Building Engines for War: Air-Cooled Radial Aircraft Engine Production in Britain and America in World War II

Building Engines for War: Air-Cooled Radial Aircraft Engine Production in Britain and America in World War II

EDWARD M. YOUNG

SAE INTERNATIONAL®

Warrendale, Pennsylvania, USA

400 Commonwealth Drive
Warrendale, PA 15096-0001 USA
E-mail: CustomerService@sae.org
Phone: 877-606-7323 (inside USA and Canada)
724-776-4970 (outside USA)
Fax: 724-776-0790

Library of Congress Catalog Number 2023948381
http://dx.doi.org/10.4271/9781468606652

ISBN-Print 978-1-4686-0664-5
ISBN-PDF 978-1-4686-0665-2
ISBN-epub 978-1-4686-0666-9

To purchase bulk quantities, please contact: SAE Customer Service

E-mail: CustomerService@sae.org
Phone: 877-606-7323 (inside USA and Canada)
724-776-4970 (outside USA)
Fax: 724-776-0790

Visit the SAE International Bookstore at books.sae.org

Publisher
Sherry Dickinson Nigam

Product Manager
Amanda Zeidan

Production and Manufacturing Associate
Michelle Silberman

This book is dedicated to my daughter, Sarah Cooper Young, and to my wife, Candis Elaine Litsey.

Contents

Foreword xii
Acknowledgments xiv

Introduction 1

Defining Mass Production 2
Modes of Production: Alternatives to Mass Production 4
The Limitations of Mass Production in Wartime 5
New Factories, New Machine Tools 6
References 8

CHAPTER 1
Building Aircraft Engines: 1929–1939 11

Introduction 11
Building Automobile Engines 12
Characteristics of the Aircraft Engine 14
Differences in Manufacturing Methods 16
How an Aircraft Engine Was Built 21
The Aircraft Engine Industry in the Interwar Period 27
Bristol, Wright, and Pratt & Whitney 30
The State of the Art in 1939: The Transformation in Engine Power 36
Conclusion 38
References 38

CHAPTER 2

Building Nine-Cylinder Engines: The Bristol Mercury and Pegasus 45

Introduction 45

Organizing Production: The First Aircraft Engine Shadow Scheme 47

The Factories 57

The Production Record 64

Production at the Shadow Factories 72

How It Was Done: The Transition to Large-Scale Production 74

Conclusion 76

References 76

CHAPTER 3

Building Nine-Cylinder Engines: The Wright R-1820 Cyclone 9 85

Introduction 85

Organizing Production 88

The Factories 98

The Studebaker Factories 101

The Production Record 104

How It Was Done: Building the Cyclone 9 at the Wright and Studebaker Factories 106

Production Engineering at Wright Aeronautical 108

Process Engineering at Wright 110

Inspection and Quality Control 119

How It Was Done at Studebaker 121

Conclusion 126

References 126

CHAPTER 4

Building 14-Cylinder Engines: The Bristol Hercules 137

Introduction 137
Organizing Production 140
The Factories 144
The Production Record 150
How It Was Done: Building the Hercules Engine 157
How It Was Done: Setting Up Quantity Production 161
The Tools 165
The Role of No. 1 Shadow Group 171
Conclusion 171
References 172

CHAPTER 5

Building 14-Cylinder Engines: The Pratt & Whitney R-1830 Twin-Wasp and the Wright R-2600 Cyclone 14 179

Introduction 179
The Engines 181
Organizing Production 183
The Factories 188
Pratt & Whitney Factories 188
The Buick and Chevrolet Factories 190
The Wright Lockland, Ohio Factory 192
The Production Record 195
The R-1830 Twin Wasp 195
Wright R-2600 Cyclone 14 196
How It Was Done: Building the Twin Wasp and the Cyclone 14 197
Building the Twin Wasp in Hartford 197
Building the Twin Wasp at Buick and Chevrolet 201
Building the Cyclone 14 at the Wright Lockland Factory 205
Conclusion 215
References 216

CHAPTER 6

Building 18-Cylinder Engines: The Pratt & Whitney R-2800 Double Wasp, Wright R-3350 Duplex Cyclone 18, and Bristol Centaurus 225

Introduction 225

The Engines 227

Organizing Production 231

The Factories 234

The Wright Aeronautical Wood-Ridge Factory 234

The Dodge-Chicago Factory 235

The Pratt & Whitney Kansas City Factory 239

The Ford Motor Company and Nash-Kelvinator Corporation Factories 240

The Bristol Engine Factory and Hawthorn Shadow Factory 241

The Production Record 242

The Pratt & Whitney R-2800 Double Wasp 242

The Wright R-3350 Duplex Cyclone 18 244

The Bristol Centaurus 245

How It Was Done 246

Building the Pratt & Whitney R-2800 246

Building the Cyclone 18 at the Dodge-Chicago Factory 251

Production Challenges at the Chicago Factory 256

Building the R-3350 at the Wright Wood-Ridge Factory 258

Building the Bristol Centaurus 258

Conclusion 260

References 261

CHAPTER 7

Comparing American and British Aircraft Engine Production 271

Introduction 271

Disparity of Resources 272

Comparing Labor Productivity at American and British Aircraft Engine Factories 274

Factories and Machine Tools 283

The Missions to Study Production 285
Holley's Dilemma of Mass Production: Quantity versus Quality 289
References 291

Bibliography 295
Index 327
About the Author 333

Foreword

Edward Young's reconstruction and analysis of the Allies' massive World War II aircraft engine programs is priceless, unique, thorough, and critical—all at once. He has revisioned this crucial wartime manufacturing initiative through political and organizational lenses while frankly documenting miraculous feats of engineering, design, construction, assembly, and testing, especially given the risks and pressures engine builders faced. In the following pages, you will visit a world of precision production on a scale never previously attempted (much less achieved), so as to deliver airplane powerplants by hundreds of thousands, ultimately overwhelming the Axis nations' aerial capabilities. Year by year, US and UK engines became ever more robust, their increased horsepower and improved performance meeting the air forces' perennial demands: faster, higher, farther. You will learn how Cyclones, Wasps, and Hercules were designed and how Pratt & Whitney, Wright, and Bristol devised novel fabrication processes, eased by erecting vast, one-story plants covering dozens of acres under a single roof. And you will rapidly recognize that British and American actors in this high-stakes drama had sharply different resource and institutional endowments, distinctive relationships with vehicle manufacturers, and separate sets of constraints and conflicts with which to contend. It's a compelling tale of continuous innovation under conditions of great stress and uncertainty, a saga not previously recounted so fully and so frankly.

Dr. Young's critical stance in this work must also be appreciated, as he demolishes the common assumption that "mass production won the war." To be sure, automotive firms shifted rapidly from building millions of cars to producing shiploads of trucks, jeeps, and tanks, but aircraft and engine manufacturers could not accept the loose tolerances, low fuel efficiencies, and persistent breakdowns that were routine in vehicles. Instead, as Young explains: "High-performance aircraft engines were at the pinnacle of precision engineering." They demanded "quality control well beyond the standards of other products produced in great volume"—hence, auto plant methods would not serve. Moreover, planes and engines alike were repeatedly redesigned while in use, as feedback from the fronts identified vulnerabilities. Production facilities had to be flexible, not fixed, so that design changes could be incorporated smoothly into assembly operations. In consequence, British and American planners created "hybrid forms of production organization," assuring the speedy flow of parts between machining stations, for example, while using tools that could be adjusted to incorporate engineering changes. Feedback redesigns, plus the ongoing search for greater power, meant that some engines went through a dozen or more model changes necessitating production reconfigurations.

A brief preview of the chapters ahead will bring these remarks to a close. The narrative commences with a review of US/UK aircraft-building practices during the prewar decade. Striking contrasts between auto and aircraft engines in design, fabrication, performance, and risk precede a revelatory, detailed reprise of the aero-engine manufacturing process at the sector's leading enterprises. Five subsequent chapters focus on a series of ever-bigger and more complex engines, from Bristol's and Wright's initial nine-cylinder designs, through a set of 14-cylinder models to several 18-cylinder giants late in the war. Chapter 2 relates the intense pressure of UK air arms for vastly expanded engine production and upgraded designs in the mid-1930s, including outreach to auto plants expected to support aircraft engine building and delicate negotiations to confirm divisions of labor and assure component quality. Similarly, in the US Wright realized that engine demand from the British and French, combined with US rearmament, necessitated not just more capacity but also a full process redesign to enable flow production, with new types of machine tools that workers having minimal shop skills could operate (Chapter 3). Engine redesigns gradually brought its hallmark Cyclones to 1,350 hp, more than double that of early 1930s models.

When introducing the 14-cylinder Hercules, which started out at 1,375 hp, Bristol largely echoed US engineering and production strategies—new greenfield plants, single-purpose (largely American) tools—but in somewhat smaller and scattered factories, due to the threat of bombing (Chapter 4). This limited flow production, yet the firm delivered over 60,000 Hercules for air service, with later versions generating nearly 1,600 hp. Some 250,000 of Pratt & Whitney and Wright's 14s flowed from purpose-built, balanced-flow plants using advanced technologies, especially automatic transfer lines and high-performance yet flexible machine tools (Chapter 5). After design upgrades, Cyclones delivered 1,900 hp. With the entry of automakers into engine assembly, veteran aircraft managers feared that prioritizing output would undercut precision, with production trumping engineering and safety. Surprisingly, auto engineers quickly accepted the centrality of precision to aircraft performance, with Buick and Studebaker plants establishing "stellar" production records. In the conflict's final years, Pratt & Whitney R-2800s and Wright R-3350s were powered by double rows with nine cylinders each, nearly 150,000 of which were installed on the air forces' dominant fighters and bombers near the war's end: P-47 Thunderbolts, F6F Hellcats, B-26s, and B-29s (Chapter 6). The Dodge Chicago Plant for Wright engines represented the world's largest manufacturing facility (nearly 7,000,000 ft^2), pioneering large-scale concrete construction, given steel's diversion to military uses. The closing chapter offers a point-by-point contrast between British and American methods, technologies, buildings, and policies during their intertwined engine programs, emphasizing rapid adaptation to changing circumstances and effective coordination, while creating hybrid forms of precision production blending high-volume output practices and sufficient flexibility to incorporate changes without loss of momentum.

In the decades ahead, this remarkable study will take its rightful place on bookshelves alongside Irving Holley's classic *Buying Aircraft: Matériel Procurement for the Army Air Forces* (1964), which emphasizes airframes, and Hermione Giffard's *Making Jet Engines in World War Two: Britain, Germany, and the United States* (2016), which previews the mid-century propulsion transition. Together, they provide a comprehensive analysis of crisis techno-politics, the intricacies of urgent engineering and production, and the powerful roles of warfare states in driving innovation. The value of such meticulous scholarship will endure, as will its contribution to understanding the war's technical and organizational foundations.

Philip Scranton
Professor Emeritus, History of Industry and Technology, Rutgers University

Acknowledgments

This book is based on my PhD dissertation for King's College London that I completed at the end of 2020. I would not have undertaken this dissertation without the enthusiastic support of my supervisor, Professor David Edgerton, Hans Rausing Professor of the History of Science and Technology and Professor of Modern British History at King's College, who believed it was a project worth undertaking, and I could not have completed it without his guidance. I am grateful for his patient direction of my research, his disciplined focus on key issues, the many stimulating intellectual discussions we had on wartime production and his care in reading the many drafts I sent him. Professor Joe Maiolo, Professor of International History at King's College, London, my second supervisor, was equally supportive and encouraging. I thank them both for their willingness to take on a mature student who had been away from academia for 40 years.

I am especially grateful to Philip Scranton, Professor Emeritus of History of Industry and Technology, Rutgers University, for agreeing to write the forward for the book. Professor Scranton's book *Endless Novelty* on methods of industrial production was fundamental to this study of aircraft engine production. I greatly benefited from several discussions we had on engine production during my research, and from his encouragement. Dr. Cristiano Ristuccia, Faculty of Economics at the University of Cambridge, gave me insightful guidance on questions of machine tool usage during World War II. He and Professor Scranton kindly agreed to be my examiners for my thesis defense. Their informed questions helped refine my thinking on aircraft engine production.

My fellow graduate students at the Centre for the Study of Science, Technology and Medicine at King's College, London, were a constant source of understanding and encouragement as we navigated the PhD process together. They were a valuable sounding board as I developed my thesis. I would particularly like to thank Tom Kelsey, Alison Skipper, Eleanor Larsson, Angelica Agredo Montealegre, Michael Charmarette, Stephen Marsh, and Andrew Erhardt for their support and assistance. I would also like to thank two fellow students from the King's College MA History of Warfare program, Rachel Blackman and David Vassallo, for their continued interest and encouragement. I benefited from the insight and sound advice of Dr. Alan James, my academic supervisor in the MA History of Warfare program. I am grateful for his on-going support of my academic interests.

I was fortunate to receive a three-year Sir Richard Trainor Fellowship jointly administrated through King's College and the Royal Air Force Museum. The Fellowship greatly facilitated my research in Britain and the US. At the Royal Air Force Museum, I would like to express my thanks to Dr. Harry Raffal and Ms. Nina Hadaway for their support. I am particularly grateful to Dr. Raffal for giving me the opportunity to present my research as part of the Royal Air Force Museum's Trenchard Lecture series. Dr. Raffal also gave me permission to use several photographs from the Museum's collection.

The research for this dissertation has taken me to libraries and archives all over Britain and the US. The staffs at these institutions were unfailingly helpful and professional, for which I am deeply grateful. I would especially like to thank Bob Hercock, a volunteer at the archives of the Rolls-Royce Heritage Trust-Bristol Branch, for his continued support. His knowledge of the Bristol Aerospace Company material in the archive was invaluable. Peter Pavey, Honorable General Secretary of the Rolls-Royce Heritage Trust-Bristol Branch, and Neil Chattle, Chief Executive of the Trust and Head of Corporate Heritage at Rolls-Royce Plc, kindly gave me permission to use photographs from the archives of the Trust's Bristol Branch. I would also like to thank Zoe Watson, formally archivist at the Aerospace Bristol Museum, for her guidance through the museum's collection of Bristol material, and Brian Riddle and Tony Pilmer at the National Aerospace Library in Britain who dug through their archives to find materials that helped my research. In Britain I am grateful for help from the staff at the Coventry City Archive, the Institution for Mechanical Engineers, the Modern Records Centre at the University of Warwick, Javis Gurr at Historic England Archives, and the National Archives at Kew. I would also like to thank Mick Donnelly, Manager of the Junction 7 Business Park in Accrington, Lancashire, for allowing me to make copies of a set of original plans of the Bristol No. 3 Shadow Factory he had in his possession.

In the US, the staffs at the Chicago, Illinois and College Park, Maryland offices of the National Archives and Records Administration, Carla Reczek at the Detroit Public Library Special Collections, and Tammy Horton at the Air Force Historical Research Agency were all extremely helpful. Andrew Beckman, archivist at the Studebaker Museum in South Bend, Indiana, helped with the archives of the Studebaker Company and gave permission to use several photos from the Museum's archive. Nicole Davis at the Museum of Flight in Seattle cheerfully responded to my many requests for photos from the Peter M. Bowers Collection. The Seattle Public Library and the Museum of Flight library held extensive collections of contemporary magazines which were invaluable. Donald Bauman, a senior architect at Albert Kahn Associates in Detroit, provided me with information on Albert Kahn's career and gave me access to the original plans for the Dodge Chicago factory the firm had designed in 1942. Brandt Rosenbusch of the Fiat Chrysler North American Archives generously shared a chronology of key documents related to the Chrysler Dodge Division's management of the Dodge Chicago factory building the Wright Cyclone 18 engine. My thanks to P.J. Muller who created the plan for the interior of the Accrington factory showing the layout of the blast walls and the comparison of the Accrington factory and the Wright Lockland factory. Kim McCutcheon, President of the Aircraft Engine Historical Society, was an enthusiastic supporter of my research and kindly provided some critical data comparing the performance of Bristol Hercules with that of the Rolls-Royce Merlin engine. I would like to thank the following individuals and their companies or institutions for obtaining photographs and permissions used in the book: Melanie Parker, Curtiss-Wright Corporation; The Chicago Tribune; Arabeth Balasko, Cincinnati Museum Center; Tasha Caswell, Connecticut Historical Society; Michael Wells, Kansas City Public Library; Debbie Petrelle, Pratt & Whitney, Raytheon Technologies.

I owe a great debt to Sherry Nigam, the Publisher at SAE International, for her willingness to publish this book and to Amanda Zeidan for cheerfully helping me through the production process. My sister, Upsana Young, generously offered her help in formatting my text for publication saving me countless hours.

My daughter, Sarah Cooper Young, encouraged me to pursue my dream of obtaining advanced degrees in history. She and her husband, Paul Worthington, were a constant source of encouragement. They generously shared their London home during my work on my PhD dissertation at King's College. Sarah's patient listening and penetrating questions helped me refine my arguments about production. I am deeply grateful to both of them. The dissertation process would have been much harder without them. My son Evan and his wife Chloë gave me every encouragement and generously accepted the long gaps in our visits while I completed my PhD.

My greatest debt, which I can never adequately repay, is to my wife Candis. She willingly agreed to disrupt our retirement in Seattle to move to London and commute back and forth between London and Seattle over four years. Her loving support was the foundation of my every effort. I could not have completed my PhD, and this book, without her.

Edward M. Young
Seattle, Washington,
Summer 2023

Introduction

Shortly after 6 p.m. on August 15, 1945, the shift-change whistle at the Chrysler Corporation's Dodge-Chicago aircraft engine factory gave a triple blast letting workers know that Japan had surrendered.[1] Within minutes, thousands of workers streamed out of the factory to begin a day of celebration. Within days, the US Army Air Force had canceled billions in aircraft and aircraft engine contracts, and within weeks, the Dodge-Chicago factory had laid off the bulk of these workers at the factory, who joined millions of others in Britain and America whose work came to an end at the close of World War II.[2] On August 29, the last of 18,413 18-cylinder Wright Aeronautical R-3350 Cyclone 18 engines built at the Dodge-Chicago factory, the most powerful and complex aircraft engine put into quantity production during the war, came off the assembly line.[3] Soon the factory became empty and quiet, with rows of stilled machine tools, engine parts stacked waiting for assembly, conveyors unmoving, and silent foundries and forges, its achievements are all but forgotten to this day.[4]

In the weeks leading up to Japan's surrender, the Dodge-Chicago factory had been a hive of activity. Each day, some 28,000 employees arrived at the factory, the largest aircraft factory in America. The main manufacturing building was at the time the largest single-story industrial building in the world. The building had a total floor area of 4,323,000 ft^2 enough to hold the Ford factory at Willow Run with room for 20 baseball fields alongside.[5] Approximately 24,700 linear feet of conveyors within the manufacturing building moved parts through machining and other processes to final assembly.[6] The Dodge-Chicago factory's magnesium and aluminum foundries, substantial buildings in their own right, and light and heavy forges provided rough castings and forgings that the more than 7,600 machine tools in the main manufacturing building machined into the 13,000 finished parts each Wright R-3350 engine required. After assembly, every single engine built at the factory went to the test cells for several hours of testing at different power settings. Every single engine was then completely disassembled, every single part inspected, and the engines reassembled and sent for a final test before packing and shipment to the factories building the Boeing B-29 Superfortress for the US Army Air Force. During July 1945, the Dodge-Chicago factory completed 1,697 Wright R-3350 engines, an average of 54 a day.

The Dodge-Chicago factory was just one part of a massive British and American Government undertaking during World War II to build airplane engines in unprecedented quantities to sustain Allied air power in the fight against the Axis.[7] During World War II, the British and American governments' commitment to air power as a key contributor to the Allied victory over the Axis powers created an unprecedented requirement for high-performance aircraft engines, creating a source of demand that justified a radical change in methods of aircraft engine production. This was not the conversion of a civilian industry to war production, but the creation of an entirely new industrial structure, with new factories far bigger than their peacetime equivalents, thousands of new machine tools, the mobilization and training of tens of thousands of workers, and new production methods. The British and American governments provided the financing for the immense amount of capital required. The scale and complexity of this effort are barely understood.

Contrary to the standard narrative on production in World War II, British and American aircraft engines were not built using mass production methods and assembly lines as commonly understood in previous histories of industrial production and that was best exemplified in the American automobile industry. High-performance aircraft engines were at the pinnacle of precision engineering. Their manufacture demanded a degree of precision and quality control well beyond the standards of other products produced in great volume. Production of aircraft engines could not be easily transferred to the methods commonly employed in the automotive industry. Instead, British and American aircraft engine manufacturers and the automotive companies who worked with them to build aircraft engines during the war created hybrid forms of production organization, selecting those automotive industry productions that addressed the unique characteristics of aircraft engines.

Defining Mass Production

Our understanding of industrial production in World War II is flawed. It is based on an incorrect and overly simplistic concept of mass production. There have been numerous studies of wartime industrial mobilization but less on the transformation in methods of production that took place during the war; the "what" has been well documented, but the "how" is less so.[8] The standard narrative of wartime production is that the mobilization of British and American industry for armaments production was straightforward, a matter of applying more money, more labor, and "mass production" to convert what were essentially civilian industries and civilian technologies to the production of war material.

An emphasis on a narrow conception of mass production focused on the assembly line as the defining aspect of wartime production conforms to "grand narratives of business and technology history that emphasize the pivotal role of large-scale corporations and mass production in American industrial development."[9] These narratives have often offered the "miracle of mass production" as an explanation for industrial production in World War II.[10] It fails to take into account how British and American industry built many technically complex weapons such as warships, aircraft, and aircraft engines in great quantities during the war. Particularly with respect to aircraft and aircraft engine production, historians "never ventured far beneath the surface of glib wartime proclamations about armament production. They repeated the motif of mass production, thus avoiding further exploration of the single most important transitional period in the aircraft industry."[11]

The term mass production has little explanatory power. Mass production simply refers to the volume of production rather than to a specific method of production. As a current text notes, "the term mass production is often loosely understood to imply a particular method of production. In fact, mass production is nothing more than production on a large scale."[12] The term mass production has come to describe a generic method of production, most closely associated with the American automotive industry and the Ford Motor Company's factory at Highland Park in Detroit, where many elements of high-volume automobile production originated. This concept of mass production incorporates five key elements:

- A division of labor into specialized tasks that can be performed by semi- or unskilled workers.
- Interchangeable parts and standardized products requiring minimal adjustment or finishing.
- Single-purpose machine tools in place of standard, universal machine tools to obtain faster production of component parts.

- Laying out machining operations and manufacturing processes in the correct sequence to ensure a continuous flow of work throughout the factory.
- Moving, mechanized assembly lines to assemble more complex products moving the material along a line of stationary workers who progressively add components.[13]

These separate elements are sometimes subsumed into and assumed to be part of the overarching term, the assembly line, seen as the defining characteristic of mass production.[14] The moving assembly line has become "central to the popular iconography of mass production" and "central to the academic historiography of Ford" as Ford's fundamental productive innovation.[15] Using these key elements to define mass production and privileging the moving assembly line "offers not analysis but description which jumbles together results, principles and techniques" and projects "specific and contingent features of Ford's practice as universal and necessary characteristics of a coherent system."[16] When applied in other contexts, however, broad definitions and generalizations about mass production methods that put primacy on the assembly line become problematic.[17] While there were many standardized products manufactured in high volumes such as automobiles whose production could, according to these commonly accepted elements, be classed as mass production, there were many that were not. Stating that production during World War II was based on mass production, as the standard narratives do, is correct in terms of production volumes, but tells us little about how armaments were actually manufactured.

Karel Williams, Colin Haslam, and John Williams have shown that at the Ford factory at Highland Park "the moving line was only one amongst several equally important instruments for moving workpieces past the stationary worker."[18] Instead, Ford's greater achievement was to significantly improve the flow of work through each stage of the manufacturing process, not just at the final assembly stage.[19] They have proposed a more encompassing concept of production that incorporates a greater variety of production methods encompassing the process of converting raw materials into finished components through machining. This they call repetitive manufacturing (ReM), the "repetition of the basic process operations so that successive workpieces travel along the same work path." ReM does not require standardized products or high production volumes, can incorporate the production of complex products with varying production times, and may or may not require assembly lines.[20] Creating what they term "the standardization of the workpath," that is, the flow of work from process to process, creates possibilities to increase production at a reduced cost by removing indirect labor as Ford had done at the Highland Park factory.[21] The key is achieving "improved flow within and between the different stages of the conversion process. Smooth, continuous flow is the basis for ReM cost reduction."[22]

Aircraft engine manufacturing is an example of repetitive manufacturing. What made the production of complex aircraft engines possible in unparalleled quantities during the war was not just larger factories, more machine tools, and more workers, but the shift to flow production on a scale that had never been attempted before. Obtaining a smooth and continuous flow of work through the production process increased production speed, lowered production times, and increased output. Continuous flow production is a concept that assumes that workpieces flow smoothly from the beginning to the end of the production process, with minimal distance between processes, and materials handling devices to minimize transfer times between operations.[23] In flow production, an assembly line might be only one component of a much larger system ensuring the progressive and continuous flow of parts through a factory.[24] The savings in time at each step in the manufacturing process could be incremental, but cumulatively significant. Different machining or processing times for different components or reliance on a timely supply of components from outside subcontractors could create interruptions. Efficient flow production minimized these irregularities and ensured simultaneous operations at all stages of the manufacturing process.[25]

Continuous flow production is a concept; line production is the method of achieving flow production. Line production has both a spatial and temporal dimension. From a spatial dimension in line production, machine tools and manufacturing processes are arranged in sequence so that the work flows successively and continuously from operation to operation, along a direct path until the component or product is completed.[26] The material being worked on must flow through the production process at a continuous but uniform rate; this is the temporal dimension of line production.[27] In other words, the manufacturing processes must be balanced so that no one operation is either significantly faster or slower than other operations; operating times are equalized throughout the manufacturing process.[28] This is referred to as line balance. The purpose of balance was "to obtain *maximum utilization of man power, maximum machine usage,* and *minimize material or time in process*."[29] Though line production was often incorporated into high-volume production methods, line production is not synonymous with mass production as high-volume production can be achieved in many different ways without the progressive arrangement of manufacturing operations that characterizes line production.[30]

The principal disadvantage of a system of line production was cost. For maximum effectiveness and economy, line production "demands a reasonably large quantity of a fairly standardized product" and predictable production volumes.[31] Setting up line production, particularly employing single-purpose machine tools and mechanical conveyor systems, could require a heavy initial capital investment. Making changes to the line could be time-consuming and costly. Manufacturing a standardized product that required no design changes maximized the economic benefits of a line production system as changes in design often required changes in machine tools or manufacturing processes, interrupting continuous flow. Freezing a design during a production run so that neither machine tools, processes, or materials needed to be changed was an accepted practice.[32] However, line production systems were not completely inflexible and could incorporate measures to allow for flexibility in design, output, layout, and equipment.[33]

Modes of Production: Alternatives to Mass Production

Mass production, that is, production on a large scale, was not suitable for every product, but there were alternatives. The large managerial firms that comprised "mass production" in America were in fact only a small proportion of the total American manufacturing sector that emerged from American industrial development in the nineteenth century. There was "a spectrum of possible approaches to manufacturing" aside from high-volume, mass production methods.[34] Philip Scranton has shown that mass production is simply one mode of production among several alternatives. He argues that in the evolution of American manufacturing into the twentieth century there were basically four approaches to manufacturing goods:

- Custom production: where a single item was manufactured to a customer's specific requirements.
- Batch production: where goods were manufactured in varied lots based on customer orders.
- Bulk production: where staple goods were manufactured in large quantities using relatively simple technology.
- Mass production: where standardized goods were manufactured in vast quantities using more capital- and technology-intensive methods to meet continuing demand.

Importantly, Scranton points out that these terms refer to approaches to manufacturing products and not to firms or industrial sectors exclusively.[35] Scranton links custom and batch production into what he terms flexible or specialty production and notes that there were numerous cases where a firm would employ various modes of production to manufacture a range of goods, what he terms "diversity within diversity."[36]

In Scranton's schema of the four approaches, the key variables between the different approaches are the nature of the product and the nature of demand. The more specialized the product, and the more uncertain and variable the demand, the more likely a firm was to adopt some forms of specialty production. Conversely, where products were standardized and demand was high, the more likely bulk or mass production would be the chosen approach.[37] These distinctions were evident to contemporary and later production engineers, if not business historians.[38] Scranton categorizes the pre-World War II airframe and aircraft engine manufacturers, for example, as using a combination of "job-shop" and "batch" production methods of manufacture and using relationships with smaller specialty firms to provide parts and subassemblies.[39] For Scranton, batch or specialty production was not a failed attempt to reach mass production, but a significant component of American industrialization.[40]

In batch production, manufacturing operations would typically be sequential, but not necessarily continuous.[41] As the name implies, batch production involved making products in batches, often of different quantities and often only to specific customer orders.[42] Often with batch production the layout of machine tools and equipment would be by function. All drilling machines, or lathes, or milling machines would be grouped together, with all drilling operations completed before the work moved on to the next type of machine. With batch orders, the work would not be passed on to the next stage of manufacturing until workers had completed operations on the entire batch.[43] This meant that a certain portion of the work would be set aside in storage until the entire batch was finished and that once it had completed its assigned tasks the operating unit could well be idle unless a follow-on batch arrived.[44] Given the unpredictability of demand, irregularities in the flow of work using batch production methods could be difficult to avoid. The time lag between the arrival of raw materials at the factory and until completion of the batch order could be considerable.

In the years before World War II, aircraft engine manufacturing, as will be made evident, was a form of batch production. The aircraft engine manufacturers built their engines in small batches according to specific orders from the military, commercial airlines, or airplane manufacturers. Rarely did the major aircraft engine firms build more than a few hundred engines a month and often considerably less. As war approached, the challenge for the British and American Governments, and the aircraft engine manufacturers, was to make the transition from batch production to quantity production rapidly and with no diminution of quality. This had never been done before on this scale, with such complex pieces of machinery, or so quickly.

The Limitations of Mass Production in Wartime

Switching from batch to quantity production of high-performance aircraft engines was not a simple matter of imposing high-volume automotive production methods onto aircraft engine manufacturing. What many standard narratives of industrial production in wartime fail to appreciate are the sometimes vast differences in the characteristics of different armaments and how these differences can determine production methods, making some methods suitable and others incompatible with mass producing a product. Moreover, military services often faced a difficult choice between obtaining weapons in large quantities and obtaining improved weapons, a dilemma of quantity or quality.[45] Mass production, building weapons in huge quantities, often required freezing a design to obtain economies of scale in production.[46] Maintaining qualitative superiority, however, requires frequent design changes that are often incompatible with high-volume production. During World War II, the race for greater engine power and greater engine reliability meant that manufacturers were constantly making design changes to existing engines and successively shifting to improved aircraft engine models. The trade-off was "between scale efficiency and flexibility: large-scale production was more efficient, but flexibility in production was essential to accommodate new weapons and modifications."[47]

Jonathan Zeitlin has argued that during the war aircraft production required not just the capacity to build aircraft in large numbers, but also the flexibility to incorporate frequent design changes to keep pace with changes in operational requirements and the constant demand for improved performance.[48] Much as Holley noted in his concept of "quantity versus quality," Zeitlin argues that "the central dilemma of wartime aircraft manufacture was the need to balance the qualitative gains obtainable through design modifications against the quantitative losses resulting from interruptions to continuous production runs."[49] As Zeitlin argues, "established mass-production methods such as those pioneered by the automobile industry, typically proved too rigid for the high level of uncertainty and rapid pace of innovation imposed by the war economy. Successful aircraft manufacturers therefore needed to find new ways of reconciling the high throughput of mass production with the adaptability of the craft workshop."[50] What emerged in both Britain and America was what Zeitlin calls "hybrid forms of productive organization" for building both aircraft and aircraft engines.[51] This was particularly true as in both countries the governments found that the automotive industry's methods and equipment were "far less appropriate" for aircraft manufacture than had been originally anticipated.[52] "Automobile Industry practice," as it turned out, "could not be directly transferred to aircraft manufacture," or, as it turned out, for producing aircraft engines.[53]

New Factories, New Machine Tools

This study of aircraft engine production in Britain and America will focus on the wartime production of high-performance, air-cooled, radial aircraft engines by three companies—the Bristol Aeroplane Company in Britain, and in America the Pratt & Whitney Aircraft Division of the United Aircraft Corporation and the Wright Aeronautical Corporation—and the automotive companies that built their engines during the war. As the study will show, wartime production of Bristol, Pratt & Whitney, and Wright aircraft engines took place in newly built, government-financed factories using newly designed machine tools specially adapted to manufacturing aircraft engine components in large quantities. Wartime aircraft engine factory design in Britain and America went through four generations of factories. The prewar Bristol, Pratt & Whitney, and Wright aircraft engine factories were built to accommodate low-volume batch production. The first generation of wartime factories, built under the British Government's first shadow factory scheme, was *smaller* than the prewar original factories and followed prewar production methods. The second generation of wartime aircraft engine factories, built under the second shadow factory scheme and as part of Wright's first expansion in capacity, was substantially bigger and was specifically designed to incorporate line production. The third generation built as American rearmament accelerated during 1940–1941 was even bigger and was exclusively American. Many were larger than prewar automotive factories and larger than any single aircraft engine factory building in Britain. The fourth generation, again built only in America, were true giants, including the largest single-story industrial buildings built up to that time (Table I.1)[54].

TABLE I.1 Four generations of factories building Bristol, Pratt & Whitney, and Wright aircraft engines.

Factory generation	Date	Machine shops size (ft²)	Employment	Batch/flow production	Automatic transfer machines
1st generation	1936–1937	100,000–175,000	1,500–5,000	Batch	No
2nd generation	1939–1940	500,000–880,000	5,000–10,700	Flow	Mixed
3rd generation	1941–1942	1,000,000–1,800,000	8,300–37,000	Flow	Yes
4th generation	1942–1943	2,900,000–4,322,000	23,000–33,000	Flow	Yes

This study will also show how the aircraft engine manufacturers worked with the British and American machine tool industries to develop new types of machine tools specifically to build aircraft engines in quantity. Machine tools underwent rapid development during the war. As the demand for all types of armaments increased, the British and American aircraft engine factories faced a shortage of skilled machinists. The only alternative was to hire semi- and unskilled workers who were available, but the tens of thousands of these workers recruited to work in the wartime aircraft engine factories were new to metalworking. They could not be expected to machine complicated components with just a few weeks of training. Instead, machines had to do the complex machining for them. In a process of deskilling, the skills of the experienced machinist had to be transferred to the machine and machining operations were broken down into simpler steps so that unskilled workers could complete the operations. Automaticity, making more and more functions of a machine tool automatic, was the only means of dealing with lower skill levels and increasing production.[55] As a result, during the war there was an increasing move to automatic machines in place of standard machine tools.[56] There were three phases of machine tool development during the war. In the first phase, from 1938 to 1940, the aircraft engine manufacturers developed jigs and fixtures for standard machine tools that enabled semi-skilled workers to operate the machines with training. In the second phase, from 1940 to 1942, the aircraft factories in Britain and America introduced semi- and automatic machine tools, especially machine tools that could perform multiple operations on a workpiece and move the workpiece from station to station in the machine with no need for intervention by an unskilled machine operator. In the third phase, from 1942 to 1944 primarily in America, Pratt & Whitney and Wright Aeronautical worked with the leading machine tool companies to develop even more advanced and more capable machine tools, specifically for manufacturing aircraft engine components that could replace multiple standard and earlier automatic machine tools.

Fundamentally, as this study will show, in making the transition from prewar low-volume, batch production to quantity production in wartime, the British and American aircraft engine manufacturers had to work out how to make large, complex aeroengines in ways that had never been attempted before without sacrificing standards of precision and quality. The key to building aircraft engines in unprecedented quantities was to shift to line production to obtain a more continuous flow during the production process, subdividing and simplifying machining operations for semi- and unskilled workers, and creating specialized, single-purpose machine tools.[57] This required careful attention to the layout of machine tools and manufacturing processes in the factories, a reconfiguring of the engine assembly process, careful control of flow through the different stages of manufacturing, and establishing balanced operations along the production line.[58]

In telling the story of these three companies and their licensees, their wartime factories, and the machine tools that built their engines, the goal of this book is to contribute to a deeper understanding of industrial production in World War II and to add a missing chapter to the history of production.

As the Harvard Business School's study of American airframe and engine production in World War II commented:

> *In summary, many of the changes in engine and airframe manufacturing processes between 1940 and 1944 could be classed as differences in kind, not just differences in degree. While techniques were borrowed from other industries, the special characteristics of airframes and engines made it impossible to adopt the established techniques of any other industry without revisions. To meet wartime production goals, the manufacturers of airframes and engines were not just forced to do, on a vastly greater scale, a job they had already been doing in peacetime. They had to do an essentially different job which neither they nor others had ever done before.*[59]

References

1 *Chicago Tribune*, August 15, 1945, p. 29.

2 *Chicago Tribune*, August 16, 1945, p. 11, August 22, 1945, p. 7; *The Wall Street Journal*, August 17, 1945, p. 5.

3 *The Wall Street Journal*, August 29, 1945, p. 10.

4 After the end of the war the Chrysler Corporation made a documentary of the Dodge-Chicago factory, titled *Engines for Super Bombers*, showing the then empty main assembly building. Available at https://www.youtube.com/watch?v=_t3akMEm9bI, accessed November 23, 2021.

5 *The Wall Street Journal*, February 25, 1943, p. 8.

6 This description of the Dodge-Chicago factory is taken from the Defense Plant Corporation pamphlet Airplane Engine Plant, Chicago, Illinois Plancor 792-Brochure A, Aluminum & Magnesium Castings Airplane Engine Plant, Chicago Illinois, Plancor 792-Brochure B, and Heat Treating and Forging Facilities Airplane Engine Plant, Chicago Illinois, Plancor 792 Brochure C, copies in Record Group 270, War Assets Administration, Real Property Disposal Case Files 1947-1951, Box 243, Chrysler Corporation, National Archives and Records Administration, Chicago office.

7 With 4,323,000 ft^2 of floor space, the Dodge-Chicago factory could contain the famous Ford Willow Run factory, often listed as the largest American industrial building built during World War II, with space for 20 baseball fields alongside. The study challenges the standard narratives on wartime industrial production and will demonstrate that the new production methods employed had little in common with the simplistic, assembly line view of mass production. *The Wall Street Journal*, February 25, 1943, p. 8.

8 See Jonathan Frear's article and Wilson, Mark R.: *Destructive Creation: American Business and the Winning of World War II*, (Philadelphia, PA, 2016), pp. 2, 294, fn. 4. Wilson's footnote to his introduction provide a useful bibliographic essay to the literature on American industrial mobilization. Other sources can be found in Edgerton, David: *Britain's War Machine: Weapons, Resources and Experts in the Second World War*, (London, 2011); Klein, Maury: *A Call to Arms: Mobilizing America for World War II* (New York, 2013).

9 Heinrich, Thomas B.: *Warship Builders: An Industrial History of U.S. Naval Shipbuilding, 1922-1945*, p. 3.

10 See for example Walton, Francis: *Miracle of World War II: How American Industry Made Victory Possible*, (New York, 1956); Kennedy, David M.: The Oxford History of the United States, Volume IX: *Freedom from Fear: The American People in Depression and War, 1929-1945*, (Oxford, 1999), p. 629; Herman, Arthur: *Freedom's Forge: How American Business Produced Victory in World War II*, (New York, 2012), p. 337; and Lacey, Jim: *Keep From All Thoughtful Men: How U.S. Economists Won World War II*, (Annapolis, MD, 2011), p. 53.

11 Ferguson, Robert G.: "One Thousand Planes a Day: Ford, Grumman, General Motors and the Arsenal of Democracy", *History and Technology*, Vol. 21, No. 2, (June 2005), p. 150.

12 Lockyer, Keith, Alan Muhlemann, and John Oakland: *Production and Operations Management*, 5th Ed., (London, 1988), p. 212.

13 This list of characteristics of mass production is drawn from Lyddon, Dave: "The Myth of Mass Production and the Mass Production of Myth", *Historical Studies of Industrial Relations*, Vol. 1, (March 1996), p. 81; Nye, David E.: *America's Assembly Line*, (Cambridge, MA, 2013), pp. 21-26; Wild, Ray: *Mass-Production Management*, (London, 1972), Chapter 1: Concepts of Mass Production.

14 Nye, *America's Assembly Line*, p. 27.

[15] Williams, Karel, Colin Haslem, John Williams, Andy Adcroft, and Sukhdeve Johal: "The Myth of the Line: Ford's Production of the Model T at Highland Park, 1909-16", *Business History*, Vol. 35, No. 3, (1993), p. 66.
[16] Ibid., pp. 67-68.
[17] Ibid., pp. 68, 78.
[18] Ibid., pp. 81-82.
[19] Ibid., p. 79.
[20] Ibid.
[21] Ibid.
[22] Ibid.
[23] Woollard, Frank G.: *Principles of Mass and Flow Production*, 55th Anniversary Special Reprint Edition, (2009), p. 48.
[24] Williams, Karel, Colin Haslem, and John Williams: "Ford versus 'Fordism': The Beginning of Mass Production?" *Work, Employment & Society*, Vol. 6, No. 4, (December 1992), p. 525.
[25] Wild, *Mass-Production Management*, p. 13.
[26] Muther, Richard: *Production Line Technique*, (New York, 1944), pp. 7-11.
[27] Ibid., p. 10.
[28] Ibid., p. 104.
[29] Ibid.
[30] Ibid., p. 11.
[31] Ibid., pp. 25, 256.
[32] Ibid., p. 255.
[33] Ibid., pp. 253-66.
[34] Scranton, Philip: *Endless Novelty: Specialty Production and American Industrialization, 1865-1925*, (Princeton, NJ, 1997), p. 8; Scranton, Philip: "Diversity in Diversity: Flexible Production and American Industrialization, 1880-1930", *Business History Review*, Vol. 65, (Spring 1991), p. 28.
[35] Scranton, *Endless Novelty*, p. 11.
[36] Scranton, *Endless Novelty*, p. 19.
[37] Scranton, *Endless Novelty*, pp. 17-19.
[38] See, for example, Northcott, Clarence H., Oliver Sheldon, J.W. Wardropper, and L. Urwick: *Factory Organization*, (London 1928); for a more contemporary view see Wild, Ray: *Mass-Production Management: The Design and Operation of Production Flow-Line Systems* (London, 1972).
[39] Scranton, *Endless Novelty*, p. 347.
[40] Scranton, "Diversity in Diversity", p. 90.
[41] Lockyer, Muhlemann, and Oakland, *Production and Operations Management*, 5th Ed., pp. 204-05.
[42] Scranton, *Endless Novelty*, p. 10.
[43] Lockyer, Muhlemann, and Oakland, *Production and Operations Management*, 5th Ed., pp. 204-05.
[44] Ibid.
[45] Holley, Irving B., Jr.: United States Army in World War II: Special Studies: *Buying Aircraft: Materiel Procurement for the Army Air Forces*, (Washington, DC, 1964), p. 512.
[46] Ibid.
[47] Edgerton, David: *Britain's War Machine: Weapons, Resources and Experts in the Second World War*, (London, 2011), pp. 207-08.
[48] Zeitlin, "Flexibility and Mass Production at War", pp. 48-49.
[49] Zeitlin, "Flexibility and Mass Production at War", p. 53.
[50] Zeitlin, "Flexibility and Mass Production at War", p. 49.
[51] Ibid.
[52] Zeitlin, "Flexibility and Mass Production at War", p. 51.
[53] Ibid.
[54] *Statistical Review 1939-1945*, (London, 1946), U.S. Department of Commerce, Civil Aeronautics Administration, Office of Aviation Information, Division of Aircraft Statistics: *U.S. Military Acceptances 1940-1945: Aircraft, Engines and Propeller Production*, (Washington, DC, 1946), Surplus War Property Division of Defense Plant Corporation: *Advance Listing of Industrial Plants and Plant Sites to be Disposed of by Defense Plant Corporation, Oct. 14, 1944*, (Washington, DC, 1944), and Surplus Property Administration: *Aircraft Plants and Facilities: Report of the Surplus Property Administration to the Congress, January 14, 1946*, (Washington, DC, 1946).
[55] Zeitlin, "Flexibility and Mass Production at War", p. 31.
[56] Ibid.
[57] "Increasing Productivity and Technological Improvements in Defense Industries", *Monthly Labor Review*, No. 34, (January 1942), pp. 34-38.
[58] Lilley, Tom, et al.: *Problems of Accelerating Aircraft Production during World War II*, (Cambridge, MA, 1947), pp. 40-41.
[59] Lilley, *Problems of Accelerating Aircraft Production during World War II*, p. 41.

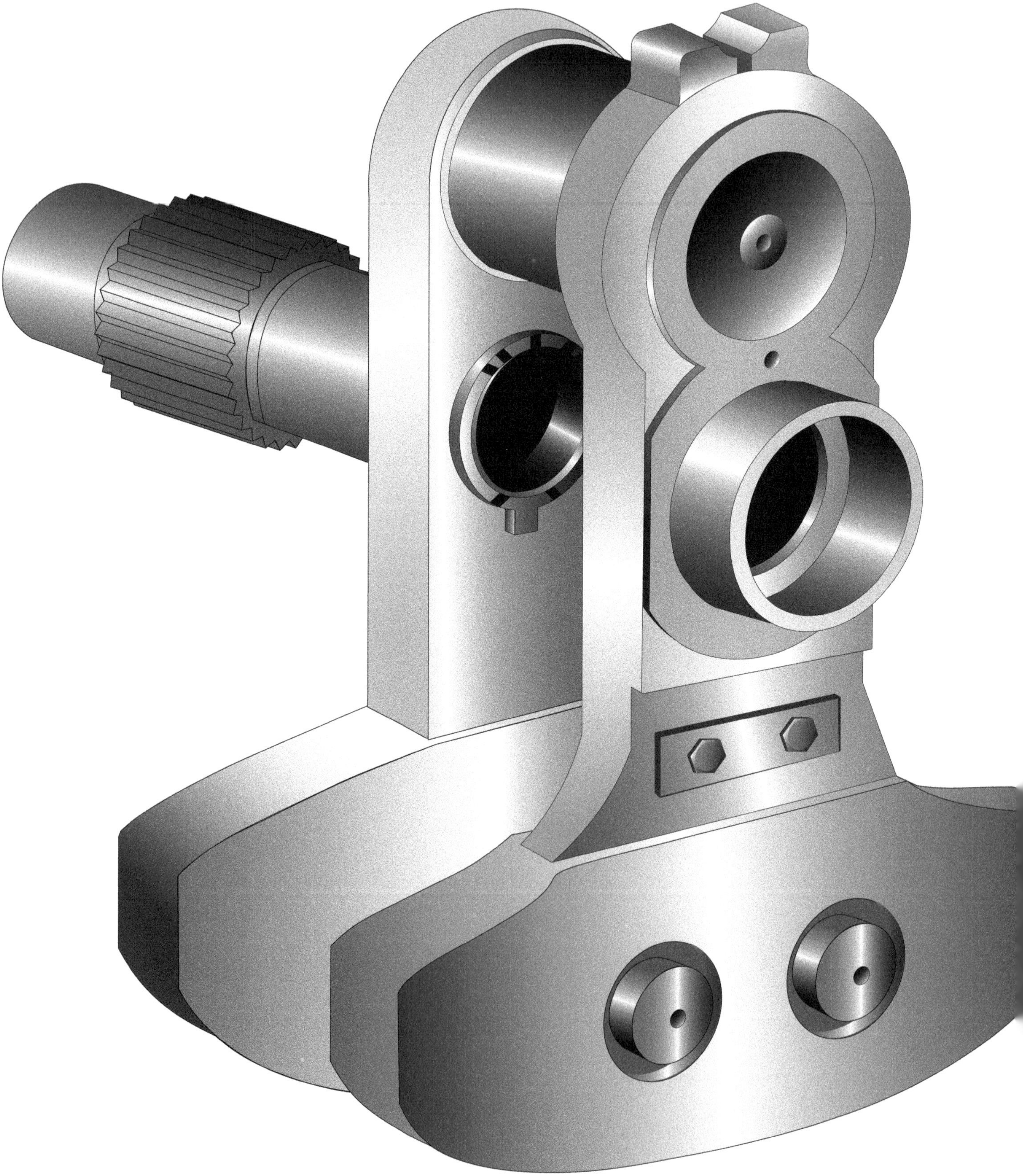

1

Building Aircraft Engines: 1929–1939

Introduction

To appreciate the magnitude of the challenges the Bristol Aeroplane Company, Pratt & Whitney, and the Wright Aeronautical Corporation faced in achieving quantity production, it is necessary to understand the complexities of building large air-cooled radial engines and the state of the companies before the war. At the time, and even subsequently, many in government and the public believed that aircraft engines could be easily manufactured in automobile factories using the same methods and machine tools used for building automobiles and automobile engines. Quantity production of aircraft engines in wartime would, therefore, be a straightforward matter of converting automobile factories to aircraft engine production. As will be shown, this proved not to be the case. The engines that Bristol, Pratt & Whitney, and Wright built stood at the pinnacle of precision engineering. These engines were far more complex machines than automobile engines and required a far higher degree of precision and quality control. Aircraft engines were made with different materials than automobile engines using different manufacturing processes. A critical difference was that aircraft engines were not built on long, continuously moving assembly lines, as were automobiles, and unlike automobiles, they were assembled, disassembled, and assembled again before shipment. These differences made building large aircraft engines in quantity challenging, particularly given the small size and limited resources of the major aircraft engine manufacturers in the interwar period.

Building Automobile Engines

To understand the complexity of an aircraft engine, one must start with its closest relative, the automobile engine. In his textbook on aircraft engines, D.R. Pye noted that in its essentials the aircraft engine was the same as any high-speed internal combustion engine.[1] Pye went on to assert, however, that there were critical qualitative differences between the two. In his view, the aircraft engine was "the aristocrat, or rather, perhaps, the athlete of the species, with no superfluous tissue and every part fine-drawn to put forth its maximum effort and to endure for long periods under those severe conditions."[2] The large high-performance aircraft engine was, as Sir Harry Ricardo claimed, at the pinnacle of precision engineering. As Ricardo said, "The modern piston aircraft-engine…represents probably the finest achievement ever reached in any sphere of mechanical engineering."[3] There are five key areas of differentiation between aircraft engines and automobile engines that need to be explored: requirements for the engine and its characteristics, the impact on design, materials used in construction, the manufacturing processes, and quality control and testing.

The basic requirements and characteristics of an automobile engine were its performance, smoothness of operation, fuel consumption, weight, and low initial cost of manufacture and installation.[4] During the 1930s, the average passenger car came with a water-cooled engine of four, six, or eight cylinders producing from 10 horsepower (hp) to 85 hp. The typical automobile engine usually runs at 22% power and only rarely above this level.[7] Thus, the engine was rarely subjected to abnormal stress. The required level of reliability was a level "deemed adequate for that particular application."[5] What was deemed adequate reliability could vary with operating conditions, but under normal operations engine failure in an automobile did not necessarily have to be life-threatening. While weight was an important consideration, as a lighter engine improved performance or allowed the weight savings to go toward other features, the power output to the weight of the engine was not a critical consideration for the average automobile engine. The automobile engines of the 1930s were straightforward machines.

The Ford Flathead V-8 engine, introduced in 1932, is a typical example. The Ford V-8 had two banks of four cylinders, set in a V form, forming the cylinder block with the upper half of the crankcase.[6] This monobloc design, with the cylinders cast integrally, provided a rigid supporting structure for each cylinder. The initial model of the Ford V-8 produced 65 hp, which later increased to 75 hp and then 95 hp by the end of the decade.[7] During this period, the Ford V-8 had only minor modifications as the horsepower increased. A complete Ford V-8 engine contained over 2,000 individual parts.[8] It did not use special high-strength steels or alloys. The cylinders and the upper half of the crankcase were cast in a single block from gray cast iron with a small amount of steel added.[9] Gray cast iron was typically used in automobile engines at the time because it was inexpensive, easy to machine, and had excellent compressive strength and vibration dampening characteristics.[10] The cylinder head attached to the top of the crankcase and the pistons were cast from aluminum alloys, while the crankshaft and camshafts were cast or forged from carbon manganese steel and then heat-treated before machining.[11] Ford used carbon manganese steel and a drop-forging process to make the connecting rods that were heat-treated after forging and a chromium and nickel alloy for the engine valves.[12] The lower half of the crankcase was a steel stamping. At 585 lb in weight, the 85 hp Ford V-8 had a weight-to-power ratio of 6.88 lb/hp.

By the late 1930s, Ford had mastered the mass production of its V-8 engines. In 1939, Ford was producing 3,600 V-8 engines a day at its River Rouge plant; a complete engine could be built in 3¼ hours.[13] The V-8 cylinder blocks were cast in a single mold at the foundry, after cooling went into an acid bath to remove sand and scale, and then washed before machining. Almost all machining operations took place on specialized machine tools, with parts moving between machines on rolling conveyors.[14] In some of the drilling, reaming, boring, and grinding operations, the machines could complete 30 to 70 cylinder blocks per hour, completing the required operation in seconds.[15] The cylinder block went through many operations in this short space of time. As an example, a Plymouth six-cylinder cylinder block required 106 operations, including multiple cleanings, painting, and inspection.[16] At the Morris Motors factory in Coventry, a cylinder block for the Morris Eight or Ten engine could be machined in approximately an hour and 53 minutes, with the longest machining operation requiring only 12 minutes to complete.[17]

After casting or forging and heat treating, the crankshaft, camshaft, pistons, connecting rods, and valves all went through similar but fewer machining operations on specialized machines. At the Morris Motors factory, a crankshaft went through 16 machining operations in 1 hour and 44 minutes.[18] Once completed, conveyor systems fed the finished parts to the assembly lines where workers assembled the complete engine. While British automobile manufacturers did not approach Ford's level of production, they still built substantial numbers of automobile engines. By 1935, using more specialized machines, moving conveyors, and line production on multiple assembly lines, the Austin Motor Company could turn out 2,000 engines a week, while Morris Motors had the capacity of producing 3,000 engines a week using similar methods.[19]

Quality control, inspection, and testing practices varied by company, but followed certain basic practices. In the production of cylinder blocks, for example, Morris Motors would cast a test bar with each pouring of molten metal from the cupolas but apparently did not test each cylinder block individually.[20] After cooling, each cylinder block would be inspected for cracks before machining. Partway through the machining process and again at the end, the cylinder block would undergo a water test for the tightness of the water jacket inside the cylinder block to reveal any cracks in the block.[21] If the cylinder block passed the water test, it went on to final assembly. After machining, a piston would be weighed to a standard with a limit of ±2 g, and its dimensions would be checked on a special measuring device.[22] Connecting rods were inspected to check that the bolt holes connecting the big end of the rod to the crankshaft were parallel, and then weighed to ensure they were within limits, if not exactly equal.[23] The crankshafts went through an inspection for dimensional accuracy and balance.[24] In American passenger automobile practice, the key measurement was the "production tolerance," the difference between the minimum and maximum dimensions of the parts that an inspector would pass, usually measured in thousandths of an inch and specified in standard tables.[25]

Once assembled, an engine went to the test department to measure its performance against specifications. At the Austin Motor Company, a test machine would drive the engine at low speeds to break it in, and then, with water, gasoline, and oil connections attached the engine's own ignition would be engaged to run the engine on its own power for up to 2 hours, including 15 minutes at full power.[26] This 2-hour engine test seems to have been common across the industry. Engines that successfully passed the tests went directly to the chassis assembly for installation in an automobile. At Vauxhall Motors in the mid-1930s, one engine in every ten was stripped down to its component parts for a more detailed inspection.[27]

The technology behind the design and construction of automobile engines did not undergo significant change during the 1930s, nor were there demands for radical increases in horsepower. The quality and reliability of these standard engines were more than adequate for the normal operations of passenger vehicles. Aircraft engines, however, were vastly different in their requirements, characteristics, and manufacturing methods.

Characteristics of the Aircraft Engine

In February 1942, the American magazine *Automobile and Aviation Industries*, the leading journal for the American automobile industry, had this to say about the differences between aircraft engines and automobile engines:

> *The fact of the matter is that the resemblance between an airplane engine and a passenger car engine is one of name only; from a manufacturing standpoint there is little resemblance. True, the component parts bear the same names—because of standardization of nomenclature—but they are made of different materials, have different form, are subject to more intensive loading, and demand an entirely different set of specifications as to tolerances and finish.*[28]

Compared to automobile engines, aircraft engines were far more powerful, more complex, more expensive, had more demanding requirements, and were more difficult to build (Figure 1.1).

FIGURE 1.1 The Wright Cyclone 9G (R-1820) introduced in 1937 offered 1,000 hp. To cope with higher temperatures, Wright added more cooling fins to the cylinders providing 2,800 in.^{2} of cooling area on each cylinder.

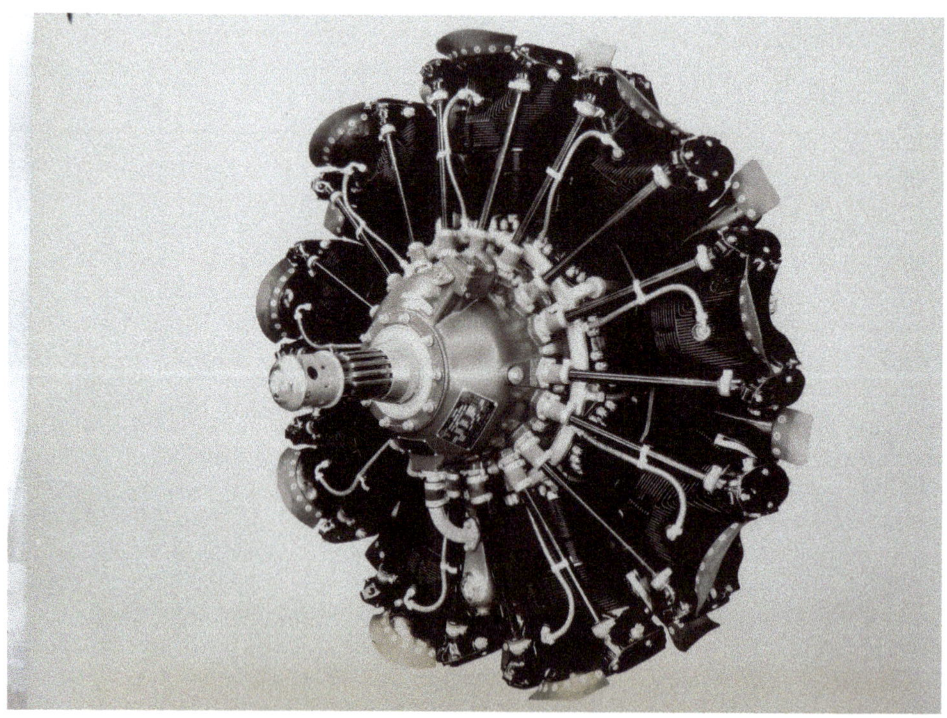

Wright Aeronautical Photo, author's collection.

Aircraft engines were very much more powerful than car engines. The Ford V-8 engine of the late 1930s produced 95 hp. In contrast, the Bristol Pegasus, Pratt & Whitney Twin Wasp, and Wright Cyclone 9 of the mid-1930s produced around 1,000 hp. To produce greater power, aircraft engines had far greater capacity. The Ford V-8 had eight cylinders with a capacity of 239 in.^{3}. The Wright Cyclone 9 and the Pegasus had nine cylinders with a displacement of 1,823 in.^{3} in the Cyclone and 1,753 in.^{3} in the Pegasus, while the Pratt & Whitney Twin Wasp had 14 cylinders and a displacement of 1,830 in.^{3}. These three engines were all radial air-cooled engines, with the cylinders arranged in a circle around the crankshaft and individually attached to the crankcase, in a single row in the Pegasus and Cyclone 9 and in a double row in the Twin Wasp. Air flowing past the cylinders provided the cooling medium, with each cylinder having fins in the cylinder head and barrel to dissipate the heat from combustion. Therefore, more cylinders and engines such as the Cyclone, Twin Wasp, and Pegasus also required more accessories than the typical automobile engine, particularly a series of gears to transfer power from the crankshaft to the propeller and a supercharger to provide more air for the air or fuel mixture at altitude. Not surprisingly, the large radial engine had more parts than an automobile engine. The Wright Cyclone 9 of the mid-1930s, for example, had 2,946 parts, and the larger double-row Wright Cyclone 14 with 14 cylinders had over 8,000 parts.

Aircraft and automobile engines differed greatly in price. In 1936, a standard Ford V-8 Tudor Sedan, Ford's most popular passenger car, was sold for $520.[29] Using a rough estimate that the V-8 engine represented 40% of the value of the automobile, the 1936 Ford V-8 would have cost around $204, or roughly $2.40 per horsepower for an 85 hp engine. In contrast, Bristol Aeroplane Company sold its Pegasus engine of 965 hp to the British Air Ministry for $8,921, or 17 times the price of the Ford Tudor (the cost to Imperial Airways, the British airline, was $9,324 per engine, while sales to the more lucrative foreign markets were $9,820 per engine).[30] Using the cost to the Air Ministry as the basis, the Pegasus cost an estimated $8.95 per horsepower. This was another indication of the greater complexity and cost of manufacture of aircraft engines compared with automobile engines.

The requirements for aircraft engines were qualitatively far more demanding. The primary requirements for an aircraft engine were performance, weight, and reliability.[31] To a far greater extent than in the design of an automobile engine, these requirements were often conflicting objectives. Maximum performance was the primary objective. The airplane designer could calculate, based on the estimated weight and drag of the airframe, the amount of horsepower required to achieve the airplane's desired performance with a given payload. The goal of the aircraft engine designer was to meet or exceed this requirement. The engine designer first chose the number of cylinders needed to generate the required horsepower and then decided on their arrangement and method of cooling. To complicate the design task, engine weight becomes a critical issue, far more critical for aircraft engines than for automobile engines.

The weight of an aircraft engine had to be kept to a minimum to increase the airplane's useful load or performance. Assuming a constant structural weight, every ounce saved in engine weight could go to payload or improved performance. The engine designer's goal was to achieve maximum horsepower for minimum weight, as measured by pounds per hp. The 85 hp Ford V-8 engine had a weight-to-power ratio of 6.88 lb/hp, while the 1,000 hp Wright Cyclone 9 engine had a weight-to-power ratio of 1.18 lb/hp. There are numerous ways of increasing engine performance, but these often involve increases in the maximum pressures and temperatures in the cylinder requiring stronger and heavier parts to cope with the higher heat and additional stress. This leads to possibly unacceptable increases in weight. The engine designer had to strike a delicate balance between strength and weight in every part of the engine. In the early 1930s, the design of aircraft engines involved "striking a judicious and well-balanced compromise between conflicting demands."[32] To compound the problem, the engine designer also had to reduce the frontal area of the engine to improve drag in flight, thereby limiting the number and placement of cylinders and compressing the space available for every other part of the engine.

Achieving maximum performance at minimum weight was challenging, but even more fundamental was engine reliability.[33] An airplane engine could have outstanding performance but would be useless if it was not dependable. Unlike its automobile counterpart, the failure of an aircraft engine could and did lead to the loss of life. All parts of the engine—valves, pistons, connecting rods, crankshaft, gears, and propeller—had to work in near-perfect harmony across the speed range of the engine. A failure of one part could cascade into a failure of the entire engine. Though difficult to achieve, the standard for reliability had to be perfection.[34] As aircraft engines increased in performance and complexity, finding a compromise between strength, weight, and reliability became more challenging. An important aspect of the pursuit of reliability was meticulous attention to detail, hence the "thousand and one steps" necessary to improve engine performance and reliability.[35]

Ensuring aircraft engine reliability was more challenging because these engines were subject to greater stress and a wider range of operating conditions than the standard automobile engine. Like all internal combustion engines, mechanical stress in an aircraft engine came from the rapid increase in pressure and heat within the cylinder during combustion and the movement of the valves, pistons, connecting rods, and crankshaft as combustion transformed heat energy into mechanical energy. At take-off and during the climb, the engine will often run at maximum power, while at cruising speed the engine operates at 50 to 60% of maximum power, compared with the typical automobile engine, which runs at 22% power and only rarely above this level.[36] Moreover, unlike other internal combustion engines, the aircraft engine must cope with a far greater range of air pressures and temperatures as it operates at different altitudes. Given the heat and mechanical stresses an engine was subject to, it was not surprising that for many years aircraft engines were subject to a long catalog of failures such as fractures in the crankshaft, broken exhaust valves, piston failures, and bearing failures.[37]

Differences in Manufacturing Methods

The requirement that aircraft engines have high performance, minimum weight, and maximum reliability under all operating conditions had important implications for the methods used in their manufacture.[38] The extent of the differences compared with the mass production methods of the automobile industry was not well understood or appreciated.[39] The methods used to manufacture aircraft engines had to incorporate certain constraints not found in most other metal machining industries: a far higher degree of precision and finishing in machining, the qualities of materials used in high-performance aircraft engines, more demanding quality control, and the need for flexibility to accommodate frequent design changes.

The need to ensure minimum weight but maximum strength of all engine components, the need to pack thousands of component parts within the smallest possible diameter, and the fact that most parts in an engine were either moving parts or in contact with other moving parts meant that all components had to be machined with far greater precision than automobile parts.[40] Similarly, these components had to fit together perfectly and all sides of a component part had to have as perfect a surface finish as possible. Any scratch or minor defect could result in the failure of the component given the intense stress aircraft engine components were subjected to.

The requirement for maximum power output with minimum weight and great reliability placed severe constraints, to a far greater extent than for automobile engines, on the materials that could be used in aircraft engine construction. For greater reliability, component parts had to be exceptionally strong to withstand the stresses of operation, but at the same time exceptionally light to minimize weight, a contradiction seemingly impossible to resolve. In addition, the engine designer had to use materials that could both withstand the heat generated in combustion and contribute to the dissipation of heat to ensure engine cooling. In designing an engine, the designer had to carefully consider the competing demands of strength, weight, and operating temperatures. Based on the required strength, the engine designer had to select appropriate materials that would meet this requirement with minimum weight, but that would also be readily adaptable to the manufacturing process.[41] As an example, the gray cast iron used in the Ford V-8 cylinder block was inexpensive, easy to cast, and easy to machine, but it was heavy and had poor thermal conductivity, making it unsuitable for aircraft engines.[42] The air-cooled aircraft engine cylinder needed lightness, rigidity, and good thermal conductivity.[43] Steel could combine rigidity with light weight, but it did not have adequate thermal conductivity. For this, engine designers had to construct the cylinder head, where temperatures from combustion were greatest, from aluminum alloy, which had far greater thermal conductivity than steel.[44]

Aircraft engine designers made extensive use of high-tensile strength steel alloys, particularly nickel–chromium–molybdenum steel alloys that could be put through a lengthy nitriding process for additional hardening.[45] For its excellent thermal conductivity, cylinder heads and pistons were made of Y-alloy aluminum, while aluminum's light weight made it suitable for radial engine crankcases. In time, stronger and lighter manganese replaced aluminum in crankcases. For valves and valve seats, especially exhaust valves that were subject to the highest temperatures during combustion, a special cobalt–chromium alloy called Stellite came into use during the 1930s. The high-performance air-cooled radial engine of the 1930s, unlike the comparable automobile engine, consisted of a variety of metals with different characteristics and manufacturing requirements. As aircraft engine power steadily increased during the 1930s, there was a constant effort to improve the quality of materials used in engine components.[46] The science of metallurgy was arguably far more important to aircraft engine development than automobile engines.

Unlike the manufacture of automobile engines, a significant amount of time and effort went into hardening components. Given the high and continuous mechanical stresses an aircraft engine was subject to, the need for engine reliability meant that engine components, even when made from high-strength steel alloys, had to go through additional hardening processes after machining. Cylinder barrels, connecting rods, crankshafts, and valves went through the nitriding process involving heating these components for a long period in an atmosphere of ammonia, in some cases for up to 72 hours, which would result in nitrogen—a component of ammonia—reacting with the steel to form a very hard compound known as nitride on the surface of the steel.[47] Smaller engine parts needing exceptional surface hardness went through a carburizing process where the components were heated to 900°C together with a carbon-containing material so that the steels could absorb the extra carbon after quenching in oil.[48] Hardening aircraft engine components through nitriding and heat treatment was more comprehensive and more time-consuming than similar processes in the automobile industry.

The greatest difference between automobile and aircraft engine practice was the far more rigorous system of quality control and inspection. With engine reliability not just a customer requirement but a vital marketing tool and potential source of reputational risk, it is not surprising that the aircraft engine manufacturers devote far more effort to quality control than their automobile counterparts. Testing and inspection were continuous throughout the production process. Every engine component had to be made with extreme accuracy to exceptionally fine tolerances and finishes to ensure minimum weight with unquestioned reliability.[49] To this end, the Bristol Aeroplane Company's Engine Department, Pratt & Whitney, and Wright Aeronautical maintained extensive engine inspection departments responsible for testing the quality of materials and the accuracy and condition of every part as it made its way through production. At the Bristol Engine Department in the mid-1930s, each engine and its component parts received over 30,000 separate inspections.[50] As an example, manufacturing a connecting rod for one type of large aircraft engine required 90 machining operations but 100 inspections, while manufacturing a connecting rod for an automobile engine required only 25 operations and 30 inspections (Figure 1.2).[51]

FIGURE 1.2 Inspecting engine parts at the Bristol Aeroplane Factory Circa 1937. In contrast to manufacturing automobile engines, every single part in an aircraft engine was subject to a meticulous inspection before assembly and then after a completed engine had undergone preliminary running. Aircraft engine parts were machined to a much higher level of precision than automotive engine parts.

Rolls-Royce Heritage Trust-Bristol Branch.

Quality control began with the arrival of raw materials and parts at the factory. Instead of taking a sample of molten metal from a cupola as in the automobile world, the Bristol Engine Department inspected and performed tests on every bar and ingot that came into the factory to ensure that all materials met official Air Ministry standards. In the production section, inspectors checked the first component from every operation and, in some cases, every component, checking for dimensional accuracy with gauges that could measure to the ten-thousandths of an inch and the quality of the finish. At the end of machining operations, every part received a final inspection before being passed, with inspectors checking for hardness and, for highly stressed components, using a Magna-Flux test machine to search for flaws in the material. The Engine Department maintained a detailed set of inspection instructions for every component, every assembly operation, and the final assembled engine. As workers combined component parts into subassemblies, inspectors examined each subassembly for tolerances and working order. The final assembly of the complete engine went through further inspections at every stage.[52]

Every single aircraft engine went through a more rigorous testing procedure than that used for automobile engines. After thorough inspection following assembly, an engine went to a test stand where it underwent an endurance test, called the "green run," running for 2 hours or more at normal power after which every engine, as opposed to one in ten, was completely stripped down to its individual parts and every part inspected for any defects. The engine was reassembled and put through a final test run for half an hour at full throttle. Additional tests determined the fuel and oil consumption and made sure that the power output conformed to standards for the specified type of engine.[53] Once all tests had been completed, there was one last full inspection and then the engine was ready for shipment. This was not a process that could be easily converted to automobile assembly line practice.

One of the hallmarks of mass production methods was the standardization of designs that allowed volume production. Maximum production came from minimal changes in design.[54] In contrast, aircraft engines were subject to frequent design changes as engineers strove for greater reliability, increased power, reduced weight, and improved production methods. In 1938, Wright Aeronautical built 1,840 aircraft engines and put through 1,032 design changes.[55] The aircraft engine industry faced three continuing conditions, which affected production: rapid engine development, pressure for greater reliability, and many variables in design, materials, and production methods.[56] These conditions apply to every part of an aircraft engine. Aircraft engine production methods had to incorporate flexibility to respond to these frequent changes in design.

Machining operations for aircraft engine components were more extensive than for automobile engine parts, partly because an aircraft engine contained considerably more components. A Wright Aeronautical study of the production of the Cyclone 9 engine determined that producing the engine's 2,946 parts required 36,857 machining operations, including 568 operations in the Wright foundry.[57] Producing one nine-cylinder Wright Cyclone 9 aircraft engine required 40 operations on each of the nine cylinder heads and 42 operations on each cylinder barrel, 57 operations on the master connecting rod, 44 operations on each articulated connecting rod, 65 operations on the front section of the crankshaft and 50 operations on the rear section, 87 operations on the supercharger housing, and 47 operations on the propeller reduction gear; this does not include machining operations on the valves, crankshaft, or pistons.[58]

In another contrast with automobile practice, machining operations on some aircraft engine components removed substantial amounts of metal. Many components started as heavy steel or aluminum forgings, made with more than the required final amount of metal to ensure the most favorable consistency or flow of the metallic grains. Finished components often weighed just 25 to 30% of the original heavy metal forging.[59] At the Bristol Aeroplane Engine Department, as an example, a forged aluminum cylinder head would arrive at the factory weighing 49 lb; 80 machining operations would reduce this to 15 lb.[60] Similarly, the master connecting rod for the Bristol Pegasus engine went from 42 lb to 9 lb 5 oz. during numerous machining operations involving successive work on a lathe, drilling machines, vertical and horizontal milling machines, and grinding machines.[61] This amount of metal had to be removed slowly and carefully, requiring more time to set up machines than in manufacturing automobile engines.[62] Automobile parts, in contrast, could be machined at higher cutting speeds and with fewer cuts.[63] Making a connecting rod for an American Cadillac automobile involved 20 machining operations taking .2 man-hours, compared with 97 operations taking 11 man-hours to make a connecting rod for the liquid-cooled Allison engine.[64] Weight reduction through machining that was common in the aircraft engine industry had never been tried in mass production.[65]

This example is indicative of another contrast with automobile practice. The machining operations in the manufacture of aircraft engine components in this period employed machine tools that were significantly different from the machine tools used in automobile factories. Removing the amount of metal typical in making aircraft engine components required machine tools that were larger and heavier than similar machine tools in the automobile industry.[66] The heavier weight provided the rigidity necessary for the finer tolerances and greater precision needed for aircraft engine components.[67] Moreover, aircraft engine manufacturers tended to use standard machine tools for their greater flexibility rather than the more specialized, single-purpose machine tools common in the automobile industry.[68] The comparatively low volume of production in the aircraft engine industry—between 50 and 200 engines a month—could not justify the heavy investment needed for specialized machine tools. In the prewar aircraft engine factories, machine tools tended to be grouped by function into separate departments rather than incorporated into a continuous process.

The volume of production was a defining characteristic that distinguished the aircraft engine industry from the automobile industry. Production in the prewar aircraft engine industry qualifies as batch manufacturing, building aircraft engines in small lots.[69] Bristol Aeroplane, Pratt & Whitney, and Wright Aeronautical built small lots of aircraft engines for specific customer orders; none of the companies built engines for inventory. In 1935, the Bristol Aeroplane Company's Engine Department produced 725 complete engines and the equivalent of 230 engines in the form of spare parts.[70] This works out to an average production of 18 engines per week. In 1938, Pratt & Whitney delivered 695 Twin Wasp engines at a rate of 13 per week, while Wright Aeronautical built 1,803 Cyclone 9 engines for the military, commercial airlines, and export at a rate of 33 engines per week.[71] This contrasts with the Ford Motor Company's production of 942,005 vehicles in 1937 at a rate of 18,115 per week.[72] The aircraft engine manufacturers also built many more models of their engines to cater to the different requirements of their military and commercial customers. Whereas the Ford Motor Company effectively produced only two models of the Ford V-8 from 1932 to 1939, there were over a dozen different models of the Bristol Mercury and Pegasus engines and the Wright Cyclone during the same period. Engine orders depended on the airframe manufacturer selecting an engine with the required horsepower, so that the aircraft engine manufacturers had to provide a range of horsepower in their engine models to meet a wider range of needs. Customers ordered specific models of the main Bristol, Pratt & Whitney, and Wright aircraft engines, for example, a Wright Cyclone 9-F-53 or Cyclone 9-G-2 or a Bristol Pegasus Mk III, which meant that it was difficult for the companies to standardize on one engine model. At this level of differentiated demand and low production volumes, there was no need and no financial incentive, to make the heavy investment in the specialized equipment, conveyor systems, and assembly lines necessary for mass production.

The mass production and assembly-line-oriented methods of automobile production were ill-suited to aircraft engine production. Unlike automobile factories, aircraft engine factories did not use continuously moving assembly lines to progressively assemble engines. Nor did the aircraft engine manufacturers use assembly lines as a means of regulating production rates.[73] As has been shown, in aircraft engine factories the main activities were machining and finishing operations, with rigorous quality control through continuous inspections. Assembly came at the end of the manufacturing process with engines assembled on movable tables. In contrast to automobile production, testing after assembly was far more rigorous. As all participants in the effort to achieve quantity production of aircraft engines would discover, "the production methods of the automobile and other similar industries had to be substantially modified before they could be used in [aircraft] engine manufacture."[74]

How an Aircraft Engine Was Built

As a further aid to appreciating the complexities of aircraft engine production, it is useful to examine how an aircraft engine was built in the interwar period in some detail. A description of building the Bristol Aeroplane Mercury engine serves as a representative example (Figure 1.3).

FIGURE 1.3 The Bristol Mercury VIS of 1933. The Bristol Aeroplane Company introduced the first nine-cylinder Mercury engine, developed from the Bristol Jupiter, in 1926. The Mercury VIS introduced in 1933 offered 605 hp and powered the Royal Air Force's Gloster Gauntlet biplane fighter.

Rolls-Royce Heritage Trust-Bristol Branch.

The manufacture of an air-cooled radial engine like the Mercury required the following steps:

- Material control: inspection of raw materials and bought-out finished parts.
- Manufacture of component parts: machining, hardening, and finishing.
- Preparation of subassemblies.
- First assembly of complete engines from components and subassemblies.
- Engine testing.
- Stripping and inspection after endurance testing.
- Final test.
- Final inspection.[75]

The component parts and subassemblies were as follows:

- Cylinder assembly, including the cylinder barrel, cylinder head, valves and rocker mechanism, and pistons.
- Crankshaft, master connecting rod, and articulated connecting rods.
- Crankcase front and rear covers.
- Blower/supercharger.
- Propeller shaft and reduction gear.[76]

From beginning to end, it took approximately 3,870 man-hours to build a Mercury engine, with 2,470 man-hours of machining, 560 man-hours of polishing and detail fitting, 420 man-hours in assembly, 150 man-hours for testing, and 250 man-hours for stripping, inspection, and rebuilding each engine.[77]

The manufacturing process began with a thorough inspection of all raw materials and all components purchased from outside specialist suppliers. Bristol maintained a Raw Material Inspection Department and a Rough View Department to carry out these inspections, checking all materials to ensure they met Air Ministry specifications, following inspection procedures laid down in the government publication *Airworthiness Handbook for Civil Aircraft*.[78] Companies supplying materials to the aircraft engine manufacturers had to use manufacturing methods that the Government's Aeronautical Inspection Department had approved and had to provide the engine companies with a release certifying that the material or part met the standards of quality.[79] A first visual inspection searched for any obvious flaws or defects in the material. Raw materials, such as steel bars or the wire used in valve springs, went through tests to determine the hardness, the presence of any internal flaws or cracks not visible, and tensile strength. Forgings for crankshafts, connecting rods, pistons and crankcases, and other components came with a test piece that could be broken off and tested to check grain flow and the orientation of metal crystals, to ensure that the flow matched the shape of the component to provide greater strength in addition to tests for hardness, tensile, and impact strength.[80]

Building the cylinders began with the cylinder barrel. The barrels arrived as rough hollow forgings of nickel–chrome–molybdenum steel, a special alloy developed to provide an extremely hard internal surface, which received the most wear, and the strength to resist distortion at the much higher power levels of large air-cooled radial engines.[81] The first operation was to bore out the inside of the rough cylinder on a turret lathe to a diameter of 5.720 in., leaving .03 in. to be removed during a later operation to obtain the correct bore of 5.75 in. for the cylinder.[82] After more work on both ends, the cylinders went for heat treatment to improve strength and hardness, being placed in an oven for 8 hours at a temperature of 500°C. When cool, cylinders went through a nitriding process to harden the internal surfaces and make them more resistant to wear. This entailed placing the cylinders in an oven at 485°C where they were bathed in ammonia gas for 72 hours, the ammonia reacting with the heated steel to leave a deposit of nitrogen on the internal surface of the cylinder.[83]

The next procedure was to machine the cooling fins on the side of the cylinder barrel. To improve the flow of air around the cooling fins, each fin was tapered to have a thickness of .062 in. at the base of the fin and .015 in. at the tip.[84] This needed a complex two-stage machining operation: The first stage was to cut the fin shape out of the bulged side of the forged cylinder, and the second stage was to taper the fins with a special type of lathe using 38 cutting tools to cut one side of the fins and another 39 tools to cut the other side. With the fins cut to the correct depth, an internal centerless grinder ground the internal surfaces of the cylinder barrel to within .002 in. of its final diameter (Figure 1.4).[85]

FIGURE 1.4 Machining cylinder heads on a Hurth table lathe. Cylinder heads required multiple operations on several different machine tools to cut and shape the cooling fins and drill holes for the intake and exhaust valves. Each machine tool worked on a single cylinder head at a time.

Rolls-Royce Heritage Trust-Bristol Branch.

The cylinder head was a drop forging, made by forcing heated metal onto a specially shaped die and then fully heat-treated. These were made from Y-alloy, an aluminum alloy containing copper, nickel, and magnesium that combined strength with resistance to high temperatures during combustion. Arriving at the engine works as a rough forging, cylinder heads went through 80 operations to remove metal and machine the work into the required size and shape, reducing the weight from 49 to 15 lb in the process.[86] Machining cylinder heads was complex due to the number of irregularly shaped cooling fins on the top and sides, the shape of the combustion chamber, and the requirement for flat spaces on either side for the two inlet and two exhaust ports. This involved careful milling to machine the correct surfaces using several different types of milling machines. Once the surfaces were prepared, additional milling machines cut out the cooling fins and then tapered them in 26 separate operations.[87] The ports for inlet and exhaust valves—two inlet and two exhaust ports on each cylinder head—were machined from the inside of the cylinder head and then valve seats where the valves rest when closed were carefully ground and holes drilled for the spark plugs.[88] Thread milling machines cut threads on the inside of the cylinder head and on the top of the cylinder barrel. The cylinder head went into an oven for an hour's heat treatment at 400°C and then had the valve seats, sparking plug adapters, and valve guides inserted while still hot, after which the cylinder head was screwed onto the cylinder barrel.[89] The final operation of this stage was to hone the inside of the cylinder barrel for an even smoother finish.

Pistons were made from R.R. 59 alloy, also known as Hiduminium, a development of the earlier Y-alloy of aluminum specifically for aircraft engine pistons that Bristol obtained from High Duty Alloys Co. Ltd. The pistons were made by stamping, where metal at room temperature as opposed to heated went under a forge or press and was forced over one or more dies to make the rough shape of the piston.[90] This 9¾ lb rough stamping went through 44 machining operations, removing 6 lb of metal.[91] The first stages of machining worked on the outside skirt of the piston to bring it close to the required dimensions and to rough out the inside of the piston. A drilling machine drilled a hole through the piston for the gudgeon pin that held the connecting rod to the piston, which was then bored to a finer finish. The outside of the piston was machined to create two recesses with flat plates for the gudgeon pin, after which a lathe cut grooves on the sides of the piston for the piston rings. As a measure of the accuracy required, the diameter of the gudgeon pinhole was bored and finished to 1.154 in., with a tolerance of +.0007 in.[92] At the end of the machining process, the piston crown and skirt underwent further surface finishing.

Inlet and exhaust valves were the final parts of the cylinder assembly. Valves had to be exceptionally strong to resist the constant hammering they received as the valve ports opened and closed and high pressures and temperatures within the combustion chamber. The valves were made from hardened steel forgings and put through a nitriding process for further hardening. Exhaust valves, which had to cope with higher temperatures, had a cavity drilled out of the stem that was filled with sodium.[93] At high temperatures, the sodium melted and moved back and forth within the cavity, helping to transfer heat down the stem of the valve. For additional protection, the valve seat at the back of the face of the valve had Stellite, a special hardening substance that resisted corrosion from the exhaust gases of higher octane fuels, welded on, while the face of the valve was covered with Brightray, another type of hardening agent.[94] The valve springs that held the valves closed were made from steel wire that was coiled to a set length, then tempered in an oven at 340°C for an hour, and afterward plated with cadmium, a corrosion-resistant coating for steel.

The crankshaft and connecting rods are described next. The crankshaft transfers the mechanical energy transmitted from the pistons, through the connecting rods, to the propeller through a shaft at the front of the crankshaft. Bristol used a two-piece crankshaft with counterweights, the front half containing the crankpin, which holds the master connecting rod and a front shaft connecting to the propeller reduction gear, while the rear half contained a shaft to drive the impellor for the blower, the two halves being clamped together with a magneton joint and magneton bolt.[95] Both parts of the crankshaft were forged from nickel–chromium–molybdenum alloy steels with high-tensile strength and received additional strengthening through nitriding. The front and rear halves of the crankshaft went through a series of similar machining operations to drill and ream holes through the crankpin and the front and rear shafts to reduce weight, drill a hole for the connecting magneton joint, and shape the counterweights and progressive grinding to ensure a smooth surface overall.[96]

Master connecting rods and the articulated connecting rods were also made from nickel–chromium–molybdenum alloy steels to better withstand torsional stresses.[97] Bristol received master connecting rods as steel stampings roughly in the correct form weighing 42 lb. After machining all surfaces in different milling machines to obtain the correct form, the finished master rod weighed 9 lb and 5½ oz.[98] Drilling machines drilled a hole at the top of the master rod for the gudgeon pin to connect with the piston and eight holes around the base to hold the articulated rods. Once drilled, these holes were bored and ground to a fine smooth finish. The articulated rods went through similar machining and drilling processes to achieve the correct form and form holes for connecting pins. After machining, both the master connecting rod and the articulated rods went through a polishing process to remove any marks that might result in fatigue fractures (Figure 1.5).[99]

FIGURE 1.5 Machining crankcases and blower (supercharger) units for Mercury engines at the Bristol Aeroplane factory in 1937. Note how the finished parts are stacked next to the machine tools until a workman with a cart can collect the parts and move them to the next phase of manufacture.

Rolls-Royce Heritage Trust-Bristol Branch.

Since they were subject to less stress, crankcases could be made from R.R. 56 aluminum alloy, another in the Hiduminium range. After stamping to obtain the rough form of the front and rear halves of the crankcase, the parts went through a heat treatment process to increase the tensile strength.[100] Once again, machining removed considerable amounts of metal, with the front half being reduced from 84 to 41 lb, while the rear half went from 75 to 42 lb. The two halves went through repeated machining operations, sometimes separately and sometimes temporarily bolted together. The center of each half was drilled out to take the front and rear shafts of the crankcase, and nine holes were cut out around the circumference of the crankcase to hold the cylinder barrels, while numerous holes were drilled on each half to bolt the halves together and to attach the cylinders and other components. After inspection and final treatment of all surfaces by hand, the crankcase was sandblasted, and then, the unprotected parts received a coating of protective enamel.[101]

The final sections to be described are the blower (supercharger) unit, the propeller shaft, and the reduction gear. Mercury engines used a gear-driven centrifugal blower.[102] The casings for the blower unit were made of lighter-weight magnesium alloy machined on both sides with holes bored in the center for the rear shaft of the crankshaft, which drove the impeller within the blower.[103] The impeller was made of much stronger nickel steel and arrived as a rough stamping. A hole was bored in the center of the stamping to connect with the crankshaft, while the individual blades of the impeller were cut out and tapered. After heat treatment to strengthen the blades, they underwent several grinding operations to obtain the correct angles. The propeller shaft and reduction gear transferred the movement of the crankshaft to the propeller but reduced the speed between the crankshaft and the propeller. The propeller shaft with the trunnions that held the reduction gear was a steel forging. Machining the propeller shaft was straightforward and could be done on a standard lathe, but the attached trunnions required special fixtures.[104] Gear shapers cut the teeth for the bevel-shaped reduction gear, which were then ground to the correct size and shape (Figure 1.6).[105]

FIGURE 1.6 Final assembly of Mercury and Pegasus engines at the Bristol Factory's East Works in May 1937. Teams of two men assembled an entire engine on a movable stand from parts held nearby. Once assembled, the engine went for testing.

Rolls-Royce Heritage Trust-Bristol Branch.

Assembly and engine testing were the final stages of the production process.[106] At Bristol, engine components underwent over 30,000 separate inspections during construction.[107] The final assembly of a Bristol aircraft engine was the responsibility of the Engine Building Department, which had three sections: fitting, subassembly, and erecting.[108] Finished components arrived at the fitting section, where they were test fitted, inspected, and then passed to the subassembly section, where, after a cleaning process to remove any traces of metal filings and further testing to ensure they met specified standards, the section assembled the separate components into subassemblies such as the crankshaft and connecting rods, the complete cylinder, and complete crankcase. To speed up the process, Bristol had divided the staff into the fitting and subassembly sections by component, so that there were teams that dealt only with the crankshaft, the crankcase, the cylinder, the supercharger, and so on. The subassembly section passed the subassemblies to the erecting section for final assembly into a finished engine. Here, the subassemblies were allocated to individual specialist assemblers who, working with one junior, would together assemble a complete engine. Neither the subassemblies nor the final engine assembly was done on a line, but on individual benches or stations with the engines mounted on erecting stands.

At this stage, subassemblies included the crankshaft and connecting rods, the cylinder unit, the complete blower unit, the propeller shaft and reduction gear, and the crankcase, with smaller components such as the pistons, rocker gear and push rods, and other engine accessories prepared in other sections. Skilled staff assembled the complete engine on special stands that allowed the engine to be rotated to progressively accept the subassemblies and other parts to build up a complete engine, checking for the alignment of all components and adjusting mechanisms before engine testing. There were no assembly lines.

Testing the assembled engine began with several hours of running-in, during which an electric motor ran the engine to check that all moving components were working correctly.[109] Next, the engine went to a dynamometer for endurance testing. With fuel and oil supplied, the engine ran for half an hour at low revolutions per minute (RPM) and then ran at progressively higher RPM and horsepower settings for another 2 hours when the engine was stopped, the oil was replaced, and the engine ran for an additional hour. When the endurance test was over, the engine went back to the inspection department where it was completely disassembled and every component closely inspected. After reassembling the engine, a final running test checked oil and fuel consumption and power output to meet Air Ministry specifications.[110] This was not a process that could be easily converted to automobile assembly line practice.

The Aircraft Engine Industry in the Interwar Period

A surprisingly small number of companies, nationally and internationally, built high-performance aircraft engines in the interwar years. There were numerous aircraft engine companies in America, Britain, and Europe that made small engines for lighter aircraft, but a smaller number of companies that built aircraft engines with a power output greater than 500 hp. Demand for high-performance aircraft engines was limited to the military services and commercial airlines. This segment of the aircraft engine industry was highly capital-intensive and required a level of specialized engineering and production knowledge that was difficult to replicate. The financial risks facing an aircraft engine manufacturer were daunting. This was due to "the enormous cost of developing an engine, the almost catastrophic consequences if the engine fails to be accepted, and the relatively small production orders in peacetime, if the engine is successful."[111] Given the high costs of entry into the industry, the huge financial risks from failure to secure orders, and the low level of demand for high-performance aircraft engines in peacetime, it is not surprising that there were few aircraft engine firms manufacturing high-performance engines and that the industry structure was something of a natural oligopoly, although highly competitive. Airframe designers typically chose a specific model of aircraft engine to power an airplane, and because engines were not interchangeable, once the airframe designer had chosen the specific engine model the aircraft engine manufacturer received an effective monopoly on providing engines for that airplane.[112] The winning aircraft engine manufacturer would receive what amounted to a guaranteed revenue stream for as long as that airplane was in service, hence the intensity of competition.

It is important to realize just how small this industry was before the beginning of rearmament. In 1936, the entire British aircraft engine industry had 14,291 employees, and the entire British aircraft industry had just 36,000 employees.[113] In contrast, the British Austin Motor Company had over 19,000 employees in 1935, while over 300,000 were employed in the automotive industry.[114] In 1936, the British aircraft engine manufacturers delivered 2,248 engines to the Royal Air Force (RAF).[115] The following year, with rearmament underway, the sector delivered 3,440 engines to the RAF and employed 16,936 people.[116] In comparison, that same year the British automobile industry produced 379,310 passenger cars and 113,946 commercial vehicles.[117] The American aviation industry was similarly minuscule in size. In 1937, the entire industry—airframe, engine, and part manufacturers—employed 30,384 people, well below 1% of total American manufacturing employment.[118] Employment in the aircraft engine sector did not go over 16,000 until February 1940.[119] In 1937, the American aircraft engine companies delivered 1,366 aircraft engines of greater than 500 hp to the military and produced an additional 2,161 in the same class for the commercial airline market.[120] That year, the Ford Motor Company alone built 942,005 vehicles.[121]

Despite fluctuations in sales to its military and commercial airline customers, during the interwar years the aircraft engine industry was decidedly military, not civilian, in its orientation.[122] All major piston engines in use in civilian airliners during the interwar period began as military engines.[123] Without a vibrant commercial airline market during the interwar period, few aircraft engine manufacturers could take on the financial risk of developing an engine purely for commercial airlines. It could take as long as four or five years to bring an engine from a prototype to full production.[124] Governments were willing to support the technical development of aircraft engines through research and development funds and military orders.[125] As a result, nearly every high-performance aircraft engine "in its prototype form and in the early production 'Marks,' generally owes its existence to the military requirements of the particular country in which it is built."[126] The military provided the funding for aircraft engine development and military orders gave the aircraft engine manufacturers the opportunity to gain experience and obtain greater reliability with their engines that would make them suitable for the commercial airline market (Figure 1.7).[127]

FIGURE 1.7 The Bristol Aeroplane Engine Department's East Works completed in early 1936. Growing demand for Mercury and Pegasus engines required Bristol to expand engine production. The new East Works factory had 200,000 ft^2 of space for manufacturing and additional facilities for engine testing.

Rolls-Royce Heritage Trust-Bristol Branch.

With large capital requirements, the need for specialist knowledge, the heavy financial risks involved, and the limited demand for aircraft engines, there were never as many aircraft engine manufacturers as there were airframe companies. In Britain, during the interwar years, the Air Ministry supported 14 airframe manufacturers but only four aircraft engine firms: Armstrong Siddeley Motors, Napier, Rolls-Royce, and the Bristol Aeroplane Company, but relied mostly on Bristol and Rolls-Royce for military aircraft engines.[128] Bristol was the dominant firm in the production of air-cooled radial engines, while Rolls-Royce's expertise was in the production of liquid-cooled inline engines. Similarly, the American Army Air Corps and US Navy used around 16 airframe manufacturers and three aircraft engine firms, none of which were particularly large.[129] Two companies, the Pratt & Whitney Aircraft Company, part of the United Aircraft Corporation, and Wright Aeronautical Corporation, part of the Curtiss-Wright Corporation, dominated the manufacture of high-performance aircraft engines. As the American military and commercial airlines had, by the 1930s, adopted the air-cooled radial engine as their preferred engine type, Pratt & Whitney and Wright Aeronautical both concentrated their efforts on radial engine technology and production.

France had four firms building engines in the greater than 500 hp class, and Italy also had four firms, while Germany had three and Japan had two.[130] Only a handful of aircraft engine manufacturers could be considered real leaders in the field based on the widespread sale of their engines to many airframe manufacturers and their granting of licenses for production to countries that lacked the resources to develop their own indigenous industry. In the mid-1930s, this list would have included Rolls-Royce and Bristol Aeroplane in Britain, Pratt & Whitney and Wright Aeronautical in America, and Gnôme-Rhône and Hispano-Suiza in France.

The aircraft engine industry was transnational in scope, in the sense that there were regular flows of information between countries and between companies. Aviation publications such as *Flight* and *The Aeroplane* in Britain, *Aero Digest* and *Aviation* in America, and other European magazines featured reports on aircraft engine developments in other countries, while industry magazines such as *Automotive Industries*, *Machinery* (London and New York), and the Journal of the Society of Automotive Engineers carried articles on aircraft engines and their manufacture. Annual yearbooks such as *The Aircraft Yearbook* series in America, the *Air Annual of the British Empire*, *Jane's All the World's Aircraft*, and *Handbuch der Luftfahrt* in Germany carried technical specifications of American, British, and European aircraft engines. The Bristol Aeroplane Company's Engine Department maintained relationships with Pratt & Whitney and Wright Aeronautical, with Bristol engineers visiting the American firms on several occasions.[131] In America, the National Advisory Committee for Aeronautics translated European technical reports on aircraft engine technology for circulation within the American aviation industry.[132] Similarly, industry magazines such as *American Machinist*, *Engineering*, *Machinery*, and *Mechanical Engineering* carried articles on developments in American and European machine tools.

Bristol, Wright, and Pratt & Whitney

Founded in 1910, the Bristol Aeroplane Company built small numbers of airplanes and ran flying schools in the years before World War I.[133] During the war, the Company built several thousand Bristol F.2B two-seat fighters for the Royal Flying Corps and the Royal Air Force. During the interwar years, Bristol built a range of aircraft for the RAF and export markets.[134] Bristol entered the aircraft engine field in 1920 when, with encouragement from the Air Ministry, it purchased the Cosmos Engineering Company, gaining the services of Roy Fedden, who would become Bristol's chief engine designer, and the design for the Jupiter nine-cylinder air-cooled engine.[135] This became the Bristol Aero Engine Department. The Jupiter was an outstanding success and saw wide service in military and commercial aircraft around the world, with several European countries building the Jupiter under license. In the late 1920s, Bristol introduced the nine-cylinder Mercury engine, which by 1935 could offer 825 hp. The larger nine-cylinder Bristol Pegasus came out in the early 1930s and by 1936 had reached 980 hp. These two engines were the backbone of Bristol's engine business during the 1930s and like the Jupiter saw widespread service with the RAF and other military services in Europe and commercial airlines.[136] Aircraft engine companies in Czechoslovakia, Finland, Italy, Poland, and Sweden built the Mercury and Pegasus under license from Bristol.[137]

The Wright Aeronautical Corporation could trace its lineage back to the company Wilbur and Orville Wright founded in 1909.[138] Merged with the Glenn L. Martin Company in 1916, during World War I the Wright-Martin Company built small numbers of Martin aircraft but thousands of Hispano-Suiza liquid-cooled engines for the US Army Air Service and the Allies.[139] Reorganized as the Wright Aeronautical Corporation in 1919, the company began working on air-cooled radial engines at the US Navy's request. In 1925, Wright introduced the seven-cylinder J-5 Whirlwind of 220 hp. This engine gained fame for powering Charles Lindbergh's "Spirit of St. Louis" on his historic New York-to-Paris flight. In the late 1920s, Wright began developing a more powerful nine-cylinder, air-cooled radial engine, the Wright Cyclone. Starting at 575 hp, by 1939 Wright had taken the power rating of the Cyclone 9 to 1,200 hp.[140] The Whirlwind and the Cyclone 9 were Wright's mainstays during the 1930s as they competed fiercely with Pratt & Whitney for military and commercial orders. Wright built two models of the Whirlwind, the R-760 (its military designation referring to a radial engine with a capacity of 760 in.^{3}) giving 285 to 350 hp and the R-975 offering 375 to 450 hp. In the Cyclone 9 G-200 series, Wright had reached 1,200 hp (the Cyclone received the military designation R-1820). Following the merger of the Curtiss and Wright groups in 1929, Wright built engines for many of the aircraft of its sister company, the Curtiss Aeroplane and Motor Company.[141] As the world's military air services began to rearm in the mid-1930s, and with the growth of commercial airlines, Wright's aircraft engines saw widespread service around the world.[142]

Frederick Rentschler established the Pratt & Whitney Aircraft Company in 1925.[143] Formerly president of Wright Aeronautical, Rentschler left Wright to establish a new aircraft engine company out of frustration with the lack of support from Wright's board of directors for additional technical development.[144] With encouragement from the US Navy, who wanted an additional source of air-cooled radial engines, the new company soon developed the 400 hp seven-cylinder Wasp engine, which was an instant success. The Navy wanted an even more powerful engine, and Pratt & Whitney followed the Wasp with the 525 hp nine-cylinder Hornet. The Wasp and the Hornet soon found a market with the Navy, the Army Air Corps, and commercial airlines and sold well abroad.[145] Like Wright Aeronautical, Pratt & Whitney became part of a larger entity, the United Aircraft Corporation, which included the Vought and Sikorsky airplane companies, which relied on Pratt & Whitney engines for their airplanes, and the Hamilton Standard Propeller Company. To compete with the increasingly powerful Wright Cyclone 9, Pratt & Whitney developed the double-row 14-cylinder Twin Wasp, which initially offered 800 hp, and by the late 1930s had reached 1,200 hp. These engines—the Wasp (R-1340), the smaller 300 hp Wasp Junior (R-985), the Hornet (R-1690), the Twin Wasp (R-1830), and the Twin Wasp Junior (R-1535)—were the mainstays of Pratt & Whitney during the 1930s. Like the Bristol and Wright engines, Pratt & Whitney's engines equipped military and commercial aircraft around the world.

In many ways in the interwar years, Bristol, Pratt & Whitney, and Wright were comparable companies. They concentrated on building air-cooled, radial aircraft engines and their primary customers were military, although Pratt & Whitney and Wright Aeronautical benefited from the much bigger and more developed American commercial airline industry where profits on engines were higher than on military sales.[146] They had a workforce of several thousand employees and small factories, and built small lots of engines for specific orders using standard, universal machine tools and highly skilled machine operators. None of the firms built more than a few hundred engines a month nor did they have any experience with large-scale production. Neither Bristol nor Pratt & Whitney nor Wright Aeronautical built engines for inventory but only for specific orders from military and civil customers. They also built many models of engines to cater to the different requirements of their military and commercial customers. Whereas the Ford Motor Company effectively produced only two models of the Ford V-8 from 1932 to 1939, there were over a dozen different models of the Bristol Mercury and Pegasus engines, the Pratt & Whitney Wasp, Hornet, and Twin Wasp, and the Wright Cyclone during the same period. Engine orders depended on the airframe manufacturer selecting an engine with the required horsepower, so the aircraft engine manufacturers had to provide a range of horsepower in their engine models to meet a wider range of needs.

In the interwar period, the aircraft engine factories were not large nor did the three companies employ many workers. When in 1936 Bristol wanted to expand production of its Mercury and Pegasus engines, the company built a new single-story factory building alongside its existing works at Filton, outside of Bristol, and increased the number of employees, adding 200,000 ft^2 of space and taking total employment to 2,800.[147] In total, the Bristol Aero Engine Department occupied around 460,000 ft^2.[148] In the new factory, building the Engine Department appears to have made some adjustments to its production methods, grouping machine tools by type of operation and organizing a modified form of line production for certain engine components.[149] In contrast to the automobile industry's methods of mass production, Bristol's Engine Department did not use specialized machine tools, instead purchasing standardized universal models that could be used to perform several different operations without the need for elaborate jigs and fixtures.[150] This flexibility was necessary when producing small batches of different models of engines at the same time. With the goal of maximizing quality and reliability, however, the Engine Department had an aim like the automobile industry to achieve the highest level of accuracy possible in order to have truly interchangeable parts and to reduce the need and cost of hand labor involved in "fitting" parts—adjusting by hand to ensure a proper fit—once completed (Figure 1.8).[151]

FIGURE 1.8 The Wright Aeronautical factory in Paterson, New Jersey, as shown in a contemporary postcard image. This four-story building had been built in 1916. Separate departments manufactured different engine parts, which had to be moved to an assembly building next door to the main structure.

Author's collection.

The layout of Wright Aeronautical's factory in Paterson, New Jersey, where Wright built all its engines for most of the 1930s, was adequate for batch production.[152] Wright's factory was a four-story reinforced concrete building built in 1916, with separate departments on three floors, each department building different engine components. The assembly building stood next door to the main building, but the Wright foundry for casting aluminum and steel components was located nearby and not next to the main factory. The main factory had no conveyor arrangements to move parts to the engine assembly building for final assembly or completed engines to the production test building next door for endurance tests. Like Bristol, Wright preferred standard universal machine tools over specialized tools to give greater flexibility as designs and models changed over time.[153] In 1937, the Wright factory had a capacity of producing around 200 engines a month, though actual production was well below this level.[154] Wright's factory was larger than Bristol's factory. By 1936, the Wright factory complex and foundry occupied some 675,000 ft^2 of floor space.[155] In 1937, to accommodate an expected increase in the production of Cyclone engines, Wright announced a $1,350,000 expansion keeping to its practice of using four-story buildings with a new wing adding another 125,000 ft^2 of space.[156] Wright's level of employment was comparable to Bristol, with an average of 2,570 employees in 1937 (Figure 1.9).[157]

FIGURE 1.9 The Pratt & Whitney aircraft engine factory in East Hartford, Connecticut, in the early 1930s. In the late 1930s, this factory would be greatly expanded to cope with increasing demand.

2002_53_12 Connecticut Historical Society.

The Pratt & Whitney factory was located in East Hartford, Connecticut, in a building comprising 400,000 ft^2, with small separate buildings for the company offices and the

Engineering Department.[158] Pratt & Whitney employed a slightly smaller number of people than Wright and Bristol, averaging 2,300 in 1937.[159] Unlike the Bristol and Wright Aeronautical factories, organized in departments by type of machining operation, Pratt & Whitney organized its factory on a completely different basis. Instead, Pratt & Whitney set up departments by type of component and by material to create a form of flow production where finished parts having arrived as raw materials moved in an uninterrupted flow down a central aisle to storerooms before final assembly.[160] One-half of the factory building worked on steel components—cylinder barrels, connecting rods, crankshaft, gears, and miscellaneous steel parts—while the other half built parts made from aluminum—the cylinder head, crankcase, and pistons. This configuration was not designed as a form of line production, but instead to accommodate the constant changes in engine design for the military, "dictated to a large extent by service conditions."[161] Each order the factory received was "custom-built, to a greater or lesser extent, although largely of standard parts."[162] The company did not believe that "the straight-line production methods and the quantity production methods used in older and larger industries" were warranted given the frequent changes in design.[163] The individual departments, however, did use a form of line production with machine tools arranged in sequence and raw materials proceeding in a straight line through the manufacturing process within the department.[164] With production revolving around several different types of engines at the same time, placing machine tools and machining processes in the departments had to incorporate a degree of flexibility. The cylinder department, for example, had to cope with 50 separate cylinder designs that might have to be manufactured concurrently depending on customer orders.[165] This form of factory organization was a precursor to what would be developed for quantity production in World War II.

During the interwar years, Bristol, Pratt & Whitney, and Wright Aeronautical relied heavily on skilled labor. Maintaining the highest level of dimensional accuracy and finish required a high level of skill and experience not only in machining operations, but also in inspection, testing, and final assembly of aircraft engines. Contemporary articles featuring aircraft engine production at Bristol, Pratt & Whitney, and Wright describe a labor-intensive process using skilled machinists operating a variety of standard universal machine tools. Photographs from the period typically show individual machinists conducting a specific operation on a single machine tool and relatively few photographs of machines carrying out the same operation on multiple workpieces at the same time. The Harvard Business School study of aircraft production during World War II characterized the prewar aircraft engine factories as "job shops," "geared to intermittent production in lots which, together with the equipment arrangement, caused the flow of production to be sluggish."[166] While length of service does not necessarily correlate with skill level, at the end of 1935 43% of Wright Aeronautical's 2,100 employees had been with the company for five years or more, including 239 employees with more than ten years of service.[167] While precise numbers are difficult to find, it is likely that Pratt & Whitney, Wright Aeronautical, and the Bristol Engine Department had only small numbers of managers and engineers running the day-to-day operations in the factories and working in the experimental departments developing and testing new models of engines. While direct comparisons of prewar labor productivity at Bristol and Wright are difficult, looking at the total number of workers per engine built per month shows that Pratt & Whitney and Wright was roughly twice as productive as Bristol, requiring 15 to 16 total workers per engine as opposed to 35 total workers per engine at Bristol.[168] Labor productivity at Wright may actually have been even greater as Wright was less reliant on subcontracting than Bristol or Pratt & Whitney.

The three companies differed in their reliance on subcontractors. The Bristol Engine Department appears to have relied on several smaller British engineering firms for forged and cast engine components, as did Pratt & Whitney, which obtained aluminum cylinder heads and steel cylinder barrels from outside suppliers, while Wright Aeronautical chose to build 90% of its engine within its own factory as it preferred to exercise rigid control over quality.[169] The three companies, however, purchased accessories such as carburetors, magnetos, valves, piston rings, and other small components from outside vendors. During World War II, arranging successful license production and building a reliable chain of subcontractors would become critical factors in expanding aircraft engine production.

The Bristol Engine Department had extensive experience in licensed production, Pratt & Whitney and Wright Aeronautical less so. During the 1920s and 1930s, Bristol had as many as 14 licensees across Europe and in Japan building the Bristol Jupiter, Mercury, and Pegasus engines. Bristol maintained a small department, the Foreign License Office, to manage its relations with its licensees. The office's functions include providing information on manufacturing methods and machine tools, complete drawings and specifications, and experienced Bristol staff to help set up production.[170] This gave the Bristol Engine Department considerable experience in helping other factories master the methods of Bristol engine production.

This, then, was the character of the aircraft engine industry for much of the interwar period, an industry that was highly concentrated, fiercely competitive, heavily reliant on government support, and government contracts, building comparatively small numbers of aircraft engines with a very small number of skilled workers using universal machine tools and working to exceptionally precise standards. It was not, at first glance, an industry that seemed capable of rapid expansion into quantity production. Factory size and plant layout, work processes, and types of machine tools and skill levels of employees were all suited to batch production as this fitted the characteristics of aircraft engines and level of demand, building small batches of engines to orders from military and commercial airline customers. When demand increased rapidly with government orders as rearmament gathered pace, Bristol, Pratt & Whitney, and Wright Aeronautical had to adapt all these factors—factory size and layout, work processes and machine tools, and worker skill levels—to increase production to unprecedented levels. During the war, the British and American aircraft engine manufacturers had to undertake major changes in process and production techniques.[171]

What compounded these challenges was the dramatic transformation in engine power that took place during the 1930s. The incessant demand for quality—improvements in performance and reliability—from commercial and military customers had, by the end of the decade, led to the development of aircraft engines with significantly greater power. The methods used to increase aircraft engine power, however, put even greater demands on aircraft engine designers and production engineers in their struggle to obtain maximum power output for minimum weight. As aircraft engines became more complicated, with more parts and using stronger materials, their production became more challenging just as demand for greater quantities began to accelerate.

The State of the Art in 1939: The Transformation in Engine Power

During the 1930s, aircraft engine manufacturers nearly quadrupled the power output of the air-cooled radial aircraft engine, an astonishing technical achievement that has received little recognition. Continued development had pushed the power output of the nine-cylinder radial air-cooled airplane engine from around 575 hp at the beginning of the 1930s to 1,000 hp or greater by the end of the decade. At the same time, new engines appeared with the potential for producing 2,000 hp and more. These developments were in response to a revolution in airplane design and performance. The all-metal monoplane with stressed-skin construction, retractable landing gear, landing flaps, and closely cowled engines became the dominant design for both military and commercial uses.[172] The transformation in aircraft performance would not have been possible without a concurrent transformation in power and reliability in the aircraft engine.[173]

By 1939, airplane engine designers and manufacturers had addressed many of the technical problems limiting power output, if not completely solving them. The most important contributors to the increase in aircraft engine power were the introduction of high-octane fuels, which may have accounted for as much as 50% of the increase, and the development of more effective superchargers.[174] Engine design, construction, and manufacturing processes had kept pace with the new high-octane aviation fuels and superchargers but increased power required stronger engine components to cope with the greater pressures and higher temperatures in the cylinder and increased stress on the moving components of the engine. As Harry Ricardo observed, "it is probably safe to say that within a period of ten years of intensive development, almost every single stressed part of every aero-engine in service has been revised, not once but many times."[175] The quality of materials in other aircraft engine components steadily improved as well, with the introduction of new steel alloys for crankcases and connecting rods with elements of nickel, chromium, and molybdenum and hardening through the nitriding process resulting in a doubling of tensile strength.[176] The most important development regarding materials was an improvement in precision engineering. Roy Fedden, who ran the Bristol Engine Department during the 1930s and into World War II, commented that "the most important factor of all, however, in connection with aircraft engine materials has been, and still is, the improvement in technique of fabrication by the introduction of high precision methods."[177]

One straightforward method of increasing engine power that came into use in the mid-1930s was to increase the number of cylinders in the engine. With the knowledge that many small cylinders were more efficient than a small number of large cylinders, the aircraft engine manufacturers began moving to the double-row air-cooled radial engine, initially with two rows of seven cylinders and later two rows of nine cylinders.[178] These double-row air-cooled radial aircraft engines produced around 1,500 hp, with 2,000 hp air-cooled radial engines on the horizon. Bristol, Pratt & Whitney, and Wright were at the forefront of this development. By the late 1930s, Bristol and Wright had 14-cylinder engines nearing production—the Hercules for Bristol and the Cyclone 14 for Wright—with the early marks of the Hercules offering 1,375 hp and the Cyclone 14 beginning at 1,600 hp. Pratt & Whitney abandoned its own effort to build a 14-cylinder engine in this class, the R-2180A, to pull ahead of the Wright Cyclone 14 with the 18-cylinder Double Wasp engine that initially offered 1,800 hp with the potential for soon reaching 2,000 hp.[179] Bristol and Wright Aeronautical also had large 18-cylinder engines in development targeting the 2,000 hp range, the Bristol Centaurus and the Wright Cyclone 18.

Beginning in the early 1930s, however, Bristol chose a different technological path from Wright, shifting from the standard poppet valves to the sleeve valve. Instead of using poppet valves to introduce the fuel–air mixture into the cylinder and expel the exhaust from combustion, a sleeve-valve aircraft engine used a cylindrical sleeve in the cylinder, placed between the side of the cylinder barrel and the piston, with ports cut in the sides of the sleeve. The sleeve valve surrounded the piston and moved up and down within the cylinder barrel in an elliptical, circular motion to enable the ports in the sleeve to line up with ports cut out of the cylinder barrel to allow the intake of the fuel–air mixture and expel the exhaust.[180] With no hot exhaust valve to trigger detonation and the premature combustion of the fuel–air mixture, a sleeve-valve engine could run at a higher compression ratio giving more power.[181] By doing away with the poppet valves and their associated parts, the sleeve-valve engine reduced the risk of failure of any of these parts and, in theory, simplified manufacturing and maintenance.[182] Having proved the concept in the earlier Perseus nine-cylinder sleeve-valve engine, Bristol chose the sleeve valve for the Hercules, Bristol's first large double-row, 14-cylinder air-cooled radial engine, and the 18-cylinder Centaurus. This was a major change, but one not followed by Pratt & Whitney and Wright, which stuck to poppet valves for their large high-performance production engines.

These large high-power, double-row radial engines were far more complex machines than the older nine-cylinder engines. Doubling the number of cylinders, adding single and two-speed superchargers, and reducing gears of various settings had all added to this complexity. The Cyclone 14, for example, contained 8,500 parts compared with the 2,946 parts in the Cyclone 9. Building the larger Cyclone 14 engine required more than 80,000 machining operations and 50,000 inspections to ensure quality.[183] The even larger 18-cylinder engines had upward of 13,000 parts.

Moving from around 3,000 parts in a nine-cylinder engine to 8,000 parts in a 14-cylinder engine required far more machining operations, more time in heat treatment, many more machine tools, many more machine operators, and more factory floor space to build. Similarly, the need to inspect and test every part, put 8,000 to 13,000 components together into a finished engine and disassemble the engine after final testing placed heavy demands on the inspection and engine assembly departments.

In 1939, Wright built only 163 Cyclone 14 engines and three of the larger Cyclone 18 engines, and Pratt & Whitney just two of its new Double Wasp engines.[184] At Bristol, the production of the Hercules was also still in its relative infancy, the company having built just 124 Hercules engines by the end of 1939, and only three Centaurus engines.[185] Soon, the two companies would face Irving Holley's "quantity versus quality" dilemma of mass production.[186] With the onset of war, the British Royal Air Force and the American Army Air Force would insist on the quantity production of all these aircraft engines in numbers undreamed of.

Conclusion

It was unlikely that the prewar batch production methods employed at Bristol, Pratt & Whitney, and Wright Aeronautical could have been replicated on a scale that would have produced the number of aircraft engines needed in the time required, although all three companies would initially employ these methods to ramp up production in the first years of the war. With the start of the war, competition for machine tools and skilled workers from other armament industries and the military services made it extremely unlikely that the aircraft engine manufacturers could have obtained the machine tools and workers they would have needed to greatly expand batch production methods. New production methods had to be devised that would meet the need for increasing quantity but with the flexibility to respond to the design changes that would come from the drive for improving quality, what the Harvard Business School study termed "a revolutionary approach to the basic methods of production."[187] How Bristol, Pratt & Whitney, and Wright Aeronautical responded to this challenge and made the transition from batch to quantity production is the subject of the following chapters.

References

1 Pye, D.R.: *The Internal Combustion Engine*, Volume II: *The Aircraft-Engine*, (London, 1934), p. 1.

2 *The Internal Combustion Engine*, p. 1.

3 Ricardo, Sir Harry R.: *The High-Speed Internal Combustion Engine*, 4th Ed., (London, 1953), p. 301.

4 Fenton, John, ed.: *Gasoline Engine Analysis for Computer Aided Design*, (London, 1986), p. 1; Heldt, P.M.: *High-Speed Combustion Engines: Design: Production: Tests*, 13th Ed., (New York, 1946), p. 62; Heywood, John B.: *Internal Combustion Engine Fundamentals*, (New York, 1988), pp. 42-43.

5 Heldt, *High-Speed Combustion Engines*, p. 62.

6 Pagé, Victor W.: *The Ford Models V8, B and A Cars: Construction-Operation-Repair*, (New York, 1933), pp. 529, 534-35.

7 Judge, Arthur W.: *Motor Manuals: Volume I: Automobile Engines in Theory, Design, Construction, Operation, Testing & Maintenance*, (London, 1942), pp. 196-97.

8 This information courtesy Patrick Collins at the National Motor Museum Trust.

9 "Machining Ford 8-h.p. Cylinder Blocks", *Machinery (London)*, Vol. 41, No. 1045, (October 20, 1932), p. 67.

10 DeGarmo, Paul, J.T. Black, Ronald A. Kosher, and Barney E. Klamecki: *Materials and Processes in Manufacturing*, 9th Ed., (Hoboken, NJ, 2003), p. 77.

11 "Machining Cast Crankshafts and Camshafts", *Machinery (London)*, Vol. 47, No. 1208, (December 5, 1935); "The Manufacture of Motor-car Engine Components", *Machinery (London)*, Vol. 47, No. 1215, (January 23, 1936).

12 Pagé, *The Ford Models V8, B and A Cars*, p. 529.

13 Bryan, Ford R.: *Rouge Pictured in its Prime*, (Dearborn, MI, 2003), p. 119.

14 "Manufacturing the Ford V-Eight", *The Automobile Engineer*, Vol. XXVI, No. 343, (March 1936), p. 96.

15 "Manufacturing the Ford V-Eight", pp. 98-99.

16 Ricardo, *High-Speed Internal Combustion Engines*, pp. 158-61.

17 "Producing Morris Eight Engines; The Foundry and Machine Methods at the Coventry Works", *The Automobile Engineer*, Vol. XXV, No. 335, (August 1935), pp. 291-296.

18 "Producing Morris Eight Engines", pp. 291-96.

19 "Austin Production Methods: An Improved Layout of the Engine Shop", *The Automobile Engineer*, Vol. XXV, No. 329, (February 1935), pp. 43-47; "Producing Morris Eight Engines", pp. 291-96.

20 "Producing Morris Eight Engines", p. 293.

21 Ricardo, *High-Speed Internal Combustion Engines*, pp. 154-55.

[22] "The Manufacture of Motor-car Engine Components", p. 491.
[23] Ricardo, *High-Speed Internal Combustion Engines*, p. 239.
[24] Ricardo, *High-Speed Internal Combustion Engines*, p. 272.
[25] Ricardo, *High-Speed Internal Combustion Engines*, p. 746.
[26] "Austin Production Methods", p. 47; "Engine-Testing Equipment: New Electrical Installation at Longbridge Works", *The Automobile Engineer*, Vol. XXV, No. 332, (May 1935), pp. 177-78.
[27] "Drilling Operations on the Vauxhall Cylinder Block", *Machinery (London)*, Vol. 43, No. 1120, (March 29, 1934), p. 771.
[28] "Keeping the Record Straight", *Automobile and Aviation Industries*, Vol. 86, No. 4, (February 15, 1942), p. 19.
[29] http://www.american-automobiles.com/Ford/1936-Ford.html, accessed 19/10/2017.
[30] Bristol Aeroplane Company Limited Financial Report for 1936, Aircraft Engine Sales for the year ending 31 December 1936, copy in the archives of the Bristol Aircraft Collection Trust, Filton.
[31] Liston, Joseph: *Aircraft Engine Design*, (New York, 1942), p. 1.
[32] Liston, *Aircraft Engine Design*, pp. 2-3.
[33] Frass, Arthur P.: *Aircraft Power Plants*, (New York, 1943), p. 4.
[34] Pye, D.R.: *The Internal Combustion Engine*, Volume II: *The Aircraft-Engine*, (London, 1934), p. 2.
[35] Ricardo, *The High-Speed Internal Combustion Engine*, p. 283.
[36] Frass, *Aircraft Power Plants*, p. 4; Gunston, Bill: *The Development of Piston Aircraft Engines*, Reprinted, (Sparkford, Somerset, UK, 1994), p. 30.
[37] Pye, *The Aircraft Engine*, Vol. II, p. 21.
[38] Fairbrother, E.: "Aircraft-Engine Production", *Aircraft Engineering*, Vol. 16, Issue 7, (July 1944), p. 209.
[39] Lilley, Tom, et al.: *Problems of Accelerating Aircraft Production During World War II*, (Boston, MA, 1947), 39.
[40] Lilley, *Problems of Accelerating Aircraft Production During World War II*, pp. 38-39.
[41] Price, Brian: "Developments of Aircraft-Engine in War 1915 to 1950: Impacts of Knowledge Management and Heuristics in the Configuration of Aircraft-Engine Architecture", in Starr, Fred, Ed Marshall, and Bryan Lawton, eds.: *The Piston Engine Revolution: Papers from a Conference on the History of Reciprocating Internal Combustion Engines held at the Museum of Science and Industry, Manchester, 14-17 April 2011*, (London, 2012), p. 403.
[42] Pye, *The Aircraft-Engine*, pp. 6-7.
[43] Pye, *The Aircraft-Engine*, p. 7.
[44] Ibid.
[45] Pye, *The Aircraft-Engine*, pp. 228, 247, 255.
[46] Ricardo, *The High-Speed Internal Combustion Engine*, p. 283.
[47] "New Heat-treatment Department at Bristol Aeroplane Co., Ltd.", *Machinery (London)*, Vol. 52, No. 1350, (August 25, 1938), p. 638; "The Manufacture of High Performance Aero Engines", p. 744.
[48] "The Manufacture of High Performance Aero Engines", p. 637.
[49] "The 'Bristol' Engine Inspection Department", *The "Bristol" Review*, Engine Issue No. 11, (November 1936), p. 30.
[50] Ibid.
[51] Lilley, *Problems of Accelerating Aircraft Production in World War II*, p. 39.
[52] "The 'Bristol' Engine Inspection Department", pp. 31-33; "Inspection: Routine Employed by the Bristol Aeroplane Co., Ltd., to Ensure Quality and Accuracy", *The Automobile Engineer*, Vol. XXIII, No. 306, (May 1933), pp. 169-70.
[53] "The "Bristol" Engine Inspection Department", pp. 30-35; Swan, Andrew: *Handbook of Aeronautics*, Volume II: *Aero-Engine Design and Practice*, (London, 1938), pp. 462-68.
[54] Holley, Irving B.: *United States Army in World War II: Special Studies: Buying Aircraft: Matériel Procurement for the Army Air Forces*, (Washington, DC, 1964).
[55] Hill, Henry C.: "Design Problems in the Quantity Production of Aircraft Engines", *S.A.E. Journal (Transactions)*, Vol. 46, No. 1, (January 1940), p. 19.
[56] Ibid., p. 21.
[57] "Wright Aero Will Expend $1,350,000 for Extensions to Factory and New Equipment", *Trade Winds*, Vol. 3, No. 7, (May 1937), p. 2.
[58] Ibid.

59 Geschelin, Joseph: "Conversion Almost Complete", *Automobile and Aviation Industries*, Vol. 87, No. 7, (October 1, 1942), p. 266.
60 "The Manufacture of High Performance Aero Engines", pp. 746-49.
61 "Machining Master Connecting Rods for Radial Aero Engines", *Machinery (London)*, Vol. 49, No. 1253, (October 15, 1936), p. 61.
62 "Keeping the Record Straight", p. 19.
63 Ibid.
64 Ibid.
65 Geschelin, "Conversion Almost Complete", p. 226.
66 "Keeping the Record Straight", p. 19.
67 Ibid.
68 Lilley, *Problems of Accelerating Aircraft Production in World War II*, p. 40.
69 Scranton, Philip: *Endless Novelty: Specialty Production and American Industrialization, 1865-1925*, (Princeton, NJ, 1997), p. 10.
70 Bristol Engine Department Engine Delivery Chart for 1935, Rolls-Royce Heritage Trust Collection.
71 Lilley, *Problems of Accelerating Aircraft Production in World War II*, Tables 17 and 18, pp. 104-105.
72 https://en.wikipedia.org/wiki/U.S._Automobile_Production_Figures, accessed 18/10/2017.
73 Wilson, Eugene E.: *Slipstream: The Autobiography of an Air Craftsman*, (New York, 1950), p. 240.
74 Lilley, *Problems of Accelerating Aircraft Production in World War II*, p. 39.
75 *History of the Production of Bristol Aero Engines by the Shadow Industry* p. 306; Fairbrother, E.: "Aero-Engine Production", *Aircraft Engineering*, Vol. 16, Issue 7, (July 1944), pp. 209-10.
76 "The New Engine Building Department", *The "Bristol" Review*, Engine Issue No. 13, (June 1938), p. 35.
77 NA, AIR 19/5, Notes on the Visit to the Works of the Bristol Aeroplane Company, 10 April 1936.
78 "The 'Bristol' Engine Inspection Department", *The "Bristol" Review*, Engine Issue No. 11, (November 1936), p. 31; Clinton, Arnold C.: "Short Survey of Aero-Engine Production", in Nelson, W/C H.: *Aero Engineering*, Vol. II, Part 1: Production, (London, 1939), p. 307.
79 Clinton, "Short Survey of Aero-Engine Production", p. 307.
80 "The 'Bristol' Engine Inspection Department", *The "Bristol" Review*, Engine Issue No. 11, (November 1936), p. 31; "Short Survey of Aero-Engine Production", p. 309; "The Manufacture of High Performance Aero Engines", *Machinery (London)*, Vol. 48, No. 1249, (September 17, 1936), p. 741.
81 "Modern Aircraft Engine Materials", *The "Bristol" Review*, Engine Issue No. 13, (June 1938), p. 28.
82 "The Manufacture of High Performance Aero Engines", pp. 743-44.
83 "The Manufacture of High Performance Aero Engines", p. 744.
84 "The Manufacture of High Performance Aero Engines", p. 745.
85 Ibid.
86 "The Manufacture of High Performance Aero Engines", p. 746.
87 "The Manufacture of High Performance Aero Engines", p. 749.
88 "The Manufacture of High Performance Aero Engines", p. 750; "Machining Operations on the Bristol 'Mercury' Engine", pp. 347, 352, 357.
89 "Machining Operations on the Bristol 'Mercury' Engine", pp. 342, 352, 354; Hesse, Herman C.: *Engineering Tools and Processes: A Study of Production Technique*, (New York, 1941), p. 487.
90 *Engineering Tools and Processes: A Study of Production Technique*, pp. 400-06, 411-15.
91 "The Manufacture of High Performance Aero Engines", p. 754; "Machining Operations on the Bristol 'Mercury' Engine", p. 354.
92 "The Manufacture of High Performance Aero Engines", p. 755.
93 "The Manufacture of High Performance Aero Engines", p. 751.
94 Ibid.; "Machining Operations on the Bristol 'Mercury' Engine", pp. 376-83.
95 "Machining Aero-Engine Crankshafts and Airscrew Shafts", *Machinery (London)*, Vol. 49, No. 1252, (October 8, 1936), p. 33; "Machining Operations on the Bristol 'Mercury' Engine", p. 364.
96 "Machining Aero-Engine Crankshafts and Airscrew Shafts", pp. 33-36.
97 "Machining Master Connecting Rods for Radial Aero Engines", *Machinery (London)*, Vol. 49, No. 1253, (October 15, 1936), pp. 61-65.
98 "The Manufacture of 'Bristol' Aero Engine Components", *The "Bristol" Review*, Engine Issue No. 8, (November 1934), p. 23.

[99] "The Manufacture of 'Bristol' Aero Engine Components", p. 24; "The Production of Aero-Engine Crankcases and Superchargers", *Machinery (London)*, Vol. 48, No. 1250, (September 24, 1936), pp. 774-78.
[100] "The Manufacture of 'Bristol' Aero Engine Components", p. 25.
[101] "The Manufacture of '"Bristol" ' Aero Engine Components", p. 26.
[102] The Bristol Aeroplane Co. Ltd.: *The "Bristol" Pegasus Aircooled Radial Engines*, Vol. III, Types XVII and XVIII, (Filton, UK, March 1940), p. 10.
[103] "The Production of Aero-Engine Crankcases and Superchargers", p. 777.
[104] "Machining Operations on the Bristol 'Mercury' Engine", p. 366.
[105] Herb, Charles O.: *Machine Tools at Work*, (New York, 1942), pp. 448-55; *Aircraft Engines*, Vol. Two, p. 275.
[106] "The Engine Fitting, Stripping and Re-Building Department", *The "Bristol" Review*, Engine Issue No. 10, (June 1936), p. 35.
[107] "The 'Bristol' Engine Inspection Department", *The "Bristol" Review*, Engine Issue No. 11, (November 1936), p. 30.
[108] See "New Engine Building Department", *Bristol Review*, Engine Issue No. 13, (June 1938), pp. 35-37.
[109] *Aircraft Engines*, Vol. Two, p. 420.
[110] "The "Bristol" Engine Inspection Department", p. 34.
[111] Banks, F.R.: "The Art of the Aviation Engine", *Journal of the Royal Aeronautical Society*, Vol. 52, Issue 453, (September 1948), p. 531.
[112] Eltscher and Young, *Curtiss-Wright: Greatness and Decline*, p. 70.
[113] Ritchie, *Industry and Air Power*, pp. 90, 138.
[114] "Austin Production Methods", *The Automobile Engineer*, Vol. XXV, No. 329, (February 1935), p. 43; Society of Motor Manufacturers and Traders: *The Motor Industry in Great Britain 1938*, (London, 1938), p. 50.
[115] Ritchie, *Industry and Air Power*, p. 138.
[116] Ibid.
[117] https://en.wikipedia.org/wiki/Automobile_industry_in_the_United_Kingdom, accessed 18/10/2017.
[118] Modley, Rudolf, ed.: *Aviation Facts and Figures 1945*, (New York, 1945), p. 18.
[119] *Aviation Facts and Figures 1945*, p. 20.
[120] *Aircraft Yearbook for 1938*, (New York, 1938), p. 446.
[121] https://en.wikipedia.org/wiki/U.S._Automobile_Production_Figures, accessed 18/10/2017.
[122] Edgerton, *England and the Aeroplane*, pp. 32-33; Vander Meulen, *The Politics of Aircraft*, pp. 3-4.
[123] Banks, "The Art of the Aviation Engine", p. 532; Schlaifer, Robert: *Development of Aero engines*, (Cambridge, MA, 1950), p. 12.
[124] Banks, "The Art of the Aviation Engine", p. 534.
[125] Banks, "The Art of the Aviation Engine", p. 532.
[126] Banks, "The Art of the Aviation Engine", p. 531.
[127] Banks, "The Art of the Aviation Engine", p. 532.
[128] Ritchie, *Industry and Air Power*, pp. 7, 18-20.
[129] This list is taken from Mingos, Howard: *Aircraft Yearbook for 1938*, (New York, 1938).
[130] Gray, C.G: *Jane's All the World's Aircraft for 1936*, (London, 1936).
[131] See United Aircraft Corporation: *The Bee Hive*, Vol. XIV, Nos. 4, 5, 6, (June 1938), which mentions a visit by Mr. Frank Nixon from the Bristol Aeroplane Company as well as a visit from Mr. Carlo F. Bona, Assistant Chief Engineer of the Italian FIAT Company's Engine Division.
[132] National Advisory Committee for Aeronautics, Technical Memorandum No. 840: "Research and Design Problems Introduced by Increased Power Output", by Oscar Kurtz, translated from *Luftwissen*, Vol. 4, No. 4, (April 1937).
[133] Barnes, C.H.: *Bristol Aircraft Since 1910*, 3rd Ed., (London, 1988), pp. 13-24.
[134] Barnes, *Bristol Aircraft Since 1910*, pp. 28-34.
[135] Lumsden, Alec: *British Piston Aero-Engines and Their Aircraft*, reprint of 1994 edition, (Ramsbury, Marlborough, UK, 2003), p. 93.
[136] Lumsden, pp. 103-11. For more background on the Bristol and Wright engines built during the war, see Appendix II in this dissertation, on p. 311.
[137] This information is from *Jane's All the World's Aircraft for 1937.*
[138] Eltscher and Young, *Curtiss-Wright: Greatness and Decline*, p. 43.
[139] Eltscher and Young, *Curtiss-Wright: Greatness and Decline*, pp. 44-47.

140 Eltscher and Young, *Curtiss-Wright: Greatness and Decline*, pp. 47-52.
141 Eltscher and Young, *Curtiss-Wright: Greatness and Decline*, pp. 69-74.
142 Eltscher and Young, *Curtiss-Wright: Greatness and Decline*, pp. 71-79, 85-88.
143 Pratt & Whitney Aircraft Division: *The Pratt & Whitney Story*, (1950), pp. 33-39.
144 Ibid., pp. 29-30.
145 Ibid., pp. 52-59.
146 Miller, Ronald, and David Sawers: *The Technical Development of Modern Aviation*, (New York, 1970), p. 87.
147 "The Manufacture of High Performance Aero Engines", p. 739.
148 NA, AVIA 10/267, See undated note "Engine Factory Floorspace."
149 "The Manufacture of High Performance Aero Engines", p. 742.
150 Ibid.
151 "The Manufacture of 'Bristol' Aero Engine Components", p. 23.
152 Faurote, Fay Leone: "Layout of a Manufacturing Plant", *The Iron Age*, (December 26, 1929), p. 1730.
153 "Wright Aero Will Expend $1,350,000 for Extensions to Factory and New Equipment", *Trade Winds*, Vol. Three, No. Seven, (May 1937), p. 2.
154 "Visit of Messrs. Evans & Mansell to U.S.A. in May-June 1937", Section B (3), p. 9.
155 "Over 900 Wright Employees Are Cited for Long Service; 52 Receive 15-year Awards", *Trade Winds*, Vol. Two, No. Seven, (October-November 1935), p. 2.
156 "Wright Aero Will Expend $1,350,000 for Extensions to Factory and New Equipment", *Trade Winds*, Vol. Three, No. Seven, (May 1937), p. 2.
157 George Haas to Secretary of the Treasury Henry Morgenthau, "Employment in the Aviation Manufacturing Industry", August 5, 1940, Volume 289, August 5-6, 1940, Henry Morgenthau, Jr. Papers, 1866-1953, Franklin D. Roosevelt Presidential Library and Museum.
158 Ward, J. Carlton, Jr.: "Plant Layout and Production Methods for Modern Aircraft Engines", *S.A.E. Journal (Transactions)*, Vol. 40, No. 5, (May 1937), p. 179.
159 Haas to Morgenthau, "Employment in the Aviation Manufacturing Industry", August 5, 1940.
160 Ward, "Plant Layout and Production Methods for Modern Aircraft Engines", pp. 179-81.
161 Ibid., p. 181.
162 Ibid.
163 Ibid.
164 Ibid., p. 182.
165 Ibid.
166 Lilley, *Problems of Accelerating Aircraft Production During World War II*, p. 40.
167 "Over 900 Wright Employees Are Cited for Long Service; 52 Receive 15-year Awards", *Trade Winds*, Vol. Two, No. Seven, (October-November 1935), p. 2.
168 Lilley, *Problems of Accelerating Aircraft Production During World War II*, p. 12; Manpower Office, Management Control, Air Technical Service Command, Army Air Forces: *Labor Statistics for the Aircraft Industry*, November 1944-July 1945, File 218.6-7 Labor Statistics, AFHRA; Mr. Haag to Secretary Morgenthau, "Employment in the Aviation Manufacturing Industry", January 14, 1941, Diaries of Henry Morgenthau, Vol. 347, January 13-14, 1941, FDR Library Digital Collections; Bristol Aeroplane Company, Summary of All Employees for 1937 and 1938, copy in the archives of Aerospace Bristol Museum.
169 "Over 36,500 Operations are Required in the Production of the Wright Cyclone", *Trade Winds*, Vol. Two, No. Six, (September 1935), p. 2; Ward, "Plant Layout and Production Methods for Modern Aircraft Engines."
170 Heaven, John: *Bristol Piston Engines Since 1914*, copy in the possession of the Rolls-Royce Heritage Trust, Filton.
171 Lilley, *Problems of Accelerating Aircraft Production During World War II*, p. 53.
172 Loftkin, Laurence, Jr.: *Quest for Performance: The Evolution of the Modern Aircraft*, (Washington, DC, 1985), Chapter 4; Miller and Sawers, *The Technical Development of Modern Aviation*, Chapter III.
173 Miller and Sawers, *The Technical Development of Modern Aviation*, p. 86.
174 Heron, S.D.: *History of the Aircraft Piston Engine*, (Detroit, MI, 1961), pp. 75, 36-37.
175 Ricardo, *The High-Speed Internal Combustion Engine*, 4th Ed., p. 303.
176 Gough, H.J.: "Materials of Aircraft Construction", Twenty-Sixth Wilbur Wright Memorial Lecture, *Royal Aeronautical Society Journal*, Vol. XLII, (October 1938), p. 941; Fedden, A.H. Roy: "Aircraft Power Plant Past and

Future", Thirty-Second Wilbur Wright Memorial Lecture, *Royal Aeronautical Society Journal*, Vol. XLVIII, (1944), pp. 419-29.

177 Fedden, "Aircraft Power Plant Past and Future", p. 420.

178 Fedden, "Aircraft Powerplant Trends", p. 457; Taylor, C. Fayette: "Engine Design in 1936", *Aviation*, Vol. 36, No. 1, (January 1937), p. 21.

179 Connors, Jack: *The Engines of Pratt & Whitney: A Technical History*, (Reston, VA, 2010), pp. 96-101.

180 Pye, *The Aero-Engine*, pp. 95-96.

181 Pye, *The Aero-Engine*, p. 96.

182 Pye, *The Aero-Engine*, p. 98.

183 "30,000 Workers in 6 Wright Plants to Build 3,780,000 HP Monthly", *Trade Winds*, (April 1941), p. 4.

184 Lilley, *Problems of Accelerating Aircraft Production*, Tables 17 and 18, pp. 104-105.

185 Table of Bristol Engines Delivered 1924-1943, copy in the Rolls-Royce Heritage Trust-Bristol Branch archives.

186 Holley, *Buying Aircraft*, p. 512.

187 Lilley, *Problems of Accelerating Aircraft Production During World War II*, p. 39.

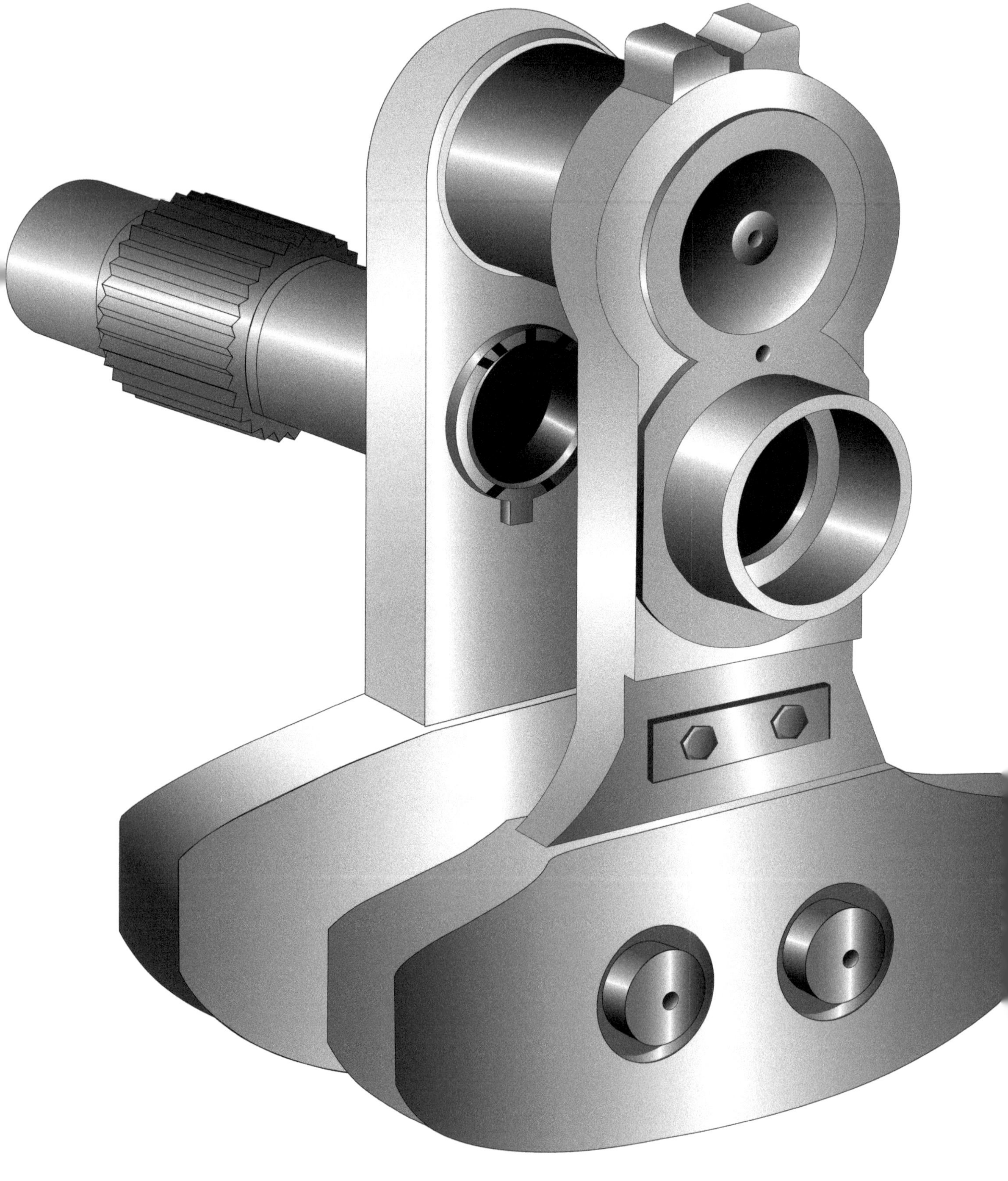

2

Building Nine-Cylinder Engines: The Bristol Mercury and Pegasus

Introduction

Since Britain began to rearm well before America, the British were the first to deal with challenges of building aircraft engines in quantity. Prior to the outbreak of war in Europe in 1939, the British Government's effort was focused on building what the Government termed war potential, the capacity to build aircraft engines in quantity in an emergency. Britain's small aviation industry did not have financial resources or the inclination to greatly expand its capacity to meet the estimated wartime requirements for what might prove to be only a short period of time.[1] The Government, therefore, devised a scheme to bring in the automobile industry to gain experience in aircraft engine production through managing small Government-owned aircraft engine factories building Bristol Mercury and Pegasus aircraft engines. It was believed at that time that with this experience, automobile manufacturers could rapidly convert their main automobile factories to aircraft engine production in an emergency so that the industry could provide a substantial potential production capacity that the Government could call on. As it turned out, this concept proved to be impractical. The factories set up to supplement Bristol's own production did manage to build the Mercury and Pegasus in quantity. They did so using existing production methods but with additional labor and machine tools.

The Mercury and the Pegasus were the Bristol Aeroplane Company's principal air-cooled radial engines in the 1930s. Bristol introduced the Mercury VI engine in 1933 as a replacement for its famous Jupiter. The Mercury offered greater horsepower than the Jupiter and was smaller in diameter, making it an ideal engine for the biplane fighters of the era. The Mercury was a nine-cylinder, poppet-valve, single-row air-cooled radial engine with a capacity of 1,520 in.^{3}, a bore of 5.75 in., and a stroke of 6.5 in.[2] A redesign of cylinder fins gave 50% more cooling area than on the Jupiter. The first production version, the Mercury IVS.2, introduced in 1932 had a normal output of 510 BHP and a maximum output of 560 BHP. Continued development led in 1934 to the Mercury VIS, which boosted the output to 605 hp.[3] In 1936, Bristol introduced the Mercury VIII, the first engine to be built at the new shadow factories. The Mercury VIII featured progressive improvements in the design of the cylinder head and barrel with a greatly increased cooling area, new pistons and valves, and strengthened components to allow for using higher octane fuels.[4] The Mercury VIII offered a significant increase in power, with a rated power output of 825 BHP and a maximum power of 840 BHP at 14,000 ft altitude. Two years later, Bristol came out with the Mercury XV, which was similar to the Mercury VIII but adapted to run on 100 octane fuel. The Mercury VIII and XV powered two of the aircraft that were important to Britain's rearmament in the late 1930s, the Bristol Blenheim light bomber and the Gloster Gladiator, the last biplane fighter in the RAF. At the beginning of the war, however, the Mercury's 840 hp was no longer deemed adequate for combat aircraft. The Mercury was also built under license in Czechoslovakia, Finland, France, Italy, Japan, and Poland (Figure 2.1).[5]

FIGURE 2.1 The nine-cylinder Bristol Pegasus engine. The Pegasus had a longer stroke and greater volume than the Bristol Mercury but produced a similar horsepower. It was much in demand for Royal Air Force bomber aircraft and civil and military flying boats.

Rolls-Royce Heritage Trust-Bristol Branch.

The Bristol Pegasus was introduced in 1933 at the same time as the Bristol Mercury. It was a larger engine intended for heavier bombers and transport aircraft. Like the Mercury, the Pegasus was a nine-cylinder, poppet-valve, single-row air-cooled radial engine with a capacity of 1,753 in.^{3}, a similar bore of 5.75 in. but a longer stroke of 7.5 in. as in the Jupiter engine.[6] The diameter of the Pegasus was about 4 in. wider than the diameter of the Mercury. With greater capacity came more power. In 1932, with various levels of supercharging, the first models of the Pegasus offered 590–635 BHP.[7] As with the Mercury, the progressive development led to engines of greater power. The Pegasus X in 1936 had a take-off power rating of 920 BHP and a maximum rated power of 875/915 BHP at 6,250 ft altitude.[8] The Pegasus was suitable for larger aircraft and in the late 1930s powered a wide range of British and European commercial and military aircraft, including the large Imperial Airways Empire class flying boats. Building the RAF's bomber force with more modern aircraft and better performance depended on the Pegasus which equipped the Handley Page Hampden and the Vickers Wellington, the principal RAF bomber in the early years of the war. The Pegasus powered several of the RAF flying boats, including the Short Sunderland and the Fairey Swordfish of the Royal Navy's Fleet Air Arm. Like the Mercury, the Pegasus was built under license in Czechoslovakia, France, Italy, Japan, and Poland.[9] While a few late model Pegasus engines could generate 1,000 hp, in the late 1930s, the Pegasus, like the Mercury, had reached the end of its potential development. More power had to come from a completely new engine.

Organizing Production: The First Aircraft Engine Shadow Scheme

Demand for the Bristol Mercury and Pegasus came from the British Government's need for thousands of these engines to power aircraft for a greatly expanded Royal Air Force. During the 1930s, the British Government saw the creation of a strong force of bombers as a form of deterrent against an air attack. Bristol Aeroplane's Mercury and Pegasus engine would power many of the new-generation all-metal, monoplane bombers that emerged in the middle years of the decade. In early 1934, in response to a recommendation from the Defence Requirements Committee, recognition of the failure of disarmament talks, and the prospect of German rearmament, the Government of Prime Minister Stanley Baldwin took the first tentative step toward expanding the Royal Air Force, promising that Britain would maintain air parity with Germany.[10] In May, the Air Ministry received approval to begin planning for an expansion program, based on the Defence Requirements Committee's proposal to add 20 squadrons to the RAF's establishment.[11] In July, the Cabinet approved Scheme A, the first of the prewar RAF expansion schemes, which called for 13 new bomber and 15 new fighter squadrons in the United Kingdom (UK) for the Metropolitan Air Force, as the RAF's UK-based force was known, and three squadrons overseas, with a complement of 1,252 aircraft of all types, an increase of 440 aircraft, to be completed by March 31, 1939.[12] Within months, British intelligence estimates of probable German aircraft production made Scheme A appear inadequate.[13] On March 4, 1935, the British Government released a Government White Paper documenting Germany's rearmament and in the House of Commons debate that followed expressed the Government's determination that British defenses should be strong enough to deter any aggressor.[14] As the House of Commons debate was about to begin, on March 10, 1935 Herman Goering announced the existence of the Luftwaffe, in violation of the Versailles Treaty.[15] Later that month, in a meeting with

Sir John Simon, Foreign Secretary, and Anthony Eden, Adolph Hitler announced that Germany had achieved parity with the Royal Air Force in the number of aircraft.[16] This led directly to the Cabinet's approval in May of Scheme C, a new expansion scheme that attempted to regain air parity with Germany by increasing the metropolitan RAF from 84 to 123 squadrons with 1,512 aircraft, a gain for the RAF of 552 aircraft with 27 new bomber and seven fighter squadrons.[17] As Britain began to accelerate rearmament in response to Germany's resurgence, the question of the nation's capacity for aircraft and aircraft engine production during the time of war became more urgent.

Even before the rearmament program began, the British Government was aware of the large gap between the existing aircraft engine capacity and probable requirements for aircraft engines during wartime. At the end of 1933, the Principal Supply Officers' Committee of the Committee of Imperial Defence reviewed a paper on the capacity for the production of aircraft engines, noting that in the first year of the war, there would be a need for 12,296 aircraft engines of all types, including 7,450 nine-cylinder Bristol aircraft engines.[18] The paper recognized that it would be difficult for Bristol to rapidly expand its production with its existing stock of machine tools and labor given the fact that aircraft engines required a certain minimum time for construction and that it would take time to acquire additional machine tools and materials and train laborers.[19] The paper estimated that Bristol could produce only around 640 aircraft engines in the first year of war. In 1931, however, Bristol had turned to the Daimler Motor Company for assistance in fulfilling a large Air Ministry contract for Bristol Jupiter engines, and Bristol engineers believed that if the entire Daimler works could be turned over to aircraft engine production, Daimler could produce 5,200 aircraft engines in a year. The paper also suggested that if other motor firms could be brought in to support Bristol's production—the paper recommended Humber Motors (Rootes), the Rover Motor Company, and the Standard Motor Company as suitable candidates—they could produce an additional 1,600 aircraft engines, bringing the total up to the required 7,450.[20] The Principal Supply Officers' Committee paper recognized that in an emergency the capacity of the main aircraft engine firms would likely be inadequate for the needs of the RAF, and as a result, the Government would have to enlist the productive capacity of the motor industry as having the nearest equivalent skills, factory space, and requisite machine tools.[21] Two years later, the estimate of total aircraft engines required in the first year of the war had increased several-fold. In December 1935, the Principal Supply Officers' Committee estimated the wartime need to be 51,106 engines of multiple types.[22]

As one of the principal suppliers of aircraft engines to the RAF, the Bristol Aeroplane Company's products and productive capacity were integral to the RAF's expansion schemes. At this time, there were 14 types of bombers, fighters, transports, and seaplanes serving in the RAF that were equipped with Bristol engines and several more Bristol-powered airplanes in development (by way of contrast, Rolls-Royce engines equipped 11 RAF aircraft). The long-serving Bristol Jupiter nine-cylinder air-cooled radial engine was nearing the end of its service life. The 600 hp of the latest Jupiter X F.B.M. series was no longer deemed adequate. In 1932, Bristol introduced the Pegasus as a replacement for the Jupiter.[23] Like the Jupiter, the Pegasus was a nine-cylinder air-cooled radial engine that incorporated improved cooling for the cylinders, with a 50% increase in the cooling area of the cylinder fins, redesigned components of greater strength, and an improved supercharger. With different levels of supercharging, the maximum power of the early Pegasus II ranged from 570 to 620 hp, with the potential for growth. That same year, Bristol came out with the Mercury, a smaller diameter nine-cylinder air-cooled radial engine intended for high-speed single-seat fighter airplanes. Like the Pegasus, the Mercury incorporated numerous improvements in design, and in the Mercury IVS had a maximum power of 560 hp. The Pegasus found ready application in RAF single- and multi-engine bombers, transports, and seaplanes, while the Mercury powered the Gloster Gauntlet fighter. In 1935, Bristol had developed the Pegasus and the Mercury to take the higher octane fuels then coming into use, significantly boosting their maximum power. The new Pegasus X offered 840–870 hp at maximum RPM, while the Mercury VI now gave 615–645 hp.[24] Bristol had also begun the production of a new family of air-cooled radial engines using a sleeve valve for the intake and exhaust stages of the combustion cycle instead of the poppet-valve system in the Mercury and Pegasus engines. The Aquila was the first of Bristol sleeve-valve engines to go into production followed by the Perseus which had a maximum power of 740–770 hp.[25] The sleeve-valve engines offered the promise of lower maintenance costs and fuel consumption, which is important for both military and civil operators alike.

Having devoted intense effort to the development of an air-cooled radial engine, the Bristol Aeroplane Company had some reservations about the ability of motor firms to manufacture aircraft engines and was confident that it could supply all the Bristol engines the Air Ministry would need from its own production. Bristol's hesitation, expressed in a response to a 1933 Air Ministry inquiry on the production of Bristol Pegasus engines in an emergency, focused on three elements of aircraft engine production: the quality of machine tools employed, inspection procedures, and engine testing.[26] Bristol expressed a concern that the machine tools commonly employed in the manufacture of automobiles might not be up to the standards of accuracy necessary for the greater precision required for machining aircraft engine components. Nor were the motor firm factories equipped with more specialized machine tools that Bristol was developing to facilitate aircraft engine production in its own aircraft engine department. Based on the company's experience working with Daimler as a subcontractor, Bristol had little doubt that machinists working in the motor firms could, with Bristol's guidance, produce parts that met Bristol's requirements for quality and accuracy, but the lack of qualified inspectors throughout the production process would be a significant constraint on production at a motor firm factory. Inspection methods common to the motor industry were inadequate for aircraft engine production. Similarly, Bristol expressed a concern that motor firms were neither equipped nor trained for more rigorous testing of aircraft engines required to meet the Air Ministry's specifications.

Under Scheme C, to be completed by March 31, 1937, the Air Ministry had ordered 784 aircraft engines from Bristol.[27] The Bristol Aero Engine Department had completed 451 new aircraft engines during 1934 and with expanded facilities was on track to build over 700 engines during 1935.[28] Writing to Sir Philip Cunliffe-Lister (Viscount Swinton), the new Secretary of State for Air, H.J. Thomas, Bristol Aeroplane Company's Assistant Managing Director, assured him that Bristol would have no difficulty completing the Air Ministry's orders for Bristol aircraft and aircraft engines. "We have every confidence," Thomas wrote, "that owing to the extensions to our establishments and the arrangements we have made, these programs will be fulfilled and perhaps exceeded in the case of the Audax machines and the Mercury and Pegasus engines, provided we have no interruptions outside our control."[29] The pace of rearmament, however, was outside the Bristol Aeroplane Company's control, and the Air Ministry was already working on revised expansion schemes that went well beyond Bristol's capacity for aircraft engine production (Figure 2.2).

FIGURE 2.2 The Bristol Blenheim light bomber was central to the Royal Air Force's expansion schemes after 1936. Scheme F, approved in February 1936, called for Bristol to build 750 Blenheims to replace the Royal Air Force's single-engine, biplane light bombers. More orders for the Blenheim would follow, creating a demand for Bristol Mercury engines in excess of what Bristol could produce on its own.

2008-03-31_image_592_01, Peter M. Bowers Collection, Museum of Flight, Seattle, WA.

The Air Ministry submitted Scheme F, its latest expansion scheme, in November 1935 against a background of continued German rearmament and the Italian invasion of Ethiopia.[30] The objective of the new scheme was to maintain parity with Germany and to build the RAF's strength overseas, adding ten new squadrons and 176 aircraft while the Metropolitan Air Force units were to receive 224 new aircraft.[31] The most important differences with earlier expansion schemes, differences that would have a direct impact on Bristol, were the transformation of the RAF's light bomber squadrons and the provision for much higher reserves of aircraft and engines.[32] In the mid-1930s, the RAF's light bomber squadrons were equipped with single-engine, fabric-covered biplanes, primarily the Hawker Hart/Audax/Hind. Scheme F called for 49 of these squadrons to be converted to medium bomber squadrons with the new all-metal monoplane Fairey Battle and Bristol's own Mercury-powered monoplane twin-engine medium bomber derived from the Bristol Type 142, Lord Rothermere's "Britain First," soon to be named the Blenheim.[33] Under Scheme F, the RAF would receive 750 Blenheims for the new medium bomber squadrons.[34] In addition, the RAF planned to add the heavier Pegasus-powered Handley Page Hampden, Vickers Wellesley, and Wellington bombers to form a more capable bomber force as a deterrent to Germany's growing air power.[35] At the recommendation of the Committee on Defence Policy and Requirements, the planned war reserve was increased to 150% of first-line strength.[36] With the re-equipment of light bomber squadrons and an increase in squadron aircraft strength from 12 to 18 aircraft for 29 of the new medium bomber squadrons and provision for 150% reserves, Scheme F called for the production of 8,009 aircraft of all types by its March 31, 1939 completion date.[37]

The number of aircraft required for Scheme F, and the deadline for completion, had an immediate impact on the demand for aircraft engines. The RAF had already placed an order with Bristol for 150 Blenheim Mark I aircraft with Bristol Mercury VIII engines and in a few months would place a second order for 434 more while arranging for expanded production.[38] In addition, the new Handley Page Hampden and Vickers Wellesley and Wellington bomber prototypes then under test, if ordered in quantity, would obviously require large numbers of Bristol Pegasus engines. It was clear to the Air Ministry that Bristol alone could not satisfy the RAF's requirements for Bristol engines.[39]

The Cabinet approved Scheme F on February 25, 1936.[40] Even before the scheme received approval, the Air Ministry realized that the existing aviation industry did not have the capacity to produce the required number of aircraft and aircraft engines in the time allowed and had decided that the motor industry had to be brought in to expand production, setting up an industry that would "shadow" the main aircraft and aircraft engine manufacturers.[41] That same month, the Committee of Defence Policy and Requirements came to the same conclusion, recommending that the Government create a shadow industry to expand the capacity for aircraft engine production and to bring at least some of this capacity into operation in peace time.[42] The objectives of the shadow scheme were to supplement production at the main aircraft engine firms to meet the requirements of Scheme F and to create a body of experience in aircraft engine production and reserve production capacity in the motor industry so that the motor firms could be rapidly converted into aircraft engine production in the event of war.[43] The Air Ministry decided to approach the main British motor firms—Austin Motors, Clement-Talbot Ltd., Daimler Motors, Rover Motors, the Humber, Clement-Talbot, and Sunbeam companies in the Rootes Securities group, Singer Motors, Standard Motors, and Wolseley Motors—to seek their assistance in setting up the shadow industry while at the same time contacting the Bristol Aeroplane Company.[44]

On March 11, 1936, a representative of the Air Ministry met with a managing director of Daimler Motors to discuss the prospect of the firm manufacturing aircraft engines.[45] The Daimler director explained that at Bristol's request, Daimler had assisted Bristol in completing a major contract, making cylinders, and connecting rods and other engine components without undue difficulty. Based on this experience, he believed that Daimler would have no problem building complete aircraft engines or engine components, though he doubted Bristol would be willing to have the motor firms build complete Bristol engines. Following this meeting, Sir Alexander Roger, Daimler's chairman, wrote to the Secretary of State for Air, now Lord Swinton, to offer the Company's technical knowledge and production experience with aircraft engines to the Air Ministry.[46] Later that month, Lord Swinton met with Captain J.P. Black, Managing Director of Standard Motors, to outline the proposed shadow scheme and to ask for his participation.[47] Like Daimler, Black raised no objections to the idea and said he would be willing to undertake the production of aircraft engines under the proposed shadow scheme. Swinton informed Black that he would soon be getting a number of motor firms together to decide on the best method of production under the shadow scheme. He asked Black if it would be preferable to have the motor firms build complete aircraft engines or to divide the production among the participating firms. Black replied that he thought the latter approach was preferable.

While reaching out to the motor firms, toward the end of February, the Air Ministry contacted the directors of the Bristol Aeroplane Company to advise them that the Ministry was considering placing substantial additional orders for Bristol aircraft engines beyond the requirements for the planned 1936–1937 expansion program, to be completed by March 1939.[48] In total, the Air Ministry had a requirement of 8,200 Bristol engines.[49] At the request of Bristol's directors, Norman Rowbotham, the manager of the Bristol Aircraft engine factory, calculated that with an expenditure of £2,372,813 ($9.5 million), Bristol could expand its production facilities to build 7,553 engines, just short of the Air Ministry's requirement.[50] On March 13, 1936, Lord Swinton, and Lord Weir, Swinton's advisor, met with Sir Stanley White, Bristol Aeroplane Company's Managing Director, and H.J. Thomas, Assistant Managing Director, to explain the Government's plans to create a shadow industry to expand aircraft engine production.[51] Lord Swinton explained that the Government intended to place educational orders with the leading motor firms for 1,500 engines to develop the technical knowledge and experience to convert the motor industry into aircraft engine production in an emergency. Swinton said that the Government was keenly interested in obtaining Bristol's new sleeve-valve engines but was concerned about the production problems Bristol was then experiencing. The Government wanted these engines but not at the cost of imperiling production. Roy Fedden, Bristol's chief aircraft engine designer who had been brought into the meeting, replied that it would be safer to build the Bristol Mercury engine.[52] While Bristol believed it could build the required number of engines by March 1939, Lord Swinton stated that it was vital that the firms outside the aircraft engine industry that would be called upon to build aircraft engines in an emergency had obtained experience beforehand.[53] This was why he was proposing giving educational orders for 1,500 engines to the motor firms. In a discussion of what the selected motor firms should undertake, Rowbotham suggested that the construction of complete engines by these firms would result in unnecessary duplication. It would be more effective, he argued, and far cheaper in terms of capital expenditure on factories, machine tools, and labor to have the firms specialize in certain components.[54]

Following his meeting with the Bristol directors and with Captain Black, Lord Swinton sent a letter on March 24 to the directors of the Austin, Daimler, Morris, Rootes, Rover, Singer, and Standard motor firms that had been allocated to the Air Ministry for war production outlining the Government's proposed shadow scheme to enlist civil industry in aircraft engine production to meet current requirements and to create the potential for large-scale production in the event of war.[55] Swinton was anxious to discuss with the motor firms the best means of achieving these objectives and determine whether large-scale production during wartime could be obtained by having each firm make complete aircraft engines or components, the same question he had put to Captain Black. He asked if the firms would be willing to send a representative to a meeting at the Air Ministry on April 7, 1936 to discuss these questions.

A week before Lord Swinton's meeting with the motor firms, the Bristol Aeroplane Company submitted a proposal to him in response to his request for ways in which Bristol could expand its own output of aircraft and aircraft engines.[56] Swinton had asked Bristol if it could expand its own facilities during the next expansion period or agree to set up a shadow factory alongside its existing works.[57] The Company responded by pointing out that it had used its own capital to finance its expansion for the 1935–1937 period and with its current facilities it could commit to produce 30 aircraft engines a week, totaling 2,910 engines by the March 31, 1939 completion date of Scheme F, though this would leave a deficiency of 2,490 engines under the Air Ministry's planned requirements for 5,200 poppet-valve engines and 1,700 sleeve-valve engines. If Bristol added an additional 260,000 ft^2 of factory space, at an estimated cost of £1,149,500, and acquired the necessary machine tools at an additional cost of £329,000 ($1.5 million), then the company could commit to increase the production to 40 poppet-valve engines a week and begin manufacturing sleeve-valve engines in October 1937 at a rate of 10 per week, growing to 30 per week by May 1938. This would total 3,700 poppet-valve engines of the Mercury or Pegasus type, and 1,700 sleeve-valve engines, closer to the Air Ministry's requirement. Bristol suggested that the cost of this expansion could come through advance payments from the Air Ministry.

However, Bristol was reluctant to use its own capital for this expansion as the company did not want to end up with excess capacity once the expansion program ended. Bristol would only go forward with this second expansion if the Government would provide reasonable compensation for redundant facilities. The company did not object to the idea of setting up a shadow factory next to its existing machine tools but expressed its general concern with sharing technical information and designs with the motor firms who might become competitors. Bristol's memorandum stated clearly that the company "was disturbed at the prospect of the Company's designs and technical assistance, incorporating years of costly research, being issued to existing large organizations which by their resources are capable of becoming competitors of the Company in future years."[58] Bristol wanted an undertaking from the Government that the motor firms participating in the shadow industry would not use any information Bristol provided to make aircraft or aircraft engines for anyone apart from the British Government. This concern was likely aimed at Morris Motors, as the Company's chairman, Lord Nuffield, owned Wolseley Aero Engines Ltd., a direct competitor. Bristol was particularly concerned about having the motor firms build complete aircraft engines and reminded the Air Ministry that it would be cheaper and more efficient to have the firms build components instead.

Lord Swinton met with Air Marshal Sir Cyril Newall, the Air Council's Air Member for Supply and Organization and the Director of Contracts for the Air Ministry on April 2 to discuss Bristol's memorandum.[59] They concluded based on Bristol's estimates of production of 2,900 aircraft engines with its current facilities that the shadow industry should manufacture 4,000 Bristol engines to make up the shortfall in the Air Ministry's requirements. Swinton and Newall were sensitive to Bristol's concerns about possible future competitors and agreed to withhold the production of the Bristol sleeve-valve engines, which was a proprietary technology, from the shadow industry firms, unless a change in the international situation made this necessary. They also agreed that the participating firms could covenant that they would not produce aircraft engines outside of the shadow scheme.

On April 7, 1936, Lord Swinton met with the representatives from six motor industry firms—Austin Motor Company, Daimler Motor Company, Rootes Securities, Rover Motor Company, Singer Motor Company, and Standard Motor Company—and despite Bristol's apparent reservations—a representative from the Wolseley Aero Engines Ltd., to explain the Government's plan for the shadow aircraft engine scheme.[60] Lord Swinton outlined the Government's plan to finance the construction of factories and machine tools for aircraft engine production and the need to complete 4,000 Bristol engines of the Mercury and Pegasus type by March 1939. He then posed two questions to the group: first, what was the most reliable method to achieve this objective, to have each of the firms build complete engines, or to have each of the seven firms build separate components of the engine and second, how rapidly could the parent motor firms turn over their existing factories and machine tools to aircraft engine production in the event of war?[61] None of the men present, all senior executives with considerable experience in automobile production, expressed any reservations about the task they were being asked to take on nor questioned the plan of converting their parent firms into aircraft engine production in an emergency. The consensus of the representatives was that it would be more efficient to have each firm build separate components of the engine, as Bristol's Rowbotham had suggested, as some firms were better equipped to undertake certain operations, and building different components might ease the demands for skilled labor, though at least one firm would have to be responsible for assembly and testing of completed engines.[62] All agreed that while some of the specialized machinery used in auto production could probably not be adapted to building aircraft engines, it would be relatively straightforward to adapt the standard machine tools in use in the motor firm factories to the task.[63] Lord Swinton suggested that the seven firms form a committee to work out the details of the shadow aircraft engine scheme with the Air Ministry.

Following Lord Swinton's suggestion, at the meeting on April 7, the participating firms formed the Aero Engine Committee to coordinate discussions among the firms and with the Air Ministry and the Bristol Aeroplane Company and appointed Sir Herbert Austin, chairman of Austin Motors, as chairman of the Committee. The Committee's first order of business was to confirm the recommendation that each of the firms should build components of the selected Bristol engine and allocate component production among them. To facilitate the Committee's deliberations, Austin decided to visit the Bristol Aero Engine Department at Filton to observe production of the Mercury and Pegasus aircraft engines and to examine the component parts of each engine in detail before the Committee's next meeting.[64]

Austin's examination of the factory layout, the machine tools employed, and machining operations for aircraft engine components reaffirmed in his mind the consensus opinion that the best way to complete the Air Ministry's order on time would be to have each participating firm in the shadow scheme build a component section of the selected engine, with the machining work divided up into equal total man hours.[65] Although Austin did not see any machining operations that he considered particularly difficult, he did note that some operations were "less easy than others and all have to be very carefully and accurately carried out—in fact, the chief point which still need [s] to be guarded is the definite necessity of reliable and painstaking manufacture and inspection."[66] Austin asked Norman Rowbotham, Bristol's production manager, to divide a Mercury or Pegasus engine into seven sections that could be assigned to each of the motor firms. When meeting with Air Marshal Sir Cyril Newall a few days later, Austin told him how impressed he was with the Bristol factory's layout and organization to the extent that he did not think that the motor firms could "materially improve the manufacturing methods adopted at present by the Bristol Company," particularly if adopting new production methods risked interfering with the interchangeability of components.[67]

At its next meeting on April 20, the Aero Engine Committee drafted a scheme for the aircraft engine shadow program calling for each of the seven firms to begin acquiring land, erecting a factory, and obtaining the necessary machine tools, jigs, and fixtures to manufacture the selected Bristol aircraft engine type, with the Air Ministry providing financing.[68] Each of the firms would manufacture one component of the engine, and the Aero Engine Committee would arrange the allocation of components based on an equal number of man-hours employed in the component's production. Two firms, Austin in Birmingham and Daimler in Coventry, would be responsible for assembling and testing complete engines. The factories in the shadow scheme would be separate from their parent firms, which would become available for aircraft engine production only in the case of a national emergency. The Government would pay each firm a management fee and a bonus for savings achieved during production. The Aero Engine Committee submitted the draft scheme to the Secretary of State for Air, who met with Sir Herbert Austin at the end of the month to review the scheme in more detail.

At this meeting, Air Marshal Newall brought up the Air Council's unease with the proposed scheme of dividing production among seven firms, the "1-7" arrangement, a point he had already raised with Austin, and that Lord Swinton shared.[69] While Lord Swinton and Air Marshal Newall recognized that the Aero Engine Committee's proposal might be the most economic and quickest means of meeting the Air Ministry's requirements by March 1939, they had concerns over the potential risks in the "1-7" arrangement and the implication for war production.[70] There was a risk in wartime that destruction of one factory would disrupt the production of complete engines or, as the Secretary of Air suggested, the same disruption could happen if one of the firms proved unable to produce its assigned component to the required standard. More importantly, there was the issue of preparing the parent motor firms for production in wartime, one of the key objectives of the aircraft engine shadow scheme. Lord Swinton questioned whether the production of a single component of an aircraft engine would provide adequate technical knowledge and experience to prepare the parent firms for the manufacture of multiple components or complete engines during the time of war. Lord Weir, who was in attendance, reminded Sir Austin of the differences between manufacturing motor engines and aircraft engines, particularly in regard to the greater need for accuracy and meticulousness in aircraft engine production. His concern, shared with Newall, was whether the "1-7" scheme would make the transition to full war production more difficult. Would it not be better, he asked Austin, to divide the firms into two or possibly three groups to make complete engines even if, as the Air Ministry recognized, this would add significantly to the cost of factories and machine tools and might delay aircraft engine production?[71]

Austin replied that he didn't think that the "1-7" arrangement would cause difficulty in wartime. He was confident that there would be no "weak link in the chain" as he put it and that if one firm should run into difficulties the other six would be available to assist. He was also confident that even if the seven firms were not building complete engines, the experience their shadow factories would collectively gain could successfully be brought to expanding wartime production at their parent firms. Apparently, Lord Swinton was still not convinced and wanted a definitive answer from all seven firms on the suitability of the "1-7" arrangement. He asked Austin to submit a questionnaire to the Aero Engine Committee asking if the Aero Engine Committee's scheme would make it easy to shift to producing multiple components during the time of war using their parent motor works or if the firms should make two or three components instead.[72]

The Aero Engine Committee responded to Lord Swinton's questions with a unanimous opinion that the proposed arrangement was the only "safe and practical scheme" to meet the Air Ministry's requirements by the set deadline.[73] Lord Swinton brought up the issue one last time at a progress meeting with the shadow firms at the end of June. While several of the firms said they would have preferred to build complete engines, all agreed that the Air Ministry's program could not be completed on time any other way, without further delays and significant costs.[74] Having teams of firms build complete engines would have required a huge increase in initial costs for factory and machine tools and would have placed a heavy burden on Bristol's ability to provide technical assistance.[75] In these series of meetings and discussions on the organization of production the shadow aircraft engine industry should employ, there was little mention of the methods of production. Only at the meeting with Lord Swinton in late June did the subject come up in discussion. Mr. Leonard Lord, the managing director of Wolseley Aero Engines Ltd., agreed that specialization was the "simplest and most direct method" to achieve the Air Ministry's goal but suggested that the Air Ministry might "break away from the methods of manufacture used by the Bristol Company and, by intensive study, design new and simpler methods which would enable production to be reached much more rapidly."[76] In response, Sir Herbert Austin argued that Lord's idea involved too great a risk and that the shadow firms should adopt the Bristol Company's designs and processes as a far more straightforward approach to the challenges of production.[77] Lord Swinton and the Air Ministry accepted the recommendation of the

Aero Engine Committee to go forward with their proposed scheme, but the Air Council's concerns over the "1-7" arrangement were not unfounded and were not forgotten.

On June 30, the day after their meeting with Lord Swinton, the members of the Aero Engine Committee with other representatives of their companies and representatives from the Air Ministry travelled to Filton for meetings with Bristol's directors and production managers.[78] There were 15 items on the agenda for the meeting, the most important of which was to determine the sections of the engine each firm would make.[79] The motor firms needed details on the size of factories and the type of machine tools required to manufacture their assigned components, sources of supply especially for the jigs and fixtures that would be needed, completion dates for factory construction, a program for engine deliveries and an agreement as to the extent of technical assistance the Bristol Aeroplane Company would provide.[80] Before the formal meeting began, the members of the Committee toured the Bristol aircraft engine works and looked at a disassembled engine to examine the components they would be making.[81] Norman Rowbotham had a Pegasus engine disassembled into five sections as an example and for each section provided the members of the Committee with a list of materials, types of machine tools, and floor space required for manufacture.[82]

At the meeting, Sir Herbert Austin told the other members of the Aero Engine Committee that Bristol had agreed to determine how to divide its engines into sections and allocate the sections to different firms. Norman Rowbotham said that Bristol could provide detailed drawings and specifications on engine components within a month of the Air Ministry deciding on which aircraft engine it wanted the shadow factories to manufacture, but he warned that detailed drawings of the jigs, fixtures, and tools required would take an additional two to three months. H.A.P. Disney, the newly appointed Director of Aeronautical Production at the Air Ministry, informed the Committee that the Ministry wanted the shadow factories to be designed to have the capacity to produce up to 65 engines per week on a day shift only, though the Committee pointed out that this would require larger factory buildings than had been originally envisioned and additional costs to the Government. This change would also delay the determination of completion dates and a production schedule. At the meeting, Sir Hubert Austin and Sir Stanley White, Bristol's Managing Director, came to an agreement on the technical assistance Bristol would provide the shadow factories and confirmed this in a joint letter to the Air Ministry. Bristol agreed to act as technical advisor to the shadow factory and to provide information on "all technical matters which are necessary for the successful manufacture of engines to the requisite standard of workmanship" but also on standards for factory and machine tools.[83] Bristol offered access to its works and its staff and agreed to provide not only full details of all components but also of all necessary jigs, fixtures and tools, and manufacturing processes. In addition, Bristol offered to set up a small staff devoted to liaison with the shadow factories to coordinate communications on all technical and production issues.

A few days after the meeting in Bristol, Disney wrote to Sir Hubert Austin to advise him that the Air Ministry had decided that the shadow factories should concentrate on the production of the Bristol Mercury VIII engine and that production of 4,000 of these engines by March 1939 remained the priority.[84] With the choice of engine decided, Bristol began working out the allocation of sections to each of the seven firms, but over the following two months, two of the firms, Singer Motors and Wolseley Aero Engines, dropped out of the shadow scheme. In the first case, the Air Ministry decided that Singer should not participate as the Ministry did not consider that the firm had an organization with the experience to undertake the project.[85] Wolseley Aero Engines went through several weeks of off-again-on-again participation in the shadow scheme until Lord Nuffield finally decided to withdraw due to his strained relations with the Air Ministry and his opposition to the shadow scheme in principle.[86] Wolseley Aero Engine's departure was a benefit to Bristol, who had objected strongly to the Air Ministry's proposal to have Wolseley, Bristol's competitor, take on the assembly and testing of complete Mercury engines.[87] But with the departure of Singer and Wolseley, the Air Ministry had to work out the allocation of production among five firms instead of seven and find a replacement for Wolseley who were also expected to manufacture the rocker gear for the Mercury engines.

Lord Swinton appealed to Bristol for help. Verdon Smith agreed that Bristol could take on assembly and testing of half the output of the remaining five shadow factories, sharing the burden with Austin, but notified Swinton that Bristol could not manufacture the Mercury rocker gear which would have to be allocated to another firm.[88] This change led to a reallocation of Mercury engine components to the participating firms with the final assignments agreed to at a meeting of the Aero Engine Committee in early October. These were (Table 2.1):

TABLE 2.1 Reallocation of mercury engine components.

	Assignment	Hours	Total hours
Austin motors	Crankshaft and reduction gear	426	676 hours
	Engine assembly and test	250	
Daimler	Crankcase front cover and oil sump	299	729 hours
	Rocker gear	430	
Rootes	Blower, crankcase rear cover	620	700 hours
	Petrol pump	80	
Rover	Connecting rods, pistons	632	715 hours
	Cam and tappet gear	83	
Standard motors	Cylinders and subassembly		676 hours
Bristol aeroplane	Engine assembly and testing		250 hours*

* Notes on a meeting of the Aero Engine Committee at the offices of the Standard Motor Company, October 8, 1936, AIR 2/2577, National Archives, Kew, UK.

This allocation of components remained in place until the Mercury engine went out of production in 1944. By the end of 1936, the terms of the agreement with the Air Ministry had been decided, a decision taken to proceed with the manufacture of the Bristol Mercury VIII engine and the component parts of the engine allocated among the five motor firms, while Bristol had started the transfer of drawings and technical information so the firms could prepare for production. The next phase was to complete the factories to produce engines.

The Factories

The nature of the product, the production processes, labor and machine tools required for its production, and the quantity desired are the key factors in determining the size of a factory.[89] During Sir Hubert Austin's first visit to the Bristol aircraft engine factory, Norman Rowbotham explained that because the planned shadow factories would be building components instead of complete engines, they would not need a large factory space. He estimated that a floor area of approximately 100,000 ft^2 would be adequate to produce around 20 engines per week, with space for machine tools, storage of materials and finished components, heat treatment and nitriding facilities, inspection, and assembly. In his letter to Sir Hubert Austin confirming the Air Ministry's decision to build the Bristol Mercury in the shadow factories, H.A.P. Disney clarified the Air Ministry's proposed output, stating that the Ministry wanted the firms to plan their factories on the basis of producing 50 sets of components per week on a single day shift.[90] In comparison, at this time the Bristol Aero Engine Department was completing 30 engines a week, primarily the Mercury and Pegasus poppet-valve engines with a small number of the new sleeve-valve Perseus engine, with a plan to expand production to 36 a week.[91] A year later, Bristol had built up production to 40 engines a week.[92] Setting a target of 50 engine sets a week for the shadow factories does not seem to have been unreasonable. As the Aero Engine Committee noted, this would increase the cost of the factories as producing the required output with a single shift would require more machine tools, thus more space. This, apparently, was acceptable to the Air Ministry, though Disney noted in his letter to Austin that the factories should be of a size sufficient to meet the Air Ministry's requirement of 4,000 engines and no larger. Room for expansion, he said, would not be necessary.[93] With the Air Ministry determining the type of aircraft engine to be built and the level of output to be achieved and with the component parts allocated to different firms, Bristol could advise each firm on the type and number of machine tools required with a suggested layout and from that determine the factory floor space needed for the machine tools and other shops (Figure 2.3).[94]

FIGURE 2.3 The Standard Motors aircraft engine shadow factory located near the main Standard Motors automobile works in Coventry. Proximity to the main Standard Motors factories was intended to facilitate their speedy conversion into aircraft engine production in the event of war.

EPW053118_The_Standard_Motor_Works_Coventry_1937, Historic England.

Easing access to labor and management at the parent motor firms, and the goal of building war potential, determined the location of the shadow factories. In 1934, the Air

Council, in considering the risk of air attacks on Britain, had divided the country into what was deemed relatively safe areas, west of a line drawn from Weston-Super-Mare in Somerset to Haltwhistle in Northumberland, and dangerous areas deemed to be in the southeast, south and eastern areas of Britain.[95] Birmingham and Coventry, the centers of the motor industry in Britain and the locations for the five planned shadow factories fell in between but were deemed "unsafe" areas.[96] Nevertheless, the need for the shadow factories to have ready access to management and labor from their parent firms was the paramount concern and outweighed the risk of placing the shadow factories next to their parent motor works. The Austin and Rover shadow factories were adjacent to their parent works in Birmingham, while the Daimler, Rootes, and Standard shadow factories were located in a ring around Coventry. The Air Ministry, too, recognized that during the time of war converting the parent motor firms into aircraft engine production would take the closest possible coordination with their shadow factories, an additional reason to have the shadow factories close by.[97] Countering the argument that a bomber aiming at one factory could hardly miss hitting another, a journalist reported that the authorities "are satisfied that the propinquity of the factories to each other is not dangerous and indeed argues that the factories are extremely difficult to identify amid the sprawling factories of the Midlands."[98]

These first aircraft engine shadow factories built for the Air Ministry were "planned to a lavishness which contrasts with the stringency that was to be the rule later" and were built using "normal peacetime standards and requirements of the managing firms."[99] Disney had requested each firm to submit as quickly as possible details of the land selected for the factory site, the number of machine tools required, and the proposed layout (Figure 2.4).[100]

FIGURE 2.4 A contemporary map showing the plot of land next to the Daimler Radford automobile works in Coventry that the Air Ministry purchased from Daimler to build the first Daimler shadow factory in 1937.

Source: Daimler Coventry: purchase site for shadow factory, AIR 2/2324: Aircraft Production: Shadow Factories (Code B, 6/2), National Archives, Kew, UK.

Daimler Motors submitted its initial proposal at the end of August, requesting financing to build a factory next to its Radford, Coventry main motor works, with 122,400 ft^2 of space (apparently later increased to 172,000 ft^2), ancillary buildings amounting to 14,000 ft^2, and offices of 8,400 ft^2 at an estimated cost, with heating, lighting, and other services of £118,000 ($472,000).[101] The Standard Motor Company shadow factory covered a total area of 230,000 ft^2, with the machine shop taking up 132,000 ft^2, while the total area of the Rootes and Rover shadow factories was 237,000 and 170,000 ft^2, respectively.[102] The machine shops at each factory varied in size based on the number and type of machine tools required for manufacturing the allocated Mercury engine component. In contrast to the dedicated buildings of the other factories, Austin Motors placed its aero engine manufacturing and assembly operations in the same 670,300 ft^2 shadow factory where it was building airframes, allocating a portion of the factory for the machine tools required for engine component production and space for the assembly of completed engines (Figure 2.5).[103]

FIGURE 2.5 A new shadow factory under construction in Coventry. These single-story structures followed American industrial building designs. The many windows to allow natural light would have to be covered over after Britain introduced a total blackout after the war had begun.

Image reproduced by kind permission of Culture Coventry Trust/Coventry Archives.

The shadow factories followed new principles of industrial factory design. After World War I, British industrial architecture adopted the American preference for larger, single-story, steel-framed factory structures instead of the previous multistory brick and reinforced concrete designs. The ideal factory building was seen as allowing "a smooth and unhampered flow" of production from raw materials to finished products.[104] This new factory design was present in the British automobile industry where several firms had adopted forms of line production, arranging machine tools in a sequence of operations rather than by type, requiring wider, more open spaces than the vertical mill factories of the past.[105] These steel-framed buildings had steel roof trusses with steel supporting columns with bays 25 ft wide, adequate for two rows of machine tools, with walls of brick.[106] The buildings typically had a saw-toothed, north-lit roof, with the north-facing slope having some clear material to let in daylight while the south-facing slope was covered with wood, asbestos, or corrugated iron.[107]

The new shadow factories were all single-story buildings with steel beam structures and brick walls, divided into separate bays, with a 6 in. steel-reinforced concrete floor to take the weight of the machine tools.[108] The factories followed "the best practice in large-scale engineering production" where the layout conformed "to straight directional work flow from the material receiving end, through the material view into the approved stores, thence to the shops and, finally, back to the finished stores and dispatch."[109] All the factories were rectangular single-story buildings built to this new factory pattern, divided into separate bays.[110] After a visit to the completed factories, a contemporary observer remarked, "the shops themselves are of the most modern design and air conditioned and have high roofs and a minimum of overhead cables and roof members. Wide concrete gangways separate the lines of machines, each of which has its separate power supply. These conditions combine to give exceptionally light, airy shops and pleasant working conditions."[111] For some of the firms, construction of their shadow factory began toward the end of August 1936, shortly after the Air Ministry approved their plans, and most had been completed by the summer of 1937. In the years leading up to the war and after, these factories would undergo expansion as the Air Ministry increased its requirements for Bristol aircraft engines (Figure 2.6).

FIGURE 2.6 Installing machine tools in a partially completed shadow factory in Coventry in August 1937. Each individual machine tool had its own separate power source. Natural light floods the work area.

Image reproduced by kind permission of Culture Coventry Trust/Coventry Archives.

The machine tools required for the shadow factories cost three to four times more than the factory buildings, though the investment varied by the factory and the number of machine tools each factory required. The number ranged from 439 machine tools at the Rootes shadow factory to 729 at the Rover shadow factory.[112] For each shadow factory, Bristol prepared a folder specifying the types of machine tools needed based on the machining times for each operation necessary for manufacturing the required component, based on Bristol's own experience in its aircraft engine works.[113] Bristol provided as well suggested layouts for machine tools based on manufacturing processes and projected output, designing the layouts for "line production" even if the initial quantities of components did not require this method of quantity production.[114] After providing information on machine tools, Bristol then prepared schedules and detailed information on the jigs, fixtures, tools, and gauges that the shadow factories would need.[115] Each of the shadow factories had a specially built standard room where the accuracy of the jigs and tools could be checked for accuracy with measuring instruments accurate to the ten-thousands of an inch required on many parts of an aircraft engine.[116] The majority of the machine tools for the first shadow factories came from British machine tool makers, and while there were delays in delivery due to the growing demand for machine tools from the breadth of rearmament underway, by October 1937, the shadow factories had most of the machine tools they needed. The Daimler shadow factory had 82% of its machine tools, Rootes had 90% and the Standard Motor Company machine tools had 92%.[117] As *Flight* magazine commented after a visit to the shadow factories that same month, "to have designed, built, and equipped in twelve months five factories with new machine tools embodying the latest practice, and this during a period of increased demand from other industries, is a feat bordering on the miraculous."[118]

With the prospect of larger orders from the Air Ministry and plans for conversion to production of sleeve-valve engines, the Bristol Aeroplane Company had been expanding its own production facilities at its own expense. In early 1936, Bristol completed a new aircraft engine factory of 200,000 ft^2, adding a second building during the first half of 1937.[119] This second building accommodated the engine fitting, subassembly and erection sections, as well as sections for manufacturing sleeve valves. Bristol financed these additions but received Government financing for a new building to assemble aircraft engines from the components produced at the shadow factories. This became Bristol Shadow Factory No. 2, next to the main aircraft engine works at the Bristol factory at Filton, costing the Government £384,023, including £100,000 for machine tools.[120] As with the shadow factories, increasing orders for Bristol aircraft engines would lead to further expansions of the Bristol aircraft engine works, financed through the Air Ministry (Table 2.2 and Figure 2.7).

TABLE 2.2 Expenditures on shadow factories 1936–1939.

	Original sanctions	Revised sanctions	Actual expenditure to 11.30.1939
Austin motors	£720,920 ($2.8M)	£701,000 ($2.8M)	£668,227 ($2.7M)
Bristol shadow no. 2	£389,786 ($1.5M)	£390,873 ($1.5)	£384,023 ($1.5M)
Daimler motor	£579,841 ($2.3M)	£764,244 ($3.0M)	£696,939 ($2.8M)
Rootes securities	£542,700 ($2.1M)	£936,959 ($3.7M)	£752,250 ($3.0M)
Rover motor	£708,221($2.8M)	£842,424 ($3.4M)	£796,945 ($3.2M)
Standard motor	£610,000 ($2.4M)	£889,804 ($3.5M)	£627,908 ($2.5M)
Totals	£3,551,468 ($14.2M)	£4,525,304 ($18.1M)	£3,926,292 ($15.7M)

Source: Statistics Relating to the Twelve Air Ministry "Shadow" Factories in Production, Memo of December 1939, "Shadow" Factories: Remuneration of Managing Firms, AIR 19/9, National Archives, Kew, UK.

FIGURE 2.7 Damage to the Rootes shadow factory in Coventry from German bombing on the night of November 14/15, 1940. The damage took some weeks to repair but production resumed fairly quickly.

The main criticism of the Government's shadow scheme was its reliance on a sole-source supply chain system. Maintaining balanced production among all the factories proved to be a challenge. Disruptions did happen during the initial years of production at the factories, as will be explained below, and following the German air raid on Coventry on the night of November 14, 1940.[121] German bombs hit the Daimler Radford works, severely damaging the nearby Daimler shadow factory and the Rootes shadow factory.[122] Fortunately, production was ahead of schedule so although it took five weeks to restore production, the total loss in output for all the shadow factories only amounted to two weeks of production.[123] Following the November raid on Coventry and the Government's push to disperse production, the automotive shadow factories acquired smaller buildings in the surrounding areas, including a carpet factory, a laundry, a mill, and several garages to use as dispersal factories.[124] As it turned out, these designated Dispersal Points did not have to be used for aero engine production, but later in the war became important sources of spare parts.[125]

The Production Record

By September 1937, the shadow factories were preparing to manufacture the Mercury Mark VIII under an initial contract for 1,500 engines and an additional 500 engines in the form of spare parts.[126] Even before the contractors had completed construction of the factories, the management began moving in machine tools and making them ready to begin production starting with small test runs. The Air Ministry hoped that the shadow factories could begin producing complete engines that month, working up to 40 engines a month by December, and then steadily increasing production to 200 complete engines a month by June 1938, but this proved to be optimistic.[127] At the Rover shadow factory, for example, Mr. S.B. Wilks, Rover's Managing Director, reported that the factory had completed its first set of components on September 15, 1937, even though the factory had only received 81% of its required machine tools and 67% of its jigs, and was still short of staff.[128] That same month, Norman Rowbotham reported to his directors that he had received the first set of components from three of the shadow factories and that he hoped to receive the remaining parts and assemble the first shadow-built Mercury engine by the end of the month, commenting on the excellent quality of the parts received.[129] Bristol had agreed to give the first engine a 100-hour type test in addition to the normal endurance test to prove its reliability, the first Mercury VIII assembled from shadow factory parts completed this test in November 1937.[130] This was an important milestone and cleared the way for quantity production.[131] By year-end, the shadow factories were ready to begin production of components, with some factories already ahead of others.[132] Rover and Rootes had run into difficulties, Rootes having problems with defective raw materials while Rover had run into problems in certain machining operations, but the Air Ministry was confident that the shadow factories could deliver 25 sets of components by December and begin full production in the spring of 1938.[133]

In January 1938, the Rover Shadow Factory reported that it now had over 1,300 staff and most of the required machine tools were in place to allow for production at a rate of 20 sets of components a week to be attained by March.[134] Production of complete Mercury aero engines from the shadow factories built up slowly. During January and February 1938, the shadow factories did not produce enough components to build complete engines but produced two complete engines in March, seven in April, and 21 in May, less than the Air Ministry's target.[135] Continued production problems at the Rootes shadow factory and a shortage of carburetors disrupted the output for July and August when no complete engines came out of shadow production, but the shadow factories finally reached and exceeded their goal of 50 Mercury engine sets a week in October, completing 252 complete engines that month.[136] Over the same period, Bristol struggled to reach its planned program of engine production. A shortage of key components and the slow rate of production of the new Hercules and Taurus sleeve-valve engines, which were effectively being built by hand, meant that by October 1938 Bristol was building less than 40 engines a week instead of the planned 48 a week.[137]

Demands from the Air Ministry for more aircraft engines were growing. On March 30, 1938, 18 days after the *Anschluss*, the Air Council informed the Aero Engine Committee that the Council had decided to go ahead with the full order of 4,000 aircraft engines from the shadow factories but had decided that these additional engines should all be the Mercury VIII, instead of a mix of Mercury and Pegasus types as had originally been proposed.[138] This decision took place against a background of intense debate over the composition and cost of the next RAF expansion scheme. Scheme L, which followed Schemes H, J, and K, received Cabinet approval on April 27, 1938 and called for an increase in total RAF aircraft strength but a reduced number of bombers in favor of more fighter aircraft.[139] But there was a dramatic shift in the timing of the expansion. Scheme L was to be an accelerated program, with the new total strength level to be completed by March 31, 1940 instead of March 31, 1941.[140] This program required a 170% increase in aircraft production over the 1937 level and called for the production of 12,000 new aircraft in two years.[141] Obviously, this would create a great demand for aircraft engines and there was now some concern in the Air Ministry to ensure that aircraft engine production kept pace with airframe production. At the end of April, Air Marshal Sir Wilfrid Freeman, Air Member for Research and Development, wrote a note to the Chief of the Air Staff expressing his concern about engine production. Freeman noted that "we have been inclined in the past to think that all was well with the engine production, but this is only so because aircraft production has been much slower than we expected. If aircraft production had been quicker the engine people would have been hard put to it to keep level. It is, in my opinion, that our engine needs should be examined and set out in the immediate future."[142] Freeman said it was obvious that an accelerated aircraft program would require an accelerated aircraft engine production and that a substantial aircraft engine production program would mean expanding production capacity at Bristol and at Rolls-Royce.[143]

H.A.P. Disney had already been in touch with Bristol about the ability of the firm to expand aircraft and engine production. Bristol had replied to his inquiry stating that even with a night shift aircraft engine production could only be increased from 40 to 44 engines per week.[144] Verdon Smith, Chairman of Bristol, met with Lord Swinton on April 28 to discuss how Bristol could accelerate aircraft production to meet the goals set for March 1940.[145] Smith explained to Swinton that this could be done but only with greater capital expenditure. Bristol was reluctant to expand its facilities using its own capital, but Smith assured Swinton that Bristol would help in every way possible if the Government would finance the factory and machine tools necessary to expand aircraft production.[146] In a follow-on letter to the Air Council Sub-Committee on Supply, Bristol pointed out that "if large numbers of additional engines will also be required, as we understand the case, it will be necessary to make considerable extensions of the engine works involving large additional capital outlays."[147]

Bristol representatives met with the Air Ministry on May 16 to review the accelerated engine program in more detail. The Air Ministry informed Bristol that there was a requirement for the following types of Bristol aircraft engines by March 31, 1940 (Figure 2.8):

- 475 Pegasus
- 900 Taurus
- 700 Hercules
- 675 Perseus
- 2,750 total

FIGURE 2.8 The Bristol Pegasus powered the Vickers Wellington Mk. I twin-engine bomber, shown here, and the twin-engine Handley Page Hampden. These two aircraft, with the Armstrong Whitworth Whitley, formed the Royal Air Force's bomber force at the beginning of the war.

2008-03-31_image_595_01, Peter M. Bowers Collection, Museum of Flight, Seattle, WA.

The Air Ministry needed more Pegasus engines for the 564 Vickers Wellington and 430 Handley Page Hampden bombers it had on order, the 900 Taurus and 675 Perseus sleeve-valve engines for the new Bristol Beaufort and Blackburn Botha General Reconnaissance aircraft, and the Hercules sleeve valve for the B.12/36 four-engine heavy bomber (the Short Stirling).[148] Mr. Daniel of Bristol remarked that this total was 1,690 more engines than the Company's planned production up to March 1940.[149] Norman Rowbotham, also in attendance, said that while it would be possible, though not without difficulty, to obtain additional facilities and machine tools, the real problem with meeting the requirement would be a shortage of labor. Air Marshal Freeman, also at the meeting, suggested to the Bristol representatives that it would be possible for the Company to make use of the shadow factories to fulfill part of the new program.[150] The Air Ministry had already been considering changes to the aircraft engine shadow factory program, including the question of shifting the shadow factories from the production of the poppet-valve Mercury and Pegasus engines to the Bristol sleeve-valve types.[151] This was partly in response to repeated requests from Lord Austin and the Aero Engine Committee who wanted to know what type of Bristol aircraft engine would follow production of the Mercury VIII and partly for the need for forward planning for aircraft engine production.[152]

Verdon Smith, Daniels, and Norman Rowbotham met Air Vice-Marshal W.L. Walsh, Air Member for Supply and Organization, Air Marshal Freeman, and other representatives of the Air Ministry on June 2 to review the revised aircraft and aircraft engine programs.[153]

Bristol had submitted a proposal to the Air Council covering the cost of expanding the aircraft engine works at Bristol, which Rowbotham estimated to be £1,926,000 ($7.7 million) for new facilities and machine tools, requesting Government financing for the expansion.[154] The directors had decided against proposing to build a new factory in the Midlands, at Government expense, that Bristol would manage as this would simply be another shadow factory which the directors found unsatisfactory. Bristol also proposed shifting production of the Pegasus engine to the shadow factories. If the shadow factories could take on the production of 642 of the Pegasus engines and an additional 353 Mercury engines on order from Bristol, this would leave Bristol with 3,686 engines to produce by March 1940, 812 engines over their current program as opposed to the 1,690 engine shortfall initially calculated.[155]

Air Marshal Freeman objected to switching the production of the Pegasus to the shadow factories, as he thought that once the production of the Pegasus had ended, the shadow factories would be left with considerable machine tools good only for manufacturing what would by that time be an obsolete engine. He argued strongly in favor of the shadow factories switching to the new and more promising Bristol Hercules. Rowbotham pointed out to Freeman that the shadow factories could not begin production of the Hercules until 1940 and that manufacture of the sleeve valve and other components for the Hercules would require a considerable amount of new machine tools and new jigs and fixtures. More importantly, the latest Hercules with a two-speed supercharger offering greater power had yet to pass its 100-hour type test.[156] The difference between the Mercury and Pegasus engines was slight, Rowbotham argued, and he fully expected that the shadow firms could take on production of the Pegasus and complete the Air Ministry's orders by December 1939. Freeman appears to have finally accepted the Bristol proposals, but asked that an "ad hoc" committee be formed of representatives of the Air Ministry, the Aero Engine Committee, and Bristol Aeroplane Company to pursue the question of shifting the shadow factories to the production of the Hercules after completion of Mercury production.[157]

H.A.P. Disney, who had attended the June 2 meeting, then called a meeting with Lord Austin and the members of the Aero Engine Committee, together with representatives of Bristol, to explain how the accelerated aircraft and aircraft engine programs would affect the shadow factories.[158] Disney explained that the Air Council had been reviewing the accelerated aircraft program and after discussions with Bristol had decided that the shadow factories should take on the production of the Bristol Pegasus XVIII engine, with an initial order for 1,000, in addition to the existing order for 4,000 Mercury engines, to be completed by the end of 1939. He asked if the Aero Engine Committee would agree in principle to take on this new order. Timing was important and the shadow firms needed to begin production as soon as possible. Disney said that Bristol would continue to manufacture Pegasus engines up to March 1939, after which it was hoped that the shadow factories could take over Pegasus production, beginning with 20 engines a month by April 1938. Lord Austin said the Committee would be willing, but there were questions on how soon Bristol could provide the necessary detailed information on the Pegasus and how different production of the Pegasus would be from the Mercury. Rowbotham and Maj. George Bulman of the Air Ministry assured the Committee that the differences between the Pegasus and the Mercury were slight, the most important difference being the two-stage blower of the Pegasus. This would put most of the burden on the Rootes shadow factory, who were making the superchargers for the Mercury. Although the meeting did not address the issue, it appears that because of the similarity of the Mercury and the Pegasus, the existing allocation of components would simply be replicated with the Pegasus. Rowbotham estimated that the cost of additional machine tools, jigs, and fixtures for the Pegasus would amount to between £70,000 and £80,000 ($280,000 and $320,000) and he saw no difficulty in producing both the Mercury and the Pegasus concurrently with minimal rearrangement of the machine tools.

The Air Ministry moved quickly on both the Bristol and shadow factory accelerated aircraft engine programs. Disney wrote to the Aero Engine Committee on June 23, confirming the order for 1,000 Pegasus XVIII engines, with planning for Pegasus production to start immediately and authorizing the shadow factories to begin ordering the necessary jigs and tools and to submit lists of new machine tools they required.[159] Disney added that an order for a small number of Hercules engines might follow. A month later, the Air Ministry approved expenditures of up to £50,000 ($200,000) for new machine tools for Pegasus production at the shadow factories. That same day, Air Council Committee on Supply met with Bristol to go over the agreed plan for expanding aircraft engine production.[160] While satisfied with Bristol's revised plans for expanding aircraft and aircraft engine production, AVM Welsh, A.M.S.O. informed Bristol that the Committee would have preferred for Bristol to set up a new factory away from Filton, but there was not time. The next day, the Air Council Committee on Supply sent a letter to Bristol saying the Committee had approved the proposed capital expenditures to allow the company to erect new factory buildings, at Government expense for aircraft and aircraft engine production.[161]

The agreed program of aircraft engine production would see Bristol gradually phase out the production of the Mercury and Pegasus poppet-valve engines to concentrate on the sleeve-valve Perseus, Taurus, and Hercules with the expectation that by June 1939 production of the Mercury and Pegasus would cease.[162] This program committed Bristol to produce 4,270 engines by March 1940. The Government would finance the construction of a new 200,000 ft^2 machine shop with 36 bays at a cost of £195,800 ($783,200), with an additional £70,600 ($306,000) for new engine test houses. The Government would commit £416,000 ($1.6 million) to purchase 500 new machine tools to equip the new factory, for a total expenditure for all items of £924,200 ($3.7 million). Freeman was correct in his assessment that the Pegasus, and the Mercury, were by this point obsolete. The final versions of the Mercury had a maximum power of 870 hp, while the last versions of the Pegasus reached 1,010 hp, but the larger and heavier aircraft coming into service needed even more power. The future of the air-cooled radial engine lay with the double-row engines of 14 and 18 cylinders, like the Hercules, with power outputs approaching 1,500 hp and greater. Although this does not seem to have been recorded in the meetings of the time, it made far more sense to have Bristol concentrate its development and production efforts on engines with the potential for greater power and leave the Mercury and the Pegasus, which continued in service until the end of the war in minor types, to the shadow factories. While Bristol received more orders for the Mercury and the Pegasus leading up to the war, the company stopped manufacturing the Mercury during 1940, continuing on with the Pegasus until 1942.[163]

The Munich Crisis in September 1938 brought calls from the Air Ministry for even greater production. At a meeting of the Aero Engine Committee on September 19, Lord Austin said that he had received a verbal notice that the Air Ministry wanted the shadow factories to push the production up to full capacity.[164] On October 7, Maj. George Bulman, recently appointed Director of Aeronautical Engine Production at the Air Ministry, told the Aero Engine Committee that the Government had decided to "double the energies in rearmament."[165] Bulman was concerned that as aircraft production ramped up the shadow factories might fall behind schedule. After some discussion, the Committee decided that 75 engine sets per week were probably the maximum balanced level of production the shadow factories could achieve, as this was the maximum number of blowers the Rootes factory could produce. Bulman replied that he was in discussion with Rootes about expanding their shadow factory. The Air Ministry was also addressing a continuing shortage of carburetors that was a drag on shadow factory production. In early November, the Air Council Committee on Supply approved the expenditure of £70,000 ($280,000) (£20,000 ($80,000) on buildings and £50,000 ($200,000) on machine tools) for the Rootes shadow factory to expand its capacity.[166] Following the Munich Crisis, the Air Ministry informed Bristol that it wanted the company to produce an additional 659 engines, 409 Pegasus types, and 250 Mercury.[167]

It was becoming clear to the Air Ministry that the initial estimates of aircraft engines required in an emergency were inadequate and would require a substantial increase to match the production of airframes.[168] More critically, the existing aircraft engine production capacity to meet this increased requirement was insufficient. It had also become apparent that the central premise of the shadow factory scheme, that in the event of an emergency the parent automotive factories could be rapidly converted into producing aircraft engines, was seriously flawed.[169] A review of automobile factories found that only 30 to 40% of machine tools in the factories could be readily converted into aircraft engine production, requiring 60 to 70% to be replaced. Not only would this be uneconomical but since the automobile factories would not be available until the start of an emergency, the time it would take to acquire and install new machine tools meant that the factories would be unable to begin aircraft engine production until several months after the outbreak of war.[170] The war potential of the automobile industry would be in the labor the industry could provide but not the factories or machine tools. Expanding the existing shadow factories for greater production would only increase their vulnerability to air attacks. The Air Ministry had no practical alternative to relying on the existing shadow factories as they were for building Mercury and Pegasus engines, expanding production by adding more machine tools, and moving to double or triple shifts in the event of war.

More orders for engines from the shadow factories and Bristol soon followed. In early October, the Air Council Committee on Supply had approved orders for an additional 550 Mercury engines and later in the month decided to divide the order between Bristol and the shadow factories.[171] This order apparently instructed the shadow factories to begin the production of a new model of the Mercury, the Mercury XV, as in January 1939 Daimler notified the Air Ministry of increased costs of tooling up for the Pegasus XVIII and the Mercury XV.[172] With the prospect of more orders to follow and the need for creating war potential, the Air Council debated spending an additional £360,000 ($1.4 million) to expand facilities at the shadow factories, with £297,000 ($1.2 million) for new machine tools.[173] By March 1939, using overtime and a partial night shift, the shadow factories had pushed their output of engine sets to 80 per week, but the Air Council estimated that the factories could not produce more than 90 engine sets a week even with three shifts due to the lack of machine tools. The additional allocation would allow the shadow factories to boost output to 120 engine sets a week in an emergency, but apparently, the Air Council deferred a decision on this point.[174] Concerned with maintaining production capacity in the shadow factories, once they had completed their initial orders for Mercury and Pegasus engines, in April, the Air Council Committee on Supply approved orders for an additional 1,036 Mercury and 330 Pegasus engines.[175] On August 1, on the eve of war, the Air Council placed more orders with the shadow firms for 1,200 Pegasus engines for Wellington bombers and for 400 Mercury engines for Westland Lysander Army Co-Operation aircraft and the Miles Master II trainer.[176] The shadow factories had completed their first 39 Pegasus engines during June and that same month completed 342 Mercury and Pegasus engines with a further 54 complete engines in the form of spare parts.[177] By the end of August 1939, the shadow factories had built 3,329 complete engines and an additional 478 as spares and were on track to complete the original order for 4,000 Mercury engines by year-end.[178]

The outbreak of war brought a surge of orders for the shadow factories (now designated No. 1 Shadow Engine Group as the Air Ministry had organized the new No. 2 Shadow Engine Group to concentrate on the Hercules as described in Chapter 4). The Air Council Committee on Supply approved an order for 1,000 Pegasus XVIII engines for 300 additional Wellington bombers ordered at the same time and an order for 1,000 Mercury XV engines on September 23.[179] To meet this demand, the shadow factories had to increase their output to 120 engine sets a week. Prior to the war, most of the factories had been on a single day shift, but now they went to a double-shift system, boosting the production to 350 complete engines in March 1940 and to 646 complete engines in May.[180] Two months after the start of the war, having advised the shadow factories to begin planning to increase their output to 150 engine sets a week, George Bulman sought approval from the Air Council Committee on Supply to allow No. 1 Shadow Group factories to purchase long lead time items to boost production to 150 engine sets a week by June/July 1940 and to plan for the production of at least 6,000 engine sets.[181] In his memo, Bulman said the authorization to support the production of 6,000 engine sets was a minimum, but he thought authorization for 10,000 engine sets would be an even better way of encouraging the shadow factories to build production.[182] On December 7, 1939, Bulman wrote a note to the Air Council stating that in his opinion, a production rate of 150 Mercury XV and Pegasus XVIII engine sets a week would not be adequate as the current aircraft program's needs for engines exceeded this level of output.[183] Bulman now sought approval to allow the shadow factories to order the machine tools they would need to reach 150 engine sets a week, but to purchase additional materials to boost production to 175 engine sets a week as well as to provide the equivalent of 20 complete engines in the form of spare parts. The Air Council approved his proposal on December 19, including the following expenditures to expand the production capacity at each of the shadow factories Table 2.3[184]:

TABLE 2.3 1939 expenditures on shadow factories.

	Machine tools	Buildings and services	Total
Austin Co.	£9,000 ($36,000)	—	£9,000 ($36,000)
Daimler Co.	£22,250 ($89,000)	£41,700 ($166,800)	£63,950 ($255,800)
Rover Co.	£131,030 ($524,120)	£57,046 ($288,184)	£188,076 ($752,304)
Rootes Securities	£208,000 ($832,000)	£60,000 ($240,000)	£268,000 ($1,072,000)
Standard Motor Co.	£37,485 ($149,940)	—	£37,485 ($149,940)
			£566,511 ($2,266,044)

Source: Aircraft Council Committee on Supply, No. 1 Shadow engine production, Memorandum by D.E.P., December 19, 1939, AIR 6/59, National Archives, Kew, UK.

Rootes required the largest investment as it was responsible for building blower units for the Mercury and the more complex two-stage blower for the Pegasus, while Bulman wanted to expand Rover's production of valves for both engines as the wastage on valves was far higher than for other engine components.[185]

As war approached, the production of the Mercury and Pegasus accelerated. By the summer of 1939, production had reached 80 engines a week. The production declined during the first few months of the war as a result of blackout restrictions, problems with transporting materials, and the call-up of workers for the services.[186] During the first half of 1940, there were production delays at the Austin and Rootes factories but Bristol and the shadow factories attained the Air Ministry's new target of 150 engine sets per week in June 1940 and reached 761 engines completed during August.[187] The new Ministry of Aircraft Production still deemed this insufficient for the number of Blenheim, Hampden, and

Wellington bombers on order that called for an output of 950 Mercury XV and Pegasus XVIII engines per month.[188] That same month, the Ministry approved a new production target for the factories of 200 complete engine sets per week, with an additional 20 sets as spares.[189] The Ministry approved additional expenditures for machine tools so the shadow factories could reach this new target.[190] To support expanded production, the Air Supply Board approved the expenditure of £55,957 ($223,828) to provide equipment to the De Havilland Aircraft Co. Ltd. to manufacture rear covers for the Mercury engines produced in No. 1 Shadow Group.[191] The Air Supply Board apparently decided to increase the rate of production in No. 1 Shadow Group even further, approving expenditures for machine tools and equipment to boost output to 280 engine sets per week.[192]

The production of the Mercury and Pegasus engines peaked during 1941 when the shadow factories and Bristol produced 9,611 complete engines, with the shadow factories achieving their highest monthly output completing 980 engines in May.[193] By the end of 1941, the aircraft types using the Mercury and Pegasus engines, principally the Blenheim, Hampden, and early Marks of the Wellington, were being progressively withdrawn from operations. The production of the two engines continued at a high rate during 1942, with a total output of 8,499 engines from Bristol and the shadow factories, but then dropped off sharply during 1943 to 1,711 engines as Bristol stopped the production of the Pegasus and demand shifted to the more powerful Hercules.[194] The production of the Pegasus ended in September 1943 while Mercury production came to an end in September 1944, after the shadow factories had shifted to producing the last versions of the Mercury, the Mercury XX, 25, and 30.[195] From June 1939 until February 1945, Bristol and the shadow factories built 16,041 Mercury engines and 15,713 Pegasus engines, with the majority completed at the shadow factories.[196] The speed with which the shadow factories increased their output, despite production problems, was impressive. During 1939, the first full year of production, the shadow factories completed 3,410 engines with a peak monthly production of 342 engines. Within two years, the shadow factories had more than doubled their monthly output to produce 9,611 complete engines, an increase of 181% (Figure 2.9).

FIGURE 2.9 Shadow factory monthly Mercury and Pegasus engine production.

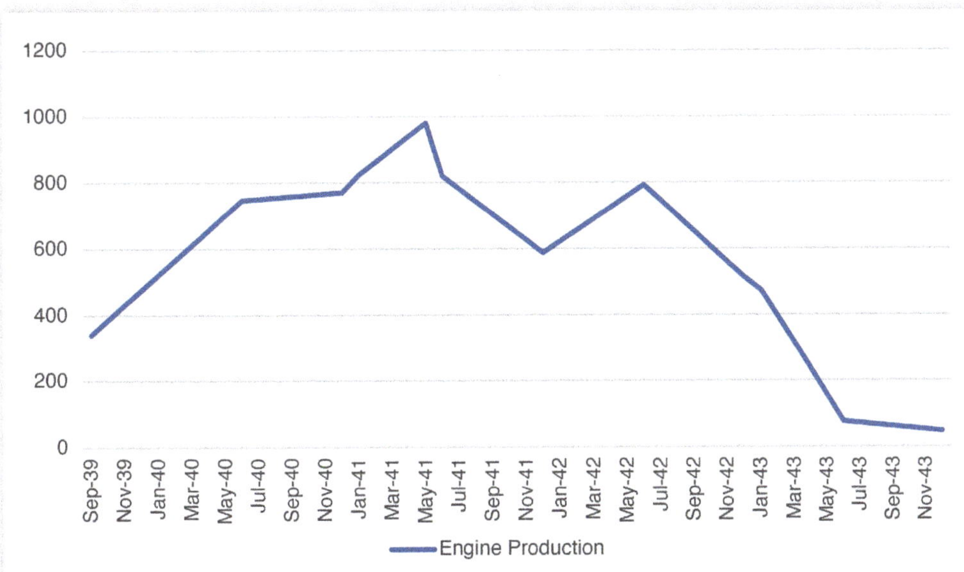

Source: Abridged Record of Engines, Spares and Accessory Output from Shadow Industry for 1938-44, "History of Production of Bristol Aero Engines by the Shadow Industry Operating Under the Joint Aero Engine Committee in Conjunction with the Bristol Aeroplane Company Limited," XC03-8876/066, Royal Air Force Museum.

Production at the Shadow Factories

Production at the shadow factories, at least until the outbreak of the war in September 1939, involved the exact replication of Bristol's prewar production methods. At the inaugural meeting between Bristol and the representatives of the shadow factories in June 1936, the automotive firms decided they would "slavishly follow Bristol Aeroplane Company production practice and improve on this if and when experience justified any changes."[197] Fortunately, Bristol had extensive experience in providing technical advice and support for manufacturing Bristol engines. For nearly a decade, Bristol had been cooperating with European firms building the Bristol Jupiter, Mercury, and Pegasus engines under license. To assist the licensees, Bristol had prepared a compilation of drawings, specifications, and schedules documenting the production in exhaustive detail known as the "Bristol Engine Bible."[198] This compilation, and Bristol's extensive experience, proved invaluable in setting up production at the shadow factories.

Soon after the inauguration of the shadow scheme, Bristol set up a special Shadow Industry Office to coordinate the supply of technical information and assistance to the factories. Over the next two years, the Office supplied more than 68 folders of detailed information and precise instructions on engine production, specifying in exact detail the types and number of machine tools to be used in each machining operation, machining times, specifications of materials, schedules of equipment for engine testing and assembly, among others.[199] Bristol based these specifications on its own experience in its aircraft engine works.[200] The factories set up a managers' committee which met frequently during this initial phase to work out common problems in cooperation with Bristol and the Air Ministry, while senior executives of the motor firms formed the Aero Engine Committee (Shadow Industry) to provide overall direction.[201]

Machine tools required for the factories cost three to four times more than factory buildings, though the investment varied by the factory and the number of machine tools each factory required. The number ranged from 439 machine tools at the Rootes factory to 729 at the Rover factory.[202] After providing information on machine tools, Bristol prepared schedules and detailed information on the jigs, fixtures, tools, and gauges that the factories would need.[203] Each of the factories had a special standard room where the jigs and tools could be checked for accuracy with measuring instruments accurate to the ten-thousands of an inch required on many parts of an aircraft engine.[204] Bristol's choice of machine tools to build the Mercury and Pegasus stressed flexibility. The Bristol machine shops did not use special-purpose machine tools that were common in the automotive industry. Instead, the company purchased standard machine tools that did not require special jigs or fixtures and that could perform several operations on the same machine tool.[205] Contemporary articles describe labor-intensive machine operations on sequences of separate and different types of machine tools, with few mentions of machine tools that could perform multiple operations on the same machine. At that time, Bristol classed over 40% of the workforce in the Aero Engine Department as skilled workers.[206] This was the operation that Sir Hubert Austin, the first chairman of the shadow factory group, observed on his visit to the Bristol engine works in June 1936, concluding that "it would certainly not be an easy matter to improve on their present practice unless special machinery was installed and a large number of engines produced."[207]

The majority of the machine tools for the first shadow factories came from British machine toolmakers, apart from small numbers of automatic machine tools and milling machines imported from America.[208] While there were delays in delivery due to the growing demand for machine tools, by October 1937, the shadow factories had most of the machine tools they needed. The Daimler shadow factory had 82% of its machine tools, Rootes had 90%, and the Standard Motor Company factory had 92%.[209]

At the time Bristol was advising the shadow factories on setting up production, the company itself was still grouping machine tools by type and function and only using line production to a limited extent.[210] This arrangement was perfectly adequate for batch production. When Bristol provided the factories with suggested layouts for machine tools, based on manufacturing processes at each factory and projected output, it provided several alternatives.[211] Bristol did offer a plan that incorporated the possibility of employing line production if the quantities of components required this method of quantity production at some later date.[212] Bristol decided, however, that the initial output of 50 complete engine sets per week did not justify setting up an extensive line production system.[213]

While following Bristol's recommendations, the automobile companies did have some leeway in choosing different approaches to production methods and laying out machine tools at their respective shadow factories. The automobile companies were familiar with line production methods and the concept of flow production. During the 1920s and 1930s, several firms reconfigured their factory layouts to improve the flow of components.[214] Despite Bristol's recommendation that a line production system was not necessary, it appears that the Rootes, Rover, and Standard Motor factories adopted some form of line production. A report on the Rootes factory noted that "the main machine shop runs in parallel lines the full length of the factory" and that the factory had taken care to "minimize the delay caused by interruption of the straight flow of machining sequences brought about by the necessity of secondary processes such as heat treatment," a good description of line production.[215] The Standard Motor shadow factory was described as "quite definitely a showplace."[216] The Standard factory had "the simplest and yet at the same time the most difficult manufacturing program."[217] The Standard factory built the cylinder head and cylinder barrel, and while machining these components was complex, it allowed the factory to adopt line production, with one line for the cylinder head and one line for the cylinder barrel, and facilitated using conveyors to move parts between machining operations.[218] How production was organized at the Austin, Daimler, and Bristol shadow factories is uncertain.

Inevitably, despite the best efforts of Bristol and the automotive firms, there were delays and disruptions in output. Before the beginning of the war, the most serious delays occurred at the Rootes Humber factory manufacturing the blower (supercharger) units for the Mercury. Apparently, Bristol had specified that the Rootes factory should use the Swiss Maag gear grinder for the blower gear wheels, but William Rootes insisted on using the British Orcutt grinding machine, only to find later that the Mercury engines equipped with the Orcutt ground gears failed their engine test, forcing a change in the gear-grinding machine tools.[219] In July 1938, Rootes was 774 units behind schedule which "caused a considerable amount of discontent among other members as bonus payments would be delayed."[220] In July-August 1938, the shadow factories failed to produce a single engine. While monthly output improved during 1939, there continued to be problems with the Austin and Rootes shadow factories meeting their production targets.[221] This became an even greater problem after the war began as the Air Ministry pushed the shadow factories to rapidly expand production. In March 1940, the Air Council Committee on Supply reported that due to difficulties in arranging the supply of materials to the Austin and Rootes shadow factories, both factories were seriously behind schedule, with Austin short 2,000 crankshafts and Rootes short 1,000 blower gear stampings.[222] This was due in part to a lack of forging capacity in Britain for forging crankshafts, forcing Bristol to develop alternative sources of supply in America.[223] This lesson was not lost on the Air Ministry.

How It Was Done: The Transition to Large-Scale Production

It took a little over a year for the shadow factories to make the transition to large-scale production. From early 1939 to late 1940, the factories expanded their weekly capacity from 50 to 200 complete engine sets a week, a four-fold increase. Few primary documents are available describing how the factories shifted to building engines in greater quantities. The Bristol history of the shadow industry refers to the "copious correspondence between the Bristol Shadow Office and the individual factories" dealing with all manner of production issues, but this correspondence does not seem to have survived.[224] However, it is possible to show that to expand production the factories are greatly increasing machine utilization by moving to double shifts, nearly doubling the number of workers at the factories with a doubling of labor productivity through the learning effect.

The transition to larger-scale production began before the start of the war in September 1939. The factories reached their initial target of 50 engines per week in August 1938. During the first half of 1939, the factories used overtime to build up to 80 engines per week, achieving this target in July.[225] This effort was part of a three-fold increase in aero engine production at all the British aero engine factories—at Bristol, the shadow factories, Rolls-Royce, Armstrong Siddeley Motors, and De Havilland—between July 1938 and July 1939.[226] A July 1939 memorandum for the Air Council from the Air Ministry's Director General of Production on the progress of production for rearmament documented how the aero engine factories had achieved this increase.[227] The memorandum stated that the increase was due to (Table 2.4):

(a) Increased machine utilization.
(b) Increased working hours.
(c) Increased production equipment.
(d) Increased labor strength.
(e) Increased area at Main Factories.
(f) New area at shadow factories.
(g) Increased subcontracting.[228]

TABLE 2.4 1940 expenditures on shadow factories.

	Number of machine tools	Estimated cost of machine tools	Estimated cost of equipment
Austin Co.	72	£81,400 ($325,600)	—
Daimler Co.	12	£11,450 ($45,800)	—
Rootes Securities	287	£341,000 ($1,364,000)	—
Rover Co.	45	£77,000 ($308,000)	£25,428 ($101,712)
Standard Motor Co.	31	£27,040 ($108,160)	£1,612 ($6,448)
Total	497	£537,890 ($2,151,560)	£27,040 ($108,160)

Source: Air Supply Board, No. 1 Shadow Group- Production of Mercury XV and Pegasus XVIII engines, August 16, 1940, Ministry of Supply: Establishment, Registered Files (Series 1). Narratives on supply of Aircraft parts, especially aero engines: source papers, AVIA 46/35, National Archives, Kew, UK.

With the exception of the Rootes factory, additions of new machine tools to the factories were modest. To achieve the target of 150 engines per week, in December 1939, the Air Ministry approved expenditures of £407,765 ($1.6 million) for more machine tools and equipment for the shadow factories and £158,746 ($635,984) for building work to provide more storage space for parts and space to assemble the increased number of engines.[229] This represented an increase of approximately 14% over the amount spent on machine tools at the factories up to November 1939.[230] The Rootes factory received half of this allocation for machine tools, representing a 40% increase over the original expenditure for the factory.[231] This was probably an effort to improve the production of blowers at the Rootes factory, which still lagged. In August 1940, a review of production determined that with more machine tools production could in fact be boosted to 200 engines a week.[232] This was done through an additional expenditure of £537,890 ($2.1 million) with £27,040 ($108,160) on other equipment, an amount roughly equal to around 20% of the original expenditure.[233] Once again Rootes received the major portion of the allocation for machine tools. These two allocations for the Rootes shadow factory nearly doubled the number of machine tools, which rose from 606 in 1939 to 1,035 at the end of 1941.[234] While the expenditure at the Rover factory represented approximately one-third of the original expenditure for the factory, the amounts allocated to the Austin, Daimler, and Standard Motor factories were less. There is no evidence that these new machine tools were different in any way from the standard machine tools already in the factories. It is highly likely, but not certain, that during this period, those factories that had not already done so shifted to line production to increase output. The directors' minutes of the Rover Company shadow factory, for example, speak of changing the layout of machine tools within the factory to increase production to 150 engine sets per week (Figure 2.10).[235]

FIGURE 2.10 Monthly engine production and total workers at the first shadow factories April 1939 and June 1941.

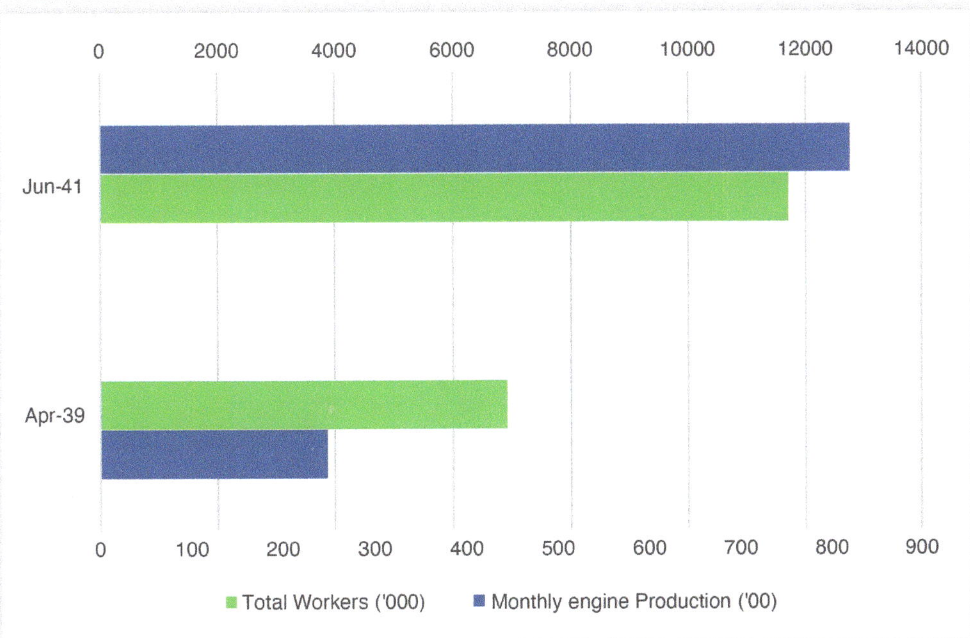

Source: Abridged Record of Engines, Spares, and Accessory Output from Shadow Industry for 1938-44, "History of Production of Bristol Aero Engines by the Shadow Industry Operating Under the Joint Aero Engine Committee in Conjunction with the Bristol Aeroplane Company Limited," XC03-8876/066, Royal Air Force Museum; Ministry of Aircraft Production, Statistical Review 1939-1945.

The most important factors in increasing production were, however, increasing the number of workers at the factories, moving to double shifts, and improving labor productivity. Between April 1939 and June 1941, the number of workers at each factory nearly doubled. More workers, enough to accommodate working double or triple shifts, allowed much greater machine tool utilization and increased component production. There is insufficient data to determine when the individual factories reached this level of employment, but it is probable that this occurred toward the end of 1940 when during October weekly production reached 200 engines per week for the first time.[236] The move to double shifts apparently took place shortly before the start of the war.[237] It is likely that as the labor force expanded, so too did the number of workers assigned to each shift (Table 2.5).

TABLE 2.5 Employment at the shadow factories 1938–1942.

	August 1938	April 1939	June 1941	March 1942
Austin	1,112	1,327	2,211	2,211 est.
Daimler	1,243	1,351	2,481	2,713
Rootes	1,306	1,542	2,873	3,413
Rover	1,320	1,425	2,891	3,429
Standard Motor	801	1,266	1,952	2,576
Total	5,782	6,911	12,408	14,342 est.

Source: 177th Progress Meeting, The Progress of Production for Air Re-armament, Appendix VI, Memorandum by D.G.P., July 12, 1939, AIR 6/58; Table Showing the strength of wage earners at certain dates of the total labor force at March 7, 1942, AVIA 10/383, National Archives, Kew, UK.

Conclusion

The first shadow scheme for aircraft engine production did not evolve as the Government had originally intended. The central premise of the shadow factory scheme that the parent automotive factories could rapidly convert into aircraft engine production proved to be flawed. The Air Ministry had no feasible alternative except to increase the number of engines that could be built in the existing factories employing more machine tools and especially more labor. While there may have been some move toward line production, the shadow factories appear to have concentrated on continuing to use existing production methods but with more workers working more shifts. Fortunately, this effort proved successful. The scheme was, as Hornby notes, "an outstanding venture into group organization for shadow industry development."[238] The transition into building the Mercury and Pegasus in greater quantities at the shadow factories after the start of the war provided Bristol, the shadow factory management, and the Air Ministry and Ministry of Aircraft Production with invaluable experience. The next challenge, the production of the larger and more complex 14-cylinder Bristol Hercules would, however, be entrusted to a new shadow factory group, set up on a completely different basis and on a completely different scale of factory size, production methods, machine tools, and output.

References

1 Smith, Malcolm: *British Air Strategy Between the Wars*, (Oxford, 1984), p. 248.

2 Lumsden, *British Piston Engines and Their Aircraft*, p. 104.

[3] "The 'Bristol' Mercury & Pegasus Engines", *The "Bristol" Review*, Engine Issue No. 4, (March 1932), pp. 6-8.
[4] "The New Series 'Bristol' Mercury and Pegasus Engines", *The "Bristol" Review*, Engine Issue No. 10, (June 1936), pp. 12-14.
[5] Heaven, John: *Bristol Piston Engines Since 1914*, unpublished manuscript, copy in the possession of the Rolls-Royce Heritage Trust, Filton.
[6] Lumsden, *British Piston Engines and Their Aircraft*, p. 108.
[7] "'Bristol' Engines for 1932", *The "Bristol" Review*, Engine Issue No. 4, (March 1932), p. 9.
[8] "The New Series 'Bristol' Mercury and Pegasus Engines", *The "Bristol" Review*, Engine Issue No. 10, (June 1936), p. 14.
[9] Heaven, *Bristol Piston Engines Since 1914*.
[10] Hyde, H. Montgomery: *British Air Policy Between the Wars 1918-1939*, (London, 1976), pp. 298-99; Dunbabin, J.P.D.: "British Rearmament in the 1930s: A Chronology and Review", *The Historical Journal*, Vol. XVIII, No. 3, (1975), pp. 589-90.
[11] Hyde, *British Air Policy Between the Wars 1918-1939*, pp. 304-05.
[12] Ibid., Appendix VII.
[13] Ibid., p. 326.
[14] Ibid., p. 333.
[15] Meredith, Richard: *Phoenix: A Complete History of the Luftwaffe 1918-1945*, Volume 2: *The Genesis of Air Power*, (Solihull, UK, 2017), p. 71.
[16] Hyde, *British Air Policy Between the Wars 1918-1939*, p. 333.
[17] Ibid., Appendix VII; Philpott, W/C I.M.: *The Royal Air Force 1930 to 1939: An Encyclopedia of the Inter-war Years*, Volume II: *Rearmament 1930 to 1939*, (Barnsley, UK, 2008), pp. 44-45.
[18] "Proposed Allocation of Capacity for the Production of Aero and Fighting Vehicle Engines", 22 December 1933, Supp 3/44, National Archives (hereinafter NA).
[19] "Proposed Allocation of Capacity for the Production of Aero and Fighting Vehicle Engines", pp. 2-3.
[20] "Proposed Allocation of Capacity for the Production of Aero and Fighting Vehicle Engines", Aircraft engine Production Table, p. 2.
[21] "Narrative: Reciprocating Aero Engines and Engine Accessories Production & Programmes", Paragraph 18, Air 46/159, NA.
[22] "Narrative: Reciprocating Aero Engines and Engine Accessories Production & Programmes", Paragraph 25.
[23] "The Bristol Mercury & Pegasus Engines", *The "Bristol" Review*, Engine Issue No. 4, (March 1932), pp. 6-8.
[24] Burge, S/L C.G.: *The Air Annual of the British Empire 1935-36*, (London, 1935), pp. 128-129.
[25] Burge, S/L C.G.: *The Air Annual of the British Empire 1935-36*, pp. 129-32.
[26] H.J. Thomas, Assistant Managing Director, Bristol Aeroplane Company, to the Secretary, Air Ministry, 13 March 1933, Air 2/629, NA.
[27] Requirements for Engines for Expansion Scheme "C", Air 6/43, NA.
[28] Bristol Aeroplane Engine Delivery Charts for 1934 and 1935, copy in the Rolls-Royce Heritage Trust Bristol Branch Archive (hereinafter RRHTBA).
[29] H.J. Thomas to Sir Philip Cunliffe-Lister, Secretary of State for Air, October 22, 1935, Air 19/4, NA.
[30] Warner, Graham: *The Bristol Blenheim: A Complete History*, (Manchester, UK, 2002), p. 59.
[31] Hyde, *British Air Policy Between the Wars 1918-1939*, p. 363, Appendix VII.
[32] Ibid., p. 363.
[33] Warner, *The Bristol Blenheim: A Complete History*, pp. 59-60; "The Expansion of the Royal Air Force 1934-1939", tables for Scheme C and Scheme F, Air 41/8, NA.
[34] Warner, *The Bristol Blenheim: A Complete History*, p. 60.
[35] Smith, *British Air Strategy Between the Wars*, pp. 233-37.
[36] Hyde, *British Air Policy Between the Wars 1918-1939*, pp. 364-65.
[37] Ibid.; "Narrative: Reciprocating Aero Engines and Engine Accessories Production & Programmes", Paragraph 26, Air 46/159, NA.
[38] Barnes, C.H.: *Bristol Aircraft Since 1910*, 3rd Ed., (London, 1988), p. 267.
[39] "Narrative: Reciprocating Aero Engines and Engine Accessories Production & Programmes", Paragraph 27.
[40] Philpott, W/C Ian M.: *The Royal Air Force: An Encyclopaedia of the Inter-war Years,* Volume II: *Re-armament 1930 to 1939*, (Barnsley, UK, 2008), p. 46.
[41] Ritchie, Sebastian: *Industry and Air Power: The Expansion of British Aircraft Production, 1935-1941*, (London, 1997), p. 58.
[42] "Narrative: Reciprocating Aero Engines and Engine Accessories Production & Programmes", Paragraph 26.
[43] "Narrative: Reciprocating Aero Engines and Engine Accessories Production & Programmes", Paragraph 27.
[44] Air Marshal Sir Cyril Newall, Air Member for Supply and Organization, to Sir Arthur Robinson, Chairman, Supply Board, 6 and 12 February 1936, Air 19/5, NA.

45 Note of Conversation with Mr. Berryman, 11.3.36, Air 19/3, NA.
46 Sir Alexander Roger to Lord Swinton, Secretary of State for Air, 19 March 1936, Air 19/3, NA.
47 33rd Progress Meeting, Interview with Standard Motor Company, 20 March 1936, Air 6/44, NA.
48 Bristol Aeroplane Company Minutes of Directors' Meeting in Committee, 25 February 1936, BAC Directors' Committees Minutes, Vol. 5, 3 January 1935 to 22 December 1936, Aerospace Bristol Archives.
49 Bristol Aeroplane Company Minutes of Directors' Meeting in Committee, 28 February 1936, BAC Directors' Committees Minutes, Vol. 5.
50 Bristol Aeroplane Company Minutes of Directors' Meeting in Committee, 28 February 1936.
51 Expansion Programme 1936-1939, Interview at the Air Ministry, Friday 13 March 1936, copy in the Bristol Shadow Factory File, RRHTA.
52 Expansion Programme 1936-1939, Interview at the Air Ministry, Friday 13 March 1936, p. 2.
53 Expansion Programme 1936-1939, Interview at the Air Ministry, Friday 13 March 1936, p. 8.
54 Expansion Programme 1936-1939, Interview at the Air Ministry, Friday 13 March 1936, p. 10.
55 Lord Swinton, Secretary of State for Air to Sir Herbert Austin, Sir Alexander Rogers, R.C. Rootes, W. Cannell, Captain R. Wilks, W.E. Bullock, Captain J.P. Black, 24 March 1936, Air 2/1738, NA.
56 Sir Stanley White, Managing Director, Bristol Aeroplane Company, to Lord Swinton, Secretary of State for Air, 31 March 1936, Air 6/44, NA.
57 Bristol Memorandum on aircraft and aircraft engine production, 31 March 1936, Air 6/44, NA.
58 Bristol Memorandum on aircraft and aircraft engine production, 31 March 1936, p. 8, Air 6/44, NA.
59 Secretary of State's 35th Progress Meeting (hereinafter SSA PM), Bristol Company Expansion Scheme, 2 April 1936, Air 19/4, NA.
60 Secretary of State's Progress Meetings, Note of a Meeting with Representatives of Automobile Firms, Air Ministry 7 April 1936, E.P.M. 39 (36), Air 19/3, National Archives.
61 Secretary of State's Progress Meetings, Note of a Meeting with Representatives of Automobile Firms, Air Ministry 7 April 1936, E.P.M. 39 (36), pp. 3-4.
62 Secretary of State's Progress Meetings, 7 April 1936, E.P.M. 39 (36), p. 3.
63 Secretary of State's Progress Meetings, 7 April 1936, E.P.M. 39 (36), p. 4.
64 Notes on a Visit to the Works of the Bristol Aeroplane Company, 10 April 1936, Air 19/4, NA.
65 Notes on the Visit to the Works of the Bristol Aeroplane Company, 10 April 1936.
66 Ibid.
67 Air Marshal Cyril Newall Minute of 20 April 1936 on meeting with Sir Herbert Austin, Air 19/5, NA.
68 Draft Scheme for the Manufacture for the Air Ministry of a Minimum of 2,000 and a Maximum of 4,000 Bristol Aero Engines to be Completed by March 1939, Air 2/1738, NA.
69 Extract from letter A.M.S.O to Sir Herbert Austin, 22 April 1936, Air 19/5, NA.
70 Secretary of State's 38th Progress Meeting, Note of a Meeting with Sir Herbert Austin, 30 April 1936, Air 2/1738, NA.
71 Secretary of State's 38th Progress Meeting, Note of a Meeting with Sir Herbert Austin, 30 April 1936; Ritchie, Sebastian: *Industry and Air Power: The Expansion of British Aircraft Production, 1935-1941*, (London, 1997), p. 122.
72 Secretary of State's 38th Progress Meeting, Note of a Meeting with Sir Herbert Austin, 30 April 1936.
73 R.E. Herrington, Secretary to the Aero Engine Committee, to Viscount Swinton, Secretary of State for Air, 7 May 1936, Air 2/1738, NA.
74 Secretary of State's Progress Meetings, Engine "Shadow Scheme" Meeting, 29 June 1936, Air 2/1738, NA.
75 Neale, M.C., ed.: *An Account of a Partnership-Industry, Government and the Aero Engine: The Memoirs of George Purvis Bulman*, Rolls-Royce Heritage Trust Historical Series No. 31, (Darby, UK, 2002), p. 224.
76 Secretary of State's Progress Meetings, Engine "Shadow Scheme" Meeting, 29 June 1936, p. 3.
77 Secretary of State's Progress Meetings, Engine "Shadow Scheme" Meeting, 29 June 1936, p. 4.
78 Notes of a Meeting of the Committee Held at the Offices of the Bristol Aeroplane Company Ltd., Bristol, 30 June 1936, Air 2/2577, NA.
79 Aero Engine Committee, Meeting to be held at the Works of the Bristol Aeroplane Company on Tuesday June 30th 1936, Agenda, Air 2/2577.
80 Ibid.
81 Notes of a Meeting of the Committee Held at the Offices of the Bristol Aeroplane Company Ltd., Bristol, 30 June 1936.
82 Neale, *An Account of a Partnership-Industry, Government and the Aero Engine: The Memoirs of George Purvis Bulman*, p. 224.
83 Notes of a Meeting of the Committee Held at the Offices of the Bristol Aeroplane Company Ltd., Bristol, 30 June 1936.
84 H.A.P. Disney, Director of Aeronautical Production, to Sir Hubert Austin, 3 July 1936, Air 2/1738, NA.

[85] Minutes of Secretary of State's 48th Expansion Progress Meeting, 28 July 1936, AVIA 46/35, NA.
[86] See Seymour, Peter: *Wolseley Radial Aero Engines: Lord Nuffield's Thwarted Venture*, (Stroud, UK, 2006), Chapter 5.
[87] Verdon Smith, Bristol Aeroplane Company, to Lord Swinton, Secretary of State for Air, 25 August 1936, Air 19/4, NA.
[88] Lord Swinton, Secretary of State for Air, to Sir Verdon Smith, Bristol Aeroplane Company, 16 September 1936, Sir Verdon Smith to Lord Swinton, 17 September 1936, Air 2/1738, NA.
[89] Anderson, Arthur G.: *Industrial Engineering and Factory Management*, (New York, 1928), pp. 120-23; Berghell, A.B.: *Production Engineering in the Aircraft Industry*, (New York, 1944), pp. 1-2.
[90] H.A.P. Disney, Director of Aeronautical Production, to Sir Hubert Austin, 3 July 1936.
[91] Bristol Aeroplane Company Minutes of Directors' Meeting in Committee, 15 and 22 September 1936, BAC Directors' Committees Minutes, Vol. 5.
[92] Bristol Aeroplane Company Minutes of Directors' Meeting in Committee, 28 September 1937, BAC Directors' Committees Minutes, Vol. 6.
[93] H.A.P. Disney, Director of Aeronautical Production, to Sir Hubert Austin, 3 July 1936.
[94] Bristol Aeroplane Company Ltd.: "History of the Production of Bristol Aero Engines by the Shadow Industry Operating under the Joint Aero Engine Committee in Conjunction with the Bristol Aeroplane Company Limited", (January 1945), X003-887/066, Royal Air Force Museum, p. 6.
[95] Hornby, William: *Factories and Machine Tools*, History of the Second World War, Civil Series, (London, 1958), p. 286.
[96] Ibid.
[97] See minute from the A.M.S.O to the Secretary of State for Air, 20 April 1936, Air 19/5, NA.
[98] James, Thurstan: "At the Shadow Factories", *The Aeroplane*, Vol. LIII, No. 1350, (3 November 1937), p. 532.
[99] Koham, C.M.: *Works and Buildings*, History of the Second World War, Civil Series, (London, 1952), p. 311.
[100] H.A.P. Disney, Director of Aeronautical Production, to Sir Hubert Austin, 3 July 1936.
[101] Treasury Inter-Service Committee, Daimler Aero Engine Factory (Shadow Industry), Air Ministry Memo 24, 7 September 1936, Air 2/1842, NA.
[102] James, "At the Shadow Factories", pp. 534-36.
[103] "Shadow Factories in Being", p. 499.
[104] Holme, C.G., Ed.: *Industrial Architecture*, (London 1935), p. 16.
[105] Winter, John: *Industrial Architecture: A Survey of Factory Building*, (London, 1970), p. 89; Collins, Paul, and Michael Stratton: *British Car Factories from 1896: A Complete Historical, Geographic, Architectural & Technological Survey*, (Goodmanstone, Dorset, 1993), p. 48.
[106] Collins and Stratton, *British Car Factories from 1896*, p. 47.
[107] Collins and Stratton, *British Car Factories from 1896*, p. 48.
[108] Daimler Motor Company Limited, Supplementary Details of Proposed Engine Factory at Radford, Coventry, 4 November 1936, Air 1/1842, NA.
[109] "The Government Aircraft 'Shadow' Scheme", p. 134.
[110] NA, AIR 2/1842, Daimler Motor Company Limited, Supplementary Details of Proposed Engine Factory at Radford, Coventry, 4 November 1936.
[111] "Shadow Factories in Being", *Flight*, (4 November 1937), p. 446.
[112] See note Rootes Securities Ltd., Aircraft Division, Engine Factory, Coventry, Air 19/7, NA; and Minutes of Management Committee Meeting 29th September 1937, Rover Company Limited (Aero Engines), Rover Company Minute Book 1937-1939, 80/148/4/67-RCO, British Motor Industry Trust.
[113] "History of the Production of Bristol Aero Engines by the Shadow Industry Operating under the Joint Aero Engine Committee in Conjunction with the Bristol Aeroplane Company Limited", p. 74.
[114] Ibid.
[115] Ibid, p. 6.
[116] "Shadow Factories in Being", p. 446.
[117] James, "At the Shadow Factories", p. 529.
[118] "Shadow Factories in Being", pp. 445-46.
[119] "A New 'Bristol' Engine Factory", *Bristol Review (Engine Issue)*, No. 11, November 1936, pp. 44-46; "Further Extensions to the 'Bristol' Engine Factory", *Bristol Review (Engine Issue)*, No. 12, June 1937, pp. 42-43.
[120] Air Ministry Memorandum No. 42, Treasury Inter-Service Committee, Bristol Engine Assembly Factory-Shadow Industry, 9 November 1936, Air 2/2366.
[121] Thoms, *War, Industry and Society*, pp. 104-112.
[122] Thoms, *War, Industry and Society*, p. 108; "History of Production of Bristol Aero Engines by the Shadow Industry Operating under the Joint Aero Engine Committee in Conjunction with the Bristol Aeroplane Company Limited", XC03-8876/066, Royal Air Force Museum, p. 65.

123 History of Production of Bristol Aero Engines by the Shadow Industry, pp. 65-66.
124 History of Production of Bristol Aero Engines by the Shadow Industry, pp. 65, 119.
125 History of Production of Bristol Aero Engines by the Shadow Industry, p. 66.
126 Ritchie, *Industry and Air Power*, p. 126.
127 Returns Showing Contractors' Actual and Estimated Deliveries of Aeroplanes and Engines, Week ending, 24 July 1937, Air 6/50, NA.
128 Minutes of Management Committee Meeting in Connection with H.M. Air Ministry Factory Held at the New Meteor Works, Coventry, on Wednesday 29th September 1937, Rover Company Minute Book No. 1, 1937-1939, L-RCO-5, British Motor Industry Heritage Trust Archive, British Motor Museum.
129 Bristol Aeroplane Company Minutes of Directors' Meeting in Committee, 21 September 1937, BAC Directors' Committees Minutes, Vol. 6.
130 Ibid.
131 Committee of Imperial Defence, Sub-Committee on Defence Policy and Requirements, Seventeenth Progress Report (October 1937) by the Air Ministry, p. 4, Air 6/51, NA.
132 Lord Austin to H.A.P. Disney, Director of Aeronautical Production, 28 September 1937, Air 2/2577, NA.
133 Committee of Imperial Defence, Sub-Committee on Defence Policy and Requirements, Eighteenth Progress Report (November 1937) by the Air Ministry, p. 3, Air 6/51, NA.
134 Minutes of Management Committee Meeting in Connection with H.M. Air Ministry Factory Held at the New Meteor Works, Coventry, on Monday 24th January 1938, Rover Company Minute Book No. 1.
135 "History of the Production of Bristol Aero Engines by the Shadow Industry Operating under the Joint Aero Engines Committee in Conjunction with the Bristol Aeroplane Company Limited", Abridged Record of Engines, Spares and Accessory Output from Shadow Industry for 1938.
136 Ibid.; Minutes of Directors Meeting, 24 May 1938, p. 1, BAC Directors Committees Minutes Vol. 6 (19 Jan '37 to 22 Feb '39).
137 Bristol Aeroplane Company Minutes of Directors' Meeting in Committee, 24 May, 14 June, 21 June, 11 October 1938, BAC Directors' Committees Minutes, Vol. 6.
138 A.H. Self, Air Ministry, to the Secretary, Aero Engine Committee, 30 March 1938, Air 2/2577, NA.
139 Hyde, *British Air Policy Between the Wars 1918-1939*, p. 417.
140 Ibid.
141 Ritchie, *Industry and Air Power: The Expansion of British Aircraft Production, 1935-1941*, p. 43.
142 Air Marshal Wilfrid Freeman, Air Member for Research and Development, to C.A.S., 29 April 1938, AVIA 10/5, NA.
143 Ibid.
144 Minutes of Directors Meeting, 22 March 1938, p. 5, BAC Directors Committees Minutes Vol. 6 (19 Jan '37 to 22 Feb '39).
145 Minutes of Directors Meeting, 3 May 1938, p. 1, BAC Directors Committees Minutes Vol. 6 (19 Jan '37 to 22 Feb '39).
146 Ibid.
147 R.D. Pearce, Secretary, Bristol Aeroplane Company, to R. Russell, Secretary, Air Council Sub-Committee, 11 May 1938, AVIA 10/151, NA.
148 Aircraft, Estimated Deliveries from 1/4/38 to 31/3/40 on Orders Already Placed, AVIA 10/151, NA.
149 Air Council Sub-Committee on Supply, Notes on Meetings Held Monday, 16th May, 1938, p. 2, AVIA 10/151, NA.
150 Ibid.
151 116th Progress Meeting, Production of Engines by the Shadow Factories, Memorandum by the A.M.S.O., 3 March 1938, Air 6/52, NA.
152 D.D.A.P. memo to D.A.P., 23 September 1937, Air 2/2577; Note of the Secretary of State's Meeting with Representatives of the Shadow Industry, 20 May 1938, Air 6/53, NA.
153 Notes on the Meeting of the Air Council Sub-Committee on Supply, Held on Thursday, 2nd June 1938, AVIA 10/152.
154 Minutes of Directors Meeting, 24 May 1938, pp. 2-3, BAC Directors Committees Minutes Vol. 6 (19 Jan '37 to 22 Feb '39).
155 Notes on the Meeting of the Air Council Sub-Committee on Supply, Held on Thursday, 2nd June 1938, AVIA 10/152.
156 Development Report for the RAF for the six months ending 31st July 1937, p. 10, Air 10/1550, NA.
157 Ibid.
158 Notes of a Meeting Held at the Austin Shadow Factory at Birmingham on the 10th June 1938, AVIA 10/152, NA.

159 H.A.P. Disney, Director of Aeronautical Production, to R.E. Herington, Secretary, Aero Engine Committee, 23 June 1938, Air 2/2577, NA.

160 Notes of the Meeting of the Air Council Committee on Supply Held on Thursday, 23rd June 1938, AVIA 10/152. NA.

161 Russell, Secretary of the Air Council Committee on Supply, to the Bristol Aeroplane Company, 24 June 1938, AVIA 10/152, NA.

162 Ibid., Appendix I (b) Programme of Production Engines.

163 Chart of New Engines Delivered 1924-1948, copy in the Rolls-Royce Heritage Trust.

164 Minutes of the Meeting of the Committee held at the offices of the Standard Motor Co., Ltd., Fletchampstead, Coventry, on Monday, 19th September 1938, Air 2/2577, NA.

165 Minutes of the Meeting of the Committee held at the offices of the Standard Motor Co., Ltd., Fletchampstead, Coventry, on Friday, 7th October 1938, Air 2/2577, NA.

166 Decisions of the Air Council Committee on Supply reached at the meeting held on Thursday 3rd November 1938, AVIA 10/156, NA.

167 Bristol Aeroplane Company Minutes of Directors' Meeting in Committee, 11 October 1938, BAC Directors' Committees Minutes, Vol. 6.

168 Air Ministry Memorandum No. 432, Treasury Inter-Service Committee, Bristol Type Engines-Provision of further capacity to provide war potential and to meet the present programme, 28 March 1939, p. 1, AVIA 10/230, NA.

169 Ibid.; Air Ministry, Progress Meetings on R.A.F. Expansion, 160th Progress Meeting, 16 March 1939, AIR 6/56.

170 Ibid.

171 Decisions of the Air Council Committee on Supply reached at the meeting held on 28th September and at an informal meetings held on 30th September and 3rd October 1938, Air 6/55; Minute E.3A, 18/10/38, AVIA 46/35, NA.

172 George Hally, Daimler Co. Ltd. (Aero Engines), to Directorate of Engine Production, 30 January 1939, Air 2/1842, NA.

173 Extension of Shadow Group Capacity for the Production of Mercury and Pegasus Engines in Emergency, 16 March 1938, Air 6/56; 160th E.P.M., Note on meeting, 20 March 1938, AVIA 46/35, NA.

174 160th E.P.M., Note on meeting, 20 March 1938, AVIA 46/35, NA.

175 41st Meeting of the Air Council Committee on Supply 19th April 1939, AVIA 46/35, NA.

176 55th Meeting of the Air Council Committee on Supply, Further Orders for Mercury and Pegasus engines from the Shadow Group, 1st August 1939, AVIA 10/161, NA.

177 "History of the Production of Bristol Aero Engines by the Shadow Industry Operating under the Joint Aero Engines Committee in Conjunction with the Bristol Aeroplane Company Limited", p. 8, Abridged Record of Engines, Spares and Accessory Output from Shadow Industry for 1939.

178 "History of the Production of Bristol Aero Engines by the Shadow Industry Operating under the Joint Aero Engines Committee in Conjunction with the Bristol Aeroplane Company Limited", Abridged Record of Engines, Spares and Accessory Output from Shadow Industry for 1939.

179 Air Council Committee on Supply, Summary of conclusions reached at the seventy-fourth meeting held on 23 September 1939, AVIA 10/162; Extract from 74th to 81st Meetings of Air Council Committee on Supply, 23rd September to 4th October 1939, AVIA 46/35, NA.

180 "History of the Production of Bristol Aero Engines by the Shadow Industry Operating under the Joint Aero Engines Committee in Conjunction with the Bristol Aeroplane Company Limited", p. 15, Abridged Record of Engines, Spares and Accessory Output from Shadow Industry for 1940.

181 Air Council Committee on Supply, note of George Bulman, D.E.P., Materials for No. 1 Shadow Engine Group, 29 November 1939, AVIA 10/164, NA.

182 Ibid.

183 Aircraft Council Committee on Supply, No. 1 Shadow engine production, 5 December 1939, AVIA 10/165, NA.

184 Aircraft Council Committee on Supply, No. 1 Shadow engine production, Memorandum by D.E.P., 19 December 1939, Air 6/59, NA.

185 Ibid.

186 Narrative-Aero Engine Production Expansion of Capacity 1935-1945, Capacity for Production of Bristol Aero Engines 1938-1945, September 1938-December 1941.

187 "History of the Production of Bristol Aero Engines by the Shadow Industry", Abridged Record of Engines, Spares and Accessory Output from Shadow Industry for 1940.

188 NA, AVIA 46/35, Air Supply Board, No. 1 Shadow Group-Production of Mercury XV and Pegasus XVIII engines, 16 August 1940.

189 Ibid.

190 Ibid.

191 Air Supply Board, Summary of Conclusions Reached at the meeting held on 27th September 1940, AVIA 10/176, NA.

192 B. Davidson to Austin Motors Co. Ltd., 2nd November 1940, AVIA 46/35, NA.

193 "History of the Production of Bristol Aero Engines by the Shadow Industry Operating under the Joint Aero Engines Committee in Conjunction with the Bristol Aeroplane Company Limited", Abridged Record of Engines, Spares and Accessory Output from Shadow Industry for 1941.

194 "History of the Production of Bristol Aero Engines by the Shadow Industry Operating under the Joint Aero Engines Committee in Conjunction with the Bristol Aeroplane Company Limited", p. 16, Abridged Record of Engines, Spares and Accessory Output from Shadow Industry for 1941-1944.

195 Ministry of Aircraft Production: *Statistical Review 1939-1945*, p. 32.

196 *Statistical Review 1939-1945*, p. 26.

197 "History of Production of Bristol Aero Engines by the Shadow Industry", p. 11.

198 Neale, M.C., ed.: *An Account of Partnership-Industry, Government and the Aero Engine: The Memoirs of George Purvis Bulman*, Rolls-Royce Heritage Trust Historical Series No. 31, (Derby, UK, 2001), p. 223.

199 "History of Production of Bristol Aero Engines by the Shadow Industry", pp. 11-14, 28-29.

200 "History of the Production of Bristol Aero Engines by the Shadow Industry", p. 74.

201 *Industry and Air Power*, p. 125.

202 See note Rootes Securities Ltd., Aircraft Division, Engine Factory, Coventry, AIR 19/7, NA; and Minutes of Management Committee Meeting 29th September 1937, Rover Company Limited (Aero Engines), Rover Company Directors Minute Book 1937-1939, 80/148/4/67-RCO, British Motor Industry Trust.

203 Muther, *Production Line Technique*, p. 6.

204 "Shadow Factories in Being", p. 446.

205 Bristol Aeroplane Company Minutes of Directors' Meeting in Committee, 11 June 1936, BAC Directors' Committees Minutes, Vol. 5.

206 See Minutes of Directors Meetings in Committee for June-September 1936, The Bristol Aeroplane Company Ltd., Minutes of Directors Committees, Vol. 5.

207 Notes on the Visit to the Works of the Bristol Aeroplane Company 10th April 1936, p. 3.

208 "The Shadow Factories: A Preliminary Survey of the Machine tools and Methods", p. 444.

209 "At the Shadow Factories", p. 529.

210 "The Manufacture of High Performance Aero Engines", p. 742.

211 "History of Production of Bristol Aero Engines by the Shadow Industry", p. 74.

212 Ibid.

213 Ibid.

214 Foreman-Peck, James, Sue Bowden, and Alan McKinlay: *The British Motor Industry*, (Manchester, UK, 1995), p. 49; "Canley Works Extensions", *The Automobile Engineer*, Vol. XXVII, No. 354, (January 1937), pp. 2-9; "The Vauxhall Works", *The Automobile Engineer*, Vol. XXVI, No. 348, (August 1936), pp. 313-17.

215 "Getting Going at the Shadow Factories", p. 530.

216 "Getting Going at the Shadow Factories", p. 534.

217 "The Shadow Factories: A Preliminary Survey of the Machine tools and Methods", p. 446.

218 Ibid.

219 Neale, M.C., ed.: *An Account of Partnership-Industry, Government and the Aero Engine: The Memoirs of George Purvis Bulman*, Rolls-Royce Heritage Trust Historical Series No. 31, (Darby, UK, 2002), pp. 227-28.

220 Aero Engine Shadow Committee Meeting Held at the Standard Aero Works 10/7/1938.

221 Minutes of Management Committee Meeting 14th November 1939, Rover Company Limited (Aero Engines), Rover Company Directors Minute Book.

222 NA, AVIA 10/169, Air Council Committee on Supply, Mercury Engines (Note by D.D.G.P.I), 1 March 1940.

223 "History of Production of Bristol Aero Engines by the Shadow Industry", p. 40.

224 "History of the Production of Bristol Aero Engines by the Shadow Industry", p. 57.

225 NA, AIR 6/58, 177th Progress Meeting, The Progress of Production for Air Re-armament, Memorandum by D.G.P., 12th July 1939, p. 16.

226 177th Progress Meeting, The Progress of Production for Air Re-armament, Memorandum by D.G.P., 12th July 1939, p. 14.

227 177th Progress Meeting, The Progress of Production for Air Re-armament, Memorandum by D.G.P., 12th July 1939, pp. 14-15.

228 Ibid.

229 NA, AVIA 10/166, Air Council Committee on Supply Meetings December 1939-January 1940, No. 1 Shadow Engine Production Memorandum by D.E.P.

230 NA, AIR 19/9, Statistics Relating to the Twelve Air Ministry "Shadow" Factories in Production, Memo of December 1939, "Shadow" Factories: Renumeration of Managing Firms. Based on Hornby's estimate of the percentages spent on buildings and machine tools at the aero engine factories. See *Factories and Machine tools*, p. 253.

231 No. 1 Shadow Engine Production Memorandum by D.E.P.

232 Narrative-Aero Engine Production Expansion of Capacity 1935-1945, Capacity for Production of Bristol Aero Engines 1938-1945, B) By Shadow Schemes.

233 Ibid.

234 Thoms, David, and Tom Donnelly: *The Motor Car Industry in Coventry Since the 1890's*, (Abingdon, Oxon., 2018), p. 128.

235 Minutes of Management Committee Meeting 17th January 1940, Rover Company Limited (Aero Engines), Rover Company Directors Minute Book.

236 Abridged Record of Engines, Spares and Accessory Output from Shadow Industry for 1940, "History of Production of Bristol Aero Engines by the Shadow Industry."

237 "History of Production of Bristol Aero Engines by the Shadow Industry", p. 15.

238 Hornby, *Factories and Machine Tools*, p. 255.

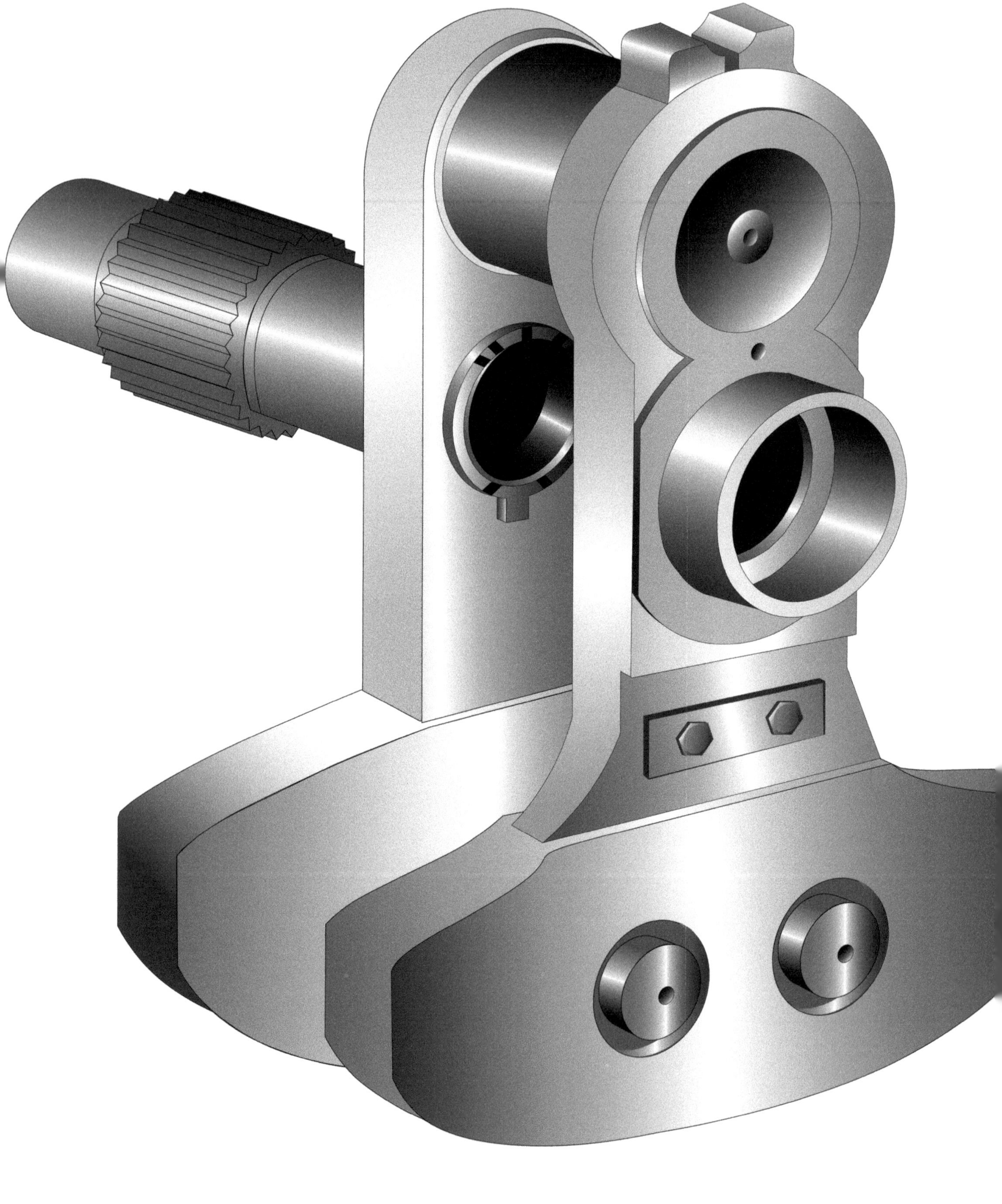

3

Building Nine-Cylinder Engines: The Wright R-1820 Cyclone 9

Introduction

As European rearmament accelerated, and as the Roosevelt administration began a belated effort to expand American air power, a principal concern was the capacity of the American aircraft engine industry to meet the demands for high-performance aircraft engines. If production capacity was insufficient, could it be expanded in time, and who would finance this expansion? These questions became more urgent during 1940 when British and French orders for American aircraft accelerated in the months leading up to the fall of France in May 1940, quickly followed by President Roosevelt's call for 50,000 aircraft for the American Army Air Corps and US Navy. The quantity of aircraft engines now required from Pratt & Whitney and Wright Aeronautical was many times greater than their annual production in the years leading up to the war. The urgency of the requirement for Wright's nine-cylinder R-1820 Cyclone 9 air-cooled radial engine forced the company to rapidly acquire, with financing initially from Britain and France, expanded production facilities, but more importantly to develop the means of shifting rapidly from the company's prewar low-volume batch production methods to high-volume quantity production. Critical to Wright's successful transition were the changes the company implemented to its production processes that incorporated the concept of continuous flow production and the realization that new types of machine tools would be required not just to increase the quantities of components but also to cope with the influx of semi- and unskilled workers who had no previous experience working with machine tools. The intensive effort Wright put into working out solutions to these problems during 1940–1941 provided a foundation for shifting to quantity production of Wright's bigger, more powerful, and more complex engines.

The Wright R-1820 engine was the Wright Aeronautical Corporation mainstay during the decade of the 1930s. During the decade, the R-1820 found wide application in American military and commercial aircraft. As aircraft designers shifted to all-metal aircraft with greater speeds, range, and payloads, Wright matched the need for increasing power with improved models of the R-1820. In the prewar years, Wright had its greatest success with multi-engine commercial airliners. Most airline operators of the Douglas DC-2/DC-3 series chose the R-1820 over Pratt & Whitney's R-1830 Twin Wasp. The R-1820 helped Wright regain a competitive position versus Pratt & Whitney.[1] The Cyclone 9 was best known as the engine that powered more than 12,000 Boeing B-17 Flying Fortress built during World War II.

In a little more than a decade, Wright Aeronautical more than doubled the power of the Cyclone 9. Development of the Cyclone 9 began in 1924, when the US Navy requested Wright to develop a new nine-cylinder air-cooled radial engine with greater power than the 200 hp Wright Whirlwind engine.[2] This in 1927 led to the first Cyclone engine, the P-2 with the military designation R-1750, which was rated at 525 hp.[3] To get more power, Wright widened the bore of the Cyclone to increase the capacity to 1,823 in.^{3}, hence its military designation of R-1820.[4] With the same overall dimensions as the R-1750, the bigger capacity of the R-1820's cylinders boosted the horsepower to 575 hp. As Figure 3.1 illustrates, following the R-1820's introduction in 1930, Wright progressively increased the Cyclone 9's power output through the R-1820F and R-1820G series, so that by 1939, the R-1820G-200 series was putting out 1,200 hp, double the power output of the first R-1820, reaching 1,350 hp in the R-1820H series later in the war.[5]

FIGURE 3.1 Increase in Wright Cyclone horsepower 1930–1943.

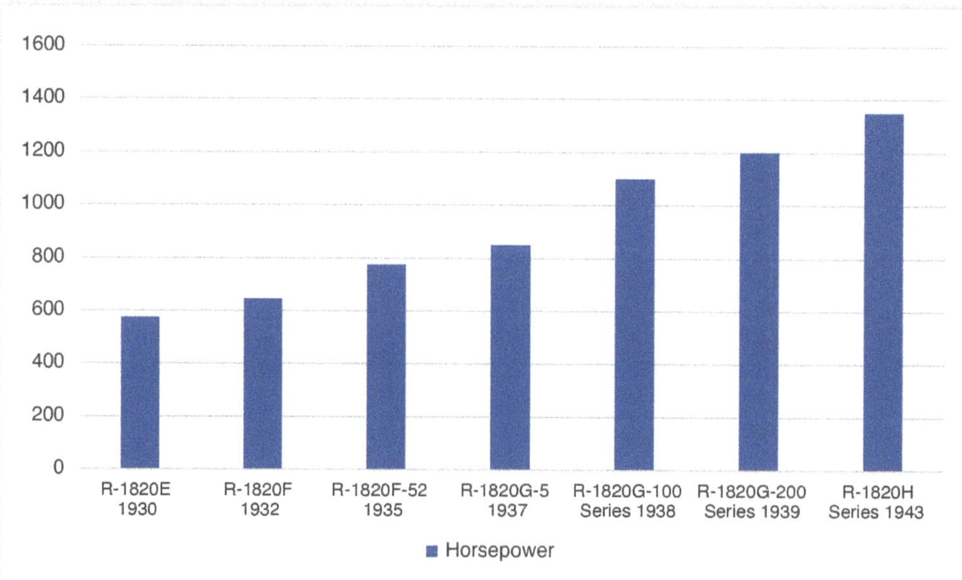

Source: Angle, Glenn: Aerosphere 1939, 1943, Jane's All the World's Aircraft 1945.

The Wright R-1820-G-200 series engine, the model built in greatest quantity during World War II in various models for the military, was a nine-cylinder, single-row, air-cooled radial engine with a dry weight of 1,310 lb. Like the Bristol Mercury and Pegasus, the R-1820 was a poppet valve engine, but with only one intake and one exhaust port in contrast to the four ports on the Mercury and the Pegasus. The G-200 series was a major redesign of the Cyclone. With the G-200, Wright decided that improved performance was more important than interchangeability of parts with earlier models of the Cyclone.[6] The cylinder heads were cast from an aluminum alloy, and to improve cooling, Wright redesigned the cylinder head to increase the cooling area by 20%, deepening the cooling fins from 1½ in. in depth to 2¼ in., although this improvement presented an even greater challenge to the Wright foundry casting the cylinder heads.[7] Wright also redesigned the intake and exhaust ports to improve their efficiency. The cylinder head was attached to a strengthened forged steel cylinder barrel which, thanks to improved machining techniques, had deeper and more closely spaced cooling fins than the G-100 series, offering 60% more cooling area.[8] The G-200 series incorporated aluminum alloy forged pistons with improved cooling and an additional piston ring for more uniform oil flow.[9]

In the R-1820-G-100 series, Wright had switched from a forged aluminum to a forged steel two-section crankcase to strengthen the engine as the power output increased. The G-200 series continued with the steel crankcase, but with an improved design to reduce weight and simplify machining operations.[10] To reduce more weight, Wright used a cast magnesium alloy for the front section of the crankcase in place of aluminum. The two-piece crankshaft for the G-200 series was made from a forged chrome–nickel steel alloy and incorporated two Wright Dynamic Dampers to reduce torsional vibration and stress to the propeller shaft.[11] The G-200 used a master rod and eight connecting rods forged from chrome–nickel–molybdenum steel alloy. Wright refined the supercharge design on the G-200 series to improve the flow of air to the intake ports, allowing a greater quantity of air to flow more directly to the impeller, and redesigned the impeller blades for greater efficiency. To save weight on the supercharger and its mounting, Wright again substituted lighter magnesium castings for the mounting and rear housing sections.[12] The propeller shaft, a hollow forging made from a steel alloy, attached to the crankshaft through the reduction gear machined from high-strength nitralloy steel.

The final wartime development of the Cyclone 9 was the R-1820H series. To obtain more power, Wright switched to using forged cylinder heads and a new form of cooling fins on the cylinder barrel, saving weight and greatly increasing the cooling area on the cylinder allowing for higher temperatures in combustion and greater power output. Designated the R-1820-56, this model of the Cyclone 9 achieved a weight to horsepower ratio of .97 lb/hp, the lowest ratio of any of the high-performance air-cooled radial engines built during World War II (Table 3.1).

TABLE 3.1 Comparative data: Wright Cyclone 9 and Bristol Pegasus XVII.

	Wright Cyclone 9	Bristol Pegasus XVII
No. of cylinders	9	9
Total cylinder volume	1,823 cu.in.	1,753 cu.in.
Diameter	55 in.	55.3 in.
Dry weight	1,310 lb	1,180 lb
Max. takeoff power	1,200 hp	1,050 hp

Source: *Jane's All the World's Aircraft 1945-46, (London, 1946).*

Organizing Production

The American aircraft engine industry was small in scale. In January 1939, the factory floor space of the entire industry, big and small firms, amounted to 1,762,000 ft^2, not significantly bigger than the factory space at the Bristol Aeroplane Company and the new shadow factories in Britain.[13] During 1938, as the shadow factories were building up to a production rate of 50 Mercury engines per week, Wright Aeronautical was building its Cyclone 9 engine at a rate of 35 per week while Pratt & Whitney was building 13 Twin Wasp engines per week. Production of these high-performance aircraft engines was growing slowly as the airlines and the military services introduced heavier and more capable airplanes. In 1936, the industry, primarily Wright Aeronautical and Pratt & Whitney, built 1,827 engines with greater than 700 hp, rising to 2,537 engines built during 1938.[14] As the world began to rearm, the demand for American aircraft and aircraft engines steadily increased. During 1937, the American aircraft engine manufacturers sold 1,048 engines abroad generating $5.9 million in sales.[15] The next year the industry sold 1,309 engines abroad for $7.9 million, and in 1939 sold 1,880 engines for $14 million, a near doubling of sales.[16] The driving force behind the increase in foreign sales were purchases of American aircraft and aircraft engines by the French and British Governments.

At the end of 1937, the French Government realized that the French Air Force and French aviation industry were woefully inadequate and incapable of matching the rapidly growing German Luftwaffe in quality and quantity of aircraft.[17] In March 1938, the French cabinet approved Plan V, a significant increase in the French Air Force's first-line strength and reserves to be achieved by March 1941.[18] In the interim, however, the Air Force would still be short of more modern aircraft. Even before Plan V received cabinet approval, the Government sent a mission to America to explore purchasing American combat airplanes, aircraft engines, and machine tools in order to "harness American industry to the French war machine" as the French representative explained.[19] The French Government hoped to purchase as many as 1,000 American aircraft, but found that there were few American military aircraft that met French requirements and that the American aviation industry was unprepared for either quantity production or quick delivery.[20] The only order that came from the mission was for 100 Curtiss-Wright P-36 fighters, designated the Hawk 75A, with Pratt & Whitney Twin Wasp engines for delivery by April 1939.[21]

The Munich Crisis in September 1938 only reinforced the reality of France's weakness in the air. With support from the Roosevelt administration, the French Government sent a second mission to America to purchase 1,000 American aircraft, particularly modern bomber aircraft. In January and February 1939, the French ordered 100 more Curtiss-Wright Hawk 75 fighters, 100 Douglas DB-7 and 110 Glenn Martin 167F bombers, 20 Chance Vought V-156 dive bombers (an additional 20 V-156 airplanes were ordered in May), and 200 North American Harvard training aircraft, all equipped with Pratt & Whitney engines.[22] In February, the French Government separately signed a contract with Pratt & Whitney for 795 Twin Wasp engines valued at $10.6 million, and a month later placed a contract for an additional 300 Twin Wasp engines.[23] These contracts were a Godsend for Pratt & Whitney which toward the end of 1938 had begun laying off employees as it completed engine contracts for the French, British, and the American Army Air Corps.[24] Over the next several months, the French Government placed additional orders with Pratt & Whitney for 1,050 Twin Wasp engines and with Glenn Martin for 150 Martin 167F bombers.[25]

British purchases of American aircraft were more restrained. In the spring of 1938, the slow pace of British aircraft production brought up the question of placing supplemental orders with the American aircraft industry.[26] Like the French Government, the British Air Ministry sent a mission to America to study the productive capacity of the American aircraft industry and identify aircraft that might be suitable for purchase for the Royal Air Force, particularly aircraft that could be used for land-based maritime reconnaissance, and training aircraft.[27] Like the French mission, the British representatives concluded that the available American military aircraft were technically behind comparable European aircraft and that it was unwise to place large orders for American aircraft, with the exception of a conversion of the Lockheed 14 Super Electra airliner for maritime reconnaissance and the North American Harvard trainer.[28] In June 1938, the Air Ministry placed orders with Lockheed for 200 Lockheed 14 aircraft, named Hudson, with an option for 50 more, and with North American for 200 Harvard trainers.[29] The British order for the Lockheed Hudson brought Wright Aeronautical an order for 400 Cyclone 9 engines at an estimated cost of $4 million.[30] A few months later, the Australian Government also placed an order with Lockheed for 50 Hudson aircraft with Wright engines.[31]

In the aftermath of Munich, the Air Ministry began to change its opinion about the possible contribution of the American aircraft industry. There was an argument that American production could augment British production to provide reserves and a flow of aircraft far from the threat of German air attack once war had begun.[32] American neutrality laws and the relatively poor performance of American military aircraft were drawbacks, but America did have huge industrial potential that could be tapped over time. This would entail a significant increase in production capacity through British capital investment to increase the flow of raw materials, machine tools, as well as aircraft and aircraft engines to Britain.[33] From this perspective, immediate purchases of aircraft and engines would contribute to building capacity for future production.

The Munich Crisis had a profound impact on President Roosevelt and his view of air power. In its aftermath, Roosevelt came to believe that air power was a force that could deter Hitler and determined to strengthen Allied and American air power to the extent he could do so within the political constraints America's isolationism imposed on his freedom of action.[34] Roosevelt "became convinced for the first time that American airplane production should be greatly stimulated with all possible speed."[35] He decided to support the French Government's purchases of American aircraft and begin an expansion of American air power.

At a conference with top government and military officials on November 14, 1938, Roosevelt stressed the need to quickly build up American air power.[36] His objectives were an Army Air Corps with 20,000 aircraft and an aircraft industry with a productive capacity of 24,000 aircraft a year, but he requested the War Department instead to draw up plans for a force of 10,000 aircraft to be built over two years.[37] In January 1939, Roosevelt sent a message to the Congress calling for an increase in national defenses and asking for $300 million in appropriations. That April Congress passed the emergency Army Defense Bill authorizing procurement of 3,251 aircraft for the Army Air Corps and an increase in the Air Corps authorized strength to 5,500 aircraft from its previously authorized but unobtained level of 2,320 aircraft.[38] With new funding, the Army Air Corps planned to solicit bids for 880 aircraft, including new two- and four-engine bombers, training planes, observation planes, and fighters; appropriations for the US Navy included provisions for 500 new aircraft.[39] On August 10, the Army Air Corps issued contracts for $78.5 million in new aircraft, including $12 million to Pratt & Whitney and $7 million to Wright Aeronautical for aircraft engines.[40] There was now a third claimant for the American aircraft industry's limited production capacity. In just six months, the industry's backlog had more than doubled, from $151 million in January 1939 to $324 million in August.[41]

Two months after the start of war in Europe, Congress amended the Neutrality Act to remove the embargo against selling arms to belligerents, allowing for "cash-and-carry," whereby countries had to pay for American materials in cash and transport them on their own ships.[42] The British and French Governments could now order airplanes, aircraft engines, and other materials without fear that they would be embargoed. The French Government did not wait for the embargo to be removed but sent a third air mission to America that arrived in Washington shortly after the start of war in Europe.[43] The two members of the mission began ordering more American aircraft and aircraft engines shortly after. To fulfill war needs for the first six months of the war, France needed 2,800 additional aircraft and 10,000 aircraft engines from all sources.[44] In a flurry of orders placed during October 1939, the new French mission ordered over 1,000 new aircraft consisting of 130 Martin 167F bombers, 530 Curtiss Hawk 75A and 100 Hawk 81 (P-40) fighters, 200 North American Harvard, and 270 Douglas DB-7 bombers, totaling $70 million.[45] At the same time, to meet a desperate need for aircraft engines to power French-built aircraft, the Government ordered 2,470 Pratt & Whitney R-1830 Twin Wasp engines for the Bloch 176 B3 bomber, 1,100 Pratt & Whitney R-985 Wasp Junior engines for the Breguet 695 attack plane, 1,440 Wright R-2600 Cyclone 14 engines for the Loiré-et-Olivier 458 B4 bomber, and 455 Wright R-1820 Cyclone 9 engines.[46] After Congress lifted the arms embargo, Britain ordered an additional 200 Hudson aircraft, 600 Harvard trainers, 120 Brewster Buffalo fighters with Wright R-1820 Cyclone 9 engines, and 50 Consolidated PBY flying boats with Pratt & Whitney R-1830 engines, creating additional engine orders for Pratt & Whitney and Wright Aeronautical.[47]

The French and British Governments now found that a real bottleneck had emerged to slow deliveries of the American airplanes they had ordered: Pratt & Whitney and Wright Aeronautical could not keep pace with the number of aircraft engines needed for the British, French, Army Air Corps, and US Navy airplanes now on order, with even larger orders likely to be placed in the future.[48] With memories of the near-collapse of the aircraft industry at the end of World War I when the American Government cancelled all contracts overnight, and the lean years of the Depression only just behind them, neither company was willing to put its own capital at risk to expand production facilities in case war could be averted and orders collapse again.[49] Depreciation under the American tax code was also a critical financial issue for the companies. At the time, the tax code did not allow rapid depreciation of factories and machinery, but only over a period of 20 years. If the aircraft engine companies built all the facilities they would need to expand production and hostilities ceased, they would be left with huge excess facilities and a depreciation expense that would cause severe financial losses.[50] This, however, was a tax matter for the Congress to resolve, which would take time. To provide immediate help to the engine companies, the French Government agreed to provide $5 million each to Pratt & Whitney and to Wright Aeronautical to double production.[51] These funds allowed the two companies to construct additional factory facilities and purchase more machine tools, increasing production capacity by 250 engines per month at Pratt & Whitney and by 200 engines per month at Wright Aeronautical.[52]

That aircraft engine production could quickly become a bottleneck to aircraft production did not come as a surprise. At a conference with Pratt & Whitney, Wright Aeronautical, and the Allison Division of General Motors to discuss their production capacity, held shortly before the outbreak of war in Europe, the representatives of the companies told Maj. Gen. Henry Arnold, Chief of the Air Corps, that even working three shifts and with expanded sources of supply, based on present and estimated requirements, there would be a shortfall in production within 9 months, and within 12 months, the shortage would amount to 450 engines a month.[53] The companies made clear that they considered any further expansion in production facilities to be "abnormal" and would want this expansion to be financed in such a way that the facilities would not become a burden to the companies at the end of any emergency.[54] They suggested that the Treasury Department agrees to complete amortization of any new facilities built to increase production in an emergency.[55]

As French and British orders for American aircraft and engines accelerated, the shortfall in engine production only worsened. In a meeting on November 9, 1939, with representatives of the War Department, the Army Air Corps, the Navy, and the Treasury and Commerce Departments, it was determined that the aircraft industry then had the capacity to produce 18,000 aircraft a year, but the aircraft engine manufacturers could only produce 12,000 engines, creating a shortfall of 6,000 engines.[56] To expand production facilities for aircraft engines, it would be necessary to address the issue of accelerated depreciation. When the issue of the disparity in production was brought to the President's attention, he wrote to Secretary of the Treasury Henry Morgenthau, Jr. that "one of the obstacles to the proper equipment of our country for national defense is the disparity between the rate of production of airplane motors and that of airplanes themselves. This must be synchronized as soon as possible."[57] The President directed Morgenthau to work out a depreciation policy whereby any facilities acquired for increased production could be fully amortized over the life of a contract or the period of emergency.[58] The President soon gave Morgenthau even greater responsibilities: to coordinate British and French aircraft purchases in America, to ensure the smooth and speedy flow of supplies to the Allies, and in effect to serve as the head of military aircraft production with a more sympathetic view of Allied purchases than that of the War Department.[59]

The main difficulty that would bedevil government planners for months was the problem of determining how many aircraft engines were required, from which manufacturer and for which type of airplane, for spares, and for airplanes to be built in Britain and France that would use American aircraft engines. Soon after the start of the war, and in expectation of large orders for American aircraft and aircraft engines, the Roosevelt administration began pressuring the British and the French to coordinate their purchases.[60] In November, the British Government set up the British Purchasing Commission in Washington to cover British purchases, and on November 25, 1939, the British and French Governments formed the Anglo-French Purchasing Commission. The Anglo-French Purchasing Commission's first task was to explore the possibility of greatly increasing purchases of American aircraft and aircraft engines.

On January 5, 1940, members of the Anglo-French Purchasing Commission met with representatives of Allison Motors, Pratt & Whitney, and Wright Aeronautical.[61] The Commission explained that the Allies were surveying the possibility of ordering 10,000 American aircraft. The members of the Commission realized that engine production would likely be a bottleneck for this aircraft program as these aircraft would require upwards of 20,000 engines, including spares, and it was hoped that these orders could be completed by the end of 1940.[62] The representatives of the engine companies withdrew for 3 hours to calculate their possible production during 1940, returning with an estimate that the three companies could probably provide only 4,000 engines, or just 20% of the number required.[63] Realizing that this was a hurried calculation, the engine companies agreed to provide the Commission with a more detailed estimate a week later.

During the meeting, the question of using the automobile industry's factories to build aircraft engines came up for discussion, but as the British Government had found when considering shifting production to the parent auto firms of the shadow factories, it was deemed equally impractical for the American automobile factories to build aircraft engines as it was for the British automobile factories. The president of the Allison Division related how Allison had examined $180,000 of machine tools in the company's sister Cadillac Division to see if these tools could be used in the Allison factory, only to find that only some 5% were suitable. The Allison president stated his belief that "no great amount of help might be expected from the automobile industry owing to the high-precision character of work done on aeronautical engines."[64] As in Britain, to build aircraft engines, the American automotive factories would have to be completely reequipped with new machine tools and their normal operations disrupted, an expensive proposition.[65] More critically, Wright Aeronautical refused to agree to using General Motors to supplement production at Wright.[66] After speaking with the president of Allison, who also described the delays Allison was encountering getting deliveries of machine tools, Secretary Morgenthau made a note saying "It is all very discouraging…The American industry will not be able to give the Allies what they want, namely 20,000 airplanes [engines] in the calendar year 1940. It just cannot be done."[67]

Because the aircraft engine companies were already working at full capacity and were having difficulties doubling capacity to meet the latest orders, it was doubtful that the companies could manufacture more than 13,000 engines by October 1941.[68] The shortage of engines meant the program could only expect to receive 8,400 airplanes, which would require 13,675 engines. By the end of 1940, the companies would have produced 24,000 engines, but after fulfilling requirements for the Army Air Corps and the Navy, the number of engines available for the Allies would not meet the target of 13,675 engines.[69] Moreover, because 5,000 of these engines were committed to airplanes that were to be built in France, the actual number available for American aircraft was around 9,000 engines, meaning that the Purchasing Commission had to cut its orders back to some 5,000 aircraft.[70]

The Anglo-French Purchasing Commission made its final recommendations on aircraft and engine purchases in late March and received French and British Government approval at the end of the month.[71] The number of aircraft to be ordered was significantly less, 10,000 aircraft, and was comprised of new aircraft types and, as a result, required a different mix of aircraft engines. The program called for purchasing 2,440 fighter aircraft and 2,260 bombers to be divided between the two countries.[72] With President Roosevelt's support, and over the objections of the War Department and the Army Air Corps, the Anglo-French Purchasing Commission had been allowed to place orders for the most advanced American airplanes, some of which had not yet entered production.[73] The Commission initially planned to place orders with the Glenn Martin Company for 700 new Model 187 twin-engine bombers (later named Baltimore) and 200 Douglas A-20s, an improved DB-7 for the Army Air Corps, and 300 dive bombers for the French, all equipped with Wright R-2600 Cyclone 14 engines, 550 Douglas DB-7s and 60 Consolidated LB-30 four-engine bombers with the Pratt & Whitney R-1830 Twin Wasp engine, and 400 Lockheed Model 37 twin-engine bombers (later named Ventura) with the Pratt & Whitney R-2800 Double Wasp engine.[74] These new orders added large numbers of Allison, Pratt & Whitney, and Wright Aeronautical engines to contracts placed earlier, taking total unfilled engine orders and options to 10,328.[75]

It was clear to the Anglo-French Purchasing Commission that timely completion of the engine contracts for these new aircraft orders would require new production capacity at Pratt & Whitney and Wright Aeronautical.[76] Since aircraft engines still presented the greatest potential bottleneck in aircraft production, the British and French Governments committed to making a substantial capital investment in new production facilities for Pratt & Whitney and Wright Aeronautical. By May 15, 1940, the British and French Governments had made capital commitments of $22.5 million to Wright Aeronautical.[77] The French

Government had previously provided $10.4 million and the British Government $4.8 million to Wright, bringing the total Allied capital commitment to the company to $37.8 million and Pratt & Whitney $8.2 million, respectively.[78] Of the total amount Wright Aeronautical received, $18.6 million went to expanding factory capacity, $7 million to personnel training, $2.4 million to support income tax payments, and $9.7 million for undesignated purposes, most likely for purchasing machine tools.[79] Allied financing was vitally important to the expansion of the American aircraft industry. As the history of the Army Air Forces in World War II noted, "it is perhaps not too much to say that the expansion financed by British and French funds in 1939 and 1940 advanced by as much as a year the time within which American aircraft production would reach its peak."[80]

Within weeks, President Roosevelt began an American rearmament that would dwarf all previous Allied aircraft orders. Following the German invasion of France and the Low Countries on May 10, 1940, the President sent a message to Congress stressing the need to build American air power and provide more airplanes to the Allies for their defense.[81] Roosevelt recommended that the country build the capacity to produce 50,000 aircraft a year and equip the Army and Navy with 50,000 aircraft.[82] Converting the President's recommendations into plans, procurement programs, and actual aircraft and engines was a massive undertaking. Planners had to determine what type of aircraft and engines would be needed, in what quantities, by what date, and from which companies and factories.[83] There were two decisions coming out of the planning process that determined the framework of American aircraft engine production during World War II: first, the realization that only the Government had the resources to finance the vast expansion in production capacity that would be necessary, and second, that the American automotive industry would have to be brought in to help expand aircraft and aircraft engine production.

Planning was the first step. To help coordinate production planning and the overall rearmament effort, the President established a new civilian agency, the Advisory Commission to the Council of National Defence, better known as the National Defence Advisory Commission (NDAC).[84] Roosevelt appointed William S. Knudsen, president of General Motors and a production expert, as commissioner for the Production Division.[85] Knudsen set up an Aeronautical Section (later the Airplane Division) to work on planning for the 50,000 aircraft program in concert with the military services, the aircraft industry, and foreign purchasers.[86] A big challenge stemming from the President's program was to "match ends with means," ensuring that America had the production facilities necessary to achieve the President's goals.[87] In the weeks following its organization, the NDAC's Aeronautics section completed several reports on the Army and Navy's plans for aircraft and aircraft engine procurement, aircraft manufacturing capacity, and aircraft engine manufacturing capacity.[88]

Report No. 7, "Report on a Study of Aircraft Engine Manufacturing Capacity," looked at the total number of engines that would be needed for the Air Corps, the Navy, and the British by June 1942, calculated the number of engines that would have to be built per month to complete this program, and then calculated the additional factory floor space that would be needed to build these engines at the specified monthly rate of production.[89] The report calculated that by June 1942, Wright Aeronautical would have to build 39,483 engines of all types, including 11,718 R-1820 Cyclone 9 engines and 23,509 R-2600 Cyclone 14 engines.[90] This was more than ten times the number of engines Wright built during 1939. Monthly production of the Cyclone 9 would have to increase from 282 engines in July 1940 to 600 a month by July 1941, while production of the Cyclone 14 would need to increase from 293 a month in June 1940 to 1,204 in July 1941 and 1,560 a month by March 1942.[91] To put these projected production levels in context, in July 1940, Wright completed 215

Cyclone 9 engines and 168 Cyclone 14 engines, short of the report's estimate of then-current production levels.[92] Wright would need to more than double Cyclone 9 production and boost Cyclone 14 production nearly ten-fold based on the report's estimates.

The report concluded, however, that it would be possible to achieve this goal by June 1942 if two conditions could be met: first, an increase in the percentage of subcontracting at each manufacturer, and second, a huge expansion in production floor space and employment.[93] At the time of the report, the percentage of work done by subcontractors was as follows:

Wright Aeronautical	38%
Pratt & Whitney	50%

The report recommended that the level of subcontracting be increased to the following levels:

Wright Aeronautical	50%
Pratt & Whitney	60%

The increase in floor area and employment required to complete the President's program was remarkable as shown in Table 3.2.

TABLE 3.2 Plant and personnel expansion: engine manufacturers.

	Production floor area (ft^2)	Present personnel	Required production floor area (ft^2)	Required productive personnel	Expansion ratio: floor area	Expansion ratio: personnel
Wright aeronautical	2,334,000	6.050	6,620,000	33,100	2.8	5.5
Pratt & Whitney	1,202,000	6,994	5,910,000	33,800	4.9	4.9
Total	4,159,000	15,394	14,120,000	74850	3.4	4.6

Source: National Defense Advisory Commission, Airplane Division Report No. 7: Report on a Study of Aircraft Engine Manufacturing Capacity, July 22, 1940, Table I, p.7, File 315.30421, War Production Board Policy Documentation File, Box 1125, File 315.30413-315.30433, Records of the War Production Board, Record Group 179, National Archives and Records Administration (NARA), Washington, DC.

Wright Aeronautical would have to expand its factory space by nearly three-fold, and its personnel more than five-fold, both figures significantly greater than the entire American aircraft engine industry's floor space and personnel in 1938. Apart from increasing subcontracting, Report No. 7 said nothing about how the engine manufacturers were to increase their monthly production to these levels, but stressed the urgency of expansion saying, "it is essential to the attainment of this program that the expansion of plant facilities be started at once to provide the full magnitude of aircraft engine production at the earliest possible date."

The size of the planned expansion and the time pressure to complete the President's program made the Army, the Navy, and the NDAC realize that the Government would have to finance most of the expansion as the aircraft industry lacked the financial resources and companies were reluctant to take on the risk of being left with excess capacity.[94] After joining the NDAC, William Knudsen proposed that the Reconstruction Finance Corporation, a Government agency established to lend money to businesses during the Depression, set up a separate subsidiary to finance factory expansion.[95] This subsidiary, the Defence Plant Corporation (DPC), provided money to cover the full cost of building a factory and equipping it with machine tools.[96] The factory was leased to the manufacturer for a set term with title reverting to the Government at the end of the lease unless the manufacturer chose to purchase the factory. From 1940 to 1945, the DPC provided $1,357,000,000 in financing for new aircraft engine factories.[97] Wright Aeronautical was one of the early beneficiaries of this new Government financing program, receiving funding to build a large new factory to build the Wright R-2600 Cyclone 14.[98]

The Aeronautical Section's program planning became more comprehensive as events brought greater integration and coordination to American and British requirements and purchasing programs. With the fall of France, on June 17, 1940, the British Purchasing Commission in Washington took over all French aircraft and aircraft engine orders amounting to $425 million.[99] With the addition of the French contracts, the British had 8,275 airplanes and 21,485 engines on order.[100] Then, on July 23, it was agreed that the British could order an additional 6,118 airplanes and exercise options for an additional 21,000 engines.[101] In September, the Army Air Corps, the Navy, and the British Purchasing Commission formed the Joint Aircraft Committee to coordinate all aircraft and engine programs.[102] From the July 23 agreement on allocations, the Aeronautical Section produced the first of a series of reports showing monthly schedules for aircraft and aircraft engines, by aircraft type, model, and manufacturer.[103] This enabled manufacturers to know exactly what was required of them in terms of production over time, greatly facilitating production planning. Report 8 covered aircraft, and from this foundation, the Aeronautical Section prepared Report 9 covering aircraft engines, followed by Report 10 covering propellers. These reports were the foundation of the aircraft program and were regularly updated. The Report 9 series used estimated production capacity compared to requirements to determine if the supply of an engine model was in surplus or deficit.[104] Concerns with a possible shortage of engines remained. The first report in the engine series, Report 9A, showed that 83,869 engines were to be built by April 1942.[105] The report concluded that "expansion of productive facilities in several specified engine types considerably beyond that already contemplated by the manufacturers would be necessary if the requirements were to be met."[106]

Recognizing that in wartime the aircraft industry was unlikely to have the capacity to manufacture all the aircraft and aircraft engines that the Air Corps would need, the Air Corps logically thought of converting the automobile industry to manufacturing aircraft and aircraft engines as the industry had vast experience in volume production.[107] The aircraft industry, however, viewed the huge and well-capitalized automobile industry as a potential competitor and was reluctant to see automotive companies enter the aircraft or aircraft engine field.[108] There were cultural differences as well between the two industries. The aircraft engine companies saw the automotive industry as obsessed with volume production, measuring products built per minute, and constantly seeking ways to speed production and cut costs which would lead, they feared, to favoring production over precision and quality.[109] The automotive industry, in turn, viewed the aircraft engine makers as having limited experience with manufacturing and too wedded to techniques that were more akin to making handcrafted products.[110] Eugene Wilson, then president of Pratt & Whitney's parent United Aircraft Corporation, commented that "to automotive men, the aircraft industry still looked something like the ancient 'carriage trade,' and they discounted our insistence that our standards of quality and precision were not easily come by in the mass production industries."[111]

In the weeks following the President's call for 50,000 aircraft, Secretary Morgenthau pushed Pratt & Whitney and Wright Aeronautical to grant the Government licenses for manufacturing their engines at different manufacturers the Government would select.[112] The companies had different attitudes toward licensing production of their engines. Pratt & Whitney recognized that its management was insufficient to oversee a great expansion in production facilities and was happy to turn over management of volume production of its standard engines to other companies.[113] While accepting the need to increase subcontracting, Wright Aeronautical was reluctant to give other companies full responsibility for assembling Wright engines, preferring to establish branch factories that it could manage on its own.[114] Nevertheless, at Secretary Morgenthau's request, in early June 1940, both Pratt & Whitney and Wright Aeronautical agreed to grant licenses to Government-designated corporations to manufacture any type of engine if the Government found it necessary to do so in the interests of national defense.[115] Pratt & Whitney and Wright agreed to provide all technical assistance, drawings, specifications, and production data to any licensee. Guy Vaughan, president of Wright Aeronautical and its parent Curtiss-Wright Corporation, had insisted on some protection for Wright against creating future competitors, so the licensing agreement with the Government provided for termination after a term of three years unless Wright and the Government mutually agreed to an extension.[116]

Despite obtaining their agreement to granting licenses to the Government and the obvious need to expand aircraft engine production, the Government did not immediately approach the automotive industry to begin license production of Pratt & Whitney and Wright engines, initially choosing to finance factory expansions at the two aircraft engine manufacturers.[117] The Government did approach the Ford Motor Company to build the Rolls-Royce Merlin engine under license from the British Government, but Henry Ford, a committed isolationist, refused to build engines for the British.[118] Knudsen went back to Ford in August 1940 and persuaded the company to manufacture the Pratt & Whitney R-2800 Double Wasp under license.[119]

The Army Air Corps' decision to greatly expand bomber production made it necessary to engage the automobile industry more fully in aircraft production. The final plan called for the Government to finance two factories in Omaha and Kansas City to build the twin-engine Martin B-26 and North American B-25 medium bombers, and two factories in Fort Worth and Tulsa to build the four-engine Consolidated B-24 bomber.[120] Plans called for the production of 100 B-26 and 100 B-25 bombers and 100 B-24 bombers a month.[121] Meeting this new demand for bombers required a large increase in aircraft engine production, above and beyond the orders already pushing Pratt & Whitney and Wright Aeronautical to a critical point.[122] The only alternative was to bring in the automobile industry. On December 21, 1940, the Studebaker Corporation received a contract to build 2,000 Wright R-2600 Cyclone 14 engines under license from Wright Aeronautical for the new B-25 medium bombers that would be coming out of the Kansas City factory, while a month later the Buick Division of General Motors was given a contract to build 2,000 Pratt & Whitney R-1830 Twin Wasp engines for the Consolidated B-24.[123]

The Studebaker Corporation, one of the so-called independent automobile companies in the automotive industry, became Wright's first automotive licensee. Although Studebaker built far fewer cars than the "Big Three" (General Motors, Ford, and Chrysler), Studebaker had a reputation for sound engineering and innovative methods of production, particularly for its automobile engines built entirely within its own factories in South Bend, Indiana.[124] Studebaker's senior management was well thought of. Having rescued Studebaker from bankruptcy during the Depression, Howard Vance, Chairman of the Board, and Paul Hoffman, president of the company, were considered to be among the best managers in the industry.[125] In July 1940, Wright Aeronautical had approached Studebaker about building parts for the Wright R-2600 engines to be built at its new factory in Ohio, proposing that Studebaker builds cylinder assemblies, connecting rods and crankshafts.[126] This led to exploratory meetings between Studebaker and Wright, so that when in September the Government requested Studebaker to consider building the Wright R-2600 engine Studebaker had an idea of what would be involved (Figure 3.2).[127]

FIGURE 3.2 The Wright Cyclone 9 R-1820 engine powered the Boeing B-17 Flying Fortress, one of the US Army Air Force's principal heavy bombers at the start of the war. Boeing and its two licensee companies, Douglas Aircraft and Lockheed-Vega, built 12,731 B-17s. After 1942, most B-17s had R-1820 engines built by Studebaker.

2008-3-31_image_594_01, Peter M. Bowers Collection, Museum of Flight, Seattle, WA.

Studebaker was well into preparations for production of the R-2600, with factories under construction and machine tools ordered, when the Air Corps changed its objectives and, as a result, its contract with Studebaker. In April 1941, General Arnold visited Britain and learned that the RAF's greatest need was for heavy bombers to attack Germany.[128] This led to a proposal to increase heavy bomber production to 500 aircraft a month, expanding production of the Consolidated B-24 and setting up a production pool with Douglas Aircraft and Vega, a subsidiary of Lockheed to build the B-17 under license from Boeing.[129] President Roosevelt approved the Heavy Bomber Program in early May.[130] The Heavy Bomber Program "necessitated a drastic reshuffling in the plans for production of engines."[131] The initial plan to build 180 B-17s a month generated a requirement for upward of 1,000 R-1820 Cyclone 9 engines, including 35% spare engines, at a time when Wright was building only 340 Cyclones a month. To meet this new demand, the Army Air Corps instructed Studebaker to drop plans to build the Wright R-2600 and switch to building the Wright R-1820, issuing Studebaker a new contract for 1,000 R-1820 engines a month.[132]

By June 1941, the framework for wartime production of the R-1820 Cyclone 9 was in place. Wright Aeronautical would build the R-1820 at its Paterson, New Jersey, factories to provide engines for other Air Corps (soon to become the Army Air Force) and Navy airplanes and would remain responsible for development work to improve the engine, while Studebaker took over quantity production of the R-1820 exclusively for the Boeing B-17. The military services' demand for the R-1820 continued until the final months of the war. In March 1945, Studebaker and Wright were still building the R-1820 at a rate of over 1,500 engines a month.[133]

The Factories

During the 1920s and 1930s, American industrial factory design shifted from the older "vertical" plane in multi-story daylight factory buildings to the "horizontal plane" in large, single-story buildings to facilitate sequencing operations and placement of machine tools for line production.[134] The idea was to "build the plant around the process."[135] The straight line is the best way of arranging line production, and a square or rectangular building allows greater flexibility in laying out production lines of machine tools, as well as more efficiency if all production processes can be accommodated within one building.[136] Building on Albert Kahn's designs for the Ford River Rouge complex, this new form of factory design incorporated five key elements:

1. A single building consisting of a series of "cell-like spaces."
2. New building materials, particularly reinforced concrete and high-strength steel to permit larger spans and larger workspaces within the factory building.
3. Larger workspaces, with wider bays and fewer supporting columns, allowed not just more machine tools to be placed within the factory, but more efficient placement of machine tools for line production.
4. The transfer of responsibility for materials handling from worker to machine, using conveyors, monorails, cranes, and motorized trucks within the factory, saving time, space, and indirect labor.
5. The switch from overhead shafting and belt-driven machine tools to electrically powered machines, allowing more efficient placement of machine tools in production lines.[137]

As America began its rearmament, the large single-story building with large unobstructed floor areas ideal for continuous quantity production became the favored design.[138] Some 80% of the new factories then under construction for defense production followed this pattern.[139] A new development that began at the end of the 1930s, and that will be seen in the Studebaker factories, was the "windowless" factory that relied on artificial florescent lighting and air-conditioning.[140] Artificial lighting improved working conditions and productivity by ensuring that all areas within the factory space had adequate lighting and when combined with air-conditioning made multiple shifts more practical. The factories building Wright engines during the war would incorporate these features.

As with aircraft engine factories in Britain, the key characteristics of factories building the Wright Cyclone 9 were location, size, and design. Between 1939 and the end of 1940, as orders for its Cyclone 9 and Cyclone 14 engines poured in from the British, French, and then American Governments, Wright had to rapidly expand its production facilities. Having decided to retain its original factory building in Paterson, New Jersey, as its center of engine production rather than building a completely new and larger factory complex, Wright had to hurriedly acquire factory space in and around Paterson. Its factories were a mix of new and older industrial building designs. As Wright built several types of aircraft engines in its

Paterson factories, these new factories became facilities for building components for multiple types of engines. The factories fed components to the main factory building for final assembly and testing. This system was similar to the first shadow factory scheme for building Bristol Mercury and Pegasus engines. Wright came to this system through happenstance and expedience as it rushed to ramp up production in response to British and French orders.

Wright's main factory building in Paterson, however, followed the older factory pattern. The Plant No. 1 building was a four-story, daylight factory built in 1916 that Wright had expanded over the years, adding three wings to increase the floor space to 675,000 ft^2.[141] In 1937, Wright hired the firm of Albert Kahn Associates, the leading industrial architectural firm in America, to design an attached four-story wing, an experimental and engine test facility, and enlarged the assembly building adjacent to the main factory.[142] The main factory's location in downtown Paterson was a constraint on its expansion. Another multi-story wing was considered the only practical option. The new wing was similar in construction to the original factory building, constructed as a flat slab, reinforced concrete structure with extensive steel sash windows to allow daylight.[143] The new wing additions to the assembly building, and the new experimental test facility increased Wright's total floor space to 880,000 ft^2.[144] While the new wing gave more space for manufacturing, the Wright factory was not ideal for quantity production. As a British observer commented at the time, "The layout of the Wright shops ensures a good flow, but the installation of its machine tools struck me as being rather crowded, and there is the added disadvantage of the fact that the building is not very modern. It is laid out on three floors, which entails a certain amount of lost work (Figure 3.3)."[145]

FIGURE 3.3 The new Wright Plant No. 2 in Paterson, New Jersey, constructed in just 57 days with funds from the French Government and completed in June 1940. Plant 2 would nearly double Wright's total production area, but would soon be insufficient to meet the demand for Cyclone 9 and Cyclone 14 engines. Wright's original factory building, which became Plant 1, can be seen center right.

General Information on Wright Aeronautical Corporation, Defense Plant Corporation: Engineers' Reports and Appendices: Plancor 10, Box 5, Reconstruction Finance Corporation, Record Group 234, National Archives and Records Administration (henceforth NARA), Washington, DC.

Eight days after receiving a large French contract in October 1939, Wright announced plans to build a new factory on a 17-acre tract of land adjacent to its original factory. The French Government placed an order for 1,440 Wright Cyclone 9 and Cyclone 14 engines in a contract worth $28.6 million, an amount greater than Wright's total sales in 1938, and agreed to provide $5 million to enable Wright to double its production capacity.[146] The new Wright factory, designated Plant No. 2, was designed for the line production of engine parts from its inception and can be classed as one of the second generation of aircraft engine factories.[147] It was a large, single-story building covering 540,000 ft^2 of factory floor space. The factory was built of reinforced steel beams and columns on concrete floors reinforced with steel mesh to take the weight of machine tools and limit vibration, with concrete and brick walls.[148] Extensive steel sash glass windows admitted natural light. Raw materials entered at one end and finished parts to exit at the other, to be trucked to the assembly building at Plant No. 1 nearby. Wider bays and fewer columns permitted 12 double lines of machine tools to cover the floor space, separated by wide aisles to allow movement of parts and materials, with three main cross aisles that served as collecting points.[149] This layout was apparently copied from examples in the automobile industry.[150] Completed in the remarkable short space of 57 days, Plant No. 2 added nearly 50% more floor space, but Wright still lacked the capacity to meet the surge in orders for Cyclone 9 and Cyclone 14.

Wright needed more capacity but was reluctant to finance additional facilities with its own limited capital. In April 1940, as part of the Anglo-French Purchasing Commission's orders for large numbers of aircraft and aircraft engines, the French and British Governments agreed to provide financing to Wright to avoid a potential bottleneck in production, granting Wright $16.9 million in capital assistance for buildings, machine tools, and employee training.[151] With this financial assistance, Wright chose the more expedient route of acquiring a disused factory in the Paterson area rather than building another new factory. The British Government purchased an idle textile mill in Fairlawn, NJ, 4 miles from Paterson, which it leased to Wright.[152] This became Wright's Plant No. 3, adding 450,000 ft^2 of floor space and several acres of additional land.[153] The complex had over a dozen buildings, including its own power and steam-generating plants. Wright converted one of the buildings to an aluminum foundry for casting aluminum cylinder heads, the first of its kind at an aircraft engine factory.[154] The factory was available, but not ideal for manufacturing aircraft engine components and had to undergo considerable renovation. Later, the American Government's Defense Plant Corporation took over this building from the British Government.

Wright's expansion continued. In September 1940, the company started building a new magnesium foundry of 110,000 ft^2 on the land next to Plant No. 3, the first foundry in America built explicitly to cast magnesium aircraft engine parts.[155] Later that fall, Wright acquired another idle factory complex, a former dying and printing plant in East Paterson, which became Plant No. 4.[156] This added 433,000 ft^2 of factory space, with two adjacent multi-story manufacturing buildings, but like Plant No. 3 not ideally suited for heavy-machining operations. Wright then acquired a smaller factory that became Plant No. 5. With the addition of Plant No. 3, Plant No. 4, and Plant No. 5, and 160,000 ft^2 of additional space added to its main factory and assembly buildings, Wright increased its total floor space from 880,000 ft^2 in July 1939 to 2.8 million ft^2 by the end of 1940.[157] Later, probably during 1943, Wright added a three-story manufacturing building with 480,000 ft^2 of floor space at Plant No. 4, taking the total floor area at the Wright's Paterson factories to over 3 million ft^2.[158]

Neither Plant No. 3 nor Plant No. 4 were easily adaptable for line production. The main manufacturing buildings were older mill buildings, made from heavy timber framing and brick walls, with concrete ground floors and wooden upper floors. They had to be renovated and converted for production of engine parts with a thicker concrete flooring to support heavier machine tools.[159] The wooden upper floors would likely not have supported the weight of the heavier, more advanced machine tools Wright was acquiring, and may have been used for servicing or administrative functions.

Each of the factories in the Paterson area built certain components for all the Cyclone models, feeding finished parts to Plant No. 1 for assembly and testing of completed engines, with the distribution as follows:

- Plant No. 1: Final assembly and miscellaneous parts.
- Plant No. 2: Crankcases, cylinder heads, barrels and cylinder assembly, crankshaft center sections, and propeller shafts.
- Plant No. 3: Casting aluminum cylinder heads, crankshaft parts and assembly, connecting rods, and magnesium castings.
- Plant No. 4: Gears and cams.
- Plant No. 5: Pistons.[160]

This factory system was not ideal for efficient line production as it required transporting parts some distance to Plant No. 1 for final assembly. It reflects the pressure Wright was under in 1939–1940 to increase production as rapidly as possible. When the American Government asked Wright to expand production of the Cyclone 14 engine, and later the Cyclone 18, the company would adopt a completely different approach to organizing production.

The Studebaker Factories

To build the Cyclone 9, the Studebaker Corporation did not convert its automobile factories to aircraft engine production as some have supposed. Studebaker's automobile plant in South Bend, Indiana, was neither suitable nor available. It is not surprising that when the American Government approached Studebaker about building aircraft engines, discussions turned to building completely new factories to build these engines. When Studebaker began planning for quantity production of Cyclone engines in late 1940s, the company had the benefit of starting with a clean slate. Studebaker could design its aircraft engine factories for quantity production of a specific level of monthly output. The factories were initially designed with the capacity to build 1,000 engines a month, but after the attack on Pearl Harbor, this was increased to 2,000 engines a month.[161]

As Studebaker was starting from scratch, with more space available around the Midwest and with Government financing, the company had the luxury of being able to build large, more modern factories designed for more efficient line production than Wright's hastily acquired factory buildings in Paterson. The principal constraint the company faced was access to labor. The limited pool of labor available in South Bend, Indiana, the site of Studebaker's main automobile factory, forced the company to plan for three separate factories in different locations to build the Cyclone 9. Studebaker built one factory in Fort Wayne, Indiana, a second in Chicago, both roughly 90 miles from South Bend, and the third in South Bend itself. The Fort Wayne and Chicago factories built parts and components for the Cyclone 9, feeding them by rail and truck to the larger, South Bend factory. This factory made other Cyclone 9 components and carried out final assembly and testing of the engines.

With a combined floor area of nearly 3 million ft^2, the three factories were nearly equal in total floor area to all the Wright factories around Paterson (Figure 3.4).

FIGURE 3.4 The Studebaker aircraft engine factory built on the outskirts of South Bend, Indiana. Most wartime aircraft engine factories adopted the form of a large rectangular shape that by the 1930s had become the standard for industrial buildings. By 1943, expansions had increased the total area of the factory to nearly 1.5 million ft^2. This "temporary" wartime building was still in use as of 2023.

Studebaker Corporation: R-2600 Aircraft Engines Construction and Production Analysis, Box 1599, Entry P-26, File 452.8, Record Group 342, NARA.

The factory at South Bend was the biggest of the three. Studebaker based the size of the factory on an estimate of the company's production capacity, the required number of machine tools to achieve the specified level of output and a belief that it should concentrate on what it did best, machining parts and assembling engines.[162] Studebaker decided to purchase all cast and forged parts from outside vendors, eliminating the need for foundries. Initially, the main machine shop building had 790,400 ft^2 of floor space. In 1943, Studebaker added another section on the east side of the machine shop, built from reinforced concrete framing with a concrete slab roof, giving an additional 166,400 ft^2 of floor space for a total area of 1.06 million ft^2 including the basement.[163] The attached engine test cell area, which covered 386,035 ft^2, brought the total area to nearly 1.5 million ft^2. This one building was bigger than all the No. 1 Shadow Group factories combined, and bigger than any of the factory buildings in No. 2 Shadow Group.

The Chicago factory was next in size, with a machine shop building having 801,191 ft^2 of floor space.[164] Responsible for manufacturing a wide range of smaller parts for the Cyclone, the Chicago factory had the most machine tools, totaling 1,638 as of October 1944. Fort Wayne was the smallest of the three, with its manufacturing building having a floor area of 374,400 ft^2 and a total area of 470,100 ft^2 accommodating 1,075 machine tools.[165] Both factories had easy access to main rail lines and highways to ship parts to the main factory at South Bend. In comparison to their British counterparts, the three Studebaker factories together had a total floor area of approximately 2.9 million ft^2.[166] The three Studebaker factories had a total floor area that was nearly as great as the five Wright factories in and around Paterson, and nearly five times the size of the factories in No. 1 Shadow Group.

These Studebaker factories were better suited to flow production than their British counterparts. All three factories had the same rectangular design, incorporating the then-standard idea of a single unit with a large, open manufacturing space that could easily accommodate machine tools for efficient line production.[167] They were windowless following the new trend seen in several airframe and aircraft engine factories then under construction.[168] Air-conditioning provided temperature control necessary for precision engineering and fluorescent lighting replaced natural light. Artificial light allowed greater flexibility in the placement of machine tools around the factory floor as well as changes in plant layout.[169] The buildings were built of structural steel framing, with brick and tile walls and a metal roof. Column spacing was 40 ft × 40 ft to allow placing more machine tools, with an 18-ft ceiling. To support heavier machine tools, the floor of each factory was a 12-in. double-reinforced concrete slab covered in 1 in. × 18 in. maple blocks, easily sanded to keep clean. Each factory had overhead traveling cranes capable of lifting up to 20 tons. None of the factories had to incorporate any protection from air attack. There were no blast walls within the machine shops to protect machine tools, the test cells were attached to the main machine shop and assembly building, and most importantly, all machining operations could be placed in one large structure for maximum efficiency.

The machine shop building at South Bend was built on a north–south axis, with parts and raw materials entering in the south end of the building and moving progressively to the engine test cell area that was attached to the machine shop at the north end. Unlike at No. 2 Shadow Group factories, where test cells were separated from final assembly buildings, this arrangement minimized the time it took to move an engine from final assembly to the test cells and back for final inspection. Machining departments began at one side of the plant with the crankshaft rear-end section, moving subsequently to the crankshaft front-end section, on to the crankshaft counterweights, crankcase front and rear sections, crankcase assembly, pistons, cylinder heads, cylinder barrels, cylinder head and barrel assembly, and finally the heat-treating department.[170] After further expansion in 1942, the engine test area had 96 production test cells. A spur line connected the factory to the main railroad, allowing delivery of raw materials and shipment of finished engines to the factories building the B-17 (Table 3.3).[171]

TABLE 3.3 Army air force industrial facilities expansions as of March 1945.

Company	Total floor area (ft^2)	Cost of construction	Cost of machinery	Total cost	Maximum monthly capacity
Wright Paterson factories	3,442,854	$9,697,496	$35,439,903	$45,262,719	1,554 Cyclone 9 Engines
Studebaker South Bend	1,536,971	$15,727,930	$34,366,128	$50,130,791	2,300 Cyclone 9 Engines
Studebaker Chicago	855,441	$7,199,182	$17,867,468	$25,318,440	Cyclone 9 Components
Studebaker Fort Wayne	505,754	$4,580,648	$12,800,233	$17,439,251	Cyclone 9 Components

Source: AAF Industrial Facilities Expansions: Status Progress Performance as of March 1945, P. 6 Engines, File 218.7-3, Air Force Historical Research Agency, Maxwell AFB, AL.

The Production Record

During the war, Wright built 35,862 Cyclone 9 engines at its Paterson factories, while also building the seven-cylinder R-975 Whirlwind, the 14-cylinder R-2600 Cyclone 14, and until 1943, the 18-cylinder R-3350 Cyclone 18.[172] Wright built 26 different versions of the Cyclone 9 engine for the military services, principally 8,746 R-1820-65 and R-1820-97 versions for the Boeing B-17, 5,399 R-1820-60 engines for the Navy's Douglas SBD Dauntless dive bomber and 6,969 R-1820-56 series for the Navy's Grumman/General Motors FM-2 Wildcat fighter.[173] Wright was fortunate that its Cyclone 9 was a mature engine; by 1939, the company had built over 9,000 Cyclone 9 engines since the early 1930s, so that the factories were familiar with its manufacture. Wright built 2,056 Cyclone 9 engines in 1939 and completed 2,235 during 1940, but with the new factories added during 1940 and new production methods, production doubled during 1941 to 4,687 engines.[174] Production doubled again during 1942 and peaked in 1943 when the Paterson factories built 9,696 Cyclone 9 engines, with a rate of just over 1,000 engines achieved during March 1943.

Production of the R-1820 at the Wright factories in Paterson resembled Wright's prewar batch production, but on a larger scale. The company manufactured specific versions of the R-1820 for specific types of aircraft, mixing large and small orders within a contract. In September 1943, the Army Air Force placed a $129 million contract with Wright Aeronautical for aircraft engines for the AAF and the US Navy which included the following orders for the R-1820:

- 500 Navy Model R-1820-54 engines
- 1500 AAF Model R-1820-56 engines with gun synchro/Impulse Generator Drive
- 3040 AAF Model R-1820-56 engines without gun synchro/Impulse Generator Drive
- 1600 Navy Model R-1820-60 engines
- 1749 AAF Model R-1820-93 engines
- 499 AAF Model R-1820-97 engines[175]

As can be seen, there were variations between the different versions of the R-1820, requiring a degree of flexibility in production to accommodate changes in component parts.

Throughout the war, Wright Aeronautical continued development work on the Cyclone 9, looking to increase performance and reliability. Performance under war conditions was considerably more demanding than in peacetime and brought to light design features that needed improvement. In late 1943, for example, the Army Air Forces Material Command reported failures in the R-1820-65 and R-1820-97 piston rings, propeller drive shaft assemblies, master rod bearings, and cam drive gears.[176] In response to this report, Wright made changes in the design of the R-1820 which required new parts and changes in production.[177] Such design changes were a common and regular occurrence during the war (Figures 3.5 and 3.6).

FIGURE 3.5 Production of principal Cyclone 9 models at the Wright Paterson factories 1941–1945.

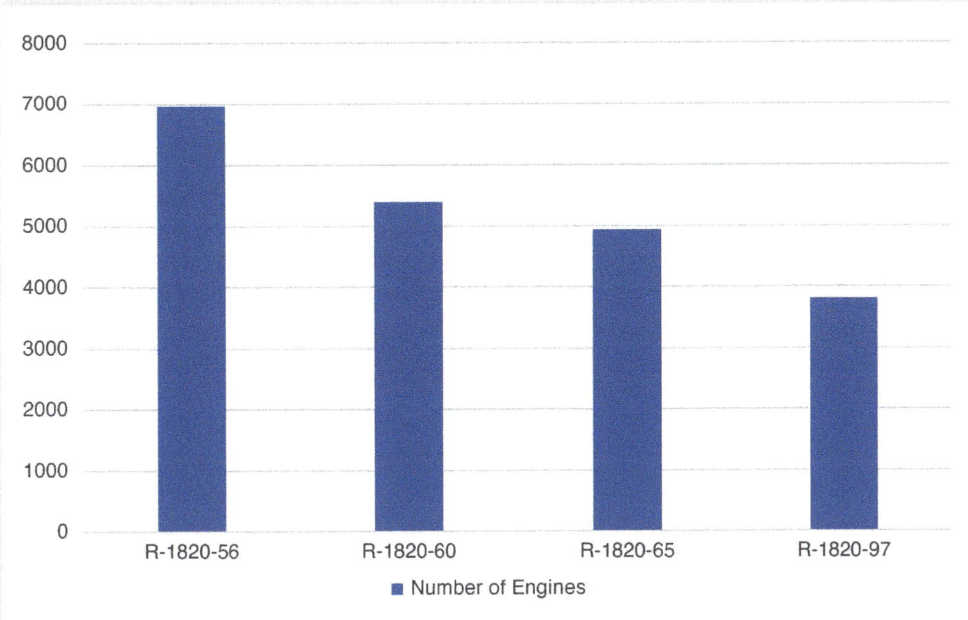

Source: Wright Aeronautical Order and Contracts Department: Summary of Engines Shipped (On an Invoice Basis) by WAD and Licensees: From 1929 thru August 31, 1960, pp. 11-12, Aircraft Engine Historical Society, http://enginehistory.org/reference.shtml#EngProd.

FIGURE 3.6 Wright Cyclone 9 production at Wright and Studebaker 1940–1945.

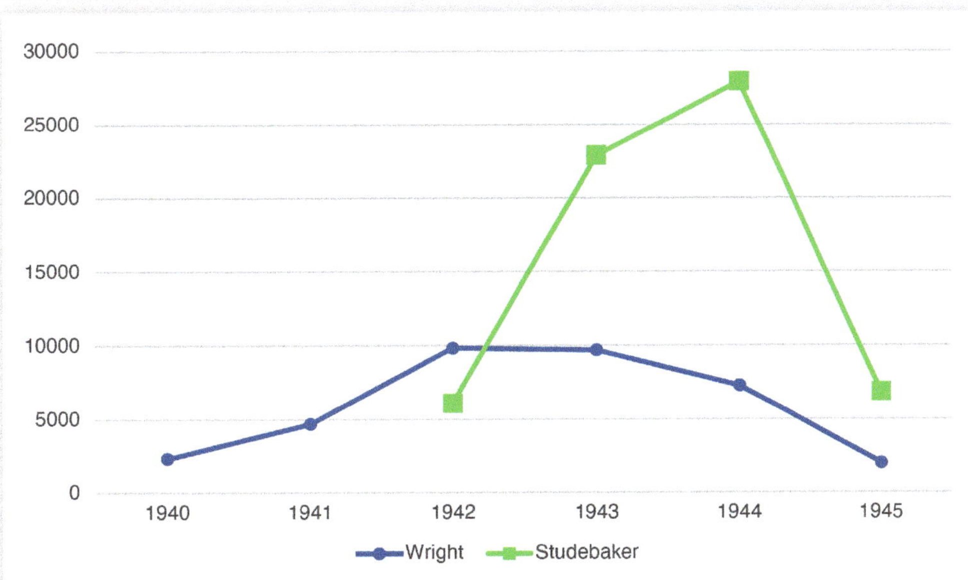

Source: US Department of Commerce, Civil Aeronautics Administration, Office of Aviation Information, Division of Aircraft Statistics: US Military Acceptances 1940-1945: Aircraft, Engines and Propeller Production, (Washington, DC, 1946).

Studebaker's contribution to Cyclone 9 production was critical. During 1943, Studebaker built more than twice as many Cyclone 9 engines as Wright, and nearly three times as many in 1944. From 1942 onward, Studebaker's entire production of the R-1820 Cyclone 9 went to the Boeing B-17. Studebaker's ability to concentrate on just one type of the Cyclone 9 engine, the -97 version, helped its production record immeasurably. In June 1941, the War Department amended Studebaker's contract to cover Cyclone 9 engines, specifying that planned production would reach a peak of 1,000 engines a month by December 1942.[178] After the attack on Pearl Harbor, the War Department ordered Studebaker to double production to 2,000 engines a month, a level of output significantly greater than that achieved by any British aircraft engine factory in World War II.[179] Studebaker began building the R-1820-65 version of the Cyclone 9 engine in early 1942, turning out four engines in February but achieving the target of 1,000 engines a month in September just eight months after starting production. Toward the end of 1942, having completed 3,504 R-1820-65 engines, Studebaker shifted to building the R-1820-97 version, which differed only in minor details from the R-1820-65, building 60,285 by the end of the war.[180] Studebaker reached the required production rate of 2,000 engines a month in May 1943, doubling its rate of production in just six months.[181] The company achieved a wartime peak in production in March 1944, completing 2,479 engines during the month.[182] From April to November 1944, Studebaker sustained a production rate of over 2,300 engines a month, producing 27,920 engines for the entire year, second only to Buick's production of 30,550 Pratt & Whitney R-1830 Twin Wasp engines.[183]

How It Was Done: Building the Cyclone 9 at the Wright and Studebaker Factories

By July 1940, the estimated number of Wright Cyclone 9 engines needed to meet Army, Navy, and British requirements by June 1942 amounted to 11,718 engines, more than five times Wright Aeronautical's production during 1939.[184] To meet this requirement, Wright would have to triple its monthly production rate, shifting from its prewar batch production methods to achieve quantity production in wartime. Compounding the problem was the shortage of skilled machinists; Wright would have to rely on employees who had little experience with machine tools, yet had to ensure no diminution of quality. Wright accomplished this transition by shifting to line production in its factories. Through improvements in production engineering and process engineering, Wright speeded the flow of parts through the factories, and by using newly developed high-production machine tools that semi- and unskilled workers could operate, cut production time even further, thereby increasing output. It took nearly two years for Wright to make the shift from batch to quantity production of the Cyclone 9 engine. During 1940, the company built an average of 186 Cyclone 9 engines a month, but more than doubled monthly production during 1941 to reach 526 engines completed in December.[185]

Prior to the war, manufacturing a Cyclone 9 engine was similar to manufacturing the Bristol Mercury and Pegasus engines described in Chapter 2. To illustrate the complexity of machining the Cyclone 9, Table 3.4 shows the number of machining operations required for the main components of the Cyclone 9 engine and the amount of metal removed from the rough forgings during the machining process.

TABLE 3.4 Machining operations on the Cyclone 9.

Part name	No. of operations	Rough weight (lb)	Finished weight (lb)
Cylinder Barrel	35	68	16
Cylinder Head	37	37.5	27.5
Cylinder Assembly	85	—	—
Total	157	105.5	43.5
Crankcase Front	45	178	40.5
Crankcase Rear	45	178	36
Crankcase Assembly	57	—	—
Total	147	356	76.5
Crankshaft Front	85	111	76.5
Crankshaft Rear	80	50	22
Counterweight Rear	41	39	19.25
Counterweight Front	36	30	15
Crankshaft Assembly	89	—	—
Total	331	230	102.75

Source: Geschelin, Joseph: "Studebaker Know-how Speeds Flying Fortress Engine Output," *Automotive and Aviation Industries,* Vol. 88, No. 7, (April 1, 1943), p. 21.

As at the Bristol Aeroplane Engine Department, the Wright Paterson factory received rough cast or forged parts from the foundry that went through successive stages of machining, heat treatment, and painting or plating before final assembly into finished engines ready for testing. As orders for its engines increased exponentially after the start of the war in Europe, Wright had to find ways to rapidly shift from building engines in small batches to continuous quantity production, increasing output while maintaining absolute adherence to its standards of quality. Initially, as factories in No. 1 Shadow Group in Britain had done, Wright expanded production by adding more workers to get higher utilization from its standard machine tools. The easiest way to increase output was to simply "...duplicate again and again any existing processes and equipment."[186] In January 1939, Wright had 3,398 employees; by the end of the year, Wright had increased employment to 5,141, and by the end of 1940 had more than doubled the number of employees to 12,408.[187]

Wright Aeronautical's problem with increasing production was now common across the American aviation industry. With their backlog growing rapidly from both foreign and domestic orders, by the end of 1939, American airframe and aircraft engine manufacturers were already close to capacity production.[188] There were now questions as to how the aircraft industry could meet the demand.[189] The answer to increasing production of aircraft and aircraft engines was, for many observers outside the industry, simple. As Henry Hill, a project engineer with Wright, noted:

> *To most people this increased production rate is merely a matter of applying the well-known methods of the automobile industry. This naïve statement has just enough truth in it to confuse the minds of many people both inside and outside the aviation industry. Between the statement and the actual fact the gulf is very wide indeed. It is true that the principles developed by the automobile people in Detroit must be applied to aircraft and engine production, but we are sure that it is equally true that these principles must be modified and further developed to suit the new set of standards and the new tempo required in the aircraft field. Whether we like it or not we must face the fact that in developing the aircraft engine we have also developed a brand-new design and manufacturing technique, which is as far removed from automobile manufacture as the automobile was from carriage building.*[190]

During 1939, a widely shared discussion began in the aircraft industry on production issues. That fall, the Society of Automotive Engineers, the professional body for automotive and aeronautical engineers, held a special conference on production problems in the industry, the National Aircraft Production Meeting.[191] Some three thousand people attended the three-day conference. Wright sent an exhibit of its engines and had Henry Hill deliver a paper on problems in the quantity production of aircraft engines.[192] Over the next two years, articles appeared regularly in major aviation and specialized industry magazines describing how airframe and aircraft engine companies were addressing the problems of increasing production, with detailed descriptions of the new methods they were using.[193]

In November 1940, the Society for Automotive Engineers held its second National Aircraft Production Meeting. For this meeting, both Pratt & Whitney and Wright exhibited aircraft engines and sent engineers to give papers.[194] A key theme of the conference was mass production and particularly the application of automobile manufacturing techniques to aircraft production.[195] Two leading engineers from the Curtiss-Wright Corporation, Wright's parent company, noted the many differences between mass production in the automobile industry and aircraft production, but asserted that despite these differences, there were opportunities to apply automobile production methods.[196] They stated that "careful analysis supports the conclusion that much can be gained through the utilization of automobile methods and practices when combined with a liberal application of common sense."[197] What this required, they argued, was careful attention to design, tool, production, and process engineering.[198] This was exactly the approach Wright took to the problem of shifting to quantity production of aircraft engines.

Production Engineering at Wright Aeronautical

Wright's problem was "to reduce the time required for all manifold operations comprising the intricate process of aircraft engine manufacture."[199] To this end, Wright's production engineers concentrated their efforts in four main areas: changing the company's policy on manufacturing engine parts, shifting to line production, reconfiguring processes to speed production, and putting much greater emphasis on materials handling. All of these efforts can be characterized as an effort to establish flow production, achieving increases in output by cutting machining and handling time wherever possible.

An observer at the November 1940 National Aircraft Production Meeting remarked on the extent of free exchange of information between companies and the general spirit of cooperation in seeking solutions to production problems.[200] Wright set up a technical production unit to examine the problems of quantity production and to make recommendations based on a study of methods in other industries.[201] The unit made a close study of automobile manufacturing methods to see what could be applied to manufacturing aircraft engines, but went beyond this to study general industrial practices looking at current production methods, new types of machine tools, and new manufacturing processes.[202] The challenge was to determine what changes to accelerate production could be justified, with the corollary demand that these changes could have no negative impact on the quality and reliability of the engines.[203] From these studies came new approaches to production engineering and a new attitude toward measuring processes, machining, and assembly operations not in terms of cost in dollars, but cost in time.[204]

For years, Wright's policy had been to manufacture nearly all the machined parts for its engines in its own factory in order to exercise complete control over quality.[205] Purchased parts made up less than 25% of the cost of a Wright engine.[206] The company now changed its policy to one of purchasing a much higher percentage of parts from outside vendors. Increasing subcontracting to speed production was also a policy that Government mobilization agencies actively encouraged. Subcontracting reduced the number of employees Wright needed to train and freed up supervisory personnel and skilled machinists for other tasks.[207]

The shift to line production was a significant change for Wright from past practices. The objective of line production was to arrange the machine tools so that the work passed sequentially from machine to machine in the proper sequence, ideally so that the work pieces could be moved a minimum distance with minimum effort and in a minimum amount of time.[208] As Wright expanded its production facilities, conversion to line production became possible within each manufacturing building, although the dispersion of Wright's factories around Paterson was not ideal. In a method that was analogous to No. 1 Shadow Group's production of the Bristol Mercury and Pegasus, Wright chose to view its Paterson factories as a single production unit, composed of greatly enlarged manufacturing operations for particular aircraft engine parts, each made at a different factory, feeding the main assembly operation in Plant No. 1.[209]

The first step in designing a layout for line production was to determine the number and type of machine tools required for each major line.[210] Analyzing the production schedule, the time required to manufacture each part and the number of working minutes each month, helped determine the required number of machine tools. Plant layout engineers then prepared a figure showing the types of machines, their placement in the sequence of machining operations, and the amount of floor space each machine tool required, making allowance for the spacing needed for the operator, proper placing between machines, storage space for finished work, and aisle space for moving rough and finished parts. This figure determined the floor space each department required, making it possible to calculate the total floor space for a new or converted building. From all this information, layout engineers prepared production strips, long pieces of black Upson Board showing the layout of a specific production line, and placing on these strips scale templates cut out in the shape of the specific machine tools to be used in their proper sequence. These production strips went on a larger board outlining the factory space available, with spaces added for other processes like heat treatment, painting, and assembly operations. Using this visual aid, production engineers could grasp the overall layout of line production and could make changes to the production strips to assess the implications of changing manufacturing processes or adding new types of machine tools. Unlike the automobile industry, where changes in design tended to be implemented when the factories were gearing up for their annual model change, design changes to aircraft engines were continuous throughout the year, thus putting a premium on building flexibility into line production.[211]

While the straight-line layout was considered the ideal for maximizing flow, under certain circumstances, this proved difficult to achieve. Developing a layout for fabrication operations was more difficult than laying out assembly operations and depended on the sequence of operations and the type of machine tools employed for each operation.[212] At Wright's Plant No. 2, the company resorted to a U-shaped line for machining cylinder barrels and cylinder heads and for the final cylinder assembly. To improve the flow of work, however, Wright laid out the sequence of machining operations so that a part never backtracked but was always moving forward.[213] Parts moved down one side of a line, would turn and move down the other side, then shift to the next line, and repeat the process until machining was complete.

Wright reorganized its foundry operations to conform to line production methods, allowing for an uninterrupted flow of operations from intake of raw materials to finished cylinder head casting.[214] At Plant No. 1 and Plant No. 2, Wright automated painting and electro-plating Cyclone engine parts, removing most handheld devices and hand operations.[215] At Plant No. 1, Wright completely revamped its assembly operation. For years, the company had used what it called a "spot assembly" system, in which four-man teams assembled an entire engine taking a full 8-hour day to complete the job.[216] Assembling a complete engine required a high degree of skill and experience as well as thorough familiarity with all the parts of a Cyclone engine. As production accelerated, Wright found that skilled assemblers were not to be had. Instead, the superintendent of the assembly department and his foremen worked out a new system that used a coordinated, progressive sequence of operations. The team broke down the process of assembling a Cyclone 9 engine into separate operations on groups of parts that took no more than 24 minutes to assemble.[217] Teams of one to three workers manned long benches set at right angles to the main assembly line. Each team built up one subassembly, each worker adding a piece or pieces as needed, the completed subassembly joining the main assembly line where it was added to the engine as it passed by.[218] The benefits to this method were the savings in assembly time and the shorter training time, as each assembler no longer had to be familiar with every part of a Cyclone engine, but only the parts that he/she would assemble.

From their study of automobile production methods, Wright's production engineers learned that "machine tools alone are not the total answer to the problem of increased output; proper material flow is of at least equal importance."[219] As Wright expanded its factories, the company devoted more attention to finding ways of moving materials by machine instead of by hand to speed production. The electro-plating operations in Plant No. 1 and Plant No. 2 involved careful cleaning of parts to be plated in various chemical solutions and then plating operations to cover parts with tin, chromium, lead, or in the case of cylinders that would be subject to saltwater corrosion, with molten aluminum.[220] Painting involved spraying parts with a primer, drying them, then spraying on two coats of enamel before baking. Previous methods involved moving parts from one operation to the next and spraying on paint manually, turning the part to paint all surfaces. In both electro-plating and painting operations, Wright installed conveyors to move individual parts through each stage of the process.[221] In its foundry operations, Wright put in automatic hoists, conveyors, and roller systems to move heavy molds, molten aluminum, and rough castings.[222] Using an overhead monorail allowed foundry workers to use a larger ladle in place of a smaller hand ladle, allowing faster and more thorough pouring into the mold.[223]

Wright progressively installed conveyor systems along its production lines so that parts moved from machine to machine along the conveyor line.[224] The conveyor systems saved floor space by eliminating the need to store work in process or finished parts alongside a machine tool. This allowed more machine tools to be placed along a production line. Conveyors replaced the hand carts and shop trucks used to move parts from station to station, removing potential bottlenecks and freeing workers for other tasks. As more and more women joined the Wright workforce, the conveyor system and hand-operated or automatic hoists reduced the need for women to use their own strength to shift heavy parts, thereby removing one of the principal objections to employing women in the factories.[225]

All these methods, such as purchasing more parts from outside vendors, shifting to line production for key engine parts, reconfiguring manufacturing processes, and expanding and improving materials handling, saved time and speeded production. This was a process of achieving incremental savings in time all along the manufacturing process, but the increments, when added together, contributed to faster rates of production. But, the company's efforts to develop high-production machine tools through process engineering were far more important to increasing Wright's production.[226]

Process Engineering at Wright

In an article written in September 1941, the *Wall Street Journal* commented that "perhaps no individual company in the aircraft industry has turned to the use of automatic single purpose machines in the same degree as the Wright Aeronautical Corporation."[227] In contrast to companies who depended on standard machine tools, the *Journal* noted that "Wright is following the automobile practice of using gigantic machines which are only good for exactly one job."[228] Wright's dilemma in 1939 was that it was facing, concurrently, a growing demand for its aircraft engines with a growing scarcity of skilled machinists.[229] The company realized that it would be impossible to train the number of workers it would need in time to meet demands for higher production.[230] The volume of production required—and the fact that the British and French Governments, and later the American Government were willing to provide financing for new machinery—now justified using expensive high-production machine tools that Wright had not found economical for batch production in place of skilled workers. The machines that Wright now required had to have two key characteristics: first, they had to be more productive than standard machine tools, producing more parts per operation and at a faster rate, and second, they had to be adapted to the limited skills and experience of the workers Wright would hire going forward. To cope with the same dilution of skill levels that Bristol faced in Britain, the skill of the machinist had to be built into the machine, the task of the process engineer.[231] Machines had to be made "practically fool proof" with operations that were "completely automatic, except for loading and unloading."[232]

When Wright began developing high-production machine tools, the company had little experience to draw on. As one Wright executive explained the problem, "there was no precedent upon which calculations could be based or any guarantee that it would be possible with automatic equipment to hold the extreme accuracy and fine finishes demanded for this class of work."[233] At first, there was considerable skepticism that automatic machine tools could be used in aircraft engine production.[234] One principal concern was whether automatic, high-production machine tools could be designed with a degree of flexibility that would accommodate the design changes that were a regular feature of making aircraft engines.[235] The Paterson factories became a laboratory for developing advanced automatic machine tools for aircraft engine production.[236]

Wright began working with the major machine tool builders to design machine tools that would meet Wright's requirements, realizing that it was presenting the machine tool companies with an "almost impossible set of problems."[237] Wright also undertook a detailed study of machine tools and methods then in use with the automobile industry to determine if there were machine tools that could be adapted to Wright's needs.[238] The American automobile industry was the leader in developing techniques of mass production and had continuously stimulated the machine tool industry to invent new types of machine tools to speed production and achieve greater precision and finer finishes at lower costs.[239]

At the same time, Wright applied the tools of process engineering to examine the fabrication of every part of a Cyclone 9 engine to see where and how a process of deskilling could be applied and a new machine tool designed for a specific operation.[240] Wright found that while the cost of a high-production machine tool was higher than an equivalent standard machine, in most cases, there was an actual saving as one high-production machine might replace several standard machine tools.[241] The difficulty Wright faced with increasing production rates using high-production machine tools was that these newly designed tools took from 6 to 15 months to build.[242] In the interim, Wright continued to use standard machine tools, but replaced them as soon as the new high-production machine tools arrived at the factories.[243]

Wright's shift to relying on more high-production machine tools went through three phases. In the first phase, beginning in 1939, Wright simply continued what it was doing, adding more standard machine tools and workers, as the shadow factories building Mercury and Pegasus engines had done. In the second phase, during 1940 and into 1941, Wright purchased semi-high-production machine tools that could be easily acquired, that could produce parts previously made on standard machine tools, and that employed semi-skilled workers.[244] It used these machines to test new production methods that it passed on to its new factory at Lockland, Ohio (see Chapter 5), to gain knowledge and experience that could be applied to the design of more advanced high-production machine tools. During this period, Wright purchased 2,000 machine tools for its Paterson factories at a cost of $28 million, although many of the high-production machine tools Wright ordered would not reach the factories until 1941, beginning the third phase of machine tool utilization (Figure 3.7).

FIGURE 3.7 Drilling 180 cylinder pad holes in the R-1820 Cyclone 9 crankcase at the Studebaker engine factory in South Bend using a specially developed Greenlee multiple drill. This special machine performed all the required operations at one time. Formed of two steel forgings weighing 300 lb, after 150 machining operations, the finished crankcase weighed just 75 lb.

AC-1027-11C, From the Collection of Studebaker National Museum, South Bend, Indiana.

Not all the high-production machine tools Wright developed in consultation with the machine tool builders were new designs. Many were derived from machine tools in use with the automobile industry, after careful study.[245] The most critical aspect of this evolution was to incorporate automaticity into the design to cope with the influx of semi- and unskilled workers.[246] As previously explained, all a semi- or unskilled operator had to do with an automatic machine tool was to load and unload the machine and check the result for accuracy with gauges provided; the machine would automatically complete all machining operations in the correct sequence with no intervention from the operator.[247]

Wright found that, contrary to what was expected, the automatic machine tools it developed were even more accurate than standard machine tools because their heavier construction limited unwanted vibrations.[248] Among the more important types of machine tools that were transferred from the automobile industry were multiple-spindle drilling, tapping and boring machines that could perform these operations in succession on multiple work pieces using a single machine tool; precision boring machines using cemented carbide tools for rough drilling and a diamond tool for fine finishing; automatic milling machines; and internal and external grinding machines and gear finishing machines.[249] Automatic multiple-spindle lathes were another important development for quantity production.[250] By mounting multiple single-point cutting tools in a specially designed jig, a multi-tool lathe could perform multiple cuts at the same time, for example, cutting all the fins on a cylinder barrel or cylinder head in one operation instead of having to shift the work piece from one machine to another for progressive operations.[251] An examination of improved production methods using high-production automatic machine tools to manufacture the main components of Cyclone 9 engines provides numerous examples of the dramatic savings in production time, space, and man power these machines delivered (Figure 3.8).

FIGURE 3.8 A line of men and women working on cylinder heads, in this case for the Pratt & Whitney R-1830 engine. Prior to the war, many operations to manufacture cylinder heads were done manually or by individual machine tools, moving the work piece from machine tool to machine tool.

na040231, Courtesy of the National Automotive History Collection, Detroit Public Library.

In manufacturing cylinder heads, machines replaced hand operations at the start of the process at the foundry. After a cast cylinder head had cooled and been removed from the mold, excess aluminum and minor irregularities had to be removed from the cast fins. This had been done manually, but Wright designed a special machine to clean up the cylinder heads using an abrasive grinding wheel and a master cam which profiled the cylinder head, moving the grinding wheel automatically.[252] Each cylinder head required a number of holes to be drilled, reamed, countersunk, and tapped, including holes for the two spark plugs, holes for the valve springs, holes for the studs holding the intake and exhaust port flanges, and holes for the intake and exhaust port rocker box covers. In the 1930s, operators used a single-spindle drill press to make all these holes in the cylinder head, one at a time, having to change the type of tool—a drill, reamer, borer, counterbore, or tap—for each operation.[253] With the need to increase production, Wright then designed semi-production machine tools, one of which could drill, ream, countersink, and tap spark plug holes in one continuous operation, automatically moving the cylinder head from station to station.[254] But the cylinder head would then have to be moved to another machine to perform these operations on the intake and exhaust ports and a third machine to machine the holes for valve springs and finally a fourth machine to drill holes for rocker box covers.[255] Wright replaced all these machines with what was no doubt the most impressive of the many high-production machine tools Wright developed during the war, the Greenlee Automatic Transfer Machine (Figure 3.9).

FIGURE 3.9 An early model of the Greenlee Automatic Transfer Machine at the Studebaker factory in South Bend. Wright Aeronautical worked with the Greenlee Brothers machine tool company to develop an automatic transfer machine that could perform all operations on a cylinder head automatically with a savings in time and labor.

na051790, Courtesy of the National Automotive History Collection, Detroit Public Library.

A transfer machine is "a multi-head automatic machine [tool] which performs a sequence of operations simultaneously on a series of product units which are automatically indexed forward from one machine station to the next."[256] In 1940, the Nash-Kelvinator Corporation, one of the small American independent car companies, and the International Harvester Company, a maker of diesel tractors and diesel tractor engines, worked with the Greenlee Brothers & Company to develop transfer machines to drill and bore holes in engine blocks.[257] The transfer machine incorporated a linked series of individual multi-spindle machine tools placed on both sides of an automatic conveyor to form a mini-production line.[258] The key feature of the transfer machine was that it "indexed," or moved, the work piece from machine to machine automatically, saving a considerable amount of time because the operator did not need to move the work piece from machine to machine. All the operator did was load the work piece at the front of the machine and unload it at the end of the machining operations.

An initial investigation of automatic machine tools indicated that there were no existing tools that could speed up cylinder head production.[259] In early 1940, a local Newark, New Jersey, company put Wright Aeronautical in touch with Greenlee Brothers, who proposed a "seemingly impossible machine" based on their earlier transfer machines for the automobile industry.[260] After lengthy discussions between Greenlee and Wright engineers, Wright ordered one automatic transfer machine from Greenlee. Working with Greenlee, Wright helped design and develop a machine that would automatically drill, ream, bore, countersink, and tap all the required holes in a cylinder head in one continuous operation.[261] This first machine, the Greenlee Automatic Transfer Machine No. 1, took nearly a year to build and was set up most likely in Plant No. 2 in Paterson in the spring of 1941.[262] This first machine was 80 ft long with 56 stations.[263] This machine required only eight men per shift, doing the work of 40 standard machines and 124 workers in a 24-hour day.[264]

The Greenlee automatic transfer machine was just one of many high-production machine tools Wright and Studebaker developed during the war to manufacture other components of the Cyclone engines as a description of manufacturing other components of the Cyclone 9 will show. The cylinder barrel for the Cyclone 9 went through two major operations, cutting the cooling fins on the outside of the barrel and creating a smooth finish on the inside. First, the cylinder barrel went through a rough finishing process to remove metal from the inside of the barrel. Multi-spindle, six-station Bullard Multi-Au-Matic boring machines replaced single-spindle boring machines.[265] After rough finishing the cylinder barrel and completing the nitriding hardening process, the cylinder barrel was transferred to a Fay automatic lathe with a special magazine tool holder that could cut all 41 cooling fins on the outside of the cylinder at one time.[266] Once the cylinder head was attached to the cylinder barrel, the complete assembly went to a specially designed Heald internal centerless grinder that automatically placed the assembly in position for machining and automatically removed the finished part and loaded a new part onto the machine.[267] A flange at the bottom of the cylinder barrel required 20 holes to be drilled, bored, reamed, and countersunk for the studs that attached the flange to the crankcase. Here again a specially designed machine replaced single drilling and reaming machines. A Baker vertical, 40-spindle, five-station machine completed this operation automatically. The cylinder assembly was loaded onto the first station and moved automatically around the machine to each of the four stations. The second station drilled ten holes in the flange, the third station drilled the remaining ten, while the fourth and fifth stations reamed and countersunk the holes to the required configuration.

The Baker machine cut the operating time from 6.5 to 1.5 minutes, replaced five standard machine tools, and reduced the operators required over three shifts from 14 to 2, saving 85 man-hours.[268]

The front and rear sections of the Cyclone 9 forged steel crankcase went through rough and finish machining, were then joined together, and placed on a Foster "Fastermatic" automatic turret lathe that finished the holes for the cylinder assemblies and the cylinder pad faces where the cylinder assembly flanges would be attached. The fixture holding the crankcase rotated automatically until the "Fastermatic" had completed machining all nine-cylinder holes and cylinder pads.[269] An Ex-cell-o automatic multiple drilling machine drilled 20 holes in the cylinder pads to anchor the cylinder assembly, drilling ten holes in two pads at a time, and then moving the part automatically to drill the second set of ten. The machining operation moved to a Baush multiple back-counterbore machine which could counterbore 16 holes at once, automatically moving the fixture holding the crankcase for the next operation. A third machine carried out tapping, finishing the threads for the bolts that would hold the cylinder flange to the crankcase.[270]

In another example, during 1941 and 1942, Wright devised a method to speed up machining the articulated rods for the Cyclone 9 at Plant No. 3 in Paterson. Previous practice had been to mill the inner section of each articulated rod separately on a Cincinnati milling machine, working on one articulated rod at a time.[271] When one side of the rod was finished, the fixture holding the rod had to be manually turned over to finish the other side. When grinding the ends of the articulated rods to the correct shape, the end of the rod had to be oscillated by hand against the grinding disc, a laborious process subject to inaccuracies. Wright engineers designed a special fixture that would fit on a standard grinding machine. An arbor, a fixture for holding multiple work pieces, held seven articulated rods, and two arbors could fit on the grinding machine. When the machine had finished grinding one side of the rods, the arbors were flipped over to grind the other side so that 14 rods could be machined at one time. The arbors were then placed on a Mattison grinding machine that automatically oscillated the rod ends while grinding them, eliminating the need for hand oscillation.[272] Two Mattison grinders replaced five standard machines, reduced the number of operators required from 14 to 4, and saved 124 man-hours in production time.[273]

David Hounshell states that the Greenlee automatic transfer machines at the Wright factory reduced production time from 59 worker-minutes to 8 worker-minutes, but there is insufficient information to estimate the relative impact on productivity of the Greenlee automatic transfer machines overall.[274] Some idea of their importance can be gleaned from a brochure Wright published late in the war documenting the high-production machine tools it put into operation in its various factories.[275] This brochure listed the savings in employees required, production hours, and standard machine tools for a range of high-production machine tools, including two large Greenlee automatic transfer machines. On a combined basis, these various high-production machine tools required 470 fewer employees, saved 3,862 production hours, and replaced 163 standard machine tools. It is interesting to note that the two Greenlee machines contributed 45% of the savings in employees required, 42% of the savings in production hours, and 46% of the savings in standard machine tools. More data is needed to determine the relative importance of the automatic transfer machines to improve productivity, particularly as the Greenlee automatic transfer machines were limited in number, while the number of other types of high-production machine tools in a factory would likely have been greater. At the Wright factories that employed them, there were only one or two Greenlee machines as these machines were costly, in some case more than 35 times the cost of other high-production machine tools. It seems likely that the factories building Wright engines had multiple high-production machine tools in their production lines (Figure 3.10).

FIGURE 3.10 Peak Cyclone 9 monthly production at Wright's Paterson factories 1940–1945.

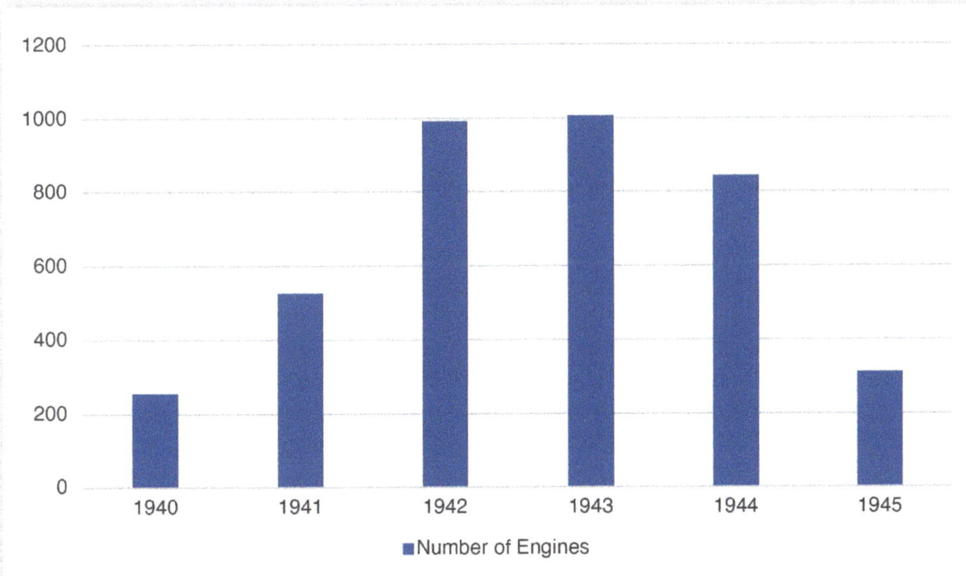

Source: US Department of Commerce, Civil Aeronautics Administration, Office of Aviation Information, Division of Aircraft Statistics: US Military Acceptances 1940-1945: Aircraft, Engines and Propeller Production, (Washington, DC, 1946).

Wright's introduction of the forged cylinder head for the Cyclone 9 in place of the cast cylinder head provides an example of how technological improvements affected production processes. This change required extensive production and process engineering to develop new methods and processes of manufacturing that were completely new parts. The benefits of using a forged cylinder head were considerable, as it enabled Wright to boost the power output of the Cyclone 9 to 1,350 hp while reducing production time.[276] By December 1941, Wright had an experimental machine shop working on production versions of the forged cylinder head and had placed orders for machine tools in expectation of beginning full-scale production in June 1942.[277] Wright believed, optimistically as it turned out, that it could produce enough forged cylinders for 100 Cyclone 9HC engines by December 1942.[278] In the event, Wright did not reach its target until October 1943.

Manufacturing a forged cylinder head proved challenging. When Wright began working on forged cylinder heads, it found that the early examples were effectively handmade and far from ready for quantity production.[279] Wright first had to work out a forging technique that could produce forged cylinders in quantity. The limited number of drop hammers in the US with the required power was only one of the many obstacles Wright had to overcome.[280] The forging process that Wright developed started with a drop hammer forcing aluminum bar stock into the general shape of the cylinder head. After heating to the proper temperature, the semi-finished work piece, or billet, went successively through three dies in a 1000-ton capacity press, the first forming the rocker boxes, the second forming the content of the cylinder head, and the third forming the combustion chamber inside the cylinder head.[281] After this, the forged cylinders had to go through a careful heat treatment process to relieve the internal stresses incurred in the forging process. This allowed the external cooling fins to be machined accurately with less wear on the cutting tools.[282]

To speed production, during 1943 Wright developed at its Paterson factories a special continuous conveyor system linked to two parallel gas–electric furnaces.[283] Twelve rough forged cylinder heads at a time went via the conveyor to the gas furnace where they were heated to 960°F and soaked for 6 hours. After cooling to 340°F, the trays carrying the cylinder heads went via the conveyor to the electric furnace where they were heat-soaked for 8 hours, followed by another cooling period and then a final 5-hour heat-soak at 450°F. Wright found that with the forging process, rejections amounted to only 5% of the total, compared to 20% with the casting process, a great savings in materials and production man-hours.[284]

Milling the cooling fins around the uneven contours of the cylinder head was the most difficult problem Wright faced with the forged cylinder head. Wright's initial attempts to mill the fins with standard milling machines proved unsuccessful.[285] Severe vibration caused fatigue cracks in the machined fins. The solution came through working with the Cincinnati Milling Machine Company to adapt one of the company's Hydro-Tel vertical profile milling machines that could mill four forged cylinder heads, using a cam and master template to guide two specially developed 13-in. cutting wheels with nine carbide-tipped teeth to cut four identical fins on the tops of all four cylinder heads at the same time.[286] A similar Hydro-Tel machine cut the fins on the sides of the cylinder head. One of these Hydro-Tel machines could work on eight forged cylinder heads at one time using two rotary fixtures.[287] Wright had the Snyder Tool & Engineering Company build a specially adapted machine tool for the other difficult machining operation, forming the valve pockets and the intake and exhaust ports on the head of the cylinder.[288] Once Wright had perfected these milling operations, a forged cylinder head could be treated much like a cast cylinder head for subsequent machining operations.

A three-section, 124-station Greenlee automatic transfer machine performed finish milling operations, and drilled, reamed, countersunk, and tapped all holes required in the cylinder head as well as counterboring seats for the valve springs, the valve guide holes, and the valve seats.[289] It took considerably more time to machine a forged cylinder head but forging eliminated the more labor-intensive and time-consuming operations casting cylinder heads at the foundry.[290]

Wright's development of the forged cylinder head illustrates the intense effort that went into process engineering. A detailed article on the process of machining fins on forged cylinders mentions no less than 16 different types of machine tools employed during the process, including two separate Greenlee automatic transfer machines.[291] All these machines had to be adapted or specially built to meet the unique requirements of machining forged cylinder heads. It took Wright and the machine tool companies months of experimentation and effort before these machine tools could be placed in operation. Wright's experience in switching to a new method of production for the forged cylinder head shows that the mass production methods Wright had adapted for aircraft engine production were not as inflexible as some observers have implied.

Inspection and Quality Control

In aircraft engine production, quality control was critical and far broader in scope than in automobile engine production. It is important to recognize that manufacturing component parts of the Cyclone 9 engine was only one part of the production process. Improvements in production engineering had to build in time and space for inspection to ensure quality control. In shifting to quantity production, with all the changes in manufacturing processes and the switch to high-production machine tools, the overriding concern of Wright's engineers was that there be no diminution in the accuracy and quality of engine parts and complete engines. As Wright's General Superintendent put it, "lower quality does not build engines—it builds trouble...for this reason, there can be no compromise with quality."[292] Quantity production vastly increased the quantities of raw materials and engine parts going through the Wright factories, compounding the problem of quality control. By the summer of 1941, when Wright was ramping up its purchases from outside vendors, Wright's Quality Control Department was inspecting 105,000 parts a day.[293] At Wright, quality control began when raw materials and rough forgings or castings arrived at the factory door, continued through the manufacturing process and only ended when the completed engine had passed its final tests.

Wright's Quality Control Department, which in August 1941 had around 1,100 inspectors on staff for the Paterson factories, contained a small group of highly experienced inspectors devoted solely to the inspection of every tool, gauge, and fixture in the factories.[294] This function was vital to maintaining accurate manufacturing but is not always considered as part of the inspection process. In addition, these inspectors performed regular checks of all the company's measuring devices using a set of master gauges as well as checking the personal measuring tools that the machinists and inspectors used in their work. The Raw Materials Inspection Department ran tests on any forgings, castings, or metal stock that came into the factories, sending a sample of each to the laboratory for testing and only releasing the materials with the laboratory's approval. In the manufacturing areas, each production department had its quota of floor inspectors who inspected the first work piece from each new setup of a machine tool for quality and accuracy. Inspectors at the Heat Treat Department tested each part after heat treatment to check for the required level of hardness. At the completion of machining operations, every component part went through a final inspection measuring dimensions, weight, and finish. The administrative load from inspection must have been considerable, as all major parts subject to high stress had their own individual record cards that were filed and maintained at the factories, even after the components had left as part of a completed engine. Inspectors put steel parts through a Magnaflux inspection to detect any minor flaws or defects undetectable by the human eye, while non-ferrous parts were X-rayed. A "profilometer" used a diamond-tipped stylus to measure the smoothness of surface finishes down to one-millionth of an inch. Parts that failed inspection went to the Salvage Department where a committee of Wright and Army Air Force engineers determined whether the part could still be used, needed to be reworked, or should be discarded. The requirements for accurate record keeping no doubt added more clerical workers to the staffs in the factories and partly helped to explain why nearly 50% of the employees never actually worked on aircraft engine production (Figure 3.11).

FIGURE 3.11 Studebaker-built Cyclone engines being assembled prior to undergoing a "green test." The worker on the right is installing the connecting rods into the crankcase while in the center workers add the cylinder assemblies.

A-1027-30C, From the Collection of Studebaker National Museum, South Bend, Indiana.

Even with monthly engine production rates increasing, Wright did not change its standard practice for the most important test of all, the production test of a complete engine. Once assembled under the eyes of inspectors, the engine went to the engine test cells where it was mounted on a test stand and connected to fuel and oil lines. During the first or "green" test, the engine was run for up to 7 hours at various speeds, with inspectors recording all data on performance. When the green test was over, the engine went back to a disassembly line, which was like the assembly line but in reverse, where every engine was taken completely apart. Subassemblies went to separate disassembly benches, where they were taken down to their component parts and each part subject to close inspection, before moving back to an assembly line for final assembly and a final, shorter test run of 15 to 30 minutes after which the engine went through one more inspection. Most of this assembly and disassembly could only be done by hand, but Wright did automate the engine cleaning process that took place after the final test run. Once laboriously done by hand, one engine at a time, Wright installed a continuous washing machine where engines attached to an overhead conveyor passed through several jets of a cleaning solution to remove all traces of dirt and oil before packing and shipment to the airframe factories.[295]

Reading about aircraft engine production during World War II, it is not readily apparent how critical the effort at quality control was to successful production, nor the volume of inspections and the sheer number of parts inspected daily, often by young women with only a few weeks of training. Like much else about wartime aircraft engine production, quality is simply taken for granted. By all accounts, Wright did an excellent job of quality control in the production of the Cyclone 9 engine at its Paterson factories. It is ironic, given the company's commitment to quality, that quality control would be the cause of so many problems and so much controversy at Wright's Lockland, Ohio and Wood-Ridge, New Jersey factories later in the war.

As with the improvements that came through production engineering, the cumulative impact of the changes in production output through process engineering and the progressive development of high-production machine tools can be seen in the increases in monthly production rates of the Cyclone 9 at Wright's Paterson factories and especially, as will be seen, at Studebaker. By the end of 1941, Wright was building Cyclone 9 engines at a rate of 500 month; 15 months later, the Paterson factories reached their peak production of 1,007 Cyclone 9 engines in a single month, more than ten times the number built in January 1940.[296] What is important to understand about Wright's successful transition to quantity production of the Cyclone 9 is that there was no one "magic bullet" that achieved this, but rather a wide range of continuous incremental gains from a change in manufacturing process or an improved machine tool that cut production time or increased the number of parts per operation. When Wright started the process, there was, as was thought at the time, no guarantee that it would be successful.

How It Was Done at Studebaker

While Wright developed the formula for shifting from batch to large-scale production, Studebaker, with the advantage of starting with a clean sheet and its own extensive experience in volume production, perfected it. Studebaker, using the knowledge Wright had accumulated and its own automotive experience, went on to establish a record of even greater output and productivity in factories that were better suited to line production and achieving improved flow. Studebaker's production of the Cyclone 9 engine integrated the size and design of factory buildings, improvements in production engineering, and more high-production machine tools. Studebaker's newer and larger factory buildings, designed from the beginning for line production, gave their production engineers greater scope for applying the lessons in production and process engineering they took from Wright (Figure 3.12).

FIGURE 3.12 Peak monthly production of the Bristol Pegasus and Wright Cyclone 9 engines 1939–1945.

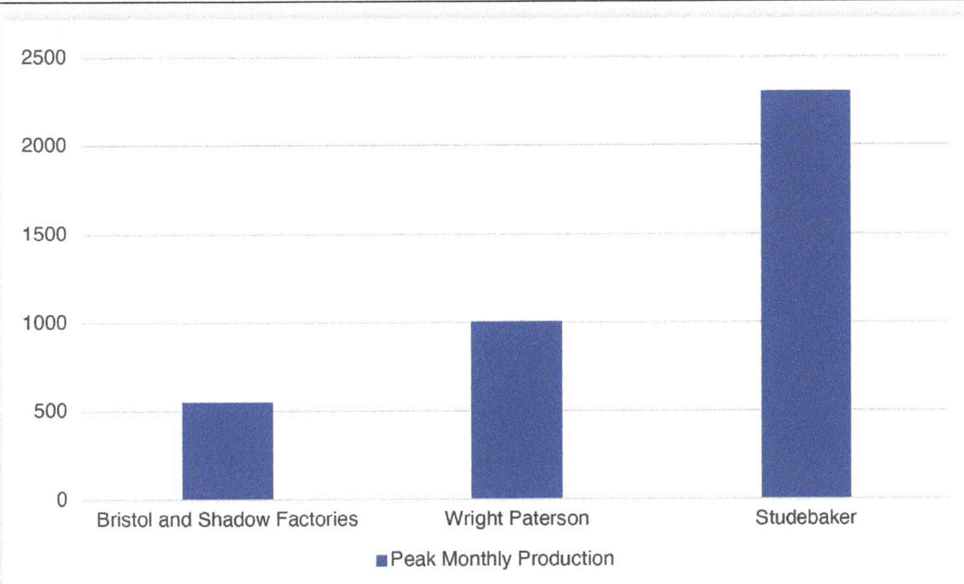

Source: US Department of Commerce, Civil Aeronautics Administration, Office of Aviation Information, Division of Aircraft Statistics: US Military Acceptances 1940-1945: Aircraft, Engines and Propeller Production, (Washington, DC, 1946); Ministry of Aircraft Production: Statistical Review 1939-45, (London 1945).

In searching for an explanation for Studebaker's wartime production record, it is necessary to start with an understanding of the knowledge and experience the company brought to manufacturing the Cyclone 9 engine. At its main South Bend automobile factory, Studebaker was using some of the most advanced machine tools in the automotive industry.[297] To manufacture engines for its new Champion series, Studebaker had several Greenlee four-station automatic transfer machines, an early version of the much larger transfer machines Wright, Greenlee, and Studebaker would develop for machining Cyclone 9 cylinder heads. Studebaker also employed several other types of Greenlee horizontal and vertical drilling machines with four work stations that could hold six engine blocks at a time.[298] Studebaker employed a variety of multi-spindle machine tools, from a Baker Brothers two-way hydraulic machine holding 30 spindles to a massive National Automatic Tool Company (NATCO) unit with 103 spindles for drilling holes in cylinder blocks.[299] The machine tools of other companies—Barnes Drill Company, Cincinnati Milling Machine Company, Norton, Ex-Cell-O, and Snyder—were also represented. The new methods and processes that Wright was developing through production and process engineering would, no doubt, have been familiar to Studebaker's own production and tooling engineers (Figure 3.13).

FIGURE 3.13 Total wartime engine production: American Automobile Licensees.

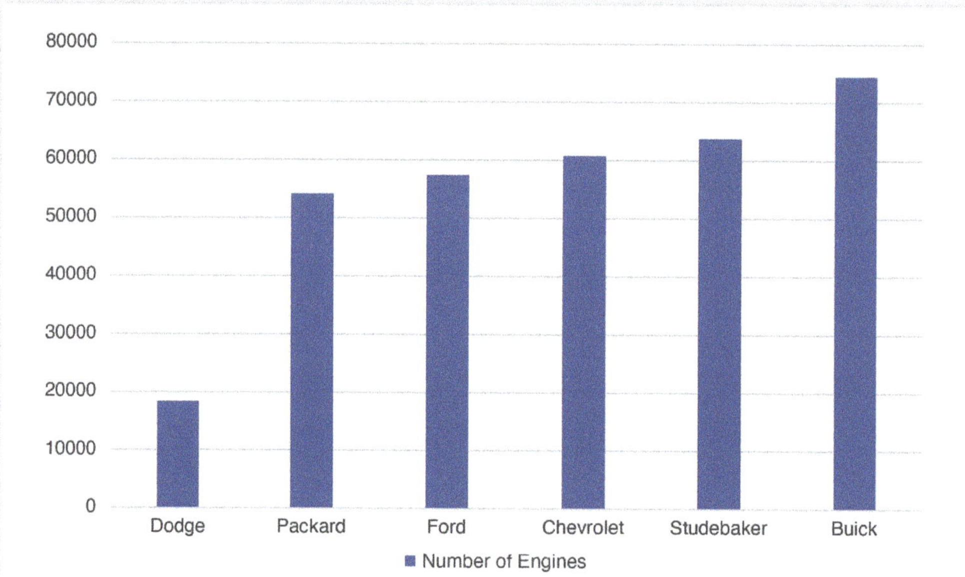

Source: US Department of Commerce, Civil Aeronautics Administration, Office of Aviation Information, Division of Aircraft Statistics: US Military Acceptances 1940-1945: Aircraft, Engines and Propeller Production, (Washington, DC, 1946).

Lists of the machine tools purchased for the South Bend and Chicago factories provide evidence supporting the argument that high-production machine tools were an important component of production at these factories.[300] Studebaker used many of the high-production automatic machine tools that Wright had developed, notably Snyder milling machines, Baker, Barnes, Ex-Cell-O and NATCO multi-spindle drilling machines, and Bullard Mult-Au-Matic machines, as well as several types of Greenlee multi-station machines, and horizontal and rotary automatic transfer machines.[301] Some of these machine tools were expensive, with a Bullard eight-spindle machine costing $38,000 (£9,500) and a NATCO multi-spindle machine $18,000 (£4,500) compared to $3,000 to $10,000 (£750 to £2,500) for standard machine tools.[302] All these specially designed machine tools replaced multiple standard machine tools. Studebaker drew on the knowledge of these machine tool companies and their experience in building machine tools for Wright's new engine factories.[303]

In 1942 Studebaker, who had also worked with the Greenlee company, added automatic transfer machines for the Cyclone 9 at its factory in South Bend.[304] The new Greenlee machine added milling operations to the drilling, reaming, countersinking, and tapping built into the earlier versions.[305] This machine was 176 ft in length, composed of three sections with 50 workstations each. In between the stations, washing machines cleaned the cylinder heads after they had gone through machining operations. At full operation, the Greenlee automatic transfer machine held 203 cylinder heads and turned out a completed head every 50 seconds (Figure 3.14).

FIGURE 3.14 An R-1820 Cyclone 9 undergoing a test run in one of the large test cells at the Studebaker factory. Every single engine built at the factory had to undergo a "green test," running for 4 hours over a normal range of speeds with several 5-minute periods of operating at maximum speeds.

A1027-36C, From the Collection of Studebaker National Museum, South Bend, Indiana.

In building the Cyclone 9, Studebaker followed Wright's methods and processes, but also incorporated its own experience in laying out a factory for efficient production.[306] Interestingly, even with the pervasive emphasis on straight-line production, Studebaker chose a different arrangement for the Fort Wayne gear factory. Manufacturing aircraft engine gears was a challenge for mass production methods as each gear presented a unique production problem due to its design, dimensions, materials, and required heat treatment.[307] As this was more akin to batch production, Studebaker arranged the Fort Wayne factory into functional departments rather than setting up progressive straight-line assembly operations, with each of the functional departments covering one type of operation—cutting, grinding, heat treatment, surface finishing—with the gears moving from department to department in the correct sequence.[308] Starting gear production completely from scratch gave Studebaker's production engineers the freedom to work out new manufacturing methods, purchase new machine tools, and develop new machine tool designs that were more efficient for machining the complicated gears.[309] Using these new methods and machine tools, the Fort Wayne factory reduced the cost of a set of gears for the Cyclone 9 engine to nearly half of Wright's original estimate (Figure 3.15).[310]

FIGURE 3.15 After completing the "green test," every engine was torn down and every part inspected. If inspectors found no defects, the engine was re-assembled and sent for a final test run before being packed for shipment.

A1027-32C, From the Collection of Studebaker National Museum, South Bend, Indiana.

At the South Bend aircraft engine factory, Studebaker paid special attention to materials handling, using mechanized and gravity conveyors and overhead monorails to move parts from machine to machine as well as hoists to lift heavier components.[311] Industrial tractors replaced hand carts to move parts around the factory. To speed assembly operations, it set up a motorized conveyor line as in its automotive plant. The conveyor line carried engines in various states of assembly on specially designed hand trucks attached to the motorized conveyor. Studebaker set up subassembly lines at right angles to the main assembly line, as Wright had done, but added motorized conveyors to some of these lines. Although Studebaker thought this assembly system acted as a constraint on its rate of production and would have preferred one continuous assembly line, with experience the assembly process turned out seven and a half engines per hour.[312]

Studebaker resorted to a high level of subcontracting. From the early stages of production planning, it had to be decided which engine parts should be made in its own factories and which should be purchased from outside vendors. Studebaker apparently went beyond Wright and purchased around 50% of the labor content of the Cyclone 9 from outside its own factories, although later in the war, the percentage of subcontracting appears to have dropped below this level.[313] Early on, Studebaker decided to purchase all steel castings for the crankshaft, crankcase, cylinder barrels, and engine gears from specialist foundry firms removing the need to build foundries at any of the three factories. Studebaker divided production of the remaining parts for the engine between its three factories, with the South Bend factory building one-half, and Chicago and Fort Wayne the rest, but in most cases, this work involved machining castings or forgings acquired from other subcontractors.[314]

Studebaker did employ more workers than the shadow factories building the Mercury and Pegasus, but these workers built proportionately more engines. In November 1943, the month of peak employment at the Studebaker factories, there were 17,551 total employees with nearly half working at the South Bend factory.[315] This compares to 11,883 workers at the shadow factories in September 1941, the year of peak Mercury and Pegasus production. This figure may not include management and engineering staff at the shadow factories, so that the actual difference between total number of employees at Studebaker and shadow factories may have been less. The important point is that with approximately 50% more workers, the Studebaker factories built 181% more engines in November 1943 than the shadow factories built in September 1941.

Studebaker's labor productivity was impressive. From June 1942 until March 1945, the rate of monthly output exceeded the increase in the total number of workers. During 1944, Studebaker built more Cyclone 9 engines while reducing the total number of workers and the total workers per engine built. As production increased and management and workers gained experience, Studebaker continually reduced the cost per Cyclone 9 engine. The cost to the Government of the first production Cyclone 9 engines was $10,061 (£2,515) per engine, but at the point of maximum production, Studebaker had reduced the cost per engine to $6,430 (£1,607), a 36% reduction in cost (Figure 3.16).[316]

FIGURE 3.16 Reduction in estimated cost per engine at the Studebaker factory.

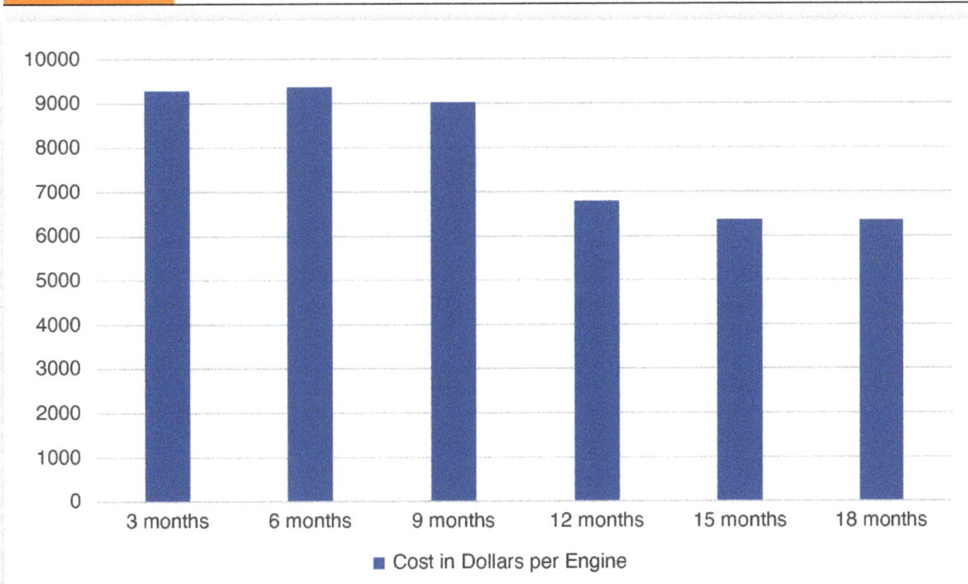

Source: The Studebaker Corporation Aviation Division, Cost of Shipments and Sales Statistics, December 31, 1943, copy in the Studebaker National Museum Archives, South Bend, IN.

Studebaker's ability to accelerate production of the Cyclone 9 was considered to be "... better and more even than that of most other manufacturers."[317] There are several reasons behind Studebaker's performance. First, as with the Mercury and Pegasus, the Cyclone 9 was a proven engine with a long production record. Technical drawings and engineering information were readily available from Wright Aeronautical, and Studebaker had the benefit of Wright's advances in production and process engineering. Second, Studebaker's production was almost entirely devoted to a single model of the Cyclone 9, limiting the disruptions from changing over to a new engine model. Thirdly, Studebaker brought to the Cyclone production program managers with years of experience in quantity production and with the strategic realism to focus on what Studebaker could do best, machining components, and assembling engines.[318]

Conclusion

The story of production of the Wright Cyclone 9 engine during World War II is a story of success that, in retrospect, appears seamless but in fact involved intense effort, ingenuity, and faith in ultimate success on the part of engineers at Wright, Studebaker, and the machine tool companies. This impressive wartime record was not a "miracle of production" but the methodical application of production and process engineering. It is important to bear in mind that in the beginning of America's expansion of airframe and aircraft engine production, there was no guarantee of success and little previous experience of quantity production to draw on. What was available was the knowledge and experience of other industries, principally the automobile and machine tool industries. These industries generously shared information with the aircraft engine manufacturers through direct exchanges, professional associations, and industry publications. Successful quantity production came through integrating factory size, production engineering, and process engineering. As in Studebaker's case, larger factory buildings gave greater scope for applying more efficient methods of production and process engineering. To build the bigger and more complex Cyclone 14 and Cyclone 18 engines in quantity would require even larger factories and more advanced high-production machine tools.

References

1 Eltscher, Louis R., and Edward M. Young: *Curtiss-Wright: Greatness and Decline*, (1998), p. 74.

2 *The Aerosphere 1939*, p. 804.

3 Ibid., p. 805.

4 Ibid., p. 813.

5 *Aerosphere 1939* provides an excellent and detailed description of the different models of the R-1820 and the changes to component parts that allowed the doubling of horsepower.

6 Hill, H.C.: "Engineering Aspects of the Single-Row Cyclone G-200 Series Engine", Wright Aeronautical Corporation, (1940), p. 2.

7 Ibid., p. 4; *G-200 Series Wright Cyclone*, Wright Aeronautical Corporation, (No Date).

8 Hill, "Engineering Aspects of the Single-Row Cyclone G-200 Series Engine", p. 4; "Wright G-200 Cyclone, Rated at 1200 H.P., Released for Both Domestic and Export Sale", *Trade Winds*, Vol. 4, No. 10, (Navy Number 1939), p. 7.

9 *G-200 Series Wright Cyclone*; "Wright G-200 Cyclone, Rated at 1200 H.P., Released for Both Domestic and Export Sale", p. 8.

10 Hill, "Engineering Aspects of the Single-Row Cyclone G-200 Series Engine", p. 5.

11 *G-200 Series Wright Cyclone*; "Wright G-200 Cyclone, Rated at 1200 H.P., Released for Both Domestic and Export Sale", p. 8.

12 *G-200 Series Wright Cyclone.*
13 Modley, Rudolf, Ed.: *Aviation Facts and Figures 1945*, (New York, 1945), p. 2.
14 Aeronautical Chamber of Commerce: *Aircraft Yearbook for 1938*, (New York, 1938), p. 446; Lilley, *Problems of Accelerating Aircraft Production During World War II*, Table 17 and Table 18, pp. 104-105.
15 Aeronautical Chamber of Commerce: *Aircraft Yearbook for 1939*, (New York, 1939), p. 512.
16 Ibid., p. 512; Aeronautical Chamber of Commerce: *Aircraft Yearbook for 1940*, (New York, 1940), p. 457.
17 Haight, John McVickar, Jr.: *American Aid to France, 1938-1940*, (New York, 1970), pp. 3-7; Baughen, Greg: *The Rise and Fall of the French Air Force: French Air Operations and Strategy 1900-1940*, (Croydon, UK, 2018), pp. 120-27.
18 Baughen, *The Rise and Fall of the French Air Force*, p. 125.
19 Memorandum of Conversation, by the Chief of the Office of Arms and Munitions Control (Green), January 18, 1938, Foreign Relations of the United States Diplomatic Papers, 1938, The British Commonwealth, Europe, Near East, and Africa, Volume II, (Washington, DC, 1955), p. 298.
20 Baughen, *The Rise and Fall of the French Air Force*, p. 127; Haight, *American Aid to France*, pp. 8-9.
21 Haight, *American Aid to France*, p. 9.
22 Ibid., pp. 24-35, 37-41, 69-75, 86-88. Correct figures for the French orders can be found in *French Air Commission: Situation des Contrats au 16 Juin 1940*, AVIA 38/426, NA.
23 *French Air Commission: Situation des Contrats au 16 Juin 1940*, AVIA 38/426, NA.
24 Wilson, *Slipstream*, pp. 218-19.
25 *French Air Commission: Situation des Contrats au 16 Juin 1940*, AVIA 38/426, NA.
26 Bailey, Gavin J.: *The Arsenal of Democracy: Aircraft Supply and the Anglo-American Alliance, 1938-1942*, (Edinburg, UK, 2013), p. 36.
27 Ibid., p. 40.
28 Hall, H. Duncan: *History of the Second World War United Kingdom Civil Series: War Production Series: North American Supply*, (London, 1955), pp. 105-06.
29 Ibid., p. 105.
30 *The Wall Street Journal*, July 16, 1938, p. 1.
31 *The Wall Street Journal*, November 3, 1938, p. 13.
32 Hall, *North American Supply*, pp. 108-109.
33 Ibid., pp. 110-11.
34 Dallek, Robert: *Franklin D. Roosevelt and American Foreign Policy, 1932-1945*, (New York, 1979), pp. 172-73.
35 Watson, Mark Skinner: *United States Army in World War II: War Department: Chief of Staff: Prewar Plans and Preparations*, (Washington, DC, 1991), p. 132.
36 Ibid., p. 137.
37 Ibid.
38 Craven, Wesley Frank and James Lea Cate: *The Army Air Forces in World War II*, Volume One: *Plans and Early Operations January 1939 to August 1942*, (Chicago, IL, 1948), p. 104; Maurer, Maurer: *Aviation in the U.S. Army, 1919-1939*, (Washington, DC, 1987), p. 352.
39 *The Wall Street Journal*, March 28, 1939, p. 11; April 26, 1939, p. 12; May 5, 1939, p. 5.
40 *The Wall Street Journal*, August 11, 1939, p. 1.
41 Ibid.
42 Dallek, *Franklin D. Roosevelt and American Foreign Policy*, pp. 204-05.
43 Haight, *American Aid to France*, p. 138.
44 Ibid., p. 133.
45 *French Air Commission: Situation des Contrats au 16 Juin 1940*, AVIA 38/426, NA.
46 Daniel, Raymond and Jean Cuny: *L'Aviation Française de Bombardement et de Renseignment (1918/1940)*, (Paris, 1978), p. 239; *French Air Commission: Situation des Contrats au 16 Juin 1940*, AVIA 38/426, NA.
47 Bailey, *The Arsenal of Democracy*, p. 52.
48 Haight, *American Aid to France*, p. 140.
49 Ibid.
50 Wilson, *Slipstream*, p. 225.
51 Haight, *American Aid to France*, p. 140.
52 Memorandum "French Purchases in the United States", October 30, 1939, Volume 220 October 30-31, 1939, p. 132B, Diaries of Henry Morgenthau, Jr., April 27, 1933-July 27, 1945, Franklin D. Roosevelt Presidential Library and Museum, Hyde Park, NY.
53 Memo from Allison Engineering Co., Wright Aeronautical Corp., Pratt & Whitney Aircraft Division, to Major General H.H. Arnold, Chief of the Air Corps, September 1, 1939, File 145.93-183 August 1939-August 1940, Air Force Historical Research Agency (AFHRA), Maxwell AFB, AL.
54 Ibid.

55 Ibid.

56 Memo from Under Secretary of the Treasury Hanes to Files, November 9, 1939, Vol. 222, November 9-16, 1939, p. 45, Diaries of Henry Morgenthau, Jr.

57 Memorandum for the Secretary of the Treasury, November 9, 1939, Vol. 222, November 9-16, 1939, p. 44, Diaries of Henry Morgenthau, Jr.

58 Ibid.

59 Holley, Irving B.: *United States Army in World War II: Special Studies: Buying Aircraft: Matériel Procurement for the Army Air Forces*, (Washington, DC, 1964), pp. 252-53; Haight, *American Aid to France*, p. 164.

60 Haight, *American Aid to France*, p. 149.

61 Memorandum for the Secretary, January 8, 1940, Vol. 233 January 1-8, 1940, p. 394, Diaries of Henry Morgenthau, Jr.

62 Ibid.

63 Ibid.

64 Ibid.

65 Mentioned in Draft Cable from Henry Self to Major Davies, 21 March 1940, AVIA 38/421, NA.

66 Ibid.

67 Note of January 4, 1940, Vol. 233, January 1-8, 1940, p. 234, Diaries of Henry Morgenthau, Jr.

68 Memorandum of January 17, 1940, Vol. 236, January 16-20, 1940, p. 90, Diaries of Henry Morgenthau, Jr. Haight, *American Aid to France*, p. 171.

69 Memorandum of January 17, 1940, Vol. 236, January 16-20, 1940, p. 90, Diaries of Henry Morgenthau, Jr.

70 Haight, *American Aid to France*, p. 187.

71 Hall, *North American Supply*, p. 121; Haight, *American Aid to France*, pp. 217-19.

72 Draft Cable No. 579 from Sir Henry Self to Major Davies, AVIA 38/421, NA.

73 Haight, *American Aid to France*, pp. 198-20, 207-10.

74 Draft Cable No 580 from Sir Henry Self to Major Davies, 21 March 1940, AVIA 38/421, NA.

75 Number of Airplane Engines Purchased by England and France, Treasury Department, Division of Research and Statistics, May 6, 1940, Vol. 260, May 4-9, 1940, p. 178, Diaries of Henry Morgenthau, Jr.

76 Draft Cable No 580 from Sir Henry Self to Major Davies, 21 March 1940, AVIA 38/421, NA.

77 Schedule A, Total Capital Commitments of the British and French Governments to the Aircraft Industry, May 15, 1940, "Capital Commitments of the British and French Governments to the Aircraft Industry, May 15, 1940", Mr. Young to Secretary Henry Morgenthau, May 16, 1940, Vol. 263, May 16-19, 1940, p. 26, Diaries of Henry Morgenthau, Jr.

78 Ibid.

79 Ibid., Schedules B-F.

80 Craven, Wesley Frank and James Lea Cate: *The Army Air Forces in World War II*, Volume Six: *Men and Planes*, (Chicago, IL, 1955), p. 301.

81 Dallek, *Franklin D. Roosevelt and American Foreign Policy*, p. 221.

82 Ibid.

83 Holley, *Buying Aircraft*, p. 229.

84 Ibid., p. 254.

85 Ibid., pp. 255-56.

86 Civilian Production Administration, Bureau of Demobilization: Historical Reports on War Administration: War Production Board: Special Study No. 21: Aircraft Production Policies under the National Defense Advisory Commission and Office of Production Management May 1940 to December 1941, (1946), pp. 4-6.

87 Holley, *Buying Aircraft*, p. 229.

88 Special Study No. 21, p. 6.

89 National Defense Advisory Commission, Airplane Division Report No. 7: Report on a Study of Aircraft Engine Manufacturing Capacity, July 22, 1940, File 315.30421, War Production Board Policy Documentation File, Box 1125, File 315.30413-315.30433, Records of the War Production Board, Record Group 179, National Archives and Records Administration (NARA), Washington, DC.

90 Ibid., Table III-D, p. 12.

91 Ibid.

92 Lilley, *Problems of Accelerating Aircraft Production*, Table 18, p. 105.

93 Report No. 7, pp. 1-7.

94 Craven and Cate, *The Army Air Forces in World War II*, Volume Six: *Men and Planes*, pp. 307-08.

95 White, Gerald T.: *Billions for Defense: Government Financing by the Defense Plant Corporation during World War II*, (University, AL, 1980), p. 19.

96 Craven and Cate *The Army Air Forces in World War II*, Volume Six: *Men and Planes*, p. 308.

97 White, *Billions for Defense*, p. 70.

[98] Craven and Cate, *The Army Air Forces in World War II*, Volume Six: *Men and Planes*, p. 309.
[99] Hall, *North American Supply*, pp. 146-147.
[100] Craven and Cate, *The Army Air Forces in World War II*, Volume Six: *Men and Planes*, p. 303.
[101] Ibid.; Report No. 7, p. 32.
[102] Report No. 7, p. 12.
[103] Ibid., p. 35.
[104] Ibid., p. 36.
[105] Special Study No. 21, p. 36.
[106] Ibid.
[107] Ibid., pp. 162-63; Lilley, *Problems Accelerating Aircraft Production*, p. 33; Holley, *Buying Aircraft*, p. 161.
[108] Holley, *Buying Aircraft*, pp. 162, 304.
[109] Walton, Francis: *Miracle of World War II: How American Industry Made Victory Possible*, (New York, 1956), pp. 280-81.
[110] Ibid., p. 281.
[111] Wilson, *Slipstream*, p. 239.
[112] See conference of May 30, 1940, Vol. 267, May 28-31, 1940, pp. 348-51.
[113] Ibid., pp. 238-39; Lilley, *Problems Accelerating Aircraft Production*, p. 33.
[114] Lilley, *Problems Accelerating Aircraft Production*, p. 33.
[115] Guy Vaughan, President, Wright Aeronautical Corporation, to Secretary of the Treasury Henry J. Morgenthau, Jr., June 5, 1940, Vol. 270, June 7-9, 1940, p. 63; E.E. Wilson, President, United Aircraft Corporation, to Secretary of the Treasury Henry J. Morgenthau, Jr., June 4, 1940, Vol. 270, June 7-9, 1940, p. 62, Diaries of Henry Morgenthau, Jr.
[116] Conference on the Air Expansion Program, May 30, 1940, Vol. 267, May 28-31, 1940, p. 349; Foley to Secretary Morgenthau, May 28, 1940, Vol. 267, May 28-31, 1940, p. 51, Diaries of Henry Morgenthau, Jr.
[117] Craven and Cate *The Army Air Forces in World War II*, Volume Six: *Men and Planes*, p. 309.
[118] Hyde, Charles K.: *Arsenal of Democracy: The American Automobile Industry in World War II*, (Detroit, MI, 2013), p. 48.
[119] Ibid., pp. 48-52.
[120] Craven and Cate *The Army Air Forces in World War II*, Volume Six: *Men and Planes*, pp. 311-12.
[121] Craven and Cate *The Army Air Forces in World War II*, Volume Six: *Men and Planes*, p. 311.
[122] Special Study No. 21, p. 66.
[123] Ibid., p. 66.
[124] Geschelin, Joseph: "Advanced Tooling Produces Newest Studebakers", *Automotive Industries*, Vol. 80, No. 13, (April 1, 1939); Studebaker production figures taken from U.S. Automobile Production Figures, https://en.wikipedia.org/wiki/U.S._Automobile_Production_Figures, accessed June 1, 2019.
[125] Bonsall, Thomas E.: *More Than They Promised: The Studebaker Story*, (Stanford, CA, 2000), pp. 208-09, 226.
[126] The Studebaker Corporation Minutes of Meetings of Directors and Shareholders, Meeting of July 26, 1940, Vol. 14-From 2-23-1940 to 12-22-1941, Studebaker Museum, South Bend, IN; *The South Bend Tribune*, August 11, 1940, p. 1.
[127] The Studebaker Corporation Minutes of Meetings of Directors and Shareholders, Meeting of September 27, 1940, Vol. 14-From 2-23-1940 to 12-22-1941.
[128] Special Study No. 21, p. 95.
[129] Ibid., p. 98.
[130] Ibid., p. 96.
[131] Ibid., p. 98.
[132] Ibid., pp. 98-99.
[133] U.S. Department of Commerce, Civil Aeronautics Administration, Office of Aviation Information, Division of Aircraft Statistics: *U.S. Military Acceptances 1940-1945: Aircraft, Engines and Propeller Production*, (Washington, DC, 1946), p. 149.
[134] Lewis, Robert: "Redesigning the Workplace: The North American Factory in the Interwar Period", *Technology and Culture*, Vol. 42, No. 4, (October 2001), p. 671.
[135] "Build the Plant around the Process", *Factory Management and Maintenance*, Vol. 97, No. 4, (April 1939), p. B-35.
[136] Muther, *Production Line Technique*, pp. 89-90.
[137] Muther, *Production Line Technique*, pp. 671-72.
[138] "What's Happening and What's Ahead in Industrial Plant Development", *Factory Management and Maintenance*, Vol. 99, No. 4, (April 1941), p. B-37, Remarks of H.K. Ferguson.
[139] "What's Happening and What's Ahead in Industrial Plant Development", p. B-38, Remarks of George Bryant.

140 Darley, Gillian: *Factory*, (London, 2003), p. 92; "What's Happening and What's Ahead in Industrial Plant Development", B-38, Remarks of George Bryant.

141 "Wright Aircraft Will Expend $1,350,000 for Extensions to Factory and New Equipment", *Trade Winds*, Vol. 3, No. 7, (May 1937), p. 2.

142 "Wright Aircraft Will Expend $1,350,000 for Extensions to Factory and New Equipment".

143 "The Wright Aeronautical Corporation Plant at Paterson, NJ", *Mill & Factory*, Vol. XXIV, No. 6, (June 1939), p. 146.

144 "Notes on General History of the Wright Aeronautical Corporation and Description of the Factory Located at Paterson, NJ, USA", copy in the Wright Aeronautical files at the Aviation Hall of Fame and Museum of New Jersey, Teterboro, NJ.

145 Waddington, J.L.: "American Methods and Modes-II", *Flight*, Vol. XXXVI, No. 1605, (September 28, 1939), p. 271.

146 Haight, John McVickar: *American Aid to France 1938-1940*, (New York, 1970), p. 140.

147 "Modern Plant Insures Clouds of Planes", *Mill & Factory*, Vol. XXVII, No. 1, (July 1940), pp. 57-58.

148 "Modern Plant Insures Clouds of Planes", pp. 56-57.

149 "Modern Plant Insures Clouds of Planes", p. 58.

150 Ibid.

151 NA, AVIA 38/429, Statement of Capital and Extraordinary Charges as at January 1, 1942, AVIA 38/489; Hornby, *Factories and Plant*, pp. 260, 263.

152 *The Wall Street Journal*, April 13, 1940, p. 11; Contract A-194, Avia 38/611, AVIA 38/611, Ministry of Supply and Ministry of Aircraft Production. North American Supply Missions, Second World War, Files. BRITISH AIR COMMISSION. Administrative files. Capital assistance. National Archives.

153 Defense Plant Corporation: *Manufacturing Plant, Fairlawn, New Jersey, Plancor 11-Plant No. 3*, Record Group 234, Records of the Reconstruction Finance Corporation: Defense Plant Corporation, Pamphlets Relating to Manufacturing Facilities, 1940-1945: Box1, 1 to 499, National Archives and Records Administration, (NARA), Washington, DC.

154 *The Morning Call*, March 21, 1941, p. 21.

155 Defense Plant Corporation: *Manufacturing Plant, Fairlawn, New Jersey, Plancor 11-Plant No. 3*; *The Wall Street Journal*, September 21, 1940, p. 7.

156 *The New York Times*, October 31, 1940, p. 7.

157 *The Wall Street Journal*, March 21, 1941, p. 1.

158 Surplus Property Administration: *Aircraft Plants and Facilities: Report of the Surplus Property Administration to the Congress, January 14, 1946*, p. 48; "Defense Plant Corporation Authorizes Expansion of Plants Coast to Coast", *American Aviation*, Vol. 6, No. 15, (January 1943), p. 50.

159 Surplus Property Administration: *Aircraft Plants and Facilities: Report of the Surplus Property Administration to the Congress, January 14, 1946*, (Washington, DC, 1946), p. 48, *The New York Times*, November 9, 1947, p. R1, "Fifteen Curtiss-Wright Factories Speed Defense Output", *Aircraft Digest*, Vol. 39, No. 4, (October, 1941), pp. 164, 166.

160 Holley, *Buying Aircraft*, pp. 161-64.

161 Summary of Engines Shipped, p. 9; Civilian Production Administration: Industrial Statistics Division: *Alphabetic Listing of Major War Supply Contracts, Cumulative June 1940 through September 1945*, Vol. 4 Rey-Z, (Washington, DC, 1946), p. 783.

162 Air Technical Service Command Plans (T-5), Logistics Planning Division: *Studebaker Corporation R-2600 Aircraft Engines Construction and Production Analysis*, pp. 3, 15.

163 Engineers' Semi-Final Report Plancor 40, South Bend & FT. Wayne, Ind & Chicago, Ill., The Studebaker Aviation Division, South Bend, Ind., RG 234, Records of the Reconstruction Finance Corporation, Defense Plant Corporation Engineers Reports and Appendices, Plancor 40, Box 30, NARA.

164 *Airplane Engine Plant, Chicago, Illinois, Plancor 40*; Engineers' Semi-Final Report Plancor 40, South Bend & FT. Wayne, Ind & Chicago, Ill., The Studebaker Aviation Division, South Bend, In.

165 *Manufacturing Plant, Fort Wayne, Indiana, Plancor 40*; Engineers' Semi-Final Report Plancor 40, South Bend & FT. Wayne, Ind & Chicago, Ill., The Studebaker Aviation Division, South Bend, In.

166 AAF Industrial Facilities Expansions: Status Progress and Performance, Data as of 31 March 1945, File 218.7-3, 31 March 1945, AFHRA.

167 Air Technical Service Command Plans (T-5), Logistics Planning Division: *Studebaker Corporation R-2600 Aircraft Engines Construction and Production Analysis*, p. 12. The description of the factories is drawn from this report and Defense Plant Corporation: *Airplane Engine Plant, Chicago, Illinois, Plancor 40; Manufacturing Plant, Fort Wayne, Indiana, Plancor 40; Manufacturing Plant, South Bend, Indiana, Plancor 40*, RG 234, Defense Plant Corporation, Pamphlets Relating to Manufacturing Facilities, 1940-1945: Box1, 1 to 499, NARA. See also *The Wall Street Journal*, March 26, 1941, p. 7.

168 "What's Happening and What's Ahead in Industrial Plant Development", B-38, Remarks of George Bryant; see also "Ford Airplane Engine Plant Has Many Distinctive Features", *Mill & Factory*, Vol. XXIX, No. 4, (October 1941), pp. 59-66.
169 Mills, Edward D.: *The Modern Factory*, (London, 1951), p. 67.
170 Geschelin, Joseph: "Studebaker Know-How Speeds Flying Fortress Engine Output", *Automotive and Aviation Industries*, Vol. 88, No. 7, (April 1, 1943), pp. 21.
171 Geschelin, "Studebaker Know-How Speeds Flying Fortress Engine Output", p. 21 and *Manufacturing Plant, South Bend, Indiana, Plancor 40*; Engineers' Semi-Final Report Plancor 40, South Bend & FT. Wayne, Ind & Chicago, Ill., The Studebaker Aviation Division, South Bend, Ind.
172 *U.S. Military Acceptances 1940-1945: Aircraft, Engines and Propeller Production*, Engine shipments for 1940-1945.
173 Wright Aeronautical Order and Contracts Department: Summary of Engines Shipped (On an Invoice Basis) by WAD and Licensees: From 1929 thru August 31, 1960, p. 7.
174 Lilley, *Problems of Accelerating Aircraft Production during World War II*, Exhibit 8, p. 12; *U.S. Military Acceptances 1940-1945: Aircraft, Engines and Propeller Production*, Engine shipments for 1940-1945.
175 Col. W.F. Volandt, Chief, Procurement Branch, Office of the Assistant Chief of Air Staff, to The Under Secretary of War, 9 September 1943, Box 1263, File 452.13 Ranger Engines to 452.13 Wright Engines July 1943-June 1944, Entry 291, Army Air Forces Central Decimal Files, General Correspondence October 1942-May 1944, RG18, NARA.
176 Col. George W. Polk, Jr., Chief, Inspection Division, Army Air Forces Material Command, to Technical Executive, AAF Material Command, 14 August 1943, File 452.13 Wright Engines July 1943-June 1944, Box 1263, AAF Central Decimal Files, Entry 291, Record Group 18, Records of the Army Air Forces, National Archives and Records Administration (NARA).
177 Maj Chester D. Seftenberg, AAF Material Command, to Chief, Development Engine Branch, Material Division, 31 August 1943, File 452.13 Wright Engines July 1943-June 1944, Box 1263, AAF Central Decimal Files, Entry 291, RG 18, (NARA).
178 *Studebaker Corporation R-2600 Aircraft Engines Construction and Production Analysis*, p. 9.
179 Summary of Engines Shipped, p. 9; Civilian Production Administration: Industrial Statistics Division: *Alphabetic Listing of Major War Supply Contracts, Cumulative June 1940 through September 1945*, Vol. 4 Rey-Z, (Washington, DC, 1946), p. 783.
180 Wright Aeronautical Order and Contracts Department: Summary of Engines Shipped (On an Invoice Basis) by WAD and Licensees: From 1929 thru August 31, 1960, p. 7, Copy in the Wright Aeronautical files, New Jersey Aviation Hall of Fame and Museum.
181 *U.S. Military Acceptances 1940-1945: Aircraft, Engines and Propeller Production*, Aircraft Engine Shipments schedules for 1942-1944.
182 Ibid.
183 *U.S. Military Acceptances 1940-1945: Aircraft, Engines and Propeller Production*, Aircraft Engine Shipments schedules for 1944.
184 Report No. 7, Table III-D, p. 12.
185 *U.S. Military Acceptances 1940-1945: Aircraft, Engines and Propeller Production*, p. 139.
186 Sutton, Kenneth E.: "Production and Quality", Wright Aeronautical Corporation (1941), p. 4.
187 Mr. Haag to Secretary Morgenthau, "Employment in the Aviation Manufacturing Industry", January 14, 1941, Morgenthau Diaries, Vol. 347, January 13-14, 1941.
188 "U.S. Manufacturers Nearing Capacity; Booked Up to Next Summer", *American Aviation*, Vol. 3, No. 12, (November 15, 1939), pp. 1, 6.
189 Wright, T.P.: "Americas Answer: Gearing Our Aviation Industry to the National Defense", *Aviation*, Vol. 38, No. 6, (June 1939), pp. 26-29, 78, 80, 82, 84; "Aircraft Expansion: Industry Studies Steps to Increase Output Should Necessity Arise", *The New York Times*, September 21, 1939, p. 1.
190 Hill, Henry C.: "More Aircraft Engines for National Defense", Paper presented at the National Aeronautic Meeting of the Society of Automotive Engineers, March 13-14, 1941, p. 4.
191 "SAE Talks Production", *Aviation*, Vol. 38, No. 11, (November 1939), pp. 32-33, 84, 86, 108.
192 "SAE Talks Production", pp. 84, 108; Hill, Henry C.: "Design Problems in the Quantity Production of Aircraft Engines", *SAE Journal (Transactions)*, Vol. 46, No. 1, (January 1940), pp. 18-24.
193 See for example "Modern Plant Insures Clouds of Planes", *Mill & Factory*, Vol. XXVII, No. 1, (July 1940), pp. 55-57; Nutt, Arthur: "Aviation Faces Biggest Production Problem", *Automotive Industries*, Vol. 83, No. 7, (October 1, 1940), pp. 321-42, 413; "Bombers in Quantity", *Aviation*, Vol. 39, No. 11, (November 1940), pp. 34-37, 108; "Unusual Processes in Aircraft Production", *Mill & Factory*, Vol. XXVIII, No. 4, (April 1941), pp. 68-74; "Modern Production at Lockheed", *Mill & Factory*, Vol. XXVII, No. 1, (May 1941), pp. 77-82.

194 "Only the Productive Can Be Strong-Only the Strong Can Be Free: A Report of the 1940 SAE National Aircraft Production Meeting, held in Los Angeles", *Aviation*, Vol. 39, No. 12, (December 1940), pp. 64, 148.
195 "Only the Productive Can Be Strong-Only the Strong Can Be Free", pp. 65, 144.
196 Berlin, Don R. and Peter F. Rossmann, "Applying Automotive Methods to Aircraft Production", *Aviation*, Vol. 40, No. 1, (January 1941), p. 42.
197 Berlin and Rossman, "Applying Automotive Methods to Aircraft Production", p. 42.
198 Ibid.
199 Ibid.
200 "Only the Productive Can Be Strong-Only the Strong Can Be Free", p. 65.
201 Holben, Martin M.: "Some Aircraft-Engine Production Methods", *SAE Journal (Transactions)*, Vol. 52, No. 10, (October 1944), p. 492.
202 Holben, "Some Aircraft-Engine Production Methods", p. 492.
203 Gagg, R.F.: "New Production Lines for Aircraft Engines", *Mechanical Engineering*, Vol. 63, No. 3, (March, 1941), p. 177.
204 Hill, "More Aircraft Engines for National Defense", p. 7.
205 Sutton, "Production and Quality", p. 2.
206 Sutton, "Production and Quality", p. 3.
207 Ibid.
208 Muther, *Production Line Technique*, p. 85.
209 Gagg, "New Production Lines for Aircraft Engines", p. 178.
210 Ibid. The description of plant layout is taken from this article.
211 Brown, P.W. and H.E. Linsley: "Engines from the Ground Up", *American Machinist*, Vol. 84, No. 11, (May 29, 1940), p. 371.
212 Muther, *Production Line Technique*, p. 87.
213 Brown and Linsley, "Engines from the Ground Up", p. 372.
214 "New Wright Line-Production Foundry Features Company's Plant No. 3 in Fair Lawn, New Jersey", *Trade Winds*, Vol. 5, No. 7, (November 1940), pp. 6-7.
215 "Wright Aero's Electro-Plating Equipment among Most Modern and Versatile in Industry", *Trade Winds*, Vol. 5, No. 7, (November 1940), pp. 10-11; "Wright Aeronautical's Line Conveyor Technique Speeds Painting of Cyclone and Whirlwind Parts", *Trade Winds*, Vol. 5, No. 7, (November 1940), pp. 14-15.
216 Sutton, Kenneth E.: "Progressive Assembly of Wright Engines", *Aircraft Digest*, Vol. 39, No. 4, (October 1941), p. 192.
217 Sutton, "Progressive Assembly of Wright Engines", p. 192.
218 Sutton, "Progressive Assembly of Wright Engines", p. 194.
219 LInsley, H.E.: "Aircraft Engines on the Production Line", *Mechanical Engineering*, Vol. 65, No. 7, (July 1943), pp. 494.
220 "Wright Aero's Electro-Plating Equipment among Most Modern and Versatile in Industry", pp. 10-11.
221 "Wright Aeronautical's Line Conveyor Technique Speeds Painting of Cyclone and Whirlwind Parts", pp. 14-15.
222 "New Wright Line-Production Foundry Features Company's Plant No. 3 in Fair Lawn, New Jersey", p. 6.
223 "New Wright Line-Production Foundry Features Company's Plant No. 3 in Fair Lawn, New Jersey", p. 7.
224 Linsley, "Aircraft Engines on the Production Line", p. 494.
225 Linsley, "Aircraft Engines on the Production Line", p. 495; Wright Aeronautical Corporation: *Application of High Production Machine Tools*, (No date), p. 3.
226 Klemin, Alexander: "A Miracle of Production", *Scientific American*, (December 1944), p. 256.
227 *Wall Street Journal*, (September 22, 1941), p. 18.
228 Ibid.
229 Holben, "Some Aircraft-Engine Production Methods", p. 492.
230 *The Morning Call*, August 27, 1943, p. 18.
231 Sutton, "Production and Quality", p. 4.
232 Sutton, "Production and Quality", p. 5.
233 Linsley, H.E.: "Mass Production for the Aircraft-Engine Industry", *Mechanical Engineering*, Vol. 64, No. 2, (February 1942), p. 100.
234 Linsley, "Aircraft Engines on the Production Line", p. 493.
235 Ibid.
236 *Wall Street Journal*, (September 22, 1941), p. 18.
237 Linsley, "Mass Production for the Aircraft-Engine Industry", p. 101.
238 Ibid.
239 Geschelin, Joseph: "Conversion for War: Influence of Automotive Mass-Production Methods", *SAE Transactions*, Vol. 50, No. 7, (July 1942), p. 278.

240 Sutton, "Production and Quality", p. 4.
241 Linsley, "Mass Production for the Aircraft-Engine Industry", p. 101.
242 Sutton, "Production and Quality", p. 2.
243 *Wall Street Journal*, (September 22, 1941), p. 18.
244 Linsley, "Mass Production for the Aircraft-Engine Industry", p. 100; "Some Aircraft-Engine Production Methods", p. 493; "Machining Methods on Wright Aero-Engines", *Machinery (London)*, Vol. 59, No. 1516, (October 30, 1940), p. 113.
245 Linsley, "Mass Production for the Aircraft-Engine Industry", p. 101.
246 Scranton, Philip: "From Depression to Globalization: Reconfiguring 20th Century American Machinery and Machine Tool Building", Practical Machinist: 06-14-2003, pp. 29-31.
247 Sutton, "Production and Quality", p. 4.
248 Linsley, "Aircraft Engines on the Production Line", p. 493.
249 "40 Years of Machine Tool Progress", *Automotive Industries*, Vol. 81, No. 7, (October 1, 1939), pp. 52-53.
250 During the war American metal working industries doubled the number of automatic multiple-spindle lathes in use. See Ristuccia, Cristiano Andrea, and J. Adam Tooze: "Machine tools and mass production in the armaments boom: Germany and the United States, 1929-44", *The Economic History Review*, Vol. 66, No. 4, (2013), pp. 963, 966.
251 "Multi-Tooling for Aero-Engine Production", *Aircraft Engineering*, Vol. XI, No. 123, (May 1939), p. 217.
252 "Wright Cyclone Production Speeded up by Special Methods", p. 173.
253 Holben, "Some Aircraft-Engine Production Methods", p. 493.
254 Holben, "Some Aircraft-Engine Production Methods", p. 494.
255 Ibid.
256 Hounshell, David A.: "Automation, Transfer Machinery and Mass Production in the U.S. Automobile Industry in the Post-World War II Era", *Enterprise & Society*, Vol. 1, No. 1, (2000), pp. 100-101.
257 Hounshell, "Automation, Transfer Machinery and Mass Production in the U.S. Automobile Industry", pp. 109-110.
258 Linsley, "Mass Production for the Aircraft-Engine Industry", p. 101.
259 *The Morning Call*, August 27, 1943, p. 18.
260 Ibid.
261 Linsley, "Mass Production for the Aircraft-Engine Industry", p. 102.
262 *The Morning Call*, August 27, 1943, p. 18.
263 *The Wall Street Journal*, 22 September 1941, p. 18.
264 Ibid.
265 Brown, P.W.: "Wright Turns to Line Production", *American Machinist*, Vol. 85, No. 13, (June 25, 1941), pp. 600, 602.
266 Brown, P.W. and H.E. Linsley: "More Horsepower for National Defense", *Aircraft Digest*, Vol. 39, No. 4, (October 1941), p. 182.
267 Ibid.
268 Linsley, H.E.: "Automatic Tooling Speeds Production in the New Wright Aircraft Plant", *Machinery (NY)*, Vol. 48, No. 6, (February 1942), p. 110; Wright Aeronautical Corporation: *Application of High Production Machine Tools*, p. 9.
269 "Manufacturing the Cyclone, Part II", *Aircraft Production*, Vol. II, No. 5, (May 1940), p. 154.
270 Ibid., p. 155.
271 "Milling Operations on the Wright 'Cyclone'", *American Machinist*, Vol. 78, No. 7, (March 28, 1934), pp. 245-48.
272 "Mass Production Methods for Building Wright Cyclones", *Aeronautical Engineering Review*, (August 1942), p. 13; Herb, Charles O.: *Machine Tools at Work*, (New York, 1942), pp. 413-15.
273 Wright Aeronautical Corporation: *Application of High Production Machine Tools*, p. 41.
274 Hounshell, "Automation, Transfer Machinery and Mass Production in the U.S. Automobile Industry in the Post-World War II Era", p. 111.
275 Ibid. Unfortunately the brochure does not say if all these machines were in one factory or in several factories, or if there were multiple numbers of these high-production machine tools in a given factory.
276 "Special Machine Tools Boost Engine Production", *The Tool Engineer*, (November 1943), p. 77.
277 A.E. Lombard to Merrill C. Meigs, Chief, Aircraft Section, Office of War Production, December 2, 1941, Folder 315.31 Aircraft Cylinder Heads, War Production Board Policy Documentation Files 315.4 Aircraft Engines, RG 179, Entry 1, Box 1120, NARA.
278 Ibid.
279 Holben, "Some Aircraft-Engine Production Methods", p. 496.
280 Linsley, H.E.: "Forged Cylinder Heads Increase Power of Cyclone Engine", p. 142.
281 Ibid.

282 A.E. Lombard to Merrill C. Meigs, Chief, Aircraft Section, Office of War Production, December 2, 1941; Linsley, H.E: "Machining the Wright Cyclone Forged Cylinder Head", *The Iron Age*, (January 13, 1944), p. 47.
283 Linsley, "Machining the Wright Cyclone Forged Cylinder Head", p. 47.
284 Linsley, "Forged Cylinder Heads Increase Power of Cyclone Engine", p. 142.
285 Holben, "Some Aircraft-Engine Production Methods", pp. 496-97; Linsley, "Machining the Wright Cyclone Forged Cylinder Head", p. 52; "Forged Cylinder Heads Require New Technique", *Aviation*, Vol. 43, No. 6, (June 1944), pp. 142-45, 248, 251.
286 Holben, "Some Aircraft-Engine Production Methods", p. 497.
287 Linsley, "Machining the Wright Cyclone Forged Cylinder Head", p. 53.
288 Holben, "Some Aircraft-Engine Production Methods", p. 497.
289 Linsley, "Machining the Wright Cyclone Forged Cylinder Head", p. 53.
290 "Forged Cylinder Heads Require New Technique", p. 251.
291 Linsley, H.E: "Machining the Wright Cyclone Forged Cylinder Head", *The Iron Age*, (January 13, 1944), p. 47-53.
292 Sutton, "Production and Quality", p. 6.
293 Ray, Samuel: "Quality Control of Wright Engine Parts", *Automotive Industries*, Vol. 85, No. 4, (August 15, 1941), p. 14.
294 Ibid., p. 15. This paragraph is drawn from Ray's article. See also "Mass Production Brings Mass Inspection", *Western Flying*, Vol. XXII, No. 8, (get date), pp. 64-65.
295 Sutton, "Progressive Assembly of Wright Engines", p. 194; Tumser, A.A.: "Quality Control: Checking and Inspection Procedure in the Production of Wright Engines", *Aircraft Digest's Aviation Engineering*, Vol. 39, No. 5, (November 1941), p. 160.
296 *U.S. Military Acceptances 1940-1945: Aircraft, Engines and Propeller Production*, pp. 139, 141, 145.
297 Geschelin, "Advanced Tooling Produces Newest Studebakers", pp. 420-32, 438-40.
298 Geschelin, "Advanced Tooling Produces Newest Studebakers", p. 426.
299 Geschelin, "Advanced Tooling Produces Newest Studebakers", pp. 427, 432.
300 See Studebaker Corporation Appendix "A", Schedules From March 1941 Through February 1944, Box 30, Plancor 40, Reconstruction Finance Corporation, Defense Plant Corporation: Engineers Reports and Appendices, Box 30, Record Group 234, NARA.
301 Geschelin, "Studebaker Know-How Speeds Flying Fortress Engine Output", pp. 19-22, 70; Wright Aeronautical, *High Production Machine Tools.*
302 See Studebaker Corporation Appendix "A", Schedules From March 1941 Through February 1944, Box 30, Plancor 40, Reconstruction Finance Corporation, Defense Plant Corporation: Engineers Reports and Appendices, Box 30.
303 Geschelin, "Studebaker Know-How Speeds Flying Fortress Engine Output", p. 20.
304 Hounshell, "Automation, Transfer Machinery and Mass Production in the U.S. Automobile Industry in the Post-World War II Era", pp. 110-12.
305 Geschelin, Joseph: "Studebaker Know-How Speeds Flying Fortress Engine Output", p. 70.
306 Geschelin, "Studebaker Know-How Speeds Flying Fortress Engine Output", p. 19.
307 Geschelin, Joseph: "Sixty-Three Parts and Assemblies from Studebaker to Cyclones", *Automotive and Aviation Industries*, Vol. 88, No. 12, (June 15, 1943), p. 33.
308 Geschelin, "Sixty-Three Parts and Assemblies from Studebaker to Cyclones", p. 33.
309 Ibid.; *Studebaker Corporation R-2600 Aircraft Engines Construction and Production Analysis*, p. 18.
310 *Studebaker Corporation R-2600 Aircraft Engines Construction and Production Analysis*, p. 18.
311 Geschelin, "Studebaker Know-How Speeds Flying Fortress Engine Output", p. 20.
312 *Studebaker Corporation R-2600 Aircraft Engines Construction and Production Analysis*, p. 25.
313 *Studebaker Corporation R-2600 Aircraft Engines Construction and Production Analysis*, p. 36; see Air Technical Service Command: *Labor Statistics for the Aircraft Industry*, November 1944-May 1945, File 218.6-7, AFHRA.
314 *Studebaker Corporation R-2600 Aircraft Engines Construction and Production Analysis*, p. 36.
315 *U.S. Military Acceptances 1940-1945: Aircraft, Engines and Propeller Production*, p. 190.
316 The Studebaker Corporation Aviation Division, Cost of Shipments and Sales Statistics, December 31, 1943, copy in the Studebaker Museum Archives, South Bend, Indiana.
317 *Studebaker Corporation R-2600 Aircraft Engines Construction and Production Analysis*, p. 3.
318 *Studebaker Corporation R-2600 Aircraft Engines Construction and Production Analysis*, pp. 3-4.

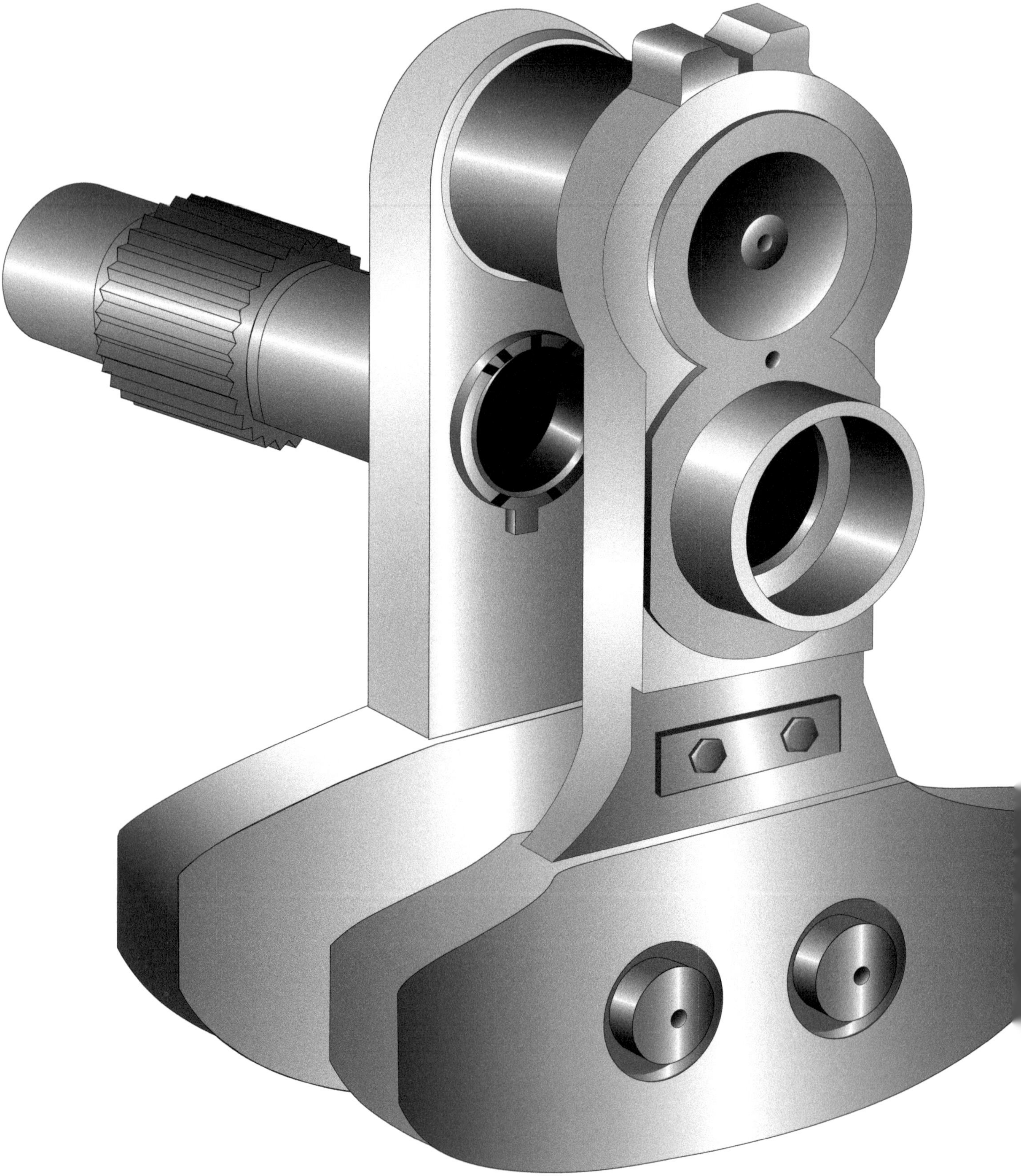

4

Building 14-Cylinder Engines: The Bristol Hercules

Introduction

The Bristol Hercules engine was second to the Rolls-Royce Merlin in terms of numbers of British aircraft engines built during World War II. The 14-cylinder Hercules was larger, more powerful, and more complicated than the Bristol Mercury and Pegasus, with a unique sleeve-valve system. It represented a new generation of British aircraft engines, with more power than the early marks of the Merlin. Yet, while the Merlin is famous, the Hercules is barely known despite the enormous scale in which it was made, and indeed the size and importance of the new factories which built it. The factories and the methods and tools used to build the Hercules were very different from those used to make the earlier generation of Bristol engines. Hercules production conforms to the second phase of wartime engine production, which was characterized by large factories, line production methods, and new types of automatic machine tools, mainly imported from the US.

Hercules production was on a completely different scale from building the Mercury and Pegasus. To build the Hercules, the Government set up a completely new shadow factory scheme, using factories with four to five times the total floor area of the first shadow factories to build complete aircraft engines, not components. The new shadow factories were built on greenfield sites, well away from their parent automotive factories. They employed new flow production methods derived from the American aircraft engine manufacturers, not from Bristol, and new types of machine tools that semi- and unskilled workers, including higher percentages of women, could use with minimum training. The new shadow factories were equipped with greater numbers of more advanced, special-purpose high-production machine tools, many imported from America, to achieve quantity production. They were thus very different from the first shadow factories. The new Bristol shadow factory at Accrington would become the model for these new methods, though the Accrington factory receives only a cursory mention in the literature.

Large-scale production of the Hercules came from advances in production engineering, specifically the effort to achieve greater flow by adopting line production, with careful attention to plant layout, placing machine tools in the proper sequence of operations. The new shadow factories also benefited from greater attention to process engineering, reconfiguring manufacturing operations for greater efficiency and higher rates of output. Facing the challenges of dilution and the need for deskilling machining operations, Bristol and the shadow factories, like Wright and Studebaker, adopted more automatic machine tools, though not to the same extent as the American companies. Yet, the Hercules factories fell short of the goal of full line production, due to constraints imposed on the design and location of the factories to minimize the risk of disruption from air attack (Figure 4.1).

FIGURE 4.1 The Hercules was Bristol's first 14-cylinder sleeve-valve engine to be built in quantity. The 1,375 hp Hercules II was the first production version, replaced in 1941 with the 1,615 hp Hercules VI, shown here, which went into production at the shadow factories. During the war, Bristol and the shadow factories built 9,206 Hercules VI engines.

Rolls-Royce Heritage Trust-Bristol Branch.

The Hercules was Bristol's response to the demand for airplane engines of ever greater power. When it was introduced in 1937, the Hercules was rated at 1,375 BHP making it at the time the most powerful British aircraft engine to have passed its type test.[1] The distinctive characteristic of the Hercules engine was the sleeve-valve mechanism which offered certain advantages over the standard poppet-valve mechanism in engines like the Mercury and Pegasus. The sleeve valve replaces the valves used in poppet-valve engines that take in the fuel–air mixture into the cylinder and eject the exhaust gases after combustion.[2] Instead, the cylinder in the sleeve-valve engines had three inlet and two exhaust ports cut into the cylinder's side. A metal sleeve, placed between the inner wall of the cylinder and the piston, moved in a reciprocating vertical, elliptical motion during the combustion cycle. As it did so, four ports cut out in the sleeve aligned with the ports on the cylinder wall to allow the fuel–air mixture to enter the cylinder and the exhaust gases to be forced out. The advantage of the sleeve valve was that it did away with the entire valve mechanism, and all its component parts, in the poppet-valve engine.[3] The removal of the valve mechanism reduced maintenance and the number of components that needed to be manufactured thereby reducing manufacturing costs, though manufacturing a cylinder with the sleeve-valve mechanism created its own unique production challenges that proved to be difficult to overcome.

The origins of the Bristol Hercules aircraft engine can be traced back to Roy Fedden's contacts with Harry Ricardo in the 1920s. Ricardo, an independent researcher and consulting engineer on internal combustion engines, had been exploring with the support of the Air Ministry the possibilities of using the single sleeve-valve mechanism to improve the efficiency of petrol engines given the poor anti-knock qualities of fuels then available and the tendency for extremely hot exhaust valves to fail, a reliable method for cooling the exhaust valve having yet to be developed.[4] Fedden thought Ricardo's work was promising, and in 1926, Bristol began an intensive program to develop an aircraft engine employing the single sleeve valve, also with financial support from the Air Ministry.[5] In 1934, after eight years of research and thousands of hours of testing, Bristol introduced the Perseus, a nine-cylinder air-cooled sleeve-valve aircraft engine offering maximum power of 760 hp.[6] At the time of the Perseus' debut, Bristol claimed that not only the cost of manufacture would be cheaper than comparable poppet-valve engines, but also, interestingly, that "Greater scope is afforded for large production with unskilled labor."[7] Bristol followed the Perseus with the Aquila, a smaller 500 hp sleeve-valve engine, but knowing that other competitors were working on more powerful double-row air-cooled radial engines to meet the never-ending demand for greater power, Fedden and his team began working on a design for a 14-cylinder two-row sleeve-valve engine using the same cylinders as in the Perseus.[8] Bristol had come to recognize that the nine-cylinder radial engines of up to 1,000 hp would no longer meet the requirements of the aircraft manufacturers who would need aircraft engines of 2,000 hp or greater.[9] By early 1935, Fedden had worked out a design for a 14-cylinder sleeve-valve engine that promised an initial power output of 1,320 hp well above the Pegasus or the Perseus nine-cylinder engines then available.[10] In November 1936, the Hercules successfully passed a 50-hour civil type test and created a sensation when it went on display at the Paris Air Show that same month.[11] The Hercules made its British debut at the Society of British Aircraft Constructors show at Hatfield in June 1937, but by then the new Bristol engine was already under consideration to power new bombers for the Royal Air Force.[12]

Organizing Production

Demand for the Hercules engine came about from a reassessment of the RAF's requirements for bomber aircraft.[13] To provide an effective deterrent against German air power, the RAF needed bombers that could carry a heavy load of bombs to Germany. This required a new class of airplane, the heavy bomber, with the speed, range, and armament necessary to fly to Germany and return. To obtain this speed and range and carry a heavier load of bombs, fuel, crew, and guns required much more powerful engines.[14] In the summer of 1936, the Air Staff issued two Operational Requirements, one for a heavy bomber under Specification B.12/36 calling for a cruising speed of 230 mph with a bomb load of 2,000 lb and a range of 1,500–2,000 miles and one for an improved multi-role medium bomber under Specification P.13/36 with similar performance.[15] In the spring of 1937, the Air Ministry ordered prototypes from two manufacturers for each specification. Short Brothers responded to the B.12/36 requirement with a proposal for a large and heavy four-engine bomber with four Bristol Hercules engines, as did Supermarine, the other chosen firm, whose four-engine design could take either the Bristol Hercules or the Rolls-Royce Merlin.[16] Named the Stirling, the Short Brothers design would enter service as the RAFs first four-engine monoplane bomber. A.V. Roe and Handley Page submitted designs for the P.13/36 requirement using two 1,800 hp Rolls-Royce Vulture engines then under development, and these would enter production as the Manchester and the Halifax.[17] Handley Page adapted the Halifax to take four Rolls-Royce Merlin engines. Later models of the Halifax would absorb much of Hercules production.

By early 1938, it was becoming apparent that the RAF would need considerable numbers of high-power aircraft engines for its new fleet of bombers and for future designs. Planners at the Air Ministry envisioned a need for 1,500 Short Stirling bombers, and this total, with an additional requirement for 1,500 Manchester and 500 Halifax bombers, became part of the RAF's expansion Scheme L in October 1938.[18] This requirement for Stirling bombers alone would require, at a minimum, 6,000 Hercules engines, while the Air Ministry was considering equipping the Vickers Wellington with the Hercules and using the Hercules in a new general reconnaissance bomber designed by Armstrong Whitworth, which would mean a need for even more Hercules engines.[19] Even with expanded facilities, it appeared doubtful that Bristol alone could produce the quantity of Hercules engines the RAF would require. The only realistic alternative was to turn to the automotive industry and the shadow factories.

To build the Hercules engine in the quantities and the time required, the Air Ministry set up a completely new shadow factory organization enlisting the automotive firms already involved in building the Bristol Mercury and Pegasus in the first shadow factory group. This second shadow factory scheme saw greater investment in larger factories, even more machine tools, and many more workers than in the first shadow factory scheme. The idea of getting the shadow factories to build the Hercules arose well before the factories had even begun to produce the Mercury engine in quantity, though the organization that finally emerged for the second shadow factory scheme was completely different from what was originally envisioned.

To build the Hercules, the Air Ministry wanted to change the structure of the shadow factory organization. At the meeting on June 10, 1938 to inform the Aircraft Engine Committee and Bristol of the Air Ministry's decision to have the shadow factories take on production of the Bristol Pegasus as well as the Mercury, H.A.P Disney, Director of Aeronautical Production, brought up the Air Ministry's intention to have the shadow factories take on production of the sleeve-valve Hercules once they had completed the contract for the 1,000 Pegasus engines.[20] Initially, the Air Ministry had decided the shadow factories should build 500 Hercules, beginning production in March 1940 or earlier at a rate of 35 Hercules engines a week based on their shop areas and using a single day shift. Concerned with the issue of vulnerability implicit in the current organizational structure where the shadow factories built component parts of the Mercury and Pegasus instead of complete engines, Disney explained that with the Hercules, the Air Ministry wanted the members of the Aircraft Engine Committee to be divided into two groups, with each group producing complete engines. He proposed the following division:

(a) Rover
Bristol
Standard Motor

(b) Rootes
Daimler
Austin

The Aircraft Engine Committee members, although all recognized the need to deal with the vulnerability issue, disagreed with the Air Ministry's proposal and proposed instead that the shadow factories keep their existing structure, but with each factory building duplicate sets of parts, so that the loss of production at one factory could be made up from another. Disney responded by saying that the Air Council Committee on Supply was "very anxious" that there should be two distinct groups but would listen to recommendations from the Aircraft Engine Committee.[21] The Aircraft Engine Committee agreed to come up with an estimate of the costs of producing the Hercules in two groups as the Air Ministry wanted and proposed meeting with Bristol to examine a Hercules engine to understand what they would be undertaking. At this meeting, the representatives of the shadow factories recognized that production of the Hercules was "a big departure from the Mercury engine."[22]

Increasing requirements for Bristol engines, a reassessment of the ability of the parent automobile firms to convert their factories to aircraft engine production in time of war, and concern with available capacity for Aircraft Engine production, against a background of a rapidly deteriorating international situation, caused the Air Ministry to change its view on the best means of organizing production of the Hercules. In early October 1938, the Air Council Committee on Supply determined a need for 4,550 Mercury engines from the shadow factories, with the order for 1,000 Pegasus engines to follow as quickly as possible.[23] The requirements for Hercules engines were even greater. The Committee discussed a plan for an immediate order for 3,000 Hercules, plus spares, to be split between Bristol and the shadow factories, and a program for 7,000 Hercules engines and spares over two years.[24] By year end, these numbers had increased again. For just the airplanes the Air Ministry had on order, there would be a need for 5,005 Mercury, 3,291 Pegasus, 1,366 Hercules, 1,354 Perseus, and 1,100 Taurus sleeve-valve engines.[25] At a meeting with the Aircraft Engine Committee on October 7, 1938, Maj. George Bulman, Director for Engine Production, had warned the Committee of the need for an accelerated program of rearmament and the concern that aircraft production might exceed aircraft engine production at the shadow factories.[26] The Aircraft Engine Committee determined that their best rate of balance production would be 75 complete engines a week, or 300 a month. At that rate, it would take over two years for the shadow factories to complete the probable number of Mercury and Pegasus engines the RAF would need.

That fall, the long-held assumption that the main automobile firms could easily convert their factories to building aircraft engines in an emergency, was called into question and found to be incorrect. At his meeting with the Aircraft Engine Committee on October 7, Bulman brought up the question of the role of the parent motor companies in time of war, suggesting that a duplicate set of jigs and tools be provided to the parent factories to manufacture the Pegasus in time of war.[27] No doubt to his astonishment, the members of the Committee replied that "such a provision would be wholly inadequate and impracticable, as duplicate jigs and tools for use in the shadow factories would be unsuited to the machinery in the parent factories and the plant available would be inadequate."[28] The Committee recommended that a thorough study should be undertaken to draw up a plan for employing the parent factories. The Air Ministry did conduct a study only to find that the majority of machine tools used to manufacture automobiles were too specialized to use in manufacturing aircraft engines and that converting these factories would be costly and time-consuming.[29] The study determined that only 30 to 40% of the machine tools in the automobile factories could be converted to producing aircraft engines, requiring the Government to finance replacing 60 to 70% of existing machinery in the factories with new machine tools, a process that would likely take months to accomplish after the start of war.[30] This would result in a huge shortfall in production. The war potential of the automotive firms was now seen as providing labor for the shadow factories, but neither factory space nor machine tools.[31] The Air Ministry now had to accept that the planned conversion of the automotive firms to producing Bristol engines in the event of war would no longer be possible, leaving the question of where and how to acquire the additional production capacity.[32]

Air Marshal Wilfrid Freeman, Air Member for Development and Production in the Air Ministry, met with representatives of Bristol on March 8 and 10, 1939 to present a proposal to increase capacity for producing Bristol engines to meet peacetime requirements and to form a base for rapid expansion of production in time of war.[33] The Air Ministry's proposal had three elements: first, to build at Government expense a new shadow factory away from Filton that Bristol would manage to produce 130 to 150 double-row engines at a cost of £2.5 million ($10 million); second, to expand the factory and plant at Filton to produce 100 to 120 single-row Mercury or Pegasus engines to cost £1 million ($4 million); and third, to engage the Standard and Rover companies to manage a new, Government-financed factory to build double-row engines at a cost of £3 million ($12 million).[34]

Standard and Rover, selected because of their record of capable management, would be invited to build 200 complete double-row engines a month, either separately or together, with a work force of 3,000 to 4,000 men. The new Bristol factory could produce 120 double-row engines a month, with 3,000 to 3,300 workers, the two new factories making up the deficiency in double-row engines. The existing shadow factories would continue with the planned conversion to Hercules production, but would reserve 20% of their capacity to continue building 40 Mercury and Pegasus engines a month. Bristol could still produce 50 Mercury and Pegasus engines at Filton, while the remaining shortfall of 110 in Mercury and Pegasus engines, Bulman suggested, could be produced at the new Bristol factory. One of the advantages of the new scheme was that the creation of more factories would, in future, make it easier to convert some factories to new types of aircraft engines without interrupting production at the others.[35] In addition, setting up new shadow factories to build complete engines instead of expanding the existing ones would reduce the vulnerability of having the shadow factories build separate components, responding to one of the main criticisms of the shadow factory scheme. Paring two firms to work together, each firm manufacturing 50% of the components of the Hercules engine would also reduce the risk of unbalanced production that had affected the first shadow factories.

After further discussions with Bristol and the automotive companies, the Air Ministry proposed that four motor firms—Standard, Daimler, Rootes, and Rover—be grouped in pairs, Standard with Daimler and Rootes with Rover.[36] Each firm in the pairings would build a factory, and each factory would manufacture 50% of the Hercules engine's components. One factory in the pairing—the Standard factory in the Standard–Daimler pairing and the Rover factory in the Rootes–Rover pairing—would take on responsibility for final assembly and testing completed engines with parts from its own production and parts transported from its partner factory. Each pair of factories was designed for an initial output of 100 complete engines per month on a day shift.[37] This was soon raised to 125 complete engines a month from each pairing, or 250 per month from the four factories.[38] The new Bristol shadow factory, to be designated Bristol No. 3 Shadow Factory, would build complete engines and was to have an initial capacity of 260 Hercules engines a month, providing a combined capacity of over 500 Hercules engines a month from all five factories.[39] This was 125 engines per week, less than the Mercury and Pegasus production program, but the Hercules was a more complicated engine to build.

With Bristol in agreement, the new shadow scheme went forward. The Treasury authorized expenditures of £6 million ($24 million) to create the new capacity for production of Bristol aircraft engines, and in June, the Air Ministry instructed Daimler, Rootes, Rover, and Standard to begin work on the four new factories and to place orders for the steel needed for construction.[40] The new factories formed No. 2 Shadow Group, with the first shadow factories becoming No. 1 Shadow Group, each with its own coordinating committee. This structure continued until May 1940, when the two committees merged to form the Joint Aircraft Engine Committee.[41] The Joint Aircraft Engine Committee would coordinate the work of the two shadow groups until the end of the war.

There was one more addition to Hercules production that took place during 1942, the delayed conversion of No. 1 Shadow Group to producing Hercules component parts to support No. 2 Shadow Group factories. The conversion had been the Air Ministry's intention, but the need for greater numbers of Mercury and Pegasus engines after the start of the war delayed the Air Ministry's plans. By mid-1941, the Air Ministry could foresee a steady decline in demand for these engines as the now-obsolete Blenheim, Hampden, and early marks of the Wellington went out of production.[42] Looking at the new generation of high-powered aircraft engines emerging from development, the Ministry of Aircraft Production's Air Supply Board, which had replaced the Air Council Committee on Supply, decided that there was insufficient production capacity for the 2,000 hp Napier Sabre engine and that No. 1 Shadow Group should convert to producing the Sabre and not the Hercules.[43] Daimler, Rootes, Rover, and Standard would make the component parts for the Sabre while Austin and Bristol's No. 2 Shadow Factory would assemble and test complete engines, the conversion to cost an estimated £3 million ($12 million).[44] By the end of the year, however, the Air Supply Board realized that continued problems with the Sabre's development would delay quantity production for some time to come.[45] In the months following the Supply Board's decision on converting No. 1 Shadow Group to Sabre production, revisions to the overall aircraft program created an unforeseen demand for an additional 2,750 Hercules engines.[46] At the end of December 1941, the Minister of Aircraft Production decided that No. 1 Shadow Group would convert to producing Hercules VI engines in cooperation with No. 2 Shadow Group, while continuing to manufacture Mercury and Pegasus engines in smaller quantities.[47] In January 1942, the Air Supply Board approved an expenditure of £3.5 million ($14 million) to cover purchase of 2,400 new machine tools for the conversion with the goal of bringing the combined production rate of No. 1 and No. 2 Shadow Groups up to 1,350 Hercules engines per month, with an additional equivalent of 250 engines as spare parts, by the end of 1943.[48] The actual expenditures under this program amounted to £4.01 ($16 million), split between No. 1 and No. 2 Shadow Groups, with No. 1 Shadow Group receiving £1.05 million ($4.2 million) and No. 2 Shadow Group £2.4 million ($9.6 million) for the new plant (Table 4.1).[49]

TABLE 4.1 Capital expenditure on no. 2 shadow group factories as of December 1942 (in millions of £).

Factory	Initial cost estimate	1942 factory	1942 machine tools	1942 total
Daimler No. 2	£1.8M ($7.2M)	£1.081M ($4.3M)	£1.865M ($7.5M)	£2.947M ($11.8M)
Rootes No. 2	£1.8M ($7.2M)	£1.212M ($4.8M)	£1.919M ($7.7M)	£3.131M ($12.5M)
Rover No. 2	£1.8M ($7.2M)	£1.574M ($6.3M)	£2.650M ($10.6M)	£4.225M ($16.9M)
Standard No. 2	£1.8M ($7.2M)	£1.805M ($7.2M)	£3.136M ($12.5M)	£4.942M ($19.8M)
Bristol No. 3 Factory	£4.026M ($16.1M)	£1.750M ($7.0M)	£4.141M ($16.5M)	£5.891M ($23.6M)

Source: CAB 102/274, National Archives, Kew, UK.

The Factories

As with the factories in the first shadow scheme, the key variables for factories built under the new scheme were location, size, and design based on required monthly output. With the contribution of the parent automotive factories now seen as providing labor and management instead of factory space and plant for aircraft engine production, the Air Ministry wanted the new shadow factories to have access to this pool of workers and managers from their parent firms.[50] But because these new factories would be substantially bigger than the No. 1 Shadow Group factories, they needed more space than was available near the parent automobile factories. The Air Ministry therefore decided that the new factories would be located in open country around Birmingham and Coventry, but not more than 3–4 miles from their parent factories to ease problems of transportation and to avoid the need to provide housing.[51] Although they would work in pairs, the factories remained some distance apart and component parts and subassemblies had to be transported from one factory to the other for final assembly, adding time to the production process.

Daimler found a site 3 miles northwest of the center of Coventry and had received instructions to locate its new factory no more than 5 miles from the factory of its partner,

Standard Motor, which was located 3½ miles to the west of the city.[52] For some reason, the same logic did not apply to the Rootes–Rover combination, as Rootes located its No. 2 Factory 17 miles away from Rover. The Rootes plant was 4½ miles southeast of Coventry at Ryton-on-Dunsmoor, while Rover's No. 2 Factory was located at Solihull, 8 miles from the center of Birmingham. For the new Bristol No. 3 Shadow Factory, the Air Ministry chose a site near the town of Accrington, northeast of Manchester, west of the line demarcating the danger zone for air attack.[53] The Ministry believed that the Accrington factory could draw labor from the area around Manchester, although this would prove difficult to do.

The factories in No. 2 Shadow Group were substantially larger than the factories in No. 1 Shadow Group. The machine shops at each of the No. 2 Shadow Group factories covered around 660,000 ft^2, with some of the assembly buildings almost as large.[54] In total area, machine shops, assembly buildings, offices, canteen, test houses, and other ancillary buildings included each of the four shadow factories covered an area of close to 1,000,000 ft^2, and in some cases more, more than five times larger than the average factory in No. 1 Shadow Group (Figure 4.2).[55]

FIGURE 4.2 An aerial view of the Standard Motors No. 2 Shadow Factory at Banner Lane, to the west of Coventry. This factory was the largest of the shadow factories building the Bristol Hercules. Paired with the Daimler No. 2 Shadow Factory, the Standard Motors No. 2 Shadow Factory built 50% of the required parts and assembled completed engines.

EAW000218_The_Standard_Motor_Company_shadow_factory_Tile_Hill_1946, Historic England.

The Standard No. 2 Shadow Factory at Banner Lane was the largest of the four factories in No. 2 Shadow Group. The complex consisted of three large parallel buildings, two for machining and one for final assembly building, ancillary buildings, and engine test stands that were separated from the final assembly building.[56] The total floor space was approximately 1,141,000 ft^2.[57] At some 925,000 ft^2, the Rover No. 2 Shadow Factory at Solihull southeast of Birmingham was three times bigger than its No. 1 Shadow Factory, and comprised one large machine shop and a separate assembly building, plus engine test stands, separate from the assembly building, and ancillary buildings.[58] Daimler and Rootes, responsible for manufacturing only components, made do with one large machine shop and a separate building for putting subassemblies together. If each of the pairings is considered as a single factory unit, then the total area they occupied was probably close to 2,000,000 ft^2 in total floor space, comparable to the new shadow factory Rolls-Royce was building at Hillington, near Glasgow, which had an initial floor area of 1,508,000 ft^2.[59] None of the new factory buildings, however, were as big as the Studebaker aircraft engine factory at South Bend, Indiana.

Bristol's No. 3 Shadow Factory at Accrington was a bit larger in total area as it was charged with building complete engines on site, like the Rolls-Royce shadow factory at Hillington. The Accrington works comprised four large square buildings, comprising three separate machine shops to manufacture components and a fourth shop for engine assembly, each 200,000 ft^2 in area (see Figure 4.3).[60] Within the machine shop buildings, an average of 130,000 ft^2 was devoted to machining operations. In the machine shops, for which accurate plans still exist, the bays appear to have been 30 ft wide, allowing ample room for machine tools, with 18 bays per shop building. Special departments for heat treating or other processes were located alongside the machining areas. Bristol added a slightly smaller building of approximately 120,000 ft^2 close by to build exhaust rings, cowlings, flame dampers, and other external parts for the Hercules. There were 24 engine test stands occupying 260,000 ft^2, placed several hundred yards from the final assembly building and protected with earthen berms, with only one entrance into the engine test stand area. Completed engines had to move from the engine assembly building by truck to the test cells and back again, which took time. With offices, canteen, and other ancillary buildings, the Accrington factory had over 1,300,000 ft^2 in total floor space.

The new factory buildings were typical of the period, large steel sheds with steel columns, placed on a thick concrete slab, supporting strong steel beams with steel trusses attached to the beams to support the roof structure.[61] Most of the buildings had a north-lit, saw-tooth roof to allow in daylight, with some additional artificial lighting, although in some factories the daylight glazing may have been blacked out at all times. Walls were of brick and were not intended to provide structural support but to keep out the weather and provide some protection from bomb fragments.[62] Additional features for air-raid protection were the principal difference between the new and the older shadow factories.

FIGURE 4.3 A plan of Bristol No. 3 Shadow Factory at Accrington. Note how the factory was divided into five separate machine shops, A to E, instead of built as one large single building. The engine test cells were to the left of the machine shops, separated by a large earthen berm and several hundred yards from Shop "A," the engine assembly building.

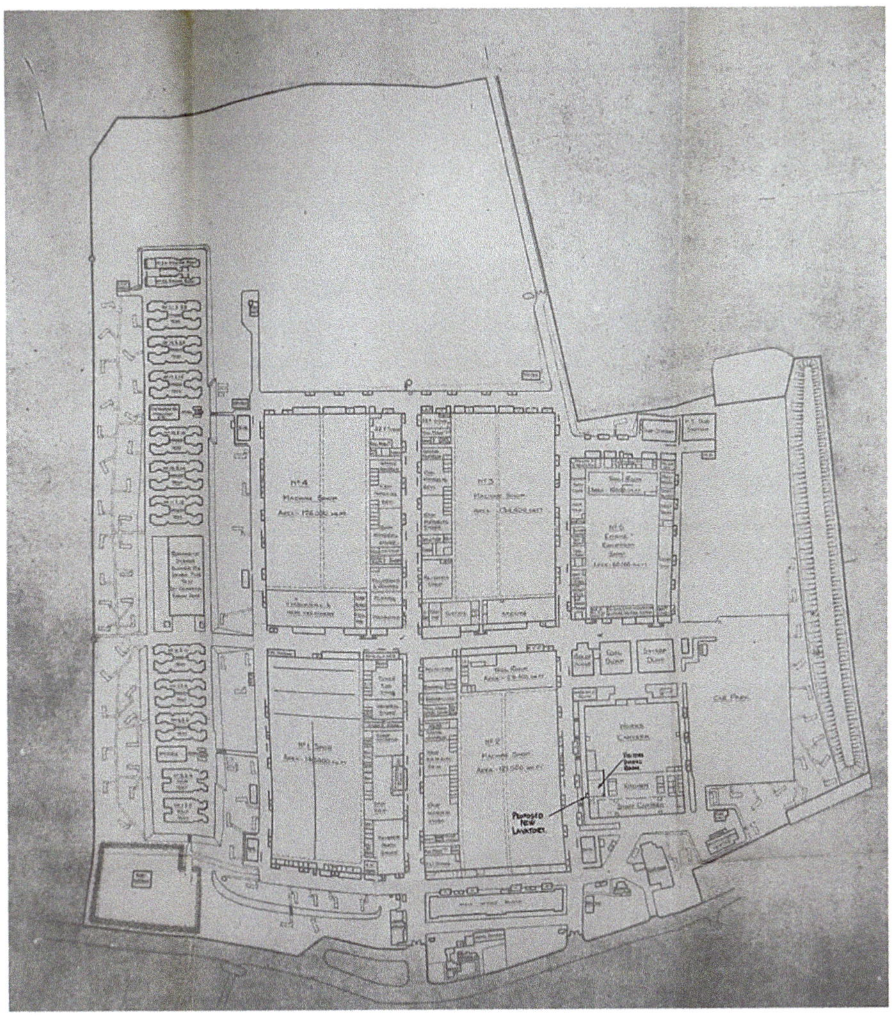

Source: AIR 2/3684, National Archives, Kew, UK.

The need for the factories in No. 2 Shadow Group and Bristol's No. 3 Shadow Factory at Accrington to take precautions externally and internally against damage from air attack had a significant impact on the design and layout of these factories. The British Government's concern with air-raid protection focused on dispersal of factory buildings and production lines "to reduce to a minimum any interference with production by bomb hits."[63] The goal was to avoid disrupting production, and a key to this was to move away from the trend toward very large industrial buildings.[64] To limit possible damage, it was deemed preferable to have several smaller buildings rather than a single large structure. A direct hit on a single large building could destroy the entire roof removing protection for the machine tools inside, but smaller buildings could, to some extent, contain an explosion thereby preventing damage to adjacent buildings.[65]

The Government recommended that factory buildings should be spaced widely apart, with a minimum of 50 ft between buildings and should not be in parallel or in a regular or symmetrical form, although the factories in No. 2 Shadow Group did not follow this prescription. This need, in turn, had a significant effect on the ability of these factories to maximize the potential for flow production. To protect machine tools in the factories, the Government recommended that blast walls be built laterally within the factory buildings.[66] The Bristol Accrington factory and the Rootes Ryton factory both incorporated blast walls, and it is likely that the other shadow factories did as well. The blast walls at the Rootes factory were "fourteen inches thick and constructed at different heights according to the configuration of the equipment being protected."[67] Up until late 1943, the machine shops at the Bristol Accrington factory had extensive blast walls within the machining area for air-raid protection (see Figure 4.4). These were 1 ft 2 in. wide and 4 ft 6 in. high. The factories in No. 2 Shadow Group and the Bristol Accrington factory were a compromise between requirements for air-raid protection and production organization. Because of the risk of air attack, they were not and could not be built as large single buildings with wide open internal spaces ideally suited to flow production (Figure 4.5).

FIGURE 4.4 Protecting the shadow factories against damage from aerial bombing was a primary concern. This photograph of the Rootes No. 2 Shadow Factory shows what appear to be blast walls separating the different groups of machine tools. With no fear of air attack, American factories could arrange their machine tools for optimum productivity.

Rolls-Royce Heritage Trust-Bristol Branch.

FIGURE 4.5 This illustration shows the layout of the blast walls in the No. 4 Machine Shop at the Bristol No. 3 Shadow Factory at Accrington, taken from the original factory plans. The blast walls were apparently removed in late 1943 or possibly later. The factory management told the Ministry of Aircraft Production that the walls had "...always caused great inconvenience." Bristol Aeroplane Co. Ltd., Accrington-Engine Factory-Proposal to remove protective walls in machine shop, September 21, 1943, Air Supply Board Meeting No. 269, September 30, 1943, AVIA 10-200, NA. The Board deferred a decision out of concern that demands from other factories would require considerable construction labor, and a concern that workers would think their safety was being compromised.

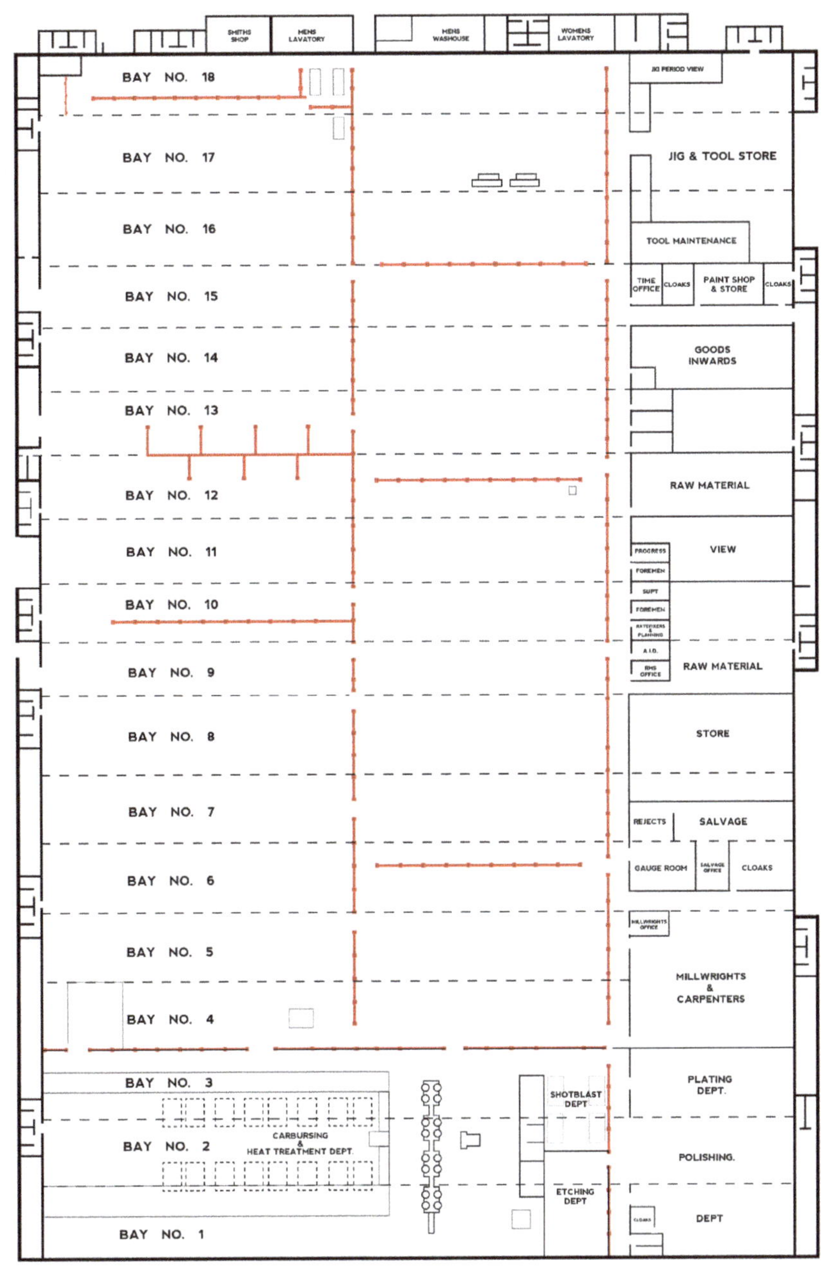

Derived from original factory plans in the possession of Michael Donnelly, Junction 7 Business Park, Clayton Le Moors, Accrington, Lancashire, BB5 5WJ, UK.

Although the factories of No. 2 Shadow Group would adopt elements of line production, the restrictions air-raid precautions imposed on the factories were not ideal for this method. At these factories, the flow of work was neither continuous nor direct. The location of blast walls within the machine shops would have interfered with the placement of machine tools and mechanical conveyor systems, restricting the flow of work. Finished parts had to move between buildings at most of the factories, from the machine shops to final assembly in separate buildings. Completed engines had to be taken from the assembly building to the engine test sheds and brought back to the assembly building for final inspection. All these steps added time to the production process, invariably slowing production from what might have been achieved in a larger, single-purpose machine and assembly building. This was not true flow production.

The Production Record

The Hercules engine was in production, on a small scale, before the war began, and continued in production until the war's end. Of the 60,672 Hercules engines built during the war, the No. 1 and No. 2 Group contributed 37,884 or 62%.[68] Bristol's No. 3 Shadow Factory built 14,379 Hercules engines, making the total for the combined shadow factories 52,243 engines, equal to 86% of the total Hercules production.[69] As the new factories ramped up Hercules production, the Bristol parent factory concentrated more on engine development, coming out with improved models of the Hercules and the more powerful 18-cylinder Centaurus sleeve-valve aircraft engine so that its contribution to Hercules production declined after 1942. The Air Ministry placed its first order for Hercules engines with Bristol in June 1938 with an order for 400 engines to be delivered by March 1940.[70] By October, the number of Hercules engines the Air Ministry required had multiplied, with the Ministry now seeing a need for as many as 7,000 Hercules engines to equip the planned order for 1,600 four-engine B.12/36 Short Stirling heavy bombers which would need 6,400 Hercules, half of which would be built at Bristol's aircraft engine works with the rest at the shadow factories.[71] In November 1938, the Air Ministry added two more aircraft to the list of requirements for the Hercules, the Saro Lerwick flying boat, and the Armstrong Whitworth Albemarle twin-engine General Reconnaissance Bomber.[72] As an example of the complexities of balancing aircraft and aircraft engine programs, of these first three applications for the Hercules engine, the Lerwick turned out to be a failure and was cancelled, the Albemarle was a disappointment in its chosen role, and its production substantially cut back, while the Short Stirling, whose performance, especially at altitude, also failed to live up to expectations, was not built in the numbers originally envisioned. Other applications for the Hercules emerged in time, saving the Air Ministry from having stocks of aircraft engines with no aircraft to put them in (Figures 4.6 and 4.7).

FIGURE 4.6 The British Air Ministry initially required Bristol Hercules engines in quantities for the Short Stirling, the Royal Air Force's first four-engine bomber shown in this photograph, and for the Saro Lerwick and Armstrong Whitworth Albemarle. Ironically, only the Stirling was built in meaningful numbers.

2008-3-31_image_599_01, Peter M. Bowers Collection, Museum of Flight, Seattle, WA.

FIGURE 4.7 Comparative Production of Complete Hercules Engines 1941–1945.

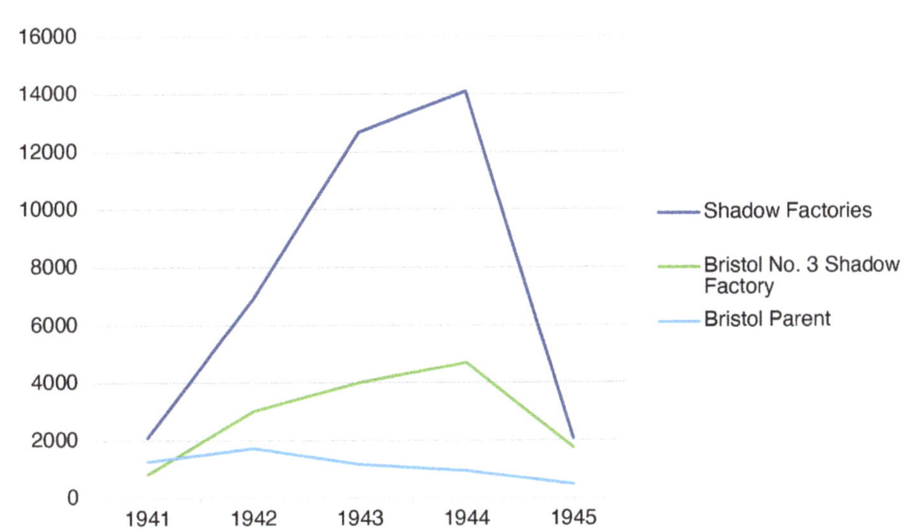

Source: Ministry of Aircraft Production Statistical Review 1939–1945; History of Production of Bristol Aircraft engines by the Shadow Industry, Abridged Record of Engines, Spares and Accessory Output from Shadow Industry 1941–1945; New Engines Delivered-Piston Engines, Rolls-Royce Heritage Trust-Bristol Branch.

Two factors created a sustained demand for the Hercules: first, the success of Bristol's twin engine, Hercules-powered Beaufighter; and second, a decision to expand the RAF's bomber force for the strategic air offensive against Germany. After the fall of France in June 1940, Prime Minister Winston Churchill wrote a minute stating his belief that the only means Britain had of defeating Germany was through "an absolutely devastating, exterminating attack by very heavy bombers from this country upon the Nazi homeland."[73] By the summer of 1941, the British aircraft industry had yet to produce bomber aircraft in the numbers Churchill believed were necessary for a strategic air offensive against Germany. On September 7, 1941, the Prime Minister requested a substantial expansion in the production of heavy and medium bombers.[74] He had been informed that to maintain a force of 4,000 bombers, the RAF would require 22,000 aircraft to be built between July 1941 and July 1943.[75] On the assumption that 5,500 aircraft could be obtained from America, that left 16,500 bombers to be built in Britain, while the forecast for British production amounted to only 11,000 bombers.[76] Churchill instructed the Ministry of Aircraft Production to arrange production of 14,500 bombers, less than required, but still 3,500 more than planned.[77] Building that many four-engine heavy bombers in a short space of time for what became the Bomber Programme would require an enormous expense for additional machine tools and materials as well as labor, all of which were in short supply. The Ministry of Aircraft Production realized that the best way to meet the Prime Minister's required number of bombers was to expand production of the twin-engine Vickers Wellington, at the time RAF Bomber Command's mainstay, and to push out the Prime Minister's time frame beyond July 1943 (Figure 4.8).[78]

FIGURE 4.8 The Bristol Hercules gave the Vickers Wellington bomber a new lease on life. Built in greater numbers than any other British bomber of World War II, the Wellington was in service in various roles until the end of the war. Out of the 11,461 Wellingtons completed during the war, the Hercules powered some 7,400.

Rolls-Royce Heritage Trust-Bristol Branch.

In 1938, the Air Ministry had instructed Vickers to develop a version of the Wellington with the Hercules III engine as the Wellington Mk III, but the difficulties with Hercules development and production delayed the Wellington Mk III's entry into service until 1941.[79] The Bomber Programme, and the Hercules engine, gave the Wellington a new lease on life. Under the Target Programme of July 1941, production of the Wellington was to begin tapering off in July 1942, declining to less than a hundred aircraft a month by January 1943, but under the Bomber Programme, Wellington production was planned to increase and expand to 300 aircraft a month by June 1943 to meet the Prime Minister's targets.[80] Production of the Wellington increased by nearly 50% from 1941 to 1942, and continued at more than 2,000 aircraft a year through 1944.[81] Ultimately, Vickers would build 1,517 Hercules-powered Wellington Mk III and 3,804 Wellington Mk X aircraft for Bomber Command and other overseas commands, and over 1,500 Hercules-powered Wellingtons for RAF Coastal Command.[82]

This created a demand for significantly more Hercules engines for the Wellington, the Short Stirling as its production expanded, a version of the AVRO Lancaster with Hercules engines, growing numbers of Bristol Beaufighters, and the Albemarle, now converted to serve as a glider tug. The Air Ministry's Engine Programme of January 1942 resulted in significant new orders for the Hercules engines from the shadow factories, with No. 1 and No. 2 Shadow Groups soon to be combined, and Bristol No. 3 Shadow Factory, with a planned combined output of 1,350 engines per month to be attained by August 1943.[83] The Engine Programme called for 5,000 sets of materials at No. 2 Shadow Group to be turned into completed Hercules engines as soon as possible, to be followed by an order for 8,000 more engines, while the Bristol No. 3 Shadow Factory was to receive an order for 4,000 Hercules engines, making a total of 17,000 engines on order.[84] This was in addition to 4,907 Hercules engines still to be delivered under previous contracts with Bristol No. 2 Shadow Group and Bristol No. 3 Shadow Factory (Figure 4.9).[85]

FIGURE 4.9 The Bristol Hercules also gave new life to the Handley Page Halifax. The Halifax with Hercules engines demonstrated improved performance over the earlier Rolls-Royce Merlin engine versions. During the war, Handley Page and its group of companies built over 3,000 Halifax bomber and transport aircraft with Hercules engines.

2008-3-31_image_600_01, Peter M. Bowers Collection, Museum of Flight, Seattle, WA.

A new source of demand for the Hercules came from the Air Ministry's decision to convert the Merlin-powered Handley Page Halifax four-engine bomber to the Hercules engine. In July 1941, the Ministry of Aircraft Production became concerned with a possible shortage of Rolls-Royce Merlin XX engines for the Halifax and the AVRO Lancaster then coming into production and requested Handley Page to fit a Halifax with the Hercules as a trial.[86] A prototype Halifax with the Hercules first flew in October 1942 and proved to have superior performance to the Merlin engine versions, which by this time had been found to be inferior to the Lancaster in terms of ceiling, bomb load, and range.[87] The Halifax and the Hercules proved to be a winning combination, giving an increase in power at takeoff compared to the Merlin, critical for heavily loaded bombers, higher maximum and cruising speeds, and an improved ceiling. Designated the Halifax Mk III, the first production version made its first flight in August 1943 and series production began soon after with a change in the engine to the Hercules XVI.[88] The Halifax Mk III was built in greater numbers than any other version of the Halifax, 2,091 being completed by the end of the war.[89]

Even better performance came from matching the Halifax with the final version of the Hercules built during the war, the Hercules 100. This version of the Hercules incorporated a major redesign of the supercharger, where the Hercules VI could produce 1,455 hp at 12,000 ft altitude and the Hercules 100 with the new supercharger design could produce 1,590 hp at 19,500 ft.[90] A Halifax with the Hercules 100 began flight tests in December 1943, with the first production model completed in October 1944 as the Halifax Mk VI.[91] With the Hercules 100, the Halifax gained an increase in maximum and cruising speeds at higher altitudes than the Hercules Mk III, and more importantly, 4,000 ft in service ceiling. Compared to the 2,000 lb bomb load the earlier Merlin-powered Halifax could carry to Berlin, the Halifax Mk VI could take 9,000 lb over the same distance.[92] It is perhaps too much to claim that the Hercules resurrected the Halifax from possible oblivion, but in giving the Halifax improved performance, the Hercules helped sustain the bombing offensive against Germany in the final year and a half of the war.

Throughout the war, the shadow factories, the Bristol No. 3 Shadow Factory and the Bristol parent factory, were continually introducing new models of the Hercules engine, responding to the RAF's requirement for ever greater engine power. The factories regularly brought a new model of an engine into production while simultaneously phasing out an older model, at times producing multiple models concurrently. In general, Bristol and the shadow factories managed this process well, sustaining production while converting to a new engine model with few disruptions.[93] The factories consistently met or came close to meeting their production targets.[94] Where there were delays, these were most often due to factors beyond their control, such as delays in the delivery of machine tools and shortages of labor (Figure 4.10).[95]

FIGURE 4.10 Production of Main Models of the Hercules 1941–1945.

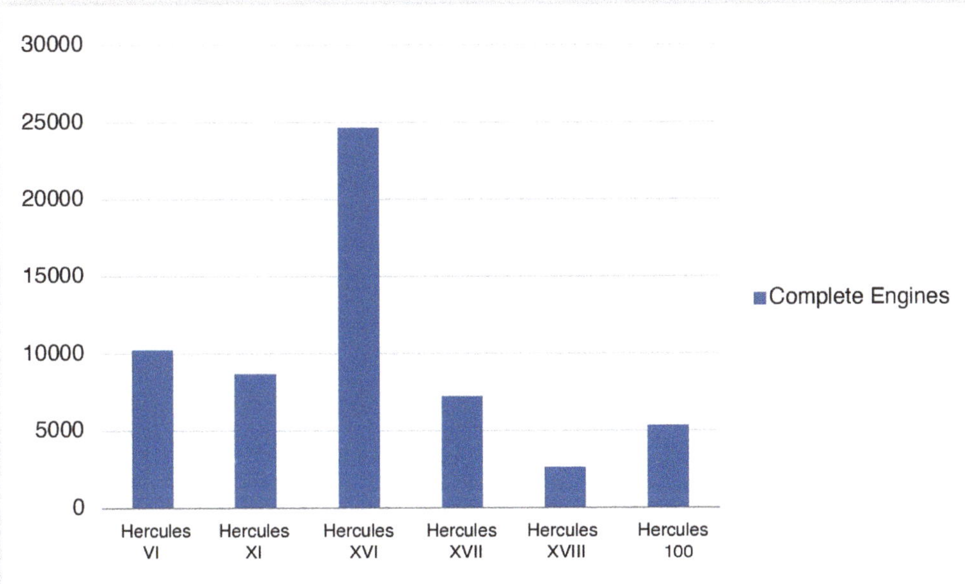

Source: Ministry of Aircraft Production Statistical Review 1939–1945, pp.;31-32.

Between 1941 and 1945, the shadow factories built six different models of the Hercules, while Bristol's No. 3 Shadow Factory built five models.[96] As an example, the Rover No. 2 Shadow Factory began manufacturing the Hercules XI in 1941, ending production in 1942 and shifting to the Hercules VI and Hercules XVI, ending production of the Hercules VI in early 1943 and shifting to the Hercules XVI, which the factory continued to produce until the end of the war. The factory started producing the Hercules XVII and XVIII at the end of 1943 and continued to produce these models until the beginning of 1945, having shifted its main effort to the Hercules 100 in 1944.[97] Transitioning from one engine model to another was not, as might be assumed, a simple matter as often component parts changed or required different operations sometimes requiring substantial re-tooling. In the case of production of the Hercules 100 at the Rover No. 2 Shadow Factory, the new model had seven completely new components to be manufactured and seven existing components that required radical modifications to their manufacture, the combination of new and modified components requiring 44 new machine tools (Figure 4.11).[98]

FIGURE 4.11 Production of Complete Hercules Engines at Rover No. 2 Shadow Factory 1941–1945.

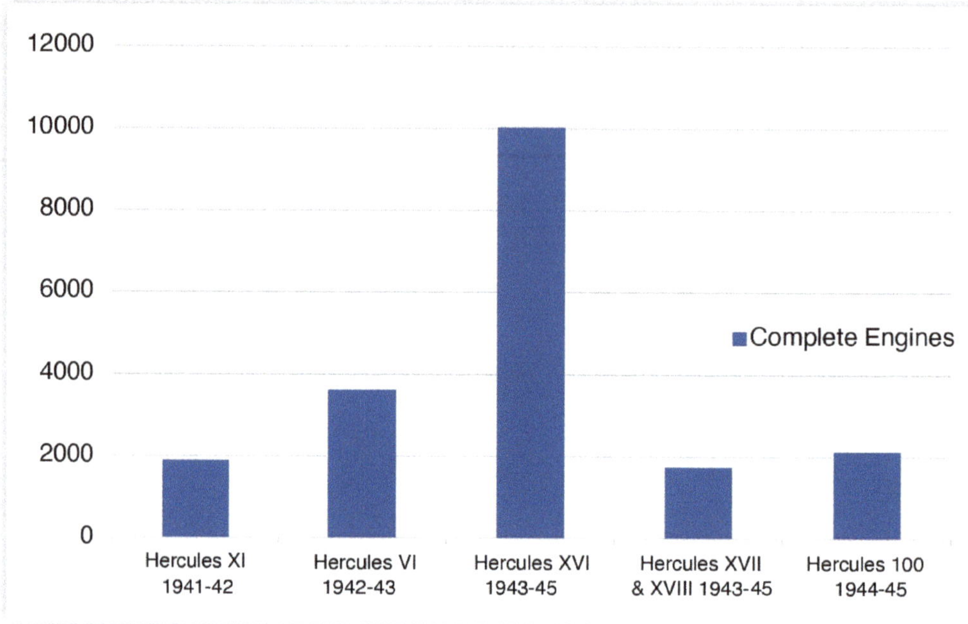

The Rover Company Limited (Aircraft engines) No. 2 Factory, Lode Lane, Solihull, Diary 1939–1945, chart of Hercules engine production. British Motor Museum Archives.

March 1944 marked the high point of Hercules production during World War II. In early 1944, the Ministry of Aircraft Production began to make reductions in the aircraft program which had an effect on the engine program. The new engine program of April 1944 called for a reversal in the decision to increase production at the shadow factories to 1,500 Hercules engines a month and at the Bristol No. 3 Shadow Factory to 425 a month to reduce planned production of all Bristol engines from 2,485 a month to 2,180 a month by January 1945.[99] More cuts in the Ministry of Aircraft Production's labor force led to further reductions in the aircraft program, particularly for Halifax bombers, leading to an amended engine program revised in November that called for further reductions in Bristol engines from all factories to 1,410 in January 1945 and down to 609 by July 1945, with output at the No. 1 and No. 2 Shadow Group factories planned to decline from 700 to 199 engines.[100] The shadow factories, the Bristol parent works, and Bristol Accrington factory began cutting back production over the summer of 1944, with production at the shadow factories dropping from 1,359 in June to 746 complete engines by December.[101]

In March 1945, with the war against Germany nearing its end, production of the Hercules at the shadow factories and at Accrington was scheduled to end in July, with the Bristol engine works at Filton taking over production on a much reduced basis, declining to 15 engines a month by early 1946.[102] Shutting down the factories took several months. The pattern at the Rover No. 2 Shadow Factory is likely to be representative of what was happening at the other shadow factories.[103] In early April 1945, officials from the Machine Tool Control office of the Ministry of Supply arrived to begin disposing of redundant machine tools, while on April 21, the factory held a "finishing up" party. At the end of May, the night shift in the machine shop ended, as did the training school, while on June 23, the machine shop ceased production. Work on the last Hercules 100 began on July 4, and the factory completed all its contracts for Hercules engines on August 31, 1945. Engine production at the Bristol No. 3 Shadow Factory ended in July, and to keep the remaining workers employed, the factory shifted to repairing Hercules engines until the war ended.

The record of Hercules aircraft engine production during World War II illustrates three important aspects common to both British and American aircraft engine manufacturing during the period: first, the critical importance of production at the automotive shadow factories; second, the challenge of managing Irving Holley's "dilemma of mass production," balancing quantity production with improvements in aircraft engine quality; and third, the equally important challenge of matching aircraft engine production programs with airframe production programs that were subject to changing operational and performance requirements.[104]

How It Was Done: Building the Hercules Engine

The Hercules was a complex piece of precision engineering with over 7,000 parts, more than double the number of parts in either the Mercury or the Pegasus, and took 4,457 man-hours to build compared to 3,662 man-hours for the Pegasus XVIII.[105] The Hercules had 14 cylinders arranged in two rows of seven and a two-speed supercharger.[106] The crankcase, made from forged aluminum alloy, was in four sections, a front cover, front section, center, and rear sections. The operating mechanism for the sleeve valves in each of the 14 cylinders was attached to the front section of the crankcase, with a series of gears driven by the crankshaft that drove 14 cranks that operated the sleeve in each cylinder. The supercharger assembly and other accessories attached to the rear section of the crankcase. As a double-row engine, the Hercules had two main journals for the two connecting rod assemblies, one for each bank of cylinders, each assembly having a master rod and six connecting rods made from forged steel alloy. The crankshaft was made of nickel–chrome Nitralloy steel, stamped to rough form, and then machined. The crankshaft drove the propeller through a series of reduction gears at the front of the crankcase and the impeller blades of the supercharger through gears attached to the rear section of the crankcase. Though there were more of them to manufacture, these components of the Hercules were basically the same as those in the Mercury and Pegasus and did not require more complex machining.[107] The Hercules cylinder and sleeve valve were not so straight-forward (Figure 4.12).

FIGURE 4.12 Manufacturing the sleeves for the Hercules was difficult. The sleeves had to be machined with exacting precision, as any distortions could cause seizure within the cylinder when the engine was operating. Bristol helped design this special machine tool to simultaneously lap the outside and the bore of the sleeve. One operator could supervise several machines.

History of Production of Bristol Aero Engines by the Shadow Industry Operating under the Joint Aero Engine Committee in Conjunction with the Bristol Aeroplane Company Limited, X003-8876/066, Royal Air Force Museum, Hendon.

The sleeve-valve cylinder barrel for the Hercules presented "many machining difficulties, particularly for line production."[108] The top and bottom thirds of the Hercules cylinder had concentric fins to aid cooling, but the middle third, termed the induction belt, had two exhaust ports and three inlet ports, with eccentric fins (i.e., not completely circular) in between.[109] The ports had a difficult profile that had to be cut to precise limits, while the cooling fins on the outside of the cylinder varied in depth adding to the difficulty of machining the cylinder barrel. Manufacturing the sleeve itself was even more challenging and resulted in "possibly more manufacturing problems than any other component."[110] To fit between the wall of the cylinder barrel and the piston, the sleeve had to be exceptionally thin, a little over 1/10 of an inch thick, machined to a tolerance of 0.0012 of an inch.[111] The thinness and hardness of the sleeve meant that distortions could occur after every machining operation.[112] These distortions could lead to seizure of the piston. Bristol spent considerable effort to find the correct material for the sleeve, finally determining that a sleeve made from austenitic nickel–chrome–tungsten steel, with nitriding, would work best.[113] In the fall of 1939, as No. 2 Shadow Group's new shadow factories were under construction, Bristol was still having problems manufacturing the Hercules.[114] It took Bristol several years to overcome the sleeve valve's production problems and establish true interchangeability.[115]

A description of manufacturing the Hercules engine begins with the cylinder. Manufacturing the cylinder barrel alone required over 100 separate machining operations.[116] The cylinder barrel was made from an aluminum alloy forging that weighed 54 lb at the start of machining and 22 lb at the end. One of the more difficult operations on the cylinder barrel was to machine the cooling cylinder fins. The Hercules cylinder had two banks of 20 concentric cooling fins on the top and bottom sections, and 19 eccentric cooling fins in the center section in between the inlet and exhaust ports. The special Maxi-cut lathe that Bristol helped to develop with the British Drummond Brothers machine tool company could cut the 40 concentric fins on the side of the cylinder barrel in one operation using 78 separate tungsten carbide cutting tools. Later in the war, a further development of this Maxi-cut machine allowed cutting either the concentric or the eccentric cooling fins, using up to 100 cutting tools in a completely automatic operation.[117] This 7-ton machine had the rigidity to enable thousands of cylinders to be cut using the same set of cutting tools. The cylinder head consisted of two light aluminum alloy die castings. A standard American Heald Borematic machine was specially adapted to automatically machine all the external surfaces of the cylinder head, the operator simply loading and unloading the workpieces onto the machine.[118]

Manufacturing the sleeve was the most difficult of all the machining operations for the Hercules engine.[119] The sleeve was a 14 in. long cylinder with a diameter of 6 and 1/8 in. and a thickness of 1/10 in., with a thicker eccentric base belt that housed a pocket for the ball joint that turned the sleeve in the cylinder. The Firth-Vickers Co. Ltd. in Sheffield provided Bristol and the shadow firms with nickel–chrome–manganese steel forged sleeves in rough form. The critical aspect of the machining process was to avoid any distortion of the sleeve which would prevent it from fitting in the cylinder barrel. Through trial and error, Bristol found ways of machining the sleeve in successive stages to reduce the risk of distortion, taking off only small amounts of excess material during each operation. British James Archdale and Co. boring machines performed a rough boring of the internal surfaces of the sleeve, while an American Cincinnati Milling centerless grinding machine finished the external surfaces. Later on in the war, a six-station American Bullard Multi-Au-Matic performed these operations, with four stations working at the same time while two were loading and unloading the workpieces.[120] Until Bristol developed a special-purpose machine for punching out the inlet and exhaust ports on the sleeve, the factories used a specially designed James Archdale Co. four-spindle profiling machine to machine the ports, working on two sleeves at a time. Grinding the base at the bottom of the sleeve took place on a special grinding machine designed to operate on seven sleeves simultaneously. After these operations, the sleeves went through a nitriding process for extra hardening, then final finishing. Strict quality control and inspection throughout the machining process ensured that the maximum ovality of the sleeve did not exceed 0.0025 in.[121]

Machining the remaining main components of the Hercules—crankcase, crankshaft, master rod and articulated rods, piston, supercharger and propeller shaft, and reduction gear—was more straightforward and followed normal practice for machining these parts, but here too Bristol and the shadow factories developed special-purpose machine tools to speed production. The Hercules crankcase was made up of four light alloy forgings, with front, center, and rear sections, and a front cover, which were bolted together to form the complete unit.[122] The front cover and the three main sections required a large hole to be bored through for the crankshaft, while the three main sections had 14 positions for attaching the cylinder barrels, with 12 holes drilled around each position to bolt the cylinder barrel to the crankcase. The front cover and center section required seven holes each for the cranks driving the sleeve-valve mechanisms. These repetitive operations provided an opportunity for machines that could perform multiple operations simultaneously or operate on several workpieces at the same time. Production engineers also broke down standard operations on the crankcase into multiple operations for unskilled workers.[123] Much of the initial finishing of the exterior of the crankcase sections took place on a British Alfred Herbert Company turret lathe equipped with multiple tools for drilling and grinding. A twin-head Archdale boring machine bored holes for the two rows of cylinder barrels using two boring tools, with the crankcase temporarily bolted together and mounted on a table, which the operator turned in sequence. To drill holes for the cranks for the sleeve-valve mechanisms, a Heald Borematic drilled all the required holes in a sequence of operations on the same machine without having to remove the crankcase.

As a double-row engine with two banks of cylinders, the Hercules required a two-throw crankshaft to attach the two master connecting rods for each bank of cylinders. The crankshaft was made of three sections, the front section, called the front web, which held a damping device and a splined shaft to connect with the propeller shaft, the center section, the central crank, and the rear web which also had an attached damping device.[124] The crankshaft sections were steel forgings that were machined to a fine finish. There appear to have been fewer opportunities to build special machine tools for multiple operations on crankshaft sections, though in a few operations two workpieces could be mounted on a single machine. The external profile of the master connecting rods, however, could be machined four at a time on the same milling machine, while another type of milling machine could perform operations on six articulated rods simultaneously.[125]

Bristol engineers developed several special-purpose machines to manufacture the supercharger for the Hercules. The volute casing (the section of the supercharger that received the fuel/air mixture after compression), an aluminum alloy casting, had a combination of boring and facing operations carried out on a single Herbert combination turret lathe, the different turret faces holding the necessary tools for each operation.[126] A single three-spindle Archdale drilling machine performed both drilling and tapping operations required on the volute casing.[127] Machining the impeller for the supercharger presented other production problems, particularly machining the complex shape of the pockets between the blades on the impeller which varied in length and depth.[128] Specially designed milling machines had to be developed to carry out this machining operation, two machines working in pairs with one machine working on the periphery of the forged alloy impeller stamping and the second machine working progressively inward to the center of the impeller. The operations on both machines were fully automatic and continuous, the workpiece attached to a special fixture. These special milling machines were modifications of another type of machine designed to mill the inlet and exhaust port contours on the sleeves and cylinders of the Hercules engine.[129]

Completed components flowed toward the assembly area where teams assembled complete Hercules engines and sent them on for rigorous testing. The assembly section had three departments, one to assemble the engine before testing, a second to strip down and inspect each engine after testing, and the third to reassemble the engines for shipment.[130] The engines moved from station to station and department to department on specially built cradles which moved along rail tracks set into the flooring. Final assemble was the closest approximation to the assembly lines familiar in automotive factories. To assemble a complete engine, the assembly team first bolted the crankcase onto the cradle which could be swiveled to the proper position for adding components or subassemblies. The cradle moved along the rail track past benches holding components and subassemblies in the proper sequence for assembly. Bristol's Shadow Industry office provided detailed instructions for assembling Hercules engines. The three sections of the crankshaft comprised a subassembly, and when assembled, the master connecting rod and articulated connecting rods were added and fitted into the crankcase. As the cradle moved along the rail track, it gained pistons, sleeves, cylinders, the many gears operating the sleeve valves, the complete supercharger assembly, and the main accessories (such as carburetors, magnetos, and oil pumps).

Every single one of the 60,672 Hercules engines built during the war underwent a thorough test before it left the factory, beginning with a 5-hour running in period during which an electric motor drove the engine to test the workings of all components together.[131] The engine then moved to a sound-proof testing station where workers inserted spark plugs and attached oil and gas lines for running the engine on its own power, first a period of running at low power then progressively increasing the throttles to run the engine at rated power for 2 hours. The engine was then run at full power for 5 minutes and then returned to the assembly shop where it was stripped down, and every part subjected to a minute inspection. The re-assembled engine was fitted to a dynamometer and run again for a set period to test that power output, fuel, and oil consumption were within set limits. Then, it was back to the test stand for more running at rated and peak power for more tests of output and fuel consumption. After this final test, the engine went back to the assembly shop for cleaning and preparation for packing and shipment. It is important to recall that, as with manufacture of the Mercury and Pegasus, inspection and quality control was continuous throughout production, from raw materials to completed engine.

How It Was Done: Setting Up Quantity Production

To produce the Hercules in the quantities the RAF needed required new methods of production. The Bristol Aeroplane Company used knowledge gained from other industries, especially from American aircraft engine manufacturers, and developed completely new production processes specifically for Hercules production. This involved an intensive effort at production engineering, setting up the factories for flow production, and process engineering, determining the manufacturing processes and sequences that would be required to build the Hercules engine and the types and numbers of machine tools that would be needed for quantity production.[132] New production processes most often required new types of machine tools, and in developing these tools, Bristol relied on the principle of the transfer of skills, from the skilled machinist to the semi- and unskilled machine operator by incorporating new jigs and fixtures for standard machine tools or by designing new types of machine tools with expanded capabilities.[133] To build the Hercules in quantity, Bristol would borrow production methods from the principal American aircraft engine manufacturers, Pratt & Whitney and Wright Aeronautical, and purchase many more machine tools from America's leading machine tool manufacturers.

In September 1939, Air Marshal Sir Wilfrid Freeman, Air Member for Research and Development at the Air Ministry, encouraged the Air Ministry to specifically request Bristol to explore "Americanising" their production practices, adopting American production methods and American machine tools where possible to attain "a standard of efficiency which would reduce the number of operations and labor requirements."[134] The Air Ministry wanted Bristol to determine if, for certain machining operations, American machine tool companies could develop one machine tool that could replace the operations of several others.[135] Bristol had already asked authorization to purchase up to 80% of the machine tools needed for the Accrington factory in America given the large backlog and long delivery dates in the British machine tool industry.[136] The demands for new plant from the factories for No. 2 Shadow Group, as well as from other sectors of the aircraft and munitions industries, had placed huge demands on the British machine tool industry to the extent that deliveries of certain types of machine tools, particularly grinding machines, would take over a year to deliver in quantity.[137] The Air Ministry agreed to Bristol's request and arranged an authorization from the Treasury for Bristol to purchase machine tools in America, with an initial limit of £1.25 million ($5 million).[138] Bristol agreed to send a production engineer to the US to investigate American methods and purchase American machine tools, selecting Frank Whitehead, Bristol's Machine Shop Superintendent, for the job.[139] The Air Council Committee on Supply met on October 3, 1939 to confirm that the Accrington factory would employ American production methods and the funds allocated to purchase American machine tools.[140]

What is important to consider are the American production methods Whitehead observed, the advances in production and process engineering the American aircraft engine manufacturers were implementing, as well as the types of American machine tools he purchased. He left England for America in early October 1939, spent four months visiting manufacturers, and returned to England in March 1940, having placed firm orders for 775 machine tools and conditional orders for an additional 137 machine tools from the leading American machine tool manufacturers, as well as jigs and fixtures.[141] Whitehead apparently wrote a full report on his visit to America, but unfortunately this report has yet to be uncovered. Whitehead visited the Pratt & Whitney aircraft engine factory in Hartford, Connecticut, and likely spent time at the Wright Aeronautical factory in Paterson, New Jersey, given Bristol's long-standing contacts with Wright, and probably visited several large machine tool companies.[142] He had visited the Pratt & Whitney factory before in 1935 and so was likely familiar with Pratt & Whitney's production methods.[143]

The factory layout and production methods at the Pratt & Whitney Hartford factory provided the model for Bristol Accrington and probably the layout of the factories in No. 2 Shadow Group, which depended on Bristol for guidance on production methods for the Hercules.[144] The advances in production engineering at the Pratt & Whitney factories were based on the flow production concept. The Hartford factory was a large single-story rectangular building 400 ft wide and 1,000 ft long, giving a total space of 400,000 ft^2.[145] The company divided this space into separate units, called manufacturing departments, each responsible for a separate component of its air-cooled radial engines, with a further division lengthwise between ferrous and non-ferrous parts. The ferrous side of the factory produced the cylinder barrel, connecting rods, crankshaft, and gears, while the non-ferrous side made the aluminum cylinder head, piston, crankcase, and miscellaneous parts.[146] Each manufacturing department made a complete component with machining operations set up on the basis of line production, with raw materials entering the department area at one side and the completed component exiting the opposite side after a thorough inspection, to be carried on trucks to the final assembly section.[147] The Pratt & Whitney factory speeded up production by arranging for the continuous forward flow of parts through successive machining operations, minimizing the time between operations, and ensuring that component parts did not need to be removed to storage between operations.[148]

During the time that Whitehead was in America and visiting Pratt & Whitney, the company completed a new 274,000-ft^2 factory building next to its original plant exclusively to build its 14-cylinder double-row Twin Wasp engine (similar but slightly smaller than the Hercules).[149] Also designed for flow production, the new building had separate production lines for different components with production flowing toward a center aisle for transfer to final assembly. There were three new elements to production that Whitehead may have observed or discussed with Pratt & Whitney factory management. First, Pratt & Whitney had conducted a detailed time study of the production of every component of an aircraft engine, adding up all the separate operations to determine the number of hours required, dividing this total by the number of working hours in a month to determine the number of parts a machine tool could produce.[150] Calculating this for all the machine tools in the factory, Pratt & Whitney could determine the number of engines it could produce per month, and if it needed to increase output, it could calculate the number of additional machine tools needed. Secondly, with an increasing number of engines on order justifying the economics, Pratt & Whitney had started developing specialized high-production machine tools to speed up production, designing one machine that could do the work of two or three standard machine tools in its older plant.[151] Thirdly, Pratt & Whitney was facing a growing scarcity of skilled machinists and was working with trade schools in the Hartford area to develop training programs for machinists and its own training programs to increase the number of machinists and tool makers, in addition to a program to train junior executives.[152]

The Bristol Accrington factory followed the Pratt & Whitney model of separate units for component production and the concept of flow production, though the factory's layout was less than optimal. Instead of one large factory building with Pratt & Whitney's straight-through production, the Accrington factory had, as noted, four separate buildings consisting of three machine shops and one assembly building for aircraft engine assembly and stripping and rebuilding after engine testing.[153] Bristol did not divide production by type of material as Pratt & Whitney had done, but did have each of the three machine shops manufacture complete components, with No. 2 Machine Shop producing crankcases and superchargers, No. 3 Machine Shop making all the cylinder barrel components, including cylinder sleeves and cylinder heads, and crankshafts and No. 4 Machine Shop building all connecting rods and all gears for operating the sleeve valves and propeller reduction gears.[154] Within each machine shop, the operations and machine tools were organized according to the principles of flow production, from raw material to final inspection before being transported to No. 1 Fitting Shop for assembly and testing, though as previously mentioned, the internal blast walls within the machine shops were an inconvenience and no doubt obstructed the uninterrupted flow of materials and parts (Figure 4.13).[155]

FIGURE 4.13 The shadow factories arranged their machine tools according to the principles of flow production as shown in this photograph of the Rootes No. 2 Shadow Factory where machines working on Hercules crankcases are grouped together. However, note how the workpieces are still stacked on the floor alongside the machine tools waiting for workers to move them to the next stage of production.

The factories in No. 2 Group also adopted the flow production concept to increase output, arranging machining operations in accordance with the line production concept, placing machine tools in the logical sequence to insure flow. In some cases, the factories took the methods and layouts that Bristol recommended through the Bristol Shadow Industry office and in other cases developing their own methods based on their experience with line production in the automotive factories.[156] Rootes Securities, for example, had also gained valuable experience in production engineering and flow production of complex machines at its shadow factory building the Bristol Blenheim aircraft.[157] This experience may well have helped Rootes in production of the Hercules. As Bristol had done with the first shadow factories, the Bristol Shadow Industry office provided detailed information to the No. 2 Group factories on building design and layout, machine tool balance, jig, fixture and tool requirements as well as detailed instructions and drawings on how to manufacture, inspect, and test components.[158] Some of the factories put in conveyor systems in parts of their production lines, but generally the factories did not use conveyors to the same extent as in the automotive factories, using instead specially designed trolleys that could protect components as they moved from machine to machine.[159] As an example of flow production, one of the shadow factories set up manufacture of the three sections of the Hercules crankcase into three separate lines with production flowing toward final assembly so that the three crankcase sections all arrived at the assembly area completely machined and prepared for assembly.[160] The layout of the Standard Motor No. 2 Shadow Factory at Banner Lane earned praise for its efficient production, helping the factory to regularly exceed its production targets and no doubt incorporating Standard's own experience in automobile production to Bristol's recommendations for Hercules production.[161] Again, the presence of blast walls within the machine shops, however, may have restricted the flow of parts.

The Tools

Whitehead's selection of American machine tools was even more important to the quantity production of Hercules engines. After reading a report from Whitehead on his progress in America, Norman Rowbotham wrote to the Air Ministry commenting that "the acquisition of plant from that country will permit a more rapid start of engine production at the Accrington factory, and will also allow it to operate with workpeople of lesser skill and numerically smaller than plant based on the processes which we are operating at Filton."[162] The machine tools that Whitehead acquired were a combination of machines designed for high levels of output and special-purpose machines that were less common in the British machine tool industry. With a much smaller automotive industry, British machine tool manufacturers had less incentive to develop the specialized and automatic machine tools required for large-scale production of automobiles.[163] Instead, British firms concentrated on building standard, universal-type machine tools, relying on imports from America and Germany for specialized machine tools.[164]

During the 1930s, the American machine tool industry developed new, improved models that offered superior performance.[165] Stronger, more rigidly built machines allowed faster cutting speeds and cemented carbide and tungsten carbide cutting tools that enabled cuts to far higher tolerances of ten thousandths of an inch. These newer cutting tools enabled machine tools to cope with the new high strength steel alloys required for high-powered aircraft engines. Precision boring machines, internal and centerless grinding machines, new types of milling machines, and surface broaching machines developed for the automotive industry produced much finer surfaces with less risk of flaws that could lead to fatigue failures. To reduce the required number of machine tools and operators, several machine tool companies developed machine tools with multiple spindles, drills, or other tools that could perform, automatically, multiple sequential operations such as drilling, boring, reaming, and tapping on a single workpiece or multiple workpieces.[166] Thus, one operator using one "automatic" machine could replace several operators using multiple machines, while improved electric motors allowed easier and more precise control. In the automotive industry, these new machine tools enabled, in some cases, a doubling of production in the same amount of time.[167]

The critical problem facing Bristol Accrington and the No. 2 Group factories was the growing shortage of skilled labor to operate machine tools. As early as May 1938, in a meeting with Air Ministry to discuss expanding Bristol's aircraft engine production, Norman Rowbotham, Bristol's head of production, had warned that while acquiring the necessary buildings, machine tools and materials would not be without difficulty, the more serious problem was the shortage of skilled labor.[168] During the rest of 1938 and into 1939, Bristol recruited more labor and maintained a high level of skilled workers, some 44% of the Bristol Engine Department being classed as skilled, but by December 1939, Rowbotham was reporting to the Bristol directors that labor shortages were holding up production.[169] The aircraft industry had expanded dramatically, growing from 40,000 employees in 1934–1936 to 370,000 by the end of 1939, but the future need was even greater.[170] In discussions with the Ministry of Labor in November 1939, the Air Ministry estimated that to fulfill the Cabinet's plan for production of 2,500 aircraft a month by July 1942, the aircraft industry would need 1,500,000 workers just as the services and other war industries were clamoring for skilled workers.[171] Supplying skilled workers for the aircraft and other munitions industries became the principal concern of planners in Government ministries (Figure 4.14).[172]

FIGURE 4.14 An example of a newly developed machine tool that incorporated the principle of deskilling with faster operation. This machine tool could drill 14 holes in the Hercules crankcase simultaneously, replacing several individual machine tools.

History of Production of Bristol Aero Engines by the Shadow Industry Operating under the Joint Aero Engine Committee In Conjunction with the Bristol Aeroplane Company Limited, X003-8876/066, Royal Air Force Museum, Hendon.

The shortage of skilled labor in Britain during the war required an intensive focus on process engineering, developing the manufacturing processes and machine tools that would accommodate the influx of less-skilled labor into the factories. Throughout the war, shortages of skilled workers, particularly toolmakers, setters for machine tools, and machinists remained chronic.[173] By early 1940, it had become apparent that the solution to the shortage of labor was dilution, the progressive deskilling of manufacturing processes through process engineering whereby the work of a skilled machinist was broken down into simpler operations that a semi- or unskilled worker could perform on standard machine tools with special jigs and fixtures and by the increasing use of machine tools that could perform multiple operations that substituted for skilled workers but that unskilled workers could operate with a modicum of training.[174] Even before the start of the war, the Air Ministry had determined that production of the Hercules would require new methods and new types of machine tools to cope with the pending shortage of skilled labor. The existing literature provides little information on how this worked in practice.

The composition of machine tools at the shadow factories changed during the war to accommodate the needs of greater production and the steady reduction in skill levels in the work force. The Bristol aircraft engine department had begun using some automatic machine tools as early as 1933, but it appears that wartime demand accelerated their use in production.[175] The productivity gains of using so-called automatics were dramatic. During the war, the British Machine Tool Control published a pamphlet demonstrating the benefits of using automatic machines that contrasted three methods of producing a single component (Table 4.2).

TABLE 4.2 Productivity gains from automatic machine tools.

	Method 1	Method 2	Method 3
Machines required	23 Centre Lathes & 10 Drilling Machines	12 Capstan Lathes	4 6-Spindle Automatic Chucking Machines
No. of operations	5	2	2 (two machines per operation)
Total operators & setters required	76 per day of 2 shifts	30 per day of 2 shifts	10 per day of 2 shifts

Source: *Intelligent Production Planning*, M.T.C. Leaflet No. 4, AVIA 22/716, National Archives, Kew, UK.

Automatic machine tools required less space, fewer operatives, and were more productive. Employing more automatic machine tools and more unskilled labor freed up the more skilled workers for other tasks better suited to their skills.

While in America, Frank Whitehead purchased multi-spindle drilling machines from Baush Machine Co., automatic vertical, multi-spindle Mult-Au-Matic lathes; boring machines from the Bullard Machine Co.; grinding machines from Brown & Sharpe and Cincinnati Milling Machine Co.; gear shaping machines from Fellows Gear Shaper Co.; turret lathes from Gisholt Machine Co.; many internal grinding machines from the Heald Machine Co.; cylindrical grinders from the Norton Co.; and turret lathes from Warner & Swasey.[176] As well as buying these machines from American machine tool manufacturers, Bristol later sent its own representatives to America to work with American machine tool companies to design machine tools, jigs, and fixtures to save time and labor in manufacturing the Hercules.[177] Whitehead ordered the machine tools of the generation that were available in late 1939, or would be available during 1940. He purchased 772 American machine tools at a cost of £1,6 million ($6.2 million) and placed orders for an additional 137 machine tools costing £349,912 ($1.4 million), exceeding the Treasury's initial allocation of £1.25 million ($5 million).[178] Most of the machine tools he ordered ranged in price from £1,250 to £3,330 ($4,000 to $13,320), with the most expensive being the Bullard Mult-Au-Matics at £5,250 ($21,000) apiece. As will be seen in Chapters 5 and 6, the more advanced machine tools that were introduced in American aircraft engine factories later in the war would be even more expensive.

Many of these machine tools featured automatic operation with multiple tools to perform several operations simultaneously with minimal setting up. In one case, Bristol worked with the Bausch Machine Tool Co. in America, a builder of multi-spindle drilling machines, to develop a machine to simultaneously drill 14 holes in the Hercules crankcase.[179] Working with the English Drummand Company, Bristol engineers built a Drummand Maxi-cut lathe to machine the external surfaces of the Hercules cylinder barrel simultaneously, while another machine, designed to cut out and profile the exhaust ports on the side of the cylinder, could perform this operation on four cylinders at the same time. Perhaps most impressive, Bristol production engineers designed a special-purpose machine to punch out the ports in the cylinder sleeve without distortion in 6 minutes.[180] The previous milling and finishing of these ports had taken 2½ hours. All these small incremental improvements, making possible multiple, simultaneous machining operations, machining multiple workpieces at the same time, reducing the need for complicated setting up of a machine tool, and cutting the time for machining operations helped speed up manufacture of Hercules components while reducing the need for additional labor. Faster machining, fewer machining operations, and less movement of parts between machine tools contributed to higher output of components per day.

While Bristol did take advantage of special-purpose, high-production machines, especially where the design of a component was less likely to change over time, a key aspect of Bristol's general policy on machine tools, which the shadow factories apparently followed, was its reliance on as many standard machine tools as possible to cope with changes in design and transition to new models of the Hercules.[181] The machine tools Bristol specified for the shadow factories comprised standard machine tools, standard tools with adaptations for less skilled workers, and some specially designed single-purpose machine tools.[182] The bulk of the machine tools used in the shadow factories and at Accrington were, apparently, standard machine tools with a mix of more specialized American machine tools.[183] While the total number of machine tools at the Accrington factory is unknown, based on Bristol's original estimate of a requirement for 2,100 machine tools to build 250 aircraft engines a month, later increased to 400 engines a month, it is possible that the number of American machine tools amounted to one-quarter to one-third of the total in the factory (by comparison, the Rover No. 2 Shadow Factory ordered 1,387 machine tools to make 50 engine sets per week, later increased to 200 per week implying an even greater number of machine tools).[184] Whereas Whitehead would have ordered American machine tools that were on offer in 1939, it also appears from descriptions in contemporary articles describing Hercules production that later American machine tools imported for Bristol and the shadow factories could be classed as part of a second generation of machine tools developed after 1939 in response to demands for higher levels of output. These same articles indicate that the factories do not appear to have used the more advanced, third-generation, high-production machine tools that could be found in the American factories building Wright and Pratt & Whitney engines later in the war.[185] In contrast to Bristol, Wright Aeronautical would rely more on replacing standard machine tools with new types of special-purpose, high-production machine tools (Figure 4.15).

FIGURE 4.15 Two key factors in expanding aircraft engine production were employing women and importing American machine tools. This photo shows a woman at one of the aircraft engine factories milling an aircraft engine piston using a vertical milling machine made by the Cincinnati Milling Machine Company in Cincinnati, Ohio.

Image taken from Women in Engineering, Ministry of Labour and National Service, copy in the archives of the Modern Records Centre, University of Warwick.

With the shortage of labor at all skill levels, simplifying manufacturing operations through process engineering was even more critical to quantity production of the Hercules yet receives barely a mention in the literature. By the time the Hercules went into production at the shadow factories, almost all the skilled workers in the engineering trades were either fully employed in the munitions industries or serving in the armed forces. The shadow factories had no option but to hire women and trainee labor, ending up with around 80% female and trainee workers.[186] By June 1942, Bristol Accrington had approximately 7,000 employees, of which 2,250, or 32%, were women.[187] At the shadow factories producing the Hercules, the average for women employed reached 34% by the middle of the war.[188] Less than 20% of the workers were classed as skilled, with Accrington having the smallest percentage of skilled workers of any engine factory in England with 13.5%.[189] The shadow factories could, and did, draw on skilled workers from No. 1 Shadow Group. These became foremen supervising teams of semi- and unskilled workers, setters who set up machine tools for less experienced machine operators, and inspectors for quality control and training staff for new employees, particularly women.[190] As several observers noted, with an acute shortage of skilled labor, it was vital to the production effort that skilled workers be used effectively.[191] It made little sense to have a skilled machinist doing work that a semi- or unskilled worker could do with proper training (Table 4.3).

TABLE 4.3 Composition of the work force at the shadow factories 1941–1942.

	Total work force 8/41	% Women in the work force	Total work force 3/42	% Women in the work force
Bristol Accrington	5,887	25%	6,745	32%
Daimler No. 2	4,275	27%	5,033	28%
Rootes No. 2	4,173	20%	5,090	27%
Rover No. 2	4,182	16%	4,690	27%
Standard No. 2	3,605	26%	5,618	32%

Source: Table of Wage Earners at Engine Factories on March 7, 1942, AVIA 10/383, National Archives, Kew, UK.

Most of the women coming to work at the shadow factories had no experience of machine tools and needed to go through a one-to-four-week training course where they were given an introduction into factory production, basic principles and mathematics of engineering, and practice on a few basic machines.[192] The factory management assigned women to different departments based on their aptitude and suitability. By the middle of the war, women workers could be found in most areas of aircraft engine production, from assembling complete engines to engine testing to inspecting parts throughout the production process, and above all operating machine tools. Women operators worked on all classes of machine tools and machine operations including boring, standard and turret lathe work, drilling, grinding, gear cutting, honing, and milling.[193] While some women advanced to higher levels of skill and more complicated machine tools, many worked on automatic machines, special-purpose machines, or standard machine tools equipped with special jigs and fixtures developed through process engineering.

The process of deskilling appears in the literature, but rarely with an explanation of what this meant in practice. Deskilling had two components, simplification of processes and specially adapted machine tools. Simplification involved the principle of "transfer of skills," getting the machine tool designer, the process engineer, the machine tool builder, and skilled machinists all working together to break down a complex machining process into simpler steps that unskilled workers could perform.[194] The basic idea was to replicate the accuracy of the skilled machinist on a standard machine tool by designing a fixture to hold the workpiece in a machine and adjusting the cutting tools so that "all that remains to be done by the operator is to insert the work in the fixture, start the machine, and then remove the work after the machining operation is completed," and in this way, "even an unskilled operator can produce accurate work."[195] This involved the extensive use of jigs and fixtures, but greatly speeded up production. As one example of process engineering at the shadow factories, production of master connecting rods was split up into 120 separate operations that unskilled or semi-skilled operators could perform.[196]

Subcontracting was a vital part of Hercules production. None of the No. 2 Shadow Group factories nor Bristol Accrington had foundries for casting aluminum or magnesium parts, or forges for stamping key components such as cylinder barrels, crankshafts, and crankcases.

Before the war, Bristol relied on specialty firms for many components of the Mercury and Pegasus engines, and this practice continued after the start of the war. Bristol selected specialty suppliers for the shadow factories, contracting with High Duty Alloys Ltd., for example, to supply forged crankcases for the Hercules. When its existing suppliers could not meet the needs of greatly expanded production, Bristol sought out new sources of supply in Britain and America, sending several Bristol representatives to work with the American suppliers.

The Role of No. 1 Shadow Group

The Air Ministry's decision to switch No. 1 Shadow Group to Hercules production as demand for the Mercury and Pegasus wound down led to converting the No. 1 Group shadow factories into component manufacturers for the factories in No. 2 Group. In effect, the two shadow groups merged into a combined manufacturing group for Hercules production, with a peak employment of 41,707 workers in November 1943, a little over half as big as the combined employment at the factories building the Rolls-Royce Merlin. The shift to Hercules production required a major investment in new machine tools for the No. 1 Group factories to build Hercules components. The Ministry of Aircraft Production's initial target, set in early 1942, to have the No. 1 Group factories build the equivalent of 500 Hercules engines a month, required an additional 2,400 machine tools costing £3.4 million ($13.6 million), with £263,930 ($1.05 million) for certain additions to buildings, though it is unclear which factories received these additions.[197] Later, when the Ministry increased the combined production target for No. 1 and No. 2 Groups to 1,500 engines a month, the Government financed an additional 393 machine tools at £517,917 and further additions to buildings and services, plant, and other equipment at the factories amounting to £404,944.[198]

The Rover No. 1 Shadow Factory gives an example of the transition to building Hercules components at the No. 1 Group factories. The transition took around a year to complete. During July 1942, the Rover No. 2 Shadow Factory transferred production of the Hercules piston to the Rover No. 1 Shadow Factory.[199] Over the next year, the transfer process continued, covering numerous small parts, sleeve-valve drive gears, and, more importantly, work on the cylinder sleeves.[200] As part of the shift, the Rover No. 2 Factory also transferred certain machine tools, such as internal centerless grinders for work on the cylinder sleeves, to the Rover No. 1 Factory. It is unclear from the remaining records how the Daimler and Rootes No. 1 Shadow Factories coordinated with their counterparts in No. 2 Shadow Group, but the assumption is that they, too, made component parts for their larger partner factories, probably in line with the type of parts they had been building. Although it had been intended to assembly Hercules engines at the Austin Shadow Factory from parts made at the No. 1 Shadow Group factories, it was determined that it was more efficient to concentrate final assembly and testing at the Rover and Standard Motor No. 2 Shadow Factories, with the Bristol No. 2 Shadow Factory as a reserve.[201]

Conclusion

The success of quantity production of the Hercules depended on the success of production engineering and process engineering, the ability to sequence operations and layout machine tools to establish flow production and the ability to deskill the manufacture of the Hercules through simplification and the adaptation of machine tools to semi- and unskilled workers. The vastly increased demand for aircraft engines from the Air Ministry created the conditions, incentives, and financing necessary for production and process engineers to develop production techniques to "achieve outputs hitherto deemed impossible."[202] It is unfortunate that the official histories of wartime production make so little mention of the importance of production and process engineering, nor the British automotive, aircraft engine, and machine tool engineers who were responsible for this success. It is doubtful that this record of production could have been achieved with the standard methods Bristol employed in the prewar years; there were simply not enough skilled machinists or machine tools in Britain to have replicated these methods on such a large scale. What is remarkable about this record is that it was so successful, and that success came quickly, came in time, and was responsive to the demands for quantity production and continuing qualitative improvement in the Hercules engine.

At the beginning of the war in 1939, Bristol and Rolls-Royce had been building aircraft engines in the 800–1,000 hp class, principally the Bristol Mercury and Pegasus, and the Rolls-Royce Merlin. By the end of the war, the companies were building aircraft engines with, in some cases, more than twice the power of the engines of 1939. Bristol was building the Centaurus engine which produced 2,500 hp and Napier had improved the Sabre engine and boosted its power output to 2,400 hp, while Rolls-Royce was building the Griffon with 1,815 hp.[203] There had also been continuous improvement in the performance of the Bristol Hercules and the Rolls-Royce Merlin. Bristol had taken the Hercules from 1,375 hp in 1939 to 1,725 hp by 1943, while Rolls-Royce had boosted the Merlin from 1,030 hp in 1939 to 1,635 hp by 1945. The British aircraft engine industry succeeded in meeting the challenge of providing both quantity and quality.

The Hercules and the Merlin together accounted for 63% of total British aircraft engine production during World War II. There were more Merlin engines built than Hercules engines, with over 105,000 built in Britain and an additional 54,000 built in America under license, more than double Hercules production. The quality of the two engines is difficult to compare. Much depends on the particular engine model, the power output, and particularly the power output at altitude. As engines for bombers, the Hercules and the Merlin were interchangeable. Both types of engines powered models of the Halifax, Lancaster, and Wellington bombers. The greater demand for the Merlin reflected the Royal Air Force's preference for liquid-cooled, in-line engines for its single-engine fighters, but also the Merlin's superior performance at altitude. The Hercules achieved its maximum power with full supercharging at medium altitudes, more than the Merlin at the same altitude, but over 20,000 ft the Merlin could produce more power than most models of the Hercules, enough to give the Merlin-powered Lancaster bomber a higher service ceiling than the Hercules-powered Halifax. Still, during the war, the Hercules powered 48% of British medium and heavy bombers, a vital contribution to the British war effort.

References

1 *Flight*, (December 2, 1937).

2 Judge, *Aircraft Engines*, Vol. 2, pp. 151-54.

3 Hassell, Patrick: "The Bristol Sleeve Valve Aircraft engines", in Starr, Fred, Edward L. Marshall, and Bryan Lawton: *The Piston Engine Revolution: Papers from a Conference on the History of the Reciprocating Internal Combustion Engines Held at the Museum of Science and Industry, Manchester, 14-17 April 2011*, (Manchester, UK, 2012), p. 120.

4 Marshal, E.L.: "A Lifelong Love Affair—Sir Harry Ricardo and the Sleeve Valve", *The International Journal for the History of Engineering and Technology*, Vol. 83, No. 1, (January 2013), p. 73.

5 Hassel, "The Bristol Sleeve Valve Aircraft Engines", p. 116.

6 Heaven, *Bristol Engines Since 1914*, pp. HIS5/3-4.

7 "The 'Bristol' Perseus Sleeve Valve Engine", *The "Bristol" Review*, Engine Issue No. 7, (July 1934), p. 18.

8 Heaven, *Bristol Engines Since 1914*, p. HIS7/1; Bristol Aeroplane Company Minutes of Directors' Meeting in Committee, 15 October 1935, BAC Directors' Committees Minutes, Vol. 5, 3 January 1935 to 22 December 1936.

9 "The New Double-Bank Sleeve-Valve Engine", *The "Bristol" Review*, Engine Issue No. 12, (June 1937), p. 15.

10 Bristol Aeroplane Company Minutes of Directors' Meeting in Committee, 26 February 1935, BAC Directors' Committees Minutes, Vol. 5, 3 January 1935 to 22 December 1936.
11 "Power at the Salon", *Flight*, (November 26, 1936), p. 579.
12 "Engines at Hatfield", *Flight*, (June 24, 1937), pp. 642-43.
13 Sinnott, Colin: *The Royal Air Force and Aircraft Design 1923-1939: Air Staff Operational Requirements*, (London, 2001), p. 157.
14 Sinnott, *The Royal Air Force and Aircraft Design 1923-1939: Air Staff Operational Requirements*, pp. 157-58; Postan, M.M., D. Hay, and J.D. Scott: *Design and Development of Weapons: Studies in Government and Industrial Organization*, (London, 1964), pp. 78-79.
15 Ibid., pp. 162-65; Bowyer, Michael J.F.: *The Stirling Bomber*, (London, 1980), p. 21.
16 Bowyer, *The Short Stirling*, p. 24.
17 Postan, Hay and Scott: *Design and Development of Weapons: Studies in Government and Industrial Organization*, p. 125.
18 Ibid., p. 125.
19 Andrews, C.F. and E.B. Morgan: *Vickers Aircraft Since 1908*, (London, 1988), pp. 326, 332; Tapper, Oliver: *Armstrong Whitworth Aircraft Since 1913*, (London, 1988), pp. 276-77.
20 Notes of a Meeting Held at the Austin Shadow Factory at Birmingham on the 10 June 1938, AVIA 10/152, NA.
21 Notes of a Meeting Held at the Austin Shadow Factory at Birmingham on the 10 June 1938, p. 5, AVIA 10/152, NA.
22 Bristol Aeroplane Company Minutes of Directors' Meeting in Committee, 21 June 1938, BAC Directors' Committees Minutes, Vol. 6.
23 Notes of an Informal meeting of the Air Council Committee on Supply held in A.M.D.P.'s room on Thursday, 6th October 1938, AVIA 10/155, NA.
24 Ibid.
25 Air Council Committee on Supply, Statement of Engine Requirements, 24 November 1938, AVIA 10/156, NA.
26 Minutes of the Meeting of the Committee held at the offices of the Standard Motor Co., Ltd., Fletchampstead, Coventry, on Friday, 7th October 1938, Air 2/2577, NA.
27 Ibid., p. 3.
28 Ibid.
29 "Narrative: Reciprocating Aircraft engines and Engine Accessories Production & Programmes", Paragraph 38, Air 46/159, NA.
30 160th Progress Meeting, War Potential-Aircraft engines (Memorandum by A.M.D.P.), 16 March 1939, Air 6/56, NA.
31 Ibid.
32 Memorandum by Air Member for Development and Production, July 1939, pp. 2-3, AVIA 10/230, NA.
33 Bristol Aeroplane Company Minutes of Directors' Meeting in Committee, 14 March 1939, BAC Directors' Committees Minutes, Vol. 7.
34 Ibid.
35 Ibid.
36 For Lord Austin's refusal, see memorandum of A.H. Self, Air Ministry, to Private Secretary, Secretary of State for Air, 12 May 1939, AVIA 10/230, NA.
37 NA, AVIA 10/230, For Lord Austin's refusal, see memorandum of A.H. Self, Air Ministry, to Private Secretary, Secretary of State for Air, 12 May 1939.
38 NA, AVIA 46/261, "Narrative: Aircraft engine Production Expansion of Capacity 1935-1945", Capacity for Production of Bristol Engines, B) Shadow Schemes, p. 4.
39 "Narrative: Aircraft engine Production Expansion of Capacity 1935-1945", Capacity for Production of Bristol Engines, B) Shadow Schemes, p. 5.
40 Notes of the Nineteenth Meeting of the D.G.P.'s Directors held on 10th July 1939, Air 10/310, NA.
41 "Narrative: Reciprocating Aircraft engines and Engine Accessories Production & Programmes", Paragraph 38.
42 Capacity for Production of Bristol Aircraft engines 1938-1945, p. 7, Narrative-Aircraft engine Production Expansion of Capacity 1935-1945; Adjustments of Aircraft Orders, Air Supply Board Meeting, S.B.M. 207/41, 22 March 1941, AVIA 10/181, NA.
43 Bristol No. 1 Shadow Group-Change Over from Bristol Mercury/Pegasus to Napier Sabre Production, Air Supply Board Meeting, S.B.M 648/41, 10 August 1941, AVIA 10/214, NA.
44 Ibid.
45 Bristol No. 1 Shadow Group-Change over from Napier Sabre planning to Bristol Hercules VI production, Air Supply Board Meeting, S.B.M. 1058/41, 21 December 1941, AVIA 10/190, NA.
46 Additional Forging Plant to meet the 11th September Programme and to provide initial capacity against the Super Programme, Air Supply Board Meeting, S.B.M. 999/41, AVIA 10/190, NA.

[47] Bristol No. 1 Shadow Group-Change over from Napier Sabre planning to Bristol Hercules VI production, Air Supply Board Meeting, S.B.M. 1058/41, 21 December 1941.

[48] Air Supply Board, Summary of Conclusions Reached at the meeting of 9th January 1942, AVIA 10/191, NA; see note in AVIA 10/616, NA.

[49] No. 1 Shadow Factories-Conversion to Hercules Manufacture, Air Supply Board Meeting, S.B.M. 20/44, 17 January 1944, AVIA 10/201, NA.

[50] 160th Progress Meeting, War Potential-Aircraft engines (Memorandum by A.M.D.P.), 16 March 1939.

[51] Capacity for Production of Bristol Aircraft engines 1938-1945, p. 5, "Narrative: Aircraft engine Production Expansion of Capacity 1935-1945."

[52] Richardson, Kenneth: *Twentieth-Century Coventry*, (Coventry, UK, 1972), pp. 66-67.

[53] NA, AIR 2/3683, A.H. Self, Air Ministry, to the Bristol Aeroplane Company, 24 September 1939.

[54] This is figure is derived from measuring post-war plans for the Daimler, Rover and Standard Motor No. 2 Shadow Factories.

[55] History of Production of Bristol Aircraft engines by the Shadow Industry, p. 119.

[56] Richardson, *Twentieth-Century Coventry*, p. 68.

[57] NA, CAB 102/274, "Government Owned Factories Operated by Contractors on Agency Terms" lists the size of this factory as 1,181,000 ft^2.

[58] Richardson, *Twentieth-Century Coventry*, p. 67.

[59] NA, AVIA 46/23, Capital Commitments as of December 1942, Schemes of Value over £800,000; History of Production of Bristol Aircraft engines by the Shadow Industry, p. 119.

[60] NA, AIR 2/3683, See Bristol Brochure of 11 September 1939, Air 2/3683 and diagrams of Accrington works in Air 2/3684.

[61] Collins and Stratton, *British Car Factories from 1896*, p. 50.

[62] "Single Storey Wartime Factory Design", p. 76.

[63] Section XI, "Standard Designs for Single Story Factories for War Industries, With Notes on Siting and Layout", *Wartime Building Construction*, First American Edition, (New York, 1942), p. 118.

[64] Ministry of Home Security, Research and Experiments Department: Bulletin C 12, "Single Storey Wartime Factory Design", *Journal of the Institution of Civil Engineers*, Vol. 16, No. 5, (March 1941), p. 73.

[65] Ministry of Home Security, "Single Storey Wartime Factory Design", p. 73.

[66] "Standard Designs for Single Story Factories for War Industries, With Notes on Siting and Layout", pp. 117-18.

[67] Thoms, *War, Industry and Society*, p. 111.

[68] Ministry of Aircraft Production *Statistical Review 1939-1945*, pp. 31-32.

[69] *Statistical Review 1939-1945*, p. 28.

[70] Air Council Committee on Supply to the Bristol Aeroplane Company, 24 June 1938, AVIA 15/3746, NA.

[71] Notes of an Informal meeting of the Air Council Committee on Supply held in A.M.D.P.'s room on Thursday, 6th October 1938, AVIA 10/155, NA; Bristol Aeroplane Company Minutes of Directors' Meeting in Committee, 11 October 1938, BAC Directors' Committees Minutes, Vol. 6.

[72] Air Council Committee on Supply statement of engine requirements, 24 November 1938, AVIA 10/156, NA.

[73] Gilbert, Martin: *Finest Hour: Winston S. Churchill 1939-1941*, (London, 1983), p. 656.

[74] Postan, M.M.: *British War Production*, (London, 1952), p. 125.

[75] Prime Minister's Personal Minute Serial No. M878.1, 7 September 1941, AVIA 9/9, NA.

[76] Postan, *British War Production*, p. 125.

[77] Ibid.

[78] Ibid., pp. 125-26.

[79] Andrews and Morgan, *Vickers Aircraft Since 1908*, pp. 332-33.

[80] Postan, *British War Production*, Table J and Table K, pp. 476-77.

[81] Ministry of Aircraft Production Statistical Review 1939-1945, pp. 5, 16-17.

[82] Ministry of Aircraft Production Statistical Review 1939-1945, pp. 5-6.

[83] "Narrative: Reciprocating Aircraft Engines and Engine Accessories Production & Programmes", Paragraphs 213-14.

[84] Air Supply Board, Review of Engine Orders, 19 February 1942, AVIA 10/617, NA.

[85] Hercules VI/XI Programme, 12 January 1942, AVIA 10/617, NA.

[86] The Handley Page Heavy Bomber, AVIA 46/112, NA.

[87] Furse, *Wilfird Freeman*, p. 322; The Handley Page Heavy Bomber, AVIA 46/112, NA.

[88] The Handley Page Heavy Bomber, AVIA 46/112, NA; Barnes, *Handley Page Aircraft Since 1907*, p. 404.

[89] Barnes, *Handley Page Aircraft Since 1907*, p. 404.

[90] Ibid., Chapter 11, p. 5.

[91] Barnes, *Handley Page Aircraft Since 1907*, p. 405.

[92] The Handley Page Heavy Bomber, AVIA 46/112, NA.

[93] Cairncross, *Planning in Wartime*, p. 87.

[94] "Narrative: Reciprocating Aircraft engines and Engine Accessories Production & Programmes", Paragraphs 235-36.
[95] Ibid.
[96] Ministry of Aircraft Production Statistical Review 1939-1945, pp. 30-31.
[97] The Rover Company Limited (Aircraft engines) No. 2 Factory, Lode Lane, Solihull, Diary 1939-1945, chart of Hercules engine production.
[98] Ibid.
[99] Ibid., Paragraphs 269, 271.
[100] Ibid., Paragraphs 276-81.
[101] History of Production of Bristol Aircraft engines by the Shadow Industry, Abridged Record of Engines, Spares and Accessory Output from Shadow Industry for 1944.
[102] Draft. Bristol Engine Programme Parent and Hawthorn, 5 April 1945, AVIA 10/261.
[103] The Rover Company Limited (Aircraft engines) No. 2 Factory, Lode Lane, Solihull, Diary 1939-1945, pp. 26-29.
[104] Holley, *Buying Aircraft: Material Procurement for the Army Air Forces*, pp. 512-13; Cairncross, Sir Alec: *Planning in Wartime: Aircraft Production in Britain, Germany and the USA*, (London, 1991), p. 15.
[105] Bristol Aeroplane Co. Ltd.: *The Power Behind Their Wings: An Account of the Part Played by the Bristol Aeroplane Company in the Development of the Air-Cooled Radial Aircraft Engine in Great Britain*, (No Date), p. 35; Table of man-hours for Aircraft and Aircraft engines, CAB 102/274, National Archives.
[106] Bristol Aeroplane Co. Ltd.: *"Bristol" Hercules VI-XI-XVI-XVII-XVIII Operator's Handbook*, (August 1945), p. 12.
[107] See "The Production of the Bristol Hercules Engine: Methods and Plant Employed in one of the Recently Built Government 'Shadow' Factories", *Machinery (London)*, Vol. 60, No. 1528, (January 22, 1942); "The Production of Aircraft engine Crankshaft's: Methods Employed at a Government 'Shadow' Factory in the Manufacture of the Bristol Hercules Engine", *Machinery (London)*, Vol. 60, No. 1534, (March 5, 1942).
[108] Oates, J. A.: "The Bristol Hercules: Part I. Sleeve Valve Design and Operation: Manufacturing the Cylinder and Sleeve", *Aircraft Production*, Vol. IV, No. 40, (February 1942), p. 178.
[109] Oates, "The Bristol Hercules: Part I", p. 178.
[110] Oates, "The Bristol Hercules: Part I", p. 183.
[111] Bristol Co., *The Power Behind Their Wings*, p. 32.
[112] Oates, "The Bristol Hercules: Part I", p. 183.
[113] Bristol Co., *The Power Behind Their* Wings, p. 31.
[114] Bristol Aeroplane Company Minutes of Directors' Meeting in Committee, 21 November, 12 December 1939, BAC Directors' Committees Minutes, Vol. 7.
[115] Neale, M.C., ed.: *An Account of a Partnership-Industry, Government and the Aircraft Engine: The Memoirs of George Purvis Bulman*, Rolls-Royce Heritage Trust Historical Series No. 31, (Darby, UK, 2002), pp. 180-81.
[116] Oates, "The Bristol Hercules: Part I", p. 177.
[117] "Machining Cylinder Fins", *Aircraft Production*, Vol. VI, No. 66, (April 1944), pp. 155-56.
[118] Oates, J.A.: "The Bristol Hercules: Part II. Production of the Master Connecting Rod, Airscrew Shaft, and Cylinder Head: The Sleeve Ball and Housing", *Aircraft Production*, Vol. IV, No. 41, (March 1942), pp. 251-52.
[119] Oates, "The Bristol Hercules: Part I", p. 183.
[120] Oates, J.A.: "Hercules Engine Components: Machining Operations on the Crankcase, Impeller, Sleeve, Cylinder Barrel and Connecting Rods", *Aircraft Production*, Vol. VI, No. 69, (July 1944), pp. 314-15.
[121] Oates, "The Bristol Hercules: Part I", p. 186.
[122] Oates, "Hercules Engine Components", p. 309.
[123] "Turret Lathe Operations on the Hercules Crank Case", *Machine Tool Review*, Vol. 31, No. 184, (March-April 1943), p. 27.
[124] Oates, J.A.: "The Hercules Crankshaft: Design Features: Machining and Assembly Operations at a Shadow Factory", *Aircraft Production*, (February 1944), pp. 59-66.
[125] Oates, "Hercules Engine Components", pp. 313-14.
[126] "The Production of the Bristol Hercules Engine: Methods and Plant Employed in one of the Recently Built Government 'Shadow' Factories", *Machinery (London)*, Vol. 60, No. 1528, (January 22, 1942), p. 9.
[127] "The Production of the Bristol Hercules Engine", p. 9.
[128] "The Production of Supercharger Impellers", *Machinery (Lon)*, Vol. 60, No. 1535, (March 12, 1942), pp. 169-70.
[129] "The Production of Supercharger Impellers", p. 173.
[130] Oates, J. A.: "The Bristol Hercules: Part III. The Gear Department: Engine Assembly and Testing: Heat Treatment and Process Sections: Laboratories and Standards Room", *Aircraft Production*, Vol. IV, No. 42, (April 1942), p. 304.
[131] Oates, "The Bristol Hercules: Part III", p. 306.
[132] Muther, *Production Line Technique*, pp. 37-38, 44-55; Lilley, *Problems of Accelerating Aircraft Production During World War II*, p. 52.

133 Hesse, Herman: *Engineering Tools and Process: A Study in Production Technique*, (New York, 1941), p. 345.
134 NA, AIR 2/3683, Notes on a Meeting at Bristol on 28th September 1939.
135 Notes on a Meeting at Bristol on 28th September 1939.
136 NA, AIR 2/3683, Air Council Committee on Supply, New Bristol Engine Shadow Factory at Accrington, Note by P.S.S. (M).
137 NA, AIR 6/58, The Progress of Production for Air Re-armament, Memorandum by D.G.P., 12 July 1939.
138 Air Council Committee on Supply, New Bristol Engine Shadow Factory at Accrington, Note by P.S.S. (M).
139 Ibid.
140 NA, AIR 2/3683, Air Council Committee on Supply, Summary of conclusions reached at the eightieth meeting held on 3rd October 1939.
141 The Bristol Aeroplane Co. Ltd, No. 3 Engine Factory, Financial Statement 13 February 1940, Air 2/3863, NA; Norman Rowbotham, Bristol Aeroplane Co. Ltd., to Brian Davidson, Air Ministry, 28 February 1940, Air 2/3863, NA.
142 A photograph of Whitehead appears in the March 1940 issue of *The Beehive*, the United Aircraft Corporation, parent of Pratt & Whitney, company magazine.
143 United Aircraft Corporation: *The Bee Hive*, Vol. IX, No. 3, (August 1935), p. 10.
144 "Herculean Development", *Aeronautical Engineering, The Aeroplane*, (November 28, 1941), p. 599.
145 "Engines for the Air Force", *American Machinist*, Vol. 83, No. 11, (May 31, 1939), p. 372.
146 "P &W Engine Production", *Aero Digest*, Vol. 34, No. 1, (January 1939), p. 58.
147 "P &W Engine Production" contains a diagram showing the flow of stores and finished parts through the factory. See also "Precision is the watchword in the Pratt & Whitney Aircraft engine plant. Exacting requirements prompt a technique unique in production methods", *Automotive Industries*, Vol. 77, No. 1, (July 3, 1937), pp. 10-12; Ward, J. Carlton, Jr.: "Plant Layout and Production Methods for Modern Aircraft Engines", *Society of Automotive Engineers Journal (Transactions)*, Vol. 40, No. 5, (May 1937), pp. 178-89.
148 "Plant Layout and Production Methods for Modern Aircraft Engines".
149 AuWerter, Jay P.: "Pratt & Whitney Expands", *Aviation*, Vol. 39, No. 7, (July 1940), pp. 38-41, 116, 118.
150 "Pratt & Whitney Expands", pp. 39-40.
151 "Pratt & Whitney Expands", p. 116.
152 "Pratt & Whitney Expands", p. 118.
153 NA, AIR 2/3684 contains plans of the Accrington factory complex.
154 "Herculean Development", p. 599.
155 Ibid.
156 "The Production of the Bristol Hercules Engine: Methods and Plant Employed in one of the Recently Built Government 'Shadow' Factories", *Machinery (London)*, Vol. 60, No. 1528, (January 22, 1942), p. 1; "The Manufacture of the New 8-HP 'Flying Standard'", *Machinery (London)*, Vol. 52, No. 1355, (September 29, 1938), pp. 789-811.
157 Foster, Bruce: "Producing the 'Shadow' Blenheim Part I. Rootes Factory in Quantity Production: Brief History of the Scheme: Methods Used on Spars, Wings and Control Surfaces", *Aircraft Production*, Vol. III, No. 27, (January 1941), pp. 7-16.
158 History of Production of Bristol Aircraft engines by the Shadow Industry, p. 20.
159 "The Production of the Bristol Hercules Engine", p. 1.
160 "The Production of Aircraft engine Crankshafts: Methods Employed at a Government 'Shadow' Factory in the Manufacture of the Bristol Hercules Engine", *Machinery (London)*, Vol. 60, No. 1534, (March 5, 1942), p. 145.
161 Thoms, *War, Industry and Society*, p. 29.
162 NA, AIR 2/3683, Norman Rowbotham, Bristol Aeroplane Company, to J.E. Keel, Air Ministry, 15 January 1940.
163 Hornby, *Factories and Plant*, p. 325.
164 Hornby, *Factories and Plant*, pp. 325-27; Lloyd-Jones, Roger and M.L. Lewis: *Alfred Herbert Ltd. and the British Machine Tool Industry, 1887-1983*, (Aldershot, UK, 2006), pp. 168-69; See for example the paper titled "Machine Tools - September 1939 to December 1940: Machine Tools for the Production of Machine Tools", no date, AVIA 10/616.
165 Miller, R.E.: "Machine Tools Developed to Meet Lower-Cost Production", *The Iron Age*, Vol. 133, No. 1, (January 4, 1934), p. 119; "40 Years of Machine Tool Progress", *Automotive Industries*, Vol. 81, No. 7, (October 1, 1939), p. 344; Oliver, F.J.: "Trends in Metal Cutting Machines and Small Tools", *The Iron Age*, Vol. 143, No. 1, (January 8, 1939), p. 115.
166 Herb, Charles O.: *Machine Tools at Work*, (New York, 1942), pp. 70-115.
167 "Increase in Production Capacity of Machine Tools in Ten Years", *Machinery (New York)*, Vol. 43, No. 5, (January 1937), pp. 325-27.
168 Air Council Sub-Committee on Supply, Notes on Meetings Held Monday, 16th May 1938, p. 2, AVIA 10/151, NA.

169 Bristol Aeroplane Company Minutes of Directors' Meeting in Committee, 21 November, 12 December 1939, BAC Directors' Committees Minutes, Vol. 7; Statement of Production and Employment in the Aircraft Industry, 1st July 1938-31st December 1938, AVIA 10/18; Statement of Labor on Airframes and Engines at Dates Given, AVIA 10/18, NA.

170 Notes of a Meeting with Representatives of the Ministry of Labor held on 10th November 1939, to Discuss Problems Involved in Securing the Labor Force Necessary to Fulfill the Aircraft Construction Programme, AVIA 10/18, NA.

171 Notes of a Meeting with Representatives of the Ministry of Labor held on 10th November 1939.

172 Hornby, *British War Production*, p. 95.

173 Hornby, *British War Production*, p. 149; Inman, P.: *Labor in the Munitions Industries*, (London, 1957), p. 81.

174 Inman, *Labor in the Munitions Industries*, p. 27.

175 "Machining Aircraft engine Parts on Automatics", *Machinery*, Vol. 41, (February 9, 1933), p. 551.

176 The Bristol Aeroplane Co. Ltd, No. 3 Engine Factory, Financial Statement 13 February 1940, Air 2/3863, NA; Albrecht, Albert B.: *The American Machine Tool Industry: Its History, Growth, Restructuring & Recovery*, 3rd Ed., (Cincinnati, OH, 2016), pp. 77-82, 145-46.

177 Albrecht, *The American Machine Tool Industry: Its History, Growth, Restructuring & Recovery*, pp. 77-82, 145-46.

178 NA, AIR 2/3683, Bristol Aeroplane Company Limited No. 3 Engine Factory Financial Statement 13 February 1940.

179 Photographs of these machine tools are in History of Production of Bristol Aircraft engines by the Shadow Industry.

180 Bristol Co., *The Power Behind Their Wings*, p. 60.

181 Bristol Co., *The Power Behind Their Wings*, p. 53; History of Production of Bristol Aircraft engines by the Shadow Industry, p. 53.

182 History of Production of Bristol Aircraft engines by the Shadow Industry, p. 74.

183 Ibid.

184 NA, AIR 2/3683, Schedule H- Machine Tools, Bristol Acceleration Programme No. 3, 21 March 1939: The Rover Company Limited (Aircraft engines) No. 2 Factory, Lode Lane, Solihull, Diary 1939-1945, Note for 1st November 1939, L-RCO-103, British Motor Museum.

185 See the series of articles in *Machinery (London)* during 1944, such as "Machining Aircraft engine Cylinder Barrels", *Machinery (Lon)*, Vol. 65, No. 1679, (December 14, 1944), pp. 645-53 and "Machining Airplane Engine Cylinder Heads", *Machinery (Lon)*, Vol. 65, No. 1681, (December 28, 1944), pp. 701-705, with "Wright Aero Produces Engines for the B-29 Superfortress", *Machinery (New York)*, Vol. 50, No. 12, (August 1944), pp. 134-44 and Holben, Martin M.: "Some Aircraft-Engine Production Methods", *SAE Journal (Transactions)*, Vol. 52, No. 10, (October 1944), pp. 492-500.

186 History of Production of Bristol Aircraft engines by the Shadow Industry, p. 19.

187 NA, AVIA 15/414, Ministry of Aircraft Production-North West Region, Bristol Aeroplane Co. Clayton-le-Mores, 2nd June 1942.

188 Ministry of Labor and National Service: *Women in Engineering*, (London, No Date) II. Aircraft engines and Accessories, p. 1.

189 Ministry of Aircraft Production-North West Region, Bristol Aeroplane Co. Clayton-le-Mores, 2nd June 1942.

190 Ibid.

191 "Labor", *Aircraft Production*, Vol. III, No. 27, (January 1941), Editorial, p. 2.

192 Ministry of Labor, *Women in Engineering*, II. Aircraft engines and Accessories, p. 5. Ministry of Labor and National Service: *Women in Industry*, (London, 1945), II. Aircraft engines and Accessories, pp. 87-88.

193 Ministry of Labor, *Women in Engineering*, II. Aircraft engines and Accessories, pp. 3-4.

194 Hesse, *Engineering Tools and Processes*, p. 345.

195 Ibid.

196 Oates, "The Bristol Hercules: Part II", p. 243.

197 Narrative-Aircraft engine Production Expansion of Capacity 1935-1945, Capacity for Production of Bristol Engines, p. 8.

198 Ibid.

199 Appendix 3, The Rover Company Limited (Aircraft engines) No. 2 Factory, Lode Lane, Solihull, Diary 1939-1945.

200 Appendix 3, The Rover Company Limited (Aircraft engines) No. 2 Factory Diary 1939-1945.

201 NA, AVIA 10/201, No. 1 Shadow Factories-Conversion to Hercules Manufacture, 17 January 1944, in Air Supply Board Meeting No. 284, 20 January 1944.

202 "New Production Techniques", *Aircraft Production*, Vol. IV, No. 41, (March 1942), p. 221.

203 Data on these engines is taken from Lumsden, *British Piston Aircraft Engines and Their Aircraft*.

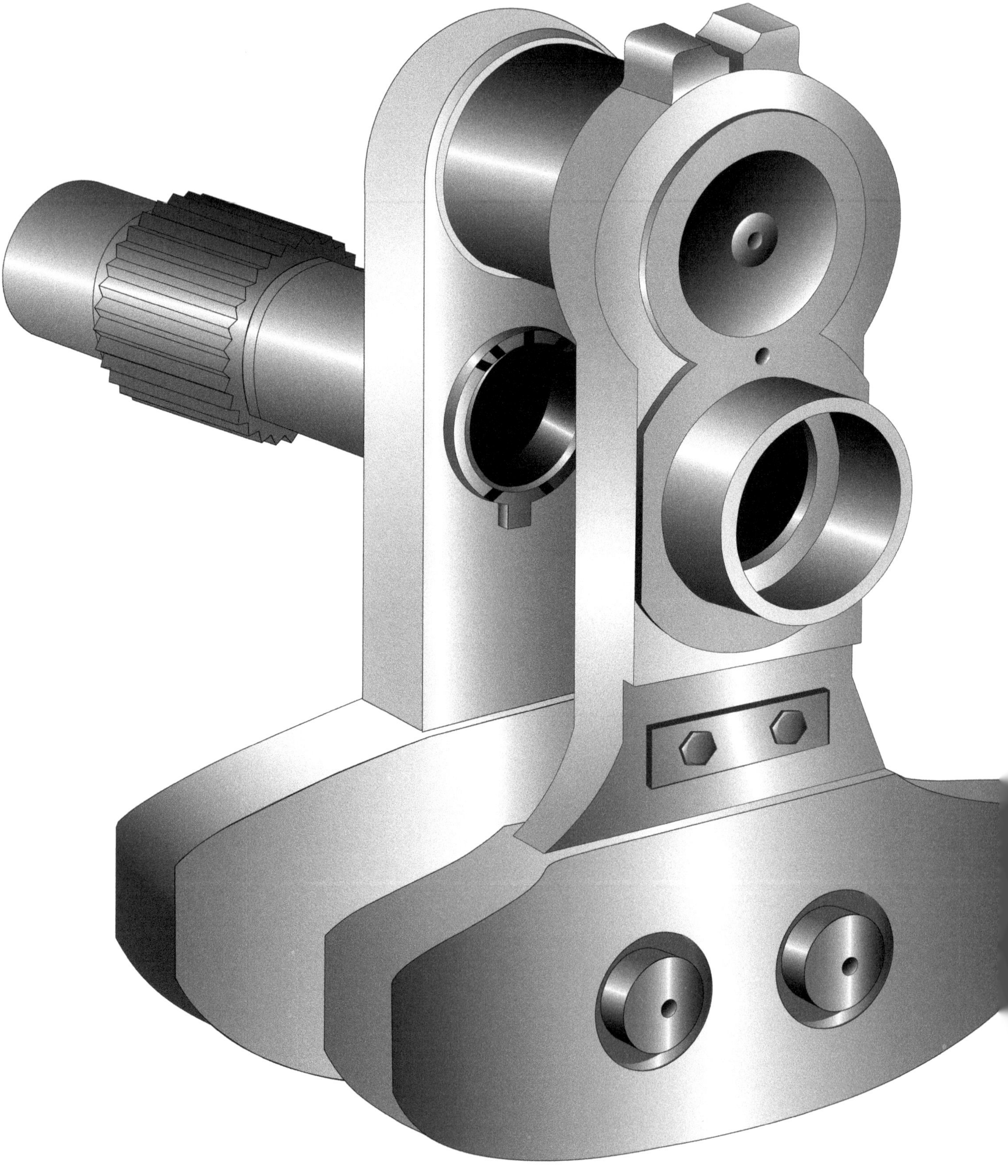

5

Building 14-Cylinder Engines: The Pratt & Whitney R-1830 Twin-Wasp and the Wright R-2600 Cyclone 14

Introduction

The Pratt & Whitney R-1830 Twin Wasp and the Wright R-2600 Cyclone 14 were built in huge numbers. During World War II, more Pratt & Whitney R-1830 Twin Wasp engines were built than any other Allied or Axis piston engine, with Pratt & Whitney and its licensees completing 171,172 engines by the war's end, compared to 160,974 Merlin engines built by Rolls-Royce and the American Packard Motor Company under license and 60,672 Hercules engines.[1] In the ranking of American and British engine production, the Wright R-2600 Cyclone 14 comes in at fifth place, just after Wright's R-1820 Cyclone 9, with 84,887 R-2600 engines completed at the Wright factories in Paterson, New Jersey, and Lockland, Ohio.[2] Production of these engines was radically different from the production of the Bristol Hercules. Pratt & Whitney, its licensees, and Wright Aeronautical built these engines on a massive scale in large factories, designed for line production, using the most advanced, automatic, high-production machine tools available. This was the precision engineering of large, complex aircraft engines on a size and scale that was unimaginable before the war.

What is astonishing about this record is not just the number of engines built during the war but the speed at which the aircraft engine factories ramped up the rate of production. It took Wright Aeronautical's Lockland, Ohio, factory just nine months from the start of production of the R-2600 Cyclone 14 to reach a rate of 500 engines per month; the Buick and Chevrolet factories building the Twin Wasp under license from Pratt & Whitney were even faster.[3] Buick turned out its first Pratt & Whitney Twin Wasp engine in February 1942 and manufactured 610 engines in April, just three months later.[4] Chevrolet took five months to boost production to over 500 engines a month.[5] It then took the Wright Lockland factory five months, the Buick factory six months and Chevrolet factories eight months to move from 500 engines a month to a level of 1,000 engines a month.[6] In contrast, it took the factories in No. 2 Shadow Group 17 months to achieve a rate of 500 Hercules engines a month.[7] Driving this rapid increase in production was the demand from the American military and Allied air forces for tens of thousands of these engines.

The history of the production of the Pratt & Whitney R-1830 Twin Wasp and the Wright R-2600 Cyclone 14 is a study of contrasting approaches to licensed production and machine tools. Pratt & Whitney willingly, and Wright Aeronautical reluctantly, agreed to license production of their aircraft engines to the automotive industry to produce the quantities of engines the Allies needed.[8] Wright preferred to maintain control over the final assembly of its engines and, as events transpired, remained the only manufacturer of its R-2600 Cyclone 14 after the Army Air Force shifted the Studebaker Corporation to building the R-1820 Cyclone 9. Wright would build the majority of its R-2600 engines in what was, at that time, the largest single-story industrial building in the world. Pratt & Whitney, while expanding its facilities around East Hartford, was happy to license production of the R-1830 Twin Wasp to the Buick Division of General Motors, and later to the Chevrolet Division. By the end of the war, 64% of all Pratt & Whitney engines were built in licensee factories, compared to 29% for Wright Aeronautical (Figure 5.1).[9]

FIGURE 5.1 Total production 1939–1945.

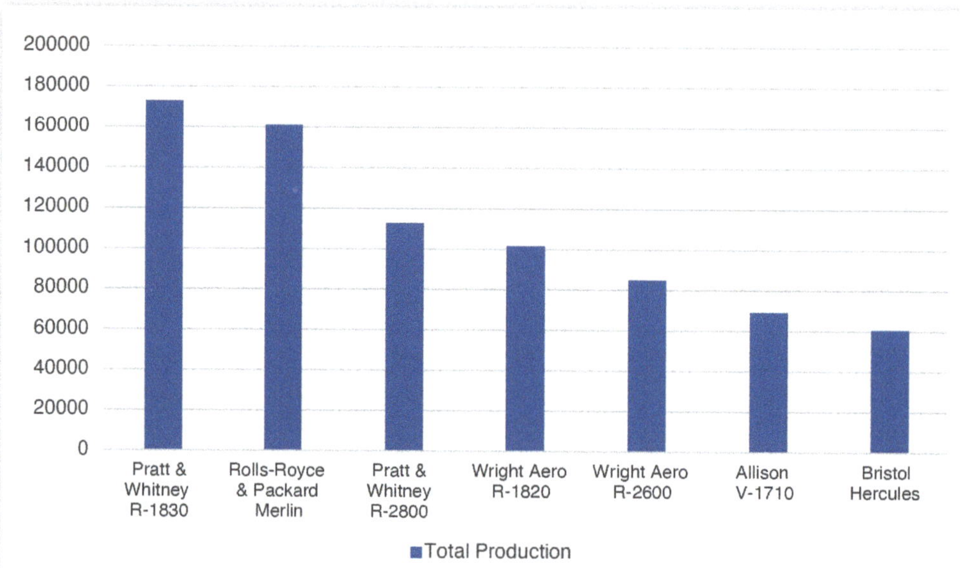

Source: Ministry of Aircraft Production: Statistical Review 1939-1945, (1946), Table 9, p.26, and US Department of Commerce, Civil Aeronautics Administration, Office of Aviation Information, Division of Aircraft Statistics: US Military Acceptances 1940-1945: Aircraft, Engines and Propeller Production, (Washington, DC, 1946).

The key to achieving quantity production at these factories were the new methods of production engineering and process engineering that Pratt & Whitney and Wright Aeronautical developed during 1940–1941 at their East Hartford and Paterson factories. Production at the Wright Lockland factory was particularly significant not simply because it was the first of the truly giant aircraft engine factories but more because the factory became renowned for its reliance on new types of advanced high-production machine tools Wright developed with the American machine tool manufacturers, notably the Greenlee automatic transfer machine. The automatic transfer machine was only one of the many advanced machine tools Wright developed specifically for aircraft engine production. The Lockland factory was at the forefront of Wright's continued development of automatic, high-production machine tools. These advanced machine tools replaced large numbers of standard machine tools, cut the number of operators, and speeded production. Buick also developed an automatic transfer machine working with the Foote Burt machine tool company. Pratt & Whitney replaced standard machine tools with automatic machine tools, but it appears to have been less reliant on the high-production machine tools that Wright and Buick developed during the war.

The Engines

Through the 1930s, Pratt & Whitney's R-1830 Twin Wasp competed directly with the Wright R-1820 Cyclone 9. Through the decade, the two engines marched, seemingly in lockstep, toward greater power output. Pratt & Whitney began the development of the Twin Wasp in 1929, choosing the double-row form with two banks of seven cylinders each.[10] In 1933, Pratt & Whitney introduced two double-row engines, the R-1830 Twin Wasp, and the smaller and less powerful R-1535 Twin Wasp Jr. The advantage of the double-row design was that it used smaller cylinders and thus the Twin Wasp was smaller in diameter than the Cyclone 9 by nearly 7 in., reducing the amount of drag while providing a similar amount of power.[11] The first model of the Twin Wasp received its Approved Type Certificate in 1933 with a rating of 830 hp.[12] With higher octane fuels, the Twin Wasp's power output reached 1,050 hp by 1936 and 1,200 hp by the start of World War II.[13] In the second half of the 1930s, Twin Wasp engines powered commercial airliners and military aircraft for both the Army Air Corps and the US Navy, particularly fighter aircraft. The Army Air Corps' Curtiss P-36 and Seversky P-35 and the Navy's F4F all used the Twin Wasp (Figure 5.2).

FIGURE 5.2 The Pratt & Whitney 14-cylinder R-1830 Twin Wasp engine. This early version of the Twin Wasp developed 830 hp. By the start of World War II, Pratt & Whitney had boosted the output of the Twin Wasp to 1,200 hp.

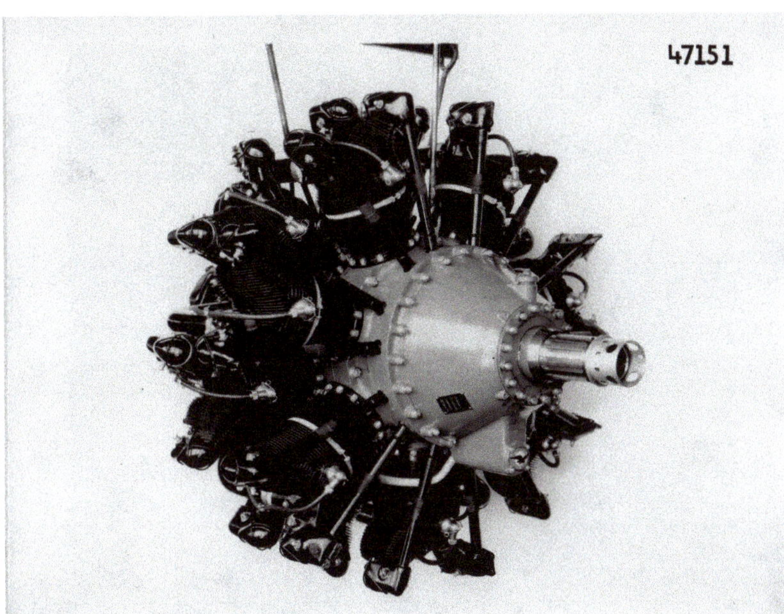

3B-47151, RG342FH, NARA.

Like the Mercury, Pegasus, and R-1820 Cyclone 9, the R-1830 Twin Wasp was a poppet-valve engine with one intake and one exhaust valve per cylinder.[14] Baffles directed the airflow to the rear bank of cylinders. The cylinder head was made up of cast aluminum fitted to a nitralloy steel cylinder barrel. Both the cylinder head and the cylinder barrel had deep, closely spaced fins for cooling. Unlike Wright Aeronautical, Pratt & Whitney continued using aluminum forgings for the three sections of the Twin Wasp crank case rather than switching to steel as Wright had done for the Cyclone 9 and Cyclone 14. In another difference from Wright's practice, the Twin Wasp used a one-piece crankshaft instead of crankshaft made up of two sections. Over time, Pratt & Whitney switched to using magnesium for certain engine parts to save weight. Many variations of the Twin Wasp came with single- or two-speed superchargers. Different design features required different manufacturing methods (Figure 5.3).

FIGURE 5.3 The Wright R-2600 Cyclone 14 was developed during the mid-1930s to power airplanes that were even larger than the Douglas DC-3. The Cyclone 14's first application was in the Boeing Model 314 Clipper for Pan-American Airways.

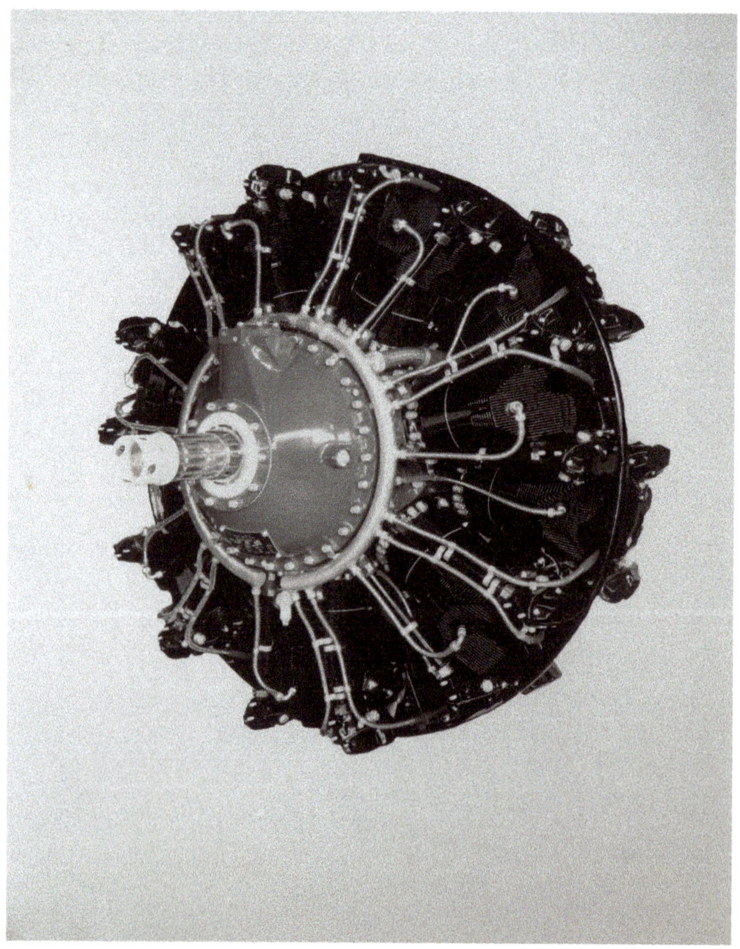

Wright Aeronautical Photo, Box 121, Motors/Engines C, Aerophile Collection, Library of Congress.

The R-2600 Cyclone 14 was Wright's first successful double-row air-cooled radial engine. By the mid-1930s, it was clear that even larger airplanes than the Cyclone 9-powered Douglas twin-engine DC-3 were on the horizon, particularly large, multi-engine flying boats to cross the Atlantic and Pacific, as well as larger bombers and patrol planes for the Army Air Corps and the Navy. To meet this expected need, Wright Aeronautical engineers decided to develop an engine that would provide 30 to 40% more power than what the progressive development of the Cyclone 9 could provide and greater power than Pratt & Whitney's Twin Wasp.[15] Initially aiming for 1,500 hp, Wright chose an engine with 14 cylinders in a double-row configuration, building on Wright's previous experience with two smaller double-row engines that had been built in limited quantities but with two and a half times the cooling area of the R-1820 made possible through improved casting techniques for the cylinder head and improved machining of longer fins on the cylinder barrel.[16]

Wright began developing the Cyclone 14 in 1935 with the engine receiving its Approved Type Certificate in June 1937.[17] The first Cyclone 14 had a take-off rating of 1,500 hp, a 50% increase in power output over the Cyclone 9 G-100 series, making it at that time the most powerful aircraft engine in production. By coincidence, the first application of the Cyclone 14, like the Bristol Hercules, was in a flying boat, in this case, the Boeing Model 314 Clipper. Although it differed in its design and construction, the Cyclone 14 was comparable in size and power output to the Hercules, the principal difference being that like all Wright air-cooled radial engines the Cyclone 14 was a poppet-valve engine while the Hercules used sleeve valves in its cylinders. From 1937 to 1945, 16 different versions of the Cyclone 14 went into production for commercial operators, the Army Air Force and the US Navy. The principal war time production models were the Cyclone 14A producing 1,600 hp, the Cyclone 14BA[18] with 1,700 hp, and the Cyclone 14BB with 1,900 hp. There were, of course, several subtypes within each major model, with odd numbers for Army Air Force engines and even numbers for Navy engines.

The Cyclone 14 drew on many design features of the Cyclone 9.[19] The cylinders were basically the same but with a greater cooling area developed for the Cyclone 9 G-200 series. The engine's 14 cylinders were arranged in two rows of seven cylinders, with the rear row slightly staggered to improve the flow of cooling air. The diameter of the Cyclone 14 remained the same as the Cyclone 9, and despite having a second row of cylinders, the engine was only a foot longer than the Cyclone 9. The Cyclone 14 was a poppet-valve engine with one intake valve and one exhaust valve in the cast aluminum cylinder head, as with the Cyclone 9. The crankcase came in three sections, bolted together to hold the two rows of cylinders. In the Cyclone 14A, the crankcase sections were made from aluminum forgings, but as the power output increased, Wright switched to using steel forgings for the crankcase. With two rows of cylinders, the Cyclone 14 needed a longer crankshaft with two sections to allow for two master connecting rods for each bank of cylinders. Wright designed a three-section crankshaft with two Wright Dynamic Dampers to reduce torsional vibration. The front section of the crankcase covering the reduction gear and the rear cover housing the supercharger were made from magnesium alloy. As a larger and more complex engine, the Cyclone 14 had many more parts than the Cyclone 9, with over 8,000 separate parts. As one would expect, machining, inspecting, and assembling an aircraft engine with 8,000 parts was a production challenge of a significantly greater magnitude than manufacturing the Cyclone 9. Production of the Cyclone 14 ended at the end of World War II.

Organizing Production

Before the outbreak of war in Europe, Pratt & Whitney received substantial orders for its R-1830 Twin Wasp engine from the French Government. The bulk of the American combat aircraft the French ordered—the Curtiss Hawk 75, Douglas DB-7, and Martin 167F—were equipped with the Twin Wasp. The French ordered additional engines as spares and replacements for the French Gnome-Rhone 14N in the Bloch 176 B3 twin-engine bomber, as the production of the Gnome-Rhone 14N was insufficient.[20] Between February and August 1939, the French Government placed orders for 2,145 Twin Wasp engines, and after the start of the war added an additional 2,850 Twin Wasps, mostly for the planned production of the Bloch 176 B3.[21] In April and May 1940, the French placed follow-on orders for 988 Twin Wasp engines.[22] French orders for the Wright R-2600 Cyclone 14 came in after the war began. In October 1939, the French Government placed an order for 1,440 R-2600 engines, again for French aircraft production, planning to use the R-2600 in the Loire-et-Olivier 458 B4 twin-engine bomber.[23] In May 1940, the French placed a follow-on order for 1,800 R-2600 engines.[24] Under the joint Anglo-French purchasing program, that same month, the Anglo-French Purchasing Commission issued letters of intent to Curtiss-Wright, parent of Wright Aeronautical, for 1,824 R-2600 engines.[25] This brought total French and Anglo-French orders to 5,651 R-1830 Twin Wasp engines costing $74 million and 5,064 R-2600 Cyclone 14 engines costing $88.5 million (Figure 5.4).[26]

FIGURE 5.4 During 1940 Wright received substantial orders from Britain and France and the US Army Air Corps and US Navy for the R-2600 engine. The R-2600 would power several light and medium bombers, including the North American B-25 shown here, and Navy carrier-based dive and torpedo bombers.

3B25546, RG342FH, NARA.

The President's 50,000 aircraft program and more British orders placed after the fall of France added to the demand for the Twin Wasp and the Cyclone 14. The NDAC Airplane

Division's Report No. 7 estimated that the Army Air Corps and the Navy would need 12,551 R-2600 engines by June 1942 in partial fulfillment of the President's program, in expectation of large orders for the Douglas A-20 (an improved version of the DB-7), North American B-25 medium bomber, and the new Curtiss CW-20 transport, while the Navy wanted the R-2600 for the Brewster SB2A dive bomber and the Martin PBM patrol bomber.[27] British orders for the Martin 187 medium bomber, Douglas Boston (equivalent to the A-20 for the Army Air Corps), and Brewster and Vultee dive bombers, in addition to previous French orders, added another 10,171 R-2600 engines for a total of 23,509, including some for commercial airliners.[28] The Army and Navy required 5,847 Twin Wasp engines primarily for the Consolidated B-24 for the Air Corps and the Navy's Grumman F4F fighter, while the British needed 7,537 engines, mostly for the French aircraft they had taken over, bringing the total to 14,592 with 1,208 engines for the commercial airlines.[29]

It was evident to the planners that neither Pratt & Whitney nor Wright Aeronautical had the production capacity to build their engines in quantities needed in the time allowed. At this point, Pratt & Whitney had orders for not just the R-1830 Twin Wasp but also for the larger 18-cylinder R-2800 Double Wasp, and the lower horsepower R-1535 Twin Wasp Jr., the R-1340 Wasp, and the R-985 Wasp Jr. In total, Pratt & Whitney had 38,832 engines on order for the Army Air Corps, the Navy, the British Empire, and commercial users.[30] Wright Aeronautical was in a similar position, with orders for its large 18-Cylinder Cyclone 18, and the lower horsepower R-975 and R-760 Whirlwind engines, in addition to the R-1820 Cyclone 9 and R-2600 Cyclone 14, taking total orders to 39,483 engines for delivery by June 1942.[31]

Initial efforts to increase production went to expanding facilities at the main engine manufacturers. As previously noted in Chapter 3, NDAC Report No. 7 had estimated that Pratt & Whitney would require a near five-fold increase in productive floor space and Wright Aeronautical a near three-fold increase.[32] With the prospect of more orders, in early 1939, Pratt & Whitney's parent United Aircraft Corporation had already begun to expand the production by rearranging space among its divisions.[33] Later in the year, Pratt & Whitney built a new addition to its East Hartford factory with financing from the French Government. After the Anglo-French Purchasing Commission placed its orders for more engines, Pratt & Whitney embarked on an $8 million expansion to the East Hartford factory, bringing the total amount of investment to $20 million over the prior 12 months, almost all financed by the French and British Governments through direct financing and payment of premiums on each engine purchased.[34] Pratt & Whitney received around $7.7 million from the French Government and around $15 million from the British Government for these additions to the East Hartford factory.[35] During the latter part of 1940 and early 1941, the American Government provided $15.5 million to Pratt & Whitney for expansion.[36] In August 1940, the US Navy, who in a division of labor with the Army Air Corps had been assigned procurement responsibility for all Pratt & Whitney engines for the two services, announced that it had issued a letter of intent to the company covering 17,000 engines with a value of $160 million.[37]

Pratt & Whitney readily agreed to expand production through licensing production of its engines to the automotive industry, first with Ford Motor Company to build the R-2800 Double Wasp engine, and then with the Buick Division to build the R-1830 Twin Wasp for the Consolidated B-24 bombers to be built in the new aircraft factories in the mid-West. In discussions with William Knudsen, Buick had agreed to build 5,000 Twin Wasp engines.[38] In January 1941, the War Department announced that Buick had been issued a contract for $36.5 million and would receive $24.3 million in Government financing to build a new factory with the capacity to build 500 engines a month.[39] By the end of the month, Buick had acquired a large track of land in Melrose Park, west of Chicago, for its new engine factory and was planning for construction (Figure 5.5).[40]

FIGURE 5.5 During World War II, the Consolidated Aircraft Corporation and its licensee companies built over 18,000 B-24 Liberator bombers. Just to get this many bombers into the air required 72,000 R-1830 engines. The R-1830 also powered the 10,174 Douglas C-47/C-53 transports built during the war.

3B-25379, RG342FH, NARA.

The President's approval of the expanded heavy bomber program in May 1941 created an even greater demand for the Twin Wasp engine with the program's substantial increase in the production of the Consolidated B-24 bomber. The result was a 100% increase in Buick's monthly rate of production to 1,000 engines a month and enlistment of the Chevrolet Division of General Motors as a second licensee to build the Twin Wasp.[41] Buick had its contract for the Twin Wasp increased to $88 million and an additional $10 million in government financing provided to expand its facilities at its new factory in Melrose Park and its existing factories in Flint, Michigan.[42] The Chevrolet Division received a contract for $89 million and a commitment of $33 million in government financing for factory buildings and machine tools at its automobile factory in Tonawanda, New York, outside of Buffalo.[43] Chevrolet would also build the Twin Wasp at a rate of 1,000 engines a month, with production scheduled to begin in 1942.[44] With the addition of Chevrolet, the organization of wartime production of the Twin Wasp was set.

Wright Aeronautical's preference to maintain control over the final assembly of its engines at its own factories or in new branch factories remained for some months after the announcement of the President's 50,000 aircraft program.[45] What Wright wanted was as much government business as it could handle in its factories before agreeing to license production to other companies, a concern that Bristol Aeroplane had as well with the shadow factories.[46] Discussing the need to increase engine production with Guy Vaughan, the president of Wright Aeronautical and Curtiss Wright, Secretary Morgenthau raised the idea of building a new factory west of the Alleghany Mountains solely to build the Wright R-2600 engine.[47] Wright Aeronautical had estimated that it would cost around $40 million to build a factory with the same capacity as the Wright factories in Paterson, New Jersey, and that this factory could build R-2600 engines at a rate of 1,000 to 1,200 a month.[48] Vaughan noted that one benefit of building a new factory to concentrate on building the R-2600 engine would be that it would free up the space at the Paterson factories for Wright's bigger R-3350 Cyclone 18 engine he knew the Army Air Corps and Navy would want in the future.[49]

Soon after joining the NDAC, William Knudsen began working on a plan to have the Government finance a new factory for Wright Aeronautical at an estimated cost of $50 million for buildings and machine tools to build 24,200 R-2600 engines at a rate of 575 engines per month.[50] What resulted was a plan for the Reconstruction Finance Corporation's subsidiary, the newly established Defense Plant Corporation, to purchase the land selected near Cincinnati, Ohio, that would be leased to a subsidiary of Wright Aeronautical.[51] The Defense Plant Corporation would provide financing to build the factory and for working capital, with additional funding available for purchasing machinery.[52] As negotiations continued, the increasing requirements for the R-2600 engine led to an adjustment in the loan amount to $92 million, which would allow the new factory to double production to 1,000 engines a month.[53] The lease with Wright Aeronautical was signed on October 19, 1940.[54] Wright broke ground for the new factory in Lockland, Ohio, on October 23, and on November 4, 1940, the War Department announced a contract with Wright Aeronautical for $119.8 million covering 6,470 R-2600 engines.[55] After Army Air Corps switched Studebaker to building the Wright R-1820 Cyclone 9, Wright Aeronautical's factories in Paterson and Lockland became the sole sources for the R-2600 Cyclone 14 and would remain so for the duration of the war.

As with the Wright R-1820 Cyclone 9, military demand for the Pratt & Whitney R-1830 Twin Wasp and the Wright R-2600 Cyclone 14 remained at a high level until the closing months of the war as indicated in Figure 5.6 which shows factory acceptances of heavy, medium, and light bombers, almost all of which used the Pratt & Whitney R-1830 Twin Wasp and the Wright Aeronautical R-1820 Cyclone 9 and R-2600 Cyclone 14, from 1941 to 1944.

FIGURE 5.6 Factory Deliveries for the Army Air Force, US Navy, and Allied Air Forces 1941–1944.

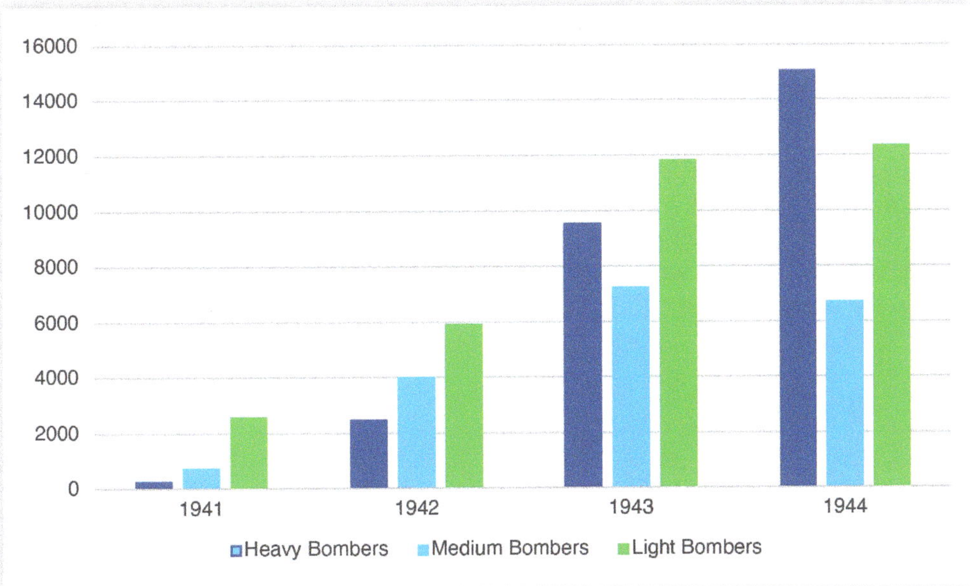

Source: Table 79, p.127. Army Air Forces Office of Statistical Control: Army Air Forces Statistical Digest World War II, (Second Printing), (Washington, DC, 1945).

The Factories

Pratt & Whitney Factories

In the spring of 1929, in the midst of an aviation industry boom, the United Aircraft and Transport Corporation announced that it had purchased 600 acres of land in East Hartford where it planned to build new factories for two of its subsidiaries, the Pratt & Whitney Company and the Chance Vought Corporation, as well as an airport for flight testing Pratt & Whitney engines and Vought airplanes.[56] Pratt & Whitney hired Albert Kahn's Detroit architectural firm to design its new factory building; Kahn also designed the new Vought factory.[57] Kahn sent a team of engineers to study Pratt & Whitney's current and future requirements that would help in the design of a factory for maximum efficiency in production.[58] Construction began on the $2 million factory buildings over the summer of 1929, and the last company unit, the offices and engineering department, moved to their new building in March 1930.[59]

Albert Kahn designed a modern, rectangular single-story building that could be easily expanded to accommodate future growth without disrupting its fundamental layout.[60] The main factory building was 400 ft wide and 1,000 ft long, providing 400,000 ft^2 of manufacturing space, with separate buildings for administration and the engineering department.[61] Lighting and ventilation were uniform within the building and there were few internal partitions. In contrast to Bristol and Wright Aeronautical, Pratt & Whitney laid out its manufacturing operations to create a smooth flow of raw materials and finished components, which may have been the first among aircraft engine manufacturers. This was not straight-line production as used in the automotive industry, but a system designed for maximum flexibility in batch production, wherein the model and number of engines in an order changed frequently. Rather than grouping machine tools by type in separate departments, Pratt & Whitney organized departments by component so that there were departments that manufactured cylinder barrels, crankshafts, crankcases, and other components.[62] The building was divided in half, with a wide central aisle running down the middle, with the departments on the north side of the building working on non-ferrous metal parts and departments on the southside working on alloy steels.[63] Raw materials came in at the west end of the factory and moved immediately to different component departments, with the finished components moving down the central aisle to the finished store area pending final assembly. A row of production test cells sat at the eastern end of the factory, just opposite the final assembly area. A section of the factory was allocated to Hamilton Standard, Pratt & Whitney's sister company that made all metal propellers. In 1936, United Aircraft built a small new factory building for Hamilton Standard next to the Pratt & Whitney factory to give Pratt & Whitney more space for aircraft engine production (Figure 5.7).[64]

FIGURE 5.7 Pratt & Whitney factory layout.

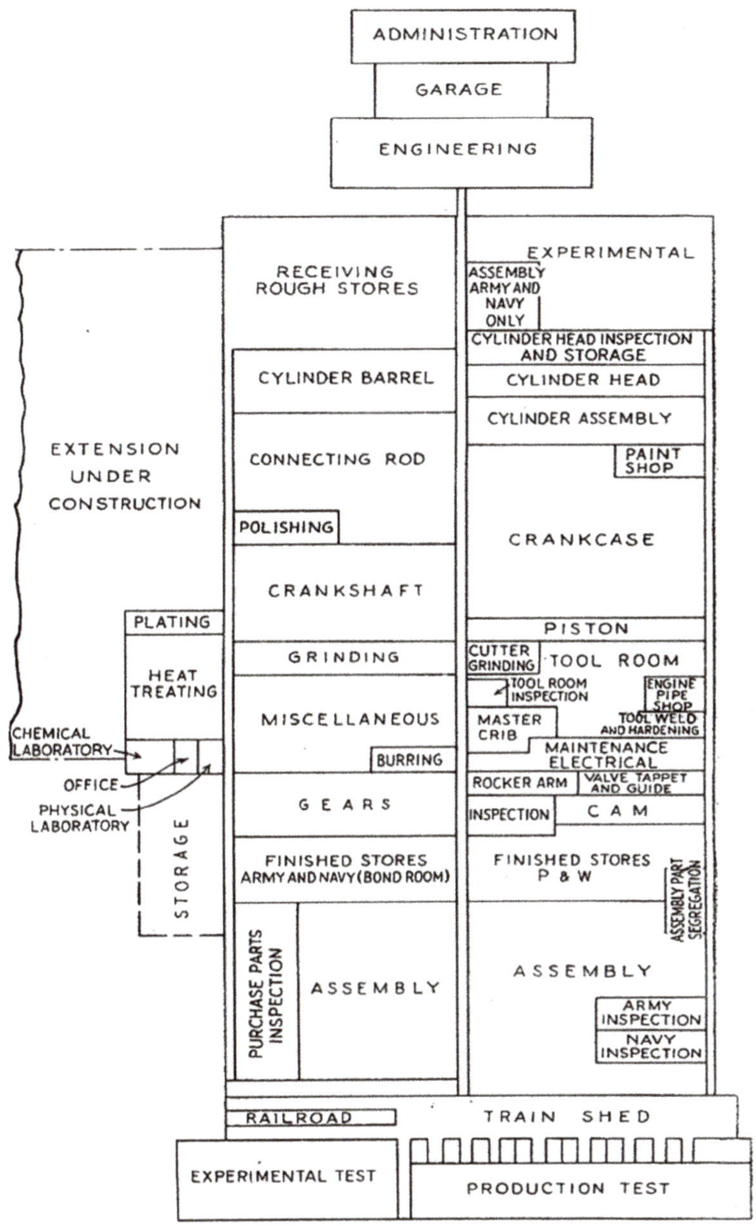

Source: Ward, J. Carlton, Jr.: "Plant Layout and Production Methods for Modern Aircraft Engines," S.A.E. Journal (Transactions), Vol. 40, No. 5, (May 1937), p. 179.

When Pratt & Whitney's order books began to fill during 1938, additional production space became imperative. United Aircraft shifted the Vought Aircraft division to Stratford, Connecticut, moved the Hamilton Standard Propeller division into the Vought factory, and had Pratt & Whitney take over the Hamilton Standard factory, providing 118,000 ft^2 of space for manufacturing.[65] This was insufficient for the flood of orders coming in during 1939, so after the start of the war in Europe, Pratt & Whitney obtained \$7.7 million in financing from the French Government to build a 274,000 ft^2 addition attached to the north of its factory that was called Plant C and known as the "French Plant."[66] Pratt & Whitney enlisted Albert Kahn once again to design this addition which was completed in three months with the first machine tools moving into the space in January 1940.[67] At the end of May 1940, United Aircraft announced plans for an $8 million addition to the Pratt & Whitney factory.[68] This addition added two factory sections, Plant D adjoining south of Plant B, the old Hamilton Standard factory section, and Plant E to the east of Plant C. These two sections, known as the British buildings from the source of their financing, added 317,000 ft^2 to the factory.[69] Finally, in August 1940, United Aircraft announced a fourth addition to the Pratt & Whitney factory, known as Plant F, that would add 400,000 ft^2 of space at a cost of $12 million devoted entirely to assembling engines.[70] Albert Kahn's firm designed all these additions, which brought Pratt & Whitney's total production area from 400,000 ft in 1937 to approximately 1,500,000 ft^2 in 18 months.[71] Plant F was devoted to engine assembly, freeing up the space in the former assembly area for manufacturing, and incorporated a new conveyor system.[72] This factory was now twice the size of the factories in No. 1 Shadow Group and larger than any two of the paired factories in No. 2 Shadow Group. None of the areas within the factory had internal walls as the No. 2 Shadow Group factories and the Bristol Accrington Factory did. In March 1941, Pratt & Whitney began the construction of the fifth addition to its East Hartford factory, adding 500,000 ft^2 of floor space.[73] Further expansion during 1942 brought the total area of the Hartford factory and testing facilities to 2,673,000 ft^2.[74]

America's entry into the war opened a floodgate of orders for airplanes and aircraft engines. Like Wright Aeronautical, requirements for all Pratt & Whitney's engines rapidly multiplied, increasing 142% from November 1941 to February 1942, from 74,118 to 179,642 engines; within eight months the total had grown to 319,206 engines.[75] Even with Buick, Chevrolet, and Ford taking some of the load, and licensing the production of Pratt & Whitney's lower-horsepower Wasp engine to Continental Motor Corporation and the Jacobs Aircraft Engine Company, Pratt & Whitney still needed more space for production.[76] During 1942, Pratt & Whitney began setting up satellite factories within a 35-mile radius of its main factory at East Hartford and moving the production of certain components to these factories. The company leased a small three-story mill building in Buckland, Connecticut to make small parts for various engines, but the main addition to capacity came from three large new factories financed through the Defense Plant Corporation under the designation Plancor 612.[77] Plant L, in Willimantic, Connecticut, manufactured crankcases and rear and nose section cases; Plant M, in East Longmeadow, Massachusetts, built crankshafts, propeller shafts, master rods, and link rods; Plant N, in Southington, Connecticut, built cylinder assemblies.[78] All these factories fed parts to the main Pratt & Whitney factory in East Hartford for the final assembly of complete engines, akin to Wright Aeronautical's factory arrangement in Paterson, New Jersey.

These factories were single-story rectangular buildings with wide open manufacturing areas, with columns spaced 40 ft by 40 ft, with walls, columns, and beams made from reinforced concrete. Plant L and M were not large, with the factory buildings and attached office building having a total area of 406,000 ft^2 at Plant L and 361,000 ft^2 at Plant M. Plant N was the largest of the three, with a total floor area of 606,400, equivalent in size to one of the factories in No. 2 Shadow Group.[79] These three factories added another 1,373,400 ft^2 to Pratt & Whitney's total floor area, bringing the total to approximately 4 million ft^2 greater than all factories in No. 1 and No. 2 Shadow Groups and the Bristol No. 3 Factory at Accrington.

The Buick and Chevrolet Factories

The Buick Division's factory in Melrose Park, Illinois, can be classed as one of the large third-generation wartime aircraft engine factories. Buick had originally intended to build the factory near its main automobile factory in Flint, Michigan, but soon realized that with expanding automobile and war production in different General Motors divisions around the area, there would likely be a shortage of labor.[80] Harlow Curtice, Buick's president and general manager, arranged, with government financing, to purchase a 125-acre tract of land in Melrose Park, west of Chicago.[81] Buick hired Albert Kahn and his firm to design the factory.[82] Designated Plancor 39 by the Defense Plant Corporation, the factory was to have four buildings, a main manufacturing building, a power house, a personnel building, and an office administrative building.[83] For the manufacturing building, Kahn followed his approach of having a large, single-story building having a wide work space with wide, open bays for maximum flexibility. The overall size of the manufacturing building was 1,305 ft by 750 ft for a total area of 978,750 ft^2 on the first floor and an additional 207,000 ft^2 in the basement below for cafeterias and equipment rooms.[84] The building had 21 bays that were 60 ft wide along its length and ten bays 75 ft wide along its width. Structural steel framing with steel trusses and columns supported the roof, while the floor was made of concrete covered in seven million wood flooring blocks with exterior walls of brick.[85] There was artificial lighting throughout the building. The manufacturing area had 18,000 ft of monorail conveyors to move component parts and 5,500 ft of monorails for electric and chain hoists. A two-story office building was attached to the south end of the manufacturing building.

Buick laid out the space within the manufacturing building by the type of component just as Pratt & Whitney had done in its Hartford factory, with sections of several bays each building the crankcase, cylinder head, cylinder barrels, pistons, gears, and miscellaneous parts.[86] Raw materials and components from subcontractors entered at the south end of the building, moving progressively down a central aisle to the final assembly area at the north end of the building. Two test cell buildings, one of 178,500 ft^2 and the other of 97,800 ft^2, were attached to the north end of the manufacturing building just off the final assembly area. The building was designed to build 500 engines a month, but before the factory was completed, the newly designated Army Air Force instructed Buick to double its rate of production to 1,000 engines a month.[87] After the attack on Pearl Harbor, the Army Air Force ordered Buick to again double production.[88] This required two extensions to the main manufacturing building, placed on the east and west sides of the building adding 261,000 ft^2 of space. The total area of the manufacturing building when completed was 1,576.00 ft^2, and, with the test cells included, amounted to 1,852,300 ft^2 or 23 acres under one roof, larger than any aircraft engine factory in Britain.

Buick had initially intended to depend on subcontractors for aluminum and steel castings and forgings, but with the planned doubling of production after America's entry into the war, Buick started making steel forgings for crankshafts, propeller shafts, connecting rods, and other steel parts at its own foundry in Flint, Michigan.[89] With $10 million in financing from the Defense Plant Corporation, Buick immediately started construction of one of the largest aluminum foundries in America to provide aluminum castings for crankcases and cylinder heads for all General Motors divisions building Pratt & Whitney engines.[90] Buick again engaged Albert Kahn's firm to design the foundry.[91] The single-story building had a floor area of 570,000 ft^2, with overhead cranes and conveyors to speed the movement of materials and castings through the foundry.[92] The foundry was designed with a capacity for sand casting 150,000 cylinder heads a month.[93]

To build the Twin Wasp, the Chevrolet Division converted several of its existing factories in the Buffalo, New York area into aircraft engine production and added new factories with financing from the Defense Plant Corporation. In 1938, Chevrolet built a new Albert Kahn-designed factory located in the Buffalo suburb of Tonawanda to make axles and automobile engines.[94] This large building had 800,000 ft^2 of floor space and became Chevrolet aircraft engine plant No. 1 building the larger components of the Twin Wasp engine, the crankcase sections, blower sections, crankshaft, reduction gear, cylinder heads, cylinder barrels, and propeller shafts.[95] Plant No. 2 was a former automobile assembly plant in Buffalo converted into manufacturing, while Plant No. 3 was the former General Motors Fisher Body factory converted into making, initially, 300 smaller parts for the Twin Wasp; these two factories added around 600,000 ft^2 of factory space.[96] The Heavy Bomber Program of May 1941 brought Chevrolet $37. 2 million in financing for a new factory located close to the original 1938 Motor and Axle factory. Confusingly, this became Plant No. 1 in Plancor 155, the Defense Plant Corporation's designation for the two factories it was financing for Chevrolet.[97] The factory comprised two joined areas, an assembly area made from reinforced concrete with steel framing and trusses that by the end of 1944 had a total floor area of 476,000 ft^2 and a test cell area with 90 production engine test cells of 240,000 ft^2, making the factory larger but comparable to the bigger factories in No. 2 Shadow Group in total area.[98] After the attack on Pearl Harbor and a requirement for more engines, the size of the factory and the test cell area were nearly doubled, with a second expansion undertaken in May 1943 and a third and final expansion began in May 1944 to support the manufacturing of the larger Pratt & Whitney R-2800-C engine.[99] The second Government-financed factory, Plancor 155 Plant No. 2, was located around 8 miles away in Buffalo. It was smaller, consisting of seven buildings, with two manufacturing buildings of 294,300 and 137,000 ft^2.[100] By 1944, the total floor area that Chevrolet had available in the Buffalo area was approximately 2,677,000 ft^2. Because Pratt & Whitney, Ford, and Buick were already taking almost all the output of Pratt & Whitney's subcontractors, Chevrolet enlisted the production facilities at other Chevrolet and General Motors factories to provide components necessary for the Twin Wasp engine.[101] Chevrolet would add its own aluminum factory as Buick had done but for forging aluminum components, not casting.[102] This forging plant was completed in seven months and went into production forging pistons, main crankcase sections, and other components for Pratt & Whitney engines, adding to production at other Chevrolet factories.[103] Chevrolet had four forging plants, two in Michigan and two in Indiana, forging parts for Pratt & Whitney engines.[104] In total, 17 Chevrolet factories around the country contributed parts for the Twin Wasp engines.[105]

The Wright Lockland, Ohio Factory

The Wright factory at Lockland, Ohio, was radically different from the factories building the Bristol Hercules in Britain. The Wright Lockland factory was huge, dwarfing the aircraft engine factories in Britain. In total floor area, the Lockland factory was larger than all the factories in No. 1 and No. 2 Shadow Groups and the Bristol No. 3 Shadow Factory at Accrington building combined. The Wright Lockland factory reflected a trend apparent in the 1930s toward "industrial giantism," a move to build truly giant factories to take maximum advantage of methods of mass production developed at Ford Motor Company.[106] The Wright factory at Lockland, Ohio was the first of the giant fourth generation of wartime aircraft engine factories to be built. In 1940, the Government preferred, for strategic reasons, that the new airframe and aircraft engine factories it was commissioning be located well away from either coast in an area between the Appalachian Mountains in the East and the Rocky Mountains in the West.[107] Wright chose to locate the new Cyclone 14 factory near the small town of Lockland, Ohio, a dozen miles from Cincinnati, a center of the machine tool industry. The Defense Plant Corporation allocated $37 million to purchase a 219-acre plot near Lockland and construct the factory, with an additional $55 million to purchase machine tools and other equipment.[108] The Lockland factory became Plancor 10. The Lockland factory would ultimately cost $141,196,000.[109]

Wright hired Albert Kahn, who had designed the new wing at the Wright Paterson factory in 1939, to design the new factory at Lockland. Albert Kahn and his firm, Albert Kahn Associated Architects and Engineers, designed all the giant aircraft engine factories building Wright and Pratt & Whitney aircraft engines. Kahn had been the lead architect of the Ford River Rouge complex. In the early 1930s, Kahn and his firm designed several giant factories as part of the Soviet Government's plans for rapid industrialization.[110] Among these were the Stalingrad tractor factory covering 960,700 ft^2 of floor space, with an assembly building of 422,000 ft^2, and the even larger tractor factory at Chelyabinsk covering 1,780,000 ft^2 with an even larger assembly building of 975,000 ft^2.[111] As the American automobile industry recovered from the depths of the Depression, Kahn received new commissions for large new factories from several automobile manufacturers, including the 1,450,000 ft^2. Ford Press Shop building at the Ford River Rouge complex built in 1939.[112] The Press Shop was the largest single-story factory building in the world.[113]

The factory Kahn and his associates designed was even bigger than Wright's original estimate. Kahn's ideal for industrial factory buildings was "...the one-story structure of incombustible materials with enormous uninterrupted floor spaces under one roof with a minimum of columns."[114] His industrial factory designs followed certain core principles: a building designed for efficient straight-line production but with the flexibility for future expansion; wide column spacing to minimize interference with production lines; proper ventilation and lighting; and optimal location of power, storage, and other ancillary buildings as well as easy access to transportation.[115] Kahn was especially concerned about flow within the factory building and the need to determine how and where raw materials would enter the building and how and where the finished product would exit (Figure 5.8).[116]

FIGURE 5.8 An aerial view of the Wright aircraft engine factory in Lockland, Ohio. At the time of its construction, this factory was the largest industrial building in the world.

SC#318 Lodder Watson Collection - #SC318-0266; Provided Courtesy of the Cincinnati Museum Center.

For Wright's factory at Lockland, Kahn designed a single-story rectangular main machine shop and assembly building with 1,468,000 ft^2 of production floor space, covering nearly 35 acres, with a more than 225,000 ft^2 basement area, and additional space for fans and air conditioning units, for a total area of over 1,800,000 ft^2.[117] At that time, the Wright Lockland factory was the largest single-story industrial building in America, eclipsing the River Rouge Press Shop. The factory was designed to produce 3,000 aircraft engines a month, tripling the initial production specification.[118] The machine shop and assembly building had structural steel framing and steel trusses with a reinforced concrete floor and 8-in. brick exterior walls with a seven-foot continuous band of steel sash windows on the sides of the building. Long roof monitors allowed in additional daylight, though there was fluorescent lighting throughout the factory. Kahn incorporated exceptionally wide bays within the factory, with column spacing of 60 ft by 75 ft to allow both uniform working areas and maximum flexibility in placing machine tools on the production lines. As with the Studebaker factories, the Lockland factory had no protections against air attack, a virtue of its location far from the coasts. There were no walls or barriers in the main machine shop and assembly building to interrupt production. A test cell building and an office building were attached to the machine shop and assembly building, with an aluminum foundry nearby. In total, the factory complex had over 2 million ft^2 of floor space, bigger than either the Standard Motor No. 2 Shadow Factory or the Bristol No. 3 Shadow Factory. During the war, this would be more than doubled.[119]

With more than 12,000 employees working at the factory complex, the movement of workers during shift changes and meal breaks could disrupt the flow of work on the production floor. Kahn's real innovation with the Lockland factory was to locate all employee services for the machine shop and assembly building in the large basement. Kahn described this as "...one of the finest features we believe we've ever hit upon in industrial plant design."[120] The basement was in the shape of a large cross underneath the main floor. Employees entered the factory by descending stairs into the basement area where they followed passageways to their lockers and to stairs leading directly to their assigned work areas.[121] These stairs and passageways also led to employee cafeterias and toilets. At the shift change, the employees working on the production floor simply reversed the process, exiting the factory building through the basement. In this way, traffic on the production floor was kept to a minimum. The area within the main machine shop and assembly building was free from obstructions, allowing optimum placement of machine tools and mechanical conveyor systems.

Contractors began laying concrete in December 1940 and erecting the steel framing beginning in January 1941.[122] The factory complex, now designated Plant No. 6, was ready on June 12, 1941, two months ahead of schedule.[123] To speed up getting into production, Kahn divided the machine shop into twelve areas. As the contractors completed each area, they enclosed it with partitions to allow Wright to start installing production machinery. Wright laid out the production lines so that the machine tools could be shifted with little difficulty when construction was completed.[124] Pilot production lines, set up to test the arrangement of machine tools and production flow, began producing parts for the Cyclone 14 in April 1941.[125] The Lockland factory completed its first Cyclone 14 engines in June and by the end of the year had sent 246 engines to the military.[126]

After America's entry into the war, Wright received approval to expand its newly completed factory at Lockland to boost the production capacity to 4,000 engines per month.[127] Two new machine shops, the South Shop Building No. 14 and the North Shop Building No. 13, went up next to the main machine shop and assembly building.[128] These buildings allowed Wright to expand the production of components for the Cyclone 14, with finished parts moving from the new shops to the main machine shop and assembly building for final assembly. The North Shop building was specifically for cylinder production.[129] The South Shop Building, like the North Shop Building, used structural steel in its construction, allowing wide bays of 60 ft by 75 ft on the main floor of 832,000 ft^2. With a large basement of 572,000 ft^2, the building had 1,478,000 ft^2 of floor space, nearly as big as the main machine and assembly building. In addition to the two new machine shops, Wright added new, large aluminum and magnesium foundries and other ancillary buildings, taking the total floor area of the Lockland factory to over 5,600,000 ft^2, bigger than the Ford Willow Run factory complex, wrongly considered the biggest factory complex in America in World War II, and bigger than the total floor space of the entire American aircraft engine industry in January 1940.[130]

The Production Record

The R-1830 Twin Wasp

The Pratt & Whitney R-1830 Twin Wasp was in great demand during World War II for the Consolidated B-24 Liberator four-engine bomber, built in greater numbers than any other American wartime aircraft, and the Douglas C-47 twin-engine transport. Between 1940 and 1945, a consortium of companies built 18,202 B-24 bombers while the Douglas Aircraft Company built 10,384 C-47 and C-53 transports. Just to fly these aircraft away from the factories required 93,576 Twin Wasp engines. From 1939 to 1945, Pratt & Whitney, Buick, and Chevrolet built 171,964 Twin Wasp engines, more than any other British or American aircraft engine built during the war.[131]

Figure 5.9 illustrates how important license production of the Twin Wasp was to meet wartime requirements. Together, Buick and Chevrolet built 80% of all the Twin Wasp engines completed between 1942 and 1945. The two General Motors divisions set records for engine production during World War II. During 1944, Buick built 30,550 Twin Wasp engines, more than any other American aircraft engine licensee or branch factory and more than double production of the Bristol Hercules at the shadow factories that year, while in November 1943, Chevrolet achieved a peak production of 3,450 engines, more than any other American aircraft engine factory during the war and again more than double the peak rate of production of the Bristol Hercules at the shadow factories.[132]

FIGURE 5.9 Pratt & Whitney R-1830 Twin Wasp Production 1939–1945.

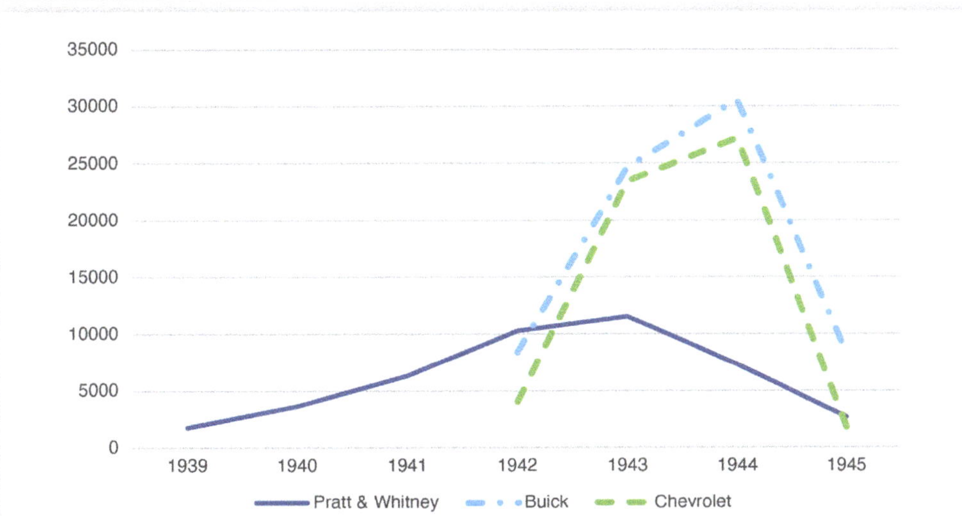

Source: US Department of Commerce, Civil Aeronautics Administration, Office of Aviation Information, Division of Aircraft Statistics: US Military Acceptances 1940-1945: Aircraft, Engines and Propeller Production, (Washington, DC, 1946).

Buick and Chevrolet both rapidly built up their monthly production rate. The first Buick-built Twin Wasp engine went to the test stand on January 23, 1942, only 243 days after the blueprints were approved.[133] Chevrolet's first engine came out nine months after the announcement of its contract to build the Twin Wasp. Like Studebaker, it took Buick just eight months from the start of production to reach a rate of 1,000 engines a month, with Chevrolet taking 13 months, compared to Ford taking 18 months to reach that rate of production of the admittedly more complicated Pratt & Whitney R-2800 Double Wasp. Remarkably, it took Chevrolet just two months to double production from 1,000 to 2,000 engines a month. Like Studebaker, Buick and Chevrolet's production rate benefited from building limited models of the Twin Wasp, principally the R-1830-43 and R-1830-65 for the Consolidated B-24 and the R-1830-90C and R-1830-92 for the Douglas C-47.

Production at Buick peaked at a rate of 3,300 Twin Wasp engines a month in November 1943, just short of Chevrolet's record, and continued at a level of over 2,700 engines a month for much of 1944.[134] In 1944, Buick began preparing to build a similar Pratt & Whitney R-2000 engine for the Douglas C-54 transport airplane, building 2,548 of these engines by the war's end.[135] When in April 1945 the Army Air Force cancelled the production of the B-24 Liberator, Buick stopped building Twin Wasp engines for the B-24 but continued building the R-1830-90C engines for the C-47 transport ending production of the Twin Wasp in September 1945, having completed 71,874 Twin Wasp engines between 1942 and 1945.[136] Similarly, Chevrolet maintained a high rate of production from the end of 1943 to the middle of 1944, building 2,600 Twin Wasp engines a month until August 1944.[137] Chevrolet also took on the production of another Pratt & Whitney engine, beginning the production of the R-2800-C Double Wasp that same month and completing 4,282 Double Wasp engines by the end of the war and 56,484 Twin Wasp engines.[138] Had the war continued, it was intended that Buick would become the sole source of Twin Wasp engines for B-24 Liberator bombers.[139]

Like Bristol Aeroplane and Wright Aeronautical, during the war Pratt & Whitney carried out both production and engine development. From the late 1930s to the end of the war, Pratt & Whitney built over 40 different models of the R-1830 Twin Wasp for many commercial and military aircraft.[140] The Hartford factory and its feeder factories built 40,456 Twin Wasp engines between 1940 and 1945, amounting to 30% of Pratt & Whitney's total wartime production at its Hartford and Kansas City factories. From the start of the war in September 1939, it took Pratt & Whitney a year to build up to a rate of 500 Twin Wasp engines a month and over two years to reach 1,000 Twin Wasp engines a month. The production of the Twin Wasp at Hartford peaked in March 1943 when the factories built 1,103 engines. The rate of production declined steadily during 1944 as Pratt & Whitney concentrated on building the more powerful R-2800 Double Wasp engine and began the production of the even bigger 28-cylinder 3,000-4,300 hp R-4360, the most powerful British or American aircraft engine to be built during the war.[141] The company continued development of the Twin Wasp and towards the end of the war introduced the R-1830-75, -94, and -98 models which boosted the power output to 1,350 hp for the Convair (formed in 1943 through a merger between Consolidated Aircraft and Vultee Aircraft) PB4Y-2 Privateer, a much-modified B-24 bomber the US Navy used as a long-range patrol aircraft.[142] The R-1830-75 incorporated a number of changes including a magnesium nose section replacing the aluminum section, a change to roller bearings in the crankshaft, different fins on the cylinder head, and aluminum fins shrunk onto the cylinder barrels instead of cut in the steel forging.[143] These changes required changes in machines and production processes.

Wright R-2600 Cyclone 14

During the war, the Lockland factory built two models of the Cyclone 14, the 1,700 hp Cyclone 14BA, and the more powerful 1,900 hp Cyclone BB, completing 60,546 Cyclone 14 engines by war's end. Following the outbreak of war in Europe, orders for the Cyclone 14 expanded rapidly. By the end of July 1940, total military orders for the Cyclone 14 amounted to 22,722, roughly 139 times the number of Cyclone 14 engines Wright had built during 1939. After the attack on Pearl Harbor, orders for the Cyclone 14 more than doubled. By the end of December 1941, total orders for the Cyclone 14 amounted to 61,000 with additional orders in later years.[144] Production of the Cyclone 14 at Lockland began in June 1941 and peaked in July 1944 when the factory completed 2,723 Cyclone 14 engines.[145] This was also the peak year of Cyclone production at Lockland, with the factory building 25,860 Cyclone 14 engines, as shown in Figure 5.10. During 1945, the Lockland factory shifted to building the Cyclone 18 but continued production of the Cyclone 14 in declining numbers until October. In total, the Lockland factory built 73% of the 84,891 Cyclone 14 engines built during the war. After the Lockland factory started production, Wright continued to build the Cyclone 14 at its Paterson factories, but as output grew at Lockland, production at the Paterson factories steadily declined. The Paterson factories built their last Cyclone 14 engine in November 1944, having completed 24,345 between 1940 and 1944.

FIGURE 5.10 Wright R-2600 Cyclone 14 production at Lockland 1941–1945.

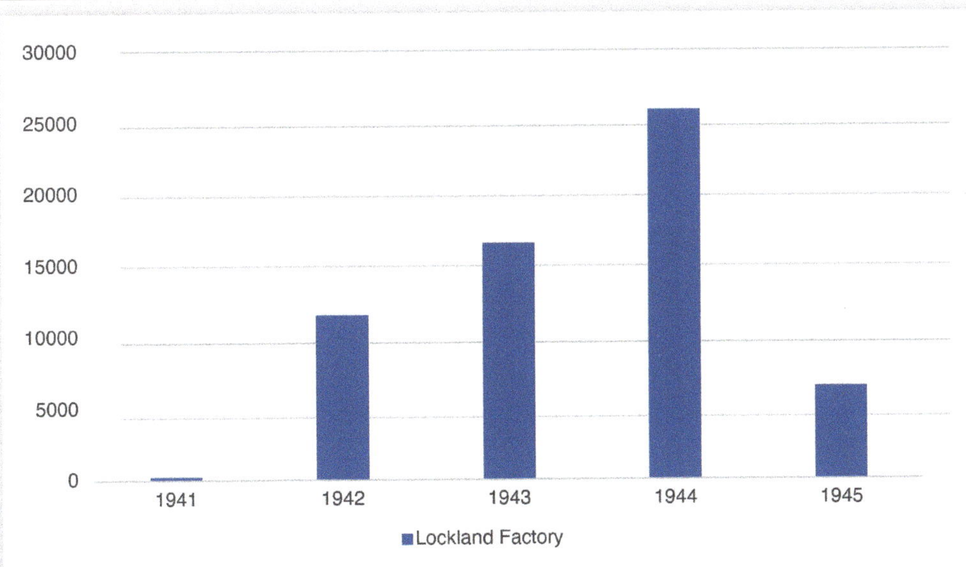

Source: US Department of Commerce, Civil Aeronautics Administration, Office of Aviation Information, Division of Aircraft Statistics: US Military Acceptances 1940-1945: Aircraft, Engines and Propeller Production, (Washington, DC, 1946).

How It Was Done: Building the Twin Wasp and the Cyclone 14

Building the Twin Wasp in Hartford

In 1940, Pratt & Whitney faced the same challenge that confronted Wright Aeronautical: how to shift from prewar batch production methods to quantity production. Pratt & Whitney's problem was compounded by the fact that the company was building seven different types of engines, compared to Wright's five engine types, and three times more lower horsepower engines than Wright was building as these engines—the Twin Wasp Jr., Wasp, and Wasp Jr.—which were in great demand for training aircraft.[146] Prior to the war, Pratt & Whitney's production methods emphasized flow and flexibility. The company had converted to a form of flow production well before Bristol Aeroplane or Wright Aeronautical. In his 1937 description of Pratt & Whitney's scheme of production in its East Hartford factory, J. Carlton Ward, Jr., the company's assistant general manager, described how raw material "flows through the various manufacturing shops to the finished store room in the rear."[147] "Theoretically, therefore," Ward wrote, "incoming raw materials should not rest but, when received, should proceed immediately through the manufacturing operations, and be delivered to the assembly floor or shipping area."[148]

Building aircraft engines in small batches among a large variety of specific engine models for the Army Air Corps, the US Navy, the American commercial airlines, and foreign customers, not all with interchangeable parts, as well as the need to incorporate frequent design changes, required flexible production methods. When asked whether the greater use of line production and quantity production methods would not be better for reducing costs, Pratt & Whitney's response was that the company continued to emphasize "flexibility of operations to meet sudden design changes rather than on the ability to manufacture quantities of a given type or even common parts (Figure 5.11)."[149]

FIGURE 5.11 The Pratt & Whitney East Hartford factory built with financing from Britain and France. The wide central aisle allowed parts to be collected on hand carts or by small tractors, but parts were still stacked close to the machine tools.

Library of Congress, Prints & Photographs Division, FSA/OWI Collection, Reproduction Number LC-USF34-080795-D.

In the new factory buildings added during 1940 with French and British financing, Pratt & Whitney ensured that the flow of materials and components was similar to the arrangement in its original factory, with a large central aisle and separate aisles running along the outside walls to serve as conduits for supplies destined for the individual production lines set up within the building.[150] Manufacturing progressed toward the center aisle where trucks carried the finished parts to store rooms or to the final assembly area. The company modified its assembly methods to speed production, setting up a form of straight-line production with lines of benches that built up subassemblies like the crankcase, crankshaft, master and connecting rods, etc. Workers attached these subassemblies to a wheeled engine stand that moved progressively down a central aisle so that by the time the engine reached the end of the assembly area, it was ready for testing in test cells.[151] When engine assembly shifted to the new Plant F, Pratt & Whitney for the first time added a conveyor system to supply parts to the assembly area, ensuring "an even flow of engines down the assembly line."[152]

With the demand for Pratt & Whitney engines accelerating to levels never previously achieved, the company had to adjust its production methods. At first, like the No. 1 Shadow Group factories and Wright Aeronautical, Pratt & Whitney hired many more workers, getting more people to do the same tasks using the same machine tools. Employment at the company doubled during 1939, and doubled again during 1940, growing from 5,642 employees in January 1940 to 11,228 in December.[153] This was about the same number of workers employed in the No. 1 Shadow Group factories by June 1941. Like all other aircraft engine manufacturers at that time, Pratt & Whitney soon ran into a shortage of skilled labor but also a need to manufacture engine parts at a faster rate than the human hand could achieve.[154] The company now confronted the necessity of deskilling, transferring the skills of the experienced machinist into a machine tool but at the same time acquiring machine tools that could speed production.[155] This necessity brought up the problem of the type of machine tools Pratt & Whitney should acquire for quantity production.

There were two opposing philosophies of production.[156] One production philosophy called for the factories to be equipped with standard machine tools that were readily available and that could be converted for higher rates of output using special jigs and fixtures. The alternative was to adopt the production methods of the automobile industry using specially designed, single-purpose machine tools that could turn out tremendous volumes of identical parts. The difficulty was that using single-purpose machine tools resulted in more rigid production and less flexibility. Despite the prospect of a vast increase in the volume of parts required, Pratt &Whitney opted for retaining the flexibility to cope with frequent design changes. Where Wright Aeronautical began working with the machine tool companies to develop a third generation of high-production machine tools, accepting the longer time it would take to get delivery, Pratt & Whitney chose to fill its factories with the second generation of machine tools, machines with more automatic features than standard machine tools but were still adaptable to modifications.

As Eugene Wilson, president of United Aircraft, explained in his postwar memoir:

> *Our production was based on the use of standard machine tools; we had few of the special single-purpose machines designed for low-cost volume production of automobiles. Our practice had been deliberately adopted for several reasons. In expanding*

for war-emergency production, we could expect machine tool builders rapidly to expand the delivery of standard machines, where special machines would require much longer. Again, whereas automotive production could be standardized for reasonable periods, in aircraft engines we must be able to incorporate every new design change without seriously retarding production. To accomplish this, we had fitted special jigs and fixtures to our standard machines; for a design change, we would scrap the fixtures but keep the machines running. No such flexibility as this could be had with the special machines of the automotive shops...[157]

Pratt & Whitney also initiated a more disciplined approach to the problem of achieving quantity production. In the summer of 1940, the company carried out a detailed time study of manufacturing operations, calculating the time required between setting up a machine tool and removing the completed part for every machining operation required.[158] From this data, the company calculated the total time required to manufacture each engine component and from this the total time required to build a complete engine. By dividing this amount of time required by the number of working hours per month, the company could calculate the rate of monthly production with the machine tools available and determine if more machine tools were needed, which they were.[159] One of the most important aspects of determining a production schedule was to ensure that the machine tools operated as continuously as possible.[160] Pratt & Whitney found that to ensure "an even flow of work along the production lines" the machine tools needed to operate at 70% capacity.[161]

Among the new automatic machine tools that Pratt & Whitney progressively acquired was a special machine designed by the Kingsbury Company that was equipped with seven tool heads for simultaneously drilling, countersinking, spot-facing, and tapping holes on the engine cylinder head.[162] Three of these machines were set up as a battery, with the first two machines drilling holes and the third machine tapping. These three machines replaced five standard machines and required fewer operators, completing the drilling and tapping operations in 5.22 minutes compared to 50 minutes on standard machines.[163] The Foote Burt Company developed a 15-ton, hydraulically activated external broaching machine, where a series of cutters were drawn across a surface, in this case, rough forgings for articulated connecting rods, to remove excess metal.[164] This machine replaced several two-spindle vertical milling machines, a horizontal milling machine, and a single-spindle vertical milling machine and could produce 300 connecting rods a day, double the number produced with the machines previously used.[165] A Heald Bore-Matic replaced a radial drill press to machine the rear crankcase cover, reducing the time needed to install the workpiece and set up tools, misalignment of tools, and producing 35 parts per 8-hour shift compared to four using a radial drill press.[166]

The other method used to increase the production at the American engine factories was, as had been done in Britain, to add workers. The total number of workers at the Pratt & Whitney factories in the Hartford area rose from 7,454 in January 1940 to a peak of 39,507 in September 1943, a nearly six-fold increase.[167] As the draft called up more and more men, Pratt & Whitney recruited a growing number of women first for inspection of parts and other routine work but later to work on machine tools of all types.[168] By November 1942, Pratt & Whitney employed 4,000 women and women represented 72% of its new hires that month and 85% of trainees at its training schools in the Hartford area.[169]

Building the Twin Wasp at Buick and Chevrolet

After William Knudsen spoke with Harlow Curtice about getting Buick to build the Pratt & Whitney R-1830 Twin Wasp engine, Curtice flew to New York to get approval from the General Motors board.[170] Alfred P. Sloan, the president and chairman of General Motors, expressed his reservations about Buick taking on this task to Secretary of the Treasury Henry Morgenthau, Jr. Despite the difficulties the Allison Engine Company, General Motors' subsidiary, was having boosting the production of its V-1710 liquid-cooled engine, Sloan was happy to support continued development and work out the production at Allison, but he was reluctant to take on a completely new endeavor. He reportedly told Morgenthau "I'm very much disturbed that you people down in Washington have asked Buick to make 5,000 of these air-cooled engines and have us get started into something we don't know anything about."[171] Sloan told Morgenthau that General Motors didn't know "a damn thing" about building air-cooled, radial aircraft engines and that the company "would have to start all over again and learn the business" and would likely have the same troubles General Motors had developing the Allison engine.[172] Buick, however, did go on to learn the business and establish a stellar record of engine production.

As a post-war history of wartime production commented, with some hyperbole, "the entry of the automotive industry into the orbit of aircraft engines seemed to portend the inevitable collision of two inimical worlds, consequent cosmic disturbances, and sparks of nothing less than meteoric size."[173]

> *The aircraft engine makers were horrified at what might befall their product in the hands of automobile people who thought of production in terms of* a thing *a minute and who were endlessly searching for shortcuts. The accepted practices of Detroit, the aircraft power plant producers insisted, could only mean an eventual ascendency of production over engineering; and the moment that occurred, they were certain, safety would vanish and engines that never whispered would be a thing of the past.*
>
> *The automobile men, initially chary of practices 'too precious for words' and inclined to think the aircraft engine makers lacked a proper realism, came on the scene with the conviction that the aircraft men had too little familiarity with 'manufacturing' and that their low production largely resulted from techniques too closely bound to handcraft.*[174]

There was some substance behind this description. Eugene Wilson felt that "to automotive men, the aircraft industry still looked like something like the ancient 'carriage trade,' and they discounted our insistence that our standards of quality and precision were not easily come by in the mass production industries."[175] There would be a process of education on both sides. Wilson recalled that when the Ford Motor Company engineers visited the Pratt & Whitney Hartford factory to familiarize themselves with aircraft engine production techniques, at first they "thought the job looked easy, but the more they had seen of our precision, the more they had become impressed with the task before them. At first, they had thought we might be overdoing the quality but after a look at the test-shed running and studying the high demands for durability and dependability, they had changed their minds."[176]

Even before signing a formal contract with the Army Air Corps, Buick sent hundreds of engineers and production experts to the Pratt & Whitney Hartford factory to learn about aircraft engine production.[177] Pratt & Whitney set up training programs for the managers, engineers, machinists, and production specialists from Buick and its other licensees, with some of the trainees working alongside Pratt & Whitney machinists in the factory.[178] In addition, Pratt & Whitney engineers provided ongoing support and responded to production problems as they arose at the licensee factories and, as production got underway, provided critical parts when shortages of supply threatened to disrupt production.[179]

In the beginning, Buick "had to start as does everyone else, by accepting the standardized procedure developed by the engine builders, using the same methods and, in general, similar equipment."[180] Because of the volume of parts required to meet production targets, the main components were manufactured in separate departments, with machine tools arranged for line production to ensure "a continuous flow of work" from the arrival of raw materials to the inspection of finished parts.[181] Approximately 60% of parts used in the Twin Wasp, including crankshafts and connecting rods, came to the Melrose factory from subcontractors or other Buick factories.[182] Buick did add features based on its automotive production experience that speeded production. At the Melrose factory and several other factories, Buick installed conveyor systems. An overhead monorail at the Melrose factory carried assembled engines to the test cells which, with other improvements, cut the time needed to set up an engine for testing from 3 to 6 hours to just 45 minutes.[183] Industrial factory trucks carried raw materials and parts arriving from subcontractors and Buick factories to the production lines and finished parts to the final assembly area.[184]

Where Buick differed from the Pratt & Whitney factories at Hartford was in the type of machine tools used in the Buick factories making the Twin Wasp. The company's initial experience of building the Twin Wasp enabled Buick management to identify new methods of increasing the pace of production.[185] An important step was the decision to employ a greater number of the latest high-production machine tools than Pratt & Whitney appeared to be using in its Hartford factories. The six-fold increase in Buick's monthly production target from 500 to 3,000 engines a month and the quantity of parts needed to meet this target now justified the use of special-purpose machine tools. These newer tools increased production, replaced standard machine tools, saved floor space, and reduced the need for skilled operators.[186] These special-purpose tools had more advanced features than the automatic tools in use at the Pratt & Whitney factories. Some of them were similar to those employed at the Wright and Studebaker factories, such as eight-station Bullard Mult-au-Matic machines incorporating 16 cutting tools with automatic indexing, Natco five- and six-way horizontal drilling machines and multiple spindle drill presses also with automatic indexing, Gisholt turret lathes, and Heald Bore-matics.[187] Buick used many of these machines to manufacture cylinder heads for the Twin Wasp (Figure 5.12).[188]

FIGURE 5.12 A line of completed cylinder barrels for the Pratt & Whitney R-1830 ready for inspection at the Buick Melrose factory.

Library of Congress, Prints & Photographs Division, FSA/OWI Collection, Reproduction Number LC-DIG-fsa-8e11114.

Seeking ways to increase cylinder head production led Buick to work with the Foote Burt Company to develop an automatic transfer machine, similar but smaller than the large Greenlee automatic transfer machines that Wright and Studebaker used in their factories. Buick production engineers noticed that operations to drill and finish the valve holes on the Twin Wasp cylinder head employed 15 different machine tools and required the cylinder heads to be moved from fixture to fixture, having to be clamped and unclamped on each machine.[189] Buick consulted Foote Burt to see if Foote Burt could make a machine that could keep the cylinder heads in a single fixture, moving the cylinder head in this single fixture automatically between sets of spindles that would also feed and return automatically.[190] Foote Burt developed a large, rectangular machine with 16 individual motorized heads set at the proper angle to perform a series of operations on the cylinder head in sequence and 18 fixtures in total.[191] The machine required just three operators, one to lift cylinder heads off a conveyor belt and place them on the machine, one to remove the finished cylinder heads and place them on another conveyor, and one operator to monitor the machine.[192] The cylinder heads moved around the rectangular machine in a clockwise direction, where the motorized heads drilled and finished the valve guide holes, roughed and finished the intake and exhaust valve seat chambers, and faced the exhaust valve guide holes.[193] At the end of 1943, Buick had two of these Foote Burt automatic transfer machines at the Melrose Park factory. These two machines released 44 men for other duties over three shifts, saving 176 man-hours, while replacing 21 machine tools.[194] Foote Burt developed two more advanced machines for Buick.

One was a ten-spindle, ten-station automatic transfer machine for machining operations on the rocker arm holes on the cylinder head which replaced 15 separate machines previously used.[195] The third machine was a multiple-spindle, vertical drill for boring holes in the rear section of the Twin Wasp crankcase. Two rear crankcase sections could be machined at one time while a third was being loaded. The machine required only one operator, replacing 6.7 operators working with other types of machines.[196]

Chevrolet faced a different set of problems getting into Twin Wasp production. As the Army Air Force doubled Chevrolet's monthly production target, then doubled it again, planning for laying out the factory and purchasing machine tools had to be substantially revised. Chevrolet found that even though the machine tools in its 1938 automobile engine and axle factory were comparatively new, only around 10% could be converted into making aircraft engine parts.[197] Chevrolet's Plant Engineering Department stripped the engine and axle factory and the Fisher Body factory bare and prepared new floor plans laying out the machine tools required to build the 300 parts needed for the Twin Wasp.[198] As more and more parts went out to subcontractors and other Chevrolet factories, the arrangement of machine tools had to be continually revised. In contrast to the Pratt & Whitney factories, Chevrolet borrowed from its automotive methods to make extensive use of equipment for handling materials, installing monorail and gravity conveyors alongside the machine tool lines, and using industrial trucks to move raw materials and parts to where they were needed.[199] By 1944, powered conveyor systems in the Chevrolet engine and axel factory had cut machining times by 5%–20%.[200] By conserving floor space and eliminating the need to load and unload the trucks as they moved from machining station to machining station, the conveyors gave the factory twice the productive capacity of a similar-size factory using only industrial trucks to move materials.[201]

With production targets similar to Buick, Chevrolet also chose to equip the factories with the latest in high-production machine tools. Descriptions of the machine tools in use in the Chevrolet factories feature types of machine tools that were also present in the Buick Melrose Park factory.[202] With experience, Chevrolet's engineers and skilled machinists found ways to improve on Pratt & Whitney's production methods to increase output. In grinding a large cam reduction gear drive for the Twin Wasp, Pratt & Whitney's process required seven steps, the gear having to be removed twice; this method produced 96 gears a day.[203] Chevrolet developed a new method of attaching the gear so that it did not have to be removed, reducing the operation to four steps and increasing production to 288 gears a day, a gain of over 200%. Tapping 224 cylinder mounting holes on the Twin Wasp crankcase required three single-spindle tapping machines.[204] Chevrolet found a way to bolt the three sections of the crankcase together and mount the crankcase on a 16-spindle machine using a new fixture. This increased production by 500% and saved two operators per shift.[205]

Chevrolet faced a completely new challenge when it had to convert its foundry for making cast-iron automobile castings into magnesium castings for aircraft engine parts. Chevrolet had to develop new methods of casting magnesium as the casting process required a high degree of care and accuracy, careful control of temperatures when pouring the molten metal, and more extensive heat treatment of finished castings.[206] Chevrolet engineers found that contrary to contemporary wisdom, magnesium castings could be poured on a moving conveyor, greatly increasing the rate of production.[207] In addition to casting magnesium, Chevrolet, like Buick, employed several factories to forge aluminum crankcase sections, pistons, and other small parts for the Twin Wasp.[208]

To assemble the Twin Wasp, Chevrolet borrowed from its automotive assembly line methods to create a modified assembly line system as Pratt & Whitney and Wright Aeronautical had done.[209] The assembly area in the engine assembly factory building was divided into sections for different subassemblies making up the Twin Wasp engine, such as the cylinder assembly, crankcase sections, crankshaft with master rod and connecting rods, and supercharger sections. The subassemblies were put together on benches that ran at right angles to a conveyor system carrying an engine on a special dolly hooked onto a moving conveyor chain built into the floor. The subassembly benches were arranged in sequence so that the finished subassembly would reach the moving engine dollies at the correct point to be added to the engine. Following the "green" engine test, every engine was taken apart and all component parts inspected, after which the parts moved to the final assembly area which mirrored the initial assembly sequence. Here, every engine was reassembled and prepared for the final acceptance test run prior to shipment.

As at the Buick factories, employment in the Chevrolet factories building the Twin Wasp grew rapidly. From 1,653 employees at the start of 1942, the number grew to a peak of 36,982 in November 1943.[210] Employment levels were comparable to the number of employees at Buick but less than at the Pratt & Whitney factories around Hartford. During 1944, employment at both Buick and Chevrolet declined by about 10,000; yet that year both firms produced more engines than the year before, likely as a result of greater production experience, higher capacity machine tools, and the innovations that each of the firms introduced based on their knowledge of automotive production.

Neither Buick nor Chevrolet made significant changes to the methods of aircraft engine production that Pratt & Whitney had established, but they did make changes. The changes were cumulative, enabling Buick and Chevrolet to attain rates of production that were two to three times greater than at the Pratt & Whitney factories. The high level of subcontracting, getting more firms and factories to produce more parts of the engine was critical to the quantity production of the Twin Wasp. A second factor was the size and design of the Buick and Chevrolet factories. At each division, the total floor space devoted to engine production was nearly two times the space available at the shadow factories building the Hercules engine. The large rectangular shape of the factories was ideally suited to line production to achieve continuous flow through the factories. The third factor was Buick and Chevrolet's use of more high-production machine tools and greater numbers of workers. At their peak levels of employment toward the end of 1943, Buick and Chevrolet had four times as many workers as the entire aircraft engine industry had at the beginning of 1940.[211]

Building the Cyclone 14 at the Wright Lockland Factory

In contrast to Wright's prewar practice of building almost all engine components in its own factories, the Lockland factory adopted a different approach to production. As America began to rearm, the National Defense Advisory Council recognized that "...dependence on subcontractors for a substantial amount of work was essential to a maximum utilization of facilities and expansion of production."[212] As a result of the Council's advice and Government urging, the Lockland factory relied far more on subcontractors than other Wright factories, even for core components of the Cyclone 14. The Lockland factory maintained a higher-than-average dependence on subcontractors until the end of the war, running at 40% of the value of the Cyclone 14 engine compared to an industry average of 29%.[213] Among the thousands of subcontractors brought into the aircraft industry were many firms that supplied parts to the automobile industry. In January 1941, Wright placed contracts with the Graham-Paige Motor Company for Cyclone 14 connecting rods, the Eaton Manufacturing Company for propeller shafts, the Ohio Crankshaft Company for Cyclone 14 crankshafts, and the Hudson Motors Car Company for Cyclone 14 pistons and rocker arms.[214] Wright also engaged the Otis Elevator Company in New Jersey to make finished crankcases for the Cyclone 14 (Figure 5.13).[215]

FIGURE 5.13 A process that was difficult to automate. Inserting headless nails by hand into one half of the sand mold for a Cyclone 14 cylinder head.

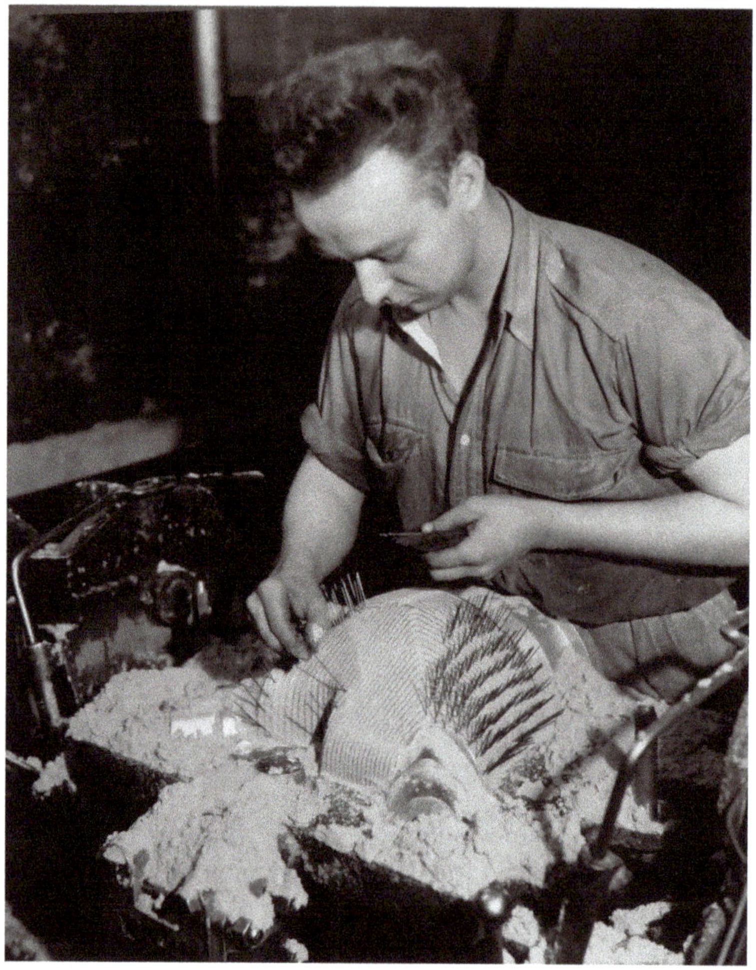

Wright Aeronautical, US Farm Security Administration/Office of War Information/Office of Emergency Management/Resettlement Administration Black & White Photographs, Prints and Photographs Reading Room, Library of Congress.

Production of the Cyclone 14 began with casting the aluminum alloy cylinder heads in the Lockland factory's aluminum foundry. While Wright had adopted line production methods for its foundry operations, making the sand molds for cylinder heads remained a labor-intensive process. Mold makers had to pack sand by hand around a bronze pattern of half the cylinder head and insert 750 headless nails in each half of the mold to support the walls of the cooling fins, the two halves being bolted together to form the finished mold.[216] To speed production, a conveyor carried the finished molds through an oven for 7 hours to harden the mold before receiving the molten aluminum. Once cooled, machines removed excess metal. Casting of magnesium sections for the crankcase nose and the supercharger housing followed a similar process of preparing sand molds, pouring molten metal, and removing the excess when the part had cooled. After rough finishing, cast parts went to the machine shop for machining and assembly.

Wright estimated that manufacturing the 8,500 parts in the Cyclone 14 engine required over 80,000 machining operations and 50,000 inspections, although there were probably fewer machining operations at the Lockland factory because of the level of subcontracting.[217] The Wright Lockland factory incorporated all the lessons Wright had learned shifting from batch to quantity production at its Paterson factories. The Lockland factory combined line production methods with high-production machine tools and a vast floor area in the main machine shop and assembly building.[218] The unobstructed floor area allowed machine tools to be placed in an optimal sequence of operations, ensuring a one-way flow of parts from rough stores to final assembly. To speed up the flow of parts, the Lockland factory complex contained over 5 miles of conveyors, mechanical tracks, and overhead lines.[219]

In its first configuration, the main machine shop and assembly building at Lockland was divided into twelve sections or manufacturing departments, each of which was a factory by itself to manufacture a particular part or component. Manufacture of complete cylinders, for example, began in a department devoted to machining cylinder barrels with a separate department for machining cylinder heads. The finished parts moved to a department that completed assembly, attaching the cylinder head to the cylinder barrel. All parts from the foundries and subcontractors entered on the west end of the machine shop and assembly building, moving steadily east to the final inspection department and finally to the final assembly area and the "green" and final engine test cells. Constructing the large South Shop Building and the smaller North Shop Building provided Wright with space to expand each manufacturing department. The final assembly of the Cyclone 14 followed the line production assembly methods developed at the Wright Paterson factories, with completed subassemblies feeding into the main assembly line at the appropriate point. Once assembled, a finished engine went through a 7-hour "green" test followed by complete disassembly and inspection of all parts, before being reassembled and sent to the final test run for an additional 5½ hours (Figure 5.14).

FIGURE 5.14 A specially designed Greenlee multistation machine tool at the Lockland factory that could automatically perform 45 boring and facing operations on the Cyclone 14 supercharger housing. This machine replaced eight separate machine tools and cut the production time from 4 hours to 24 minutes.

na051791, Courtesy of the National Automotive History Collection, Detroit Public Library.

The most important feature of the Lockland factory was its extensive use of high-production machine tools, mainly designed specifically for the factory. *The Wall Street Journal* pronounced that at the Lockland factory "breath-taking things are being done to arm America" through "the wholesale use of automatic machinery."[220] The Lockland factory benefited directly from the experimental work done at the Paterson factories on developing high-production machine tools.[221] The Lockland factory initially had over 2,000 machine tools installed; by the end of the war, the total had reached almost 6,000, a third more than the number at the Studebaker factories at Chicago, Ft. Wayne and South Bend, and double the number at the Buick Melrose Park factory.[222] The Lockland factory employed many more of the latest in automatic and semiautomatic machine tools.[223] As an example, the factory used several different types of multiple-spindle drilling machines, including a 21-spindle Foote Burt drill to drill 21 holes in the crankcase simultaneously, a Baker multiple-spindle drill that automatically drilled 24 holes in the supercharger housing and many Bullard Mult-Au-Matic machines, three of these machines replacing 18 standard machines used at the Paterson factories.[224] A reporter noted that the Lockland factory "has few single-spindle machines, whereas in Europe today there are almost no multiple-spindle machines in an aviation plant."[225] For the first time, Wright used Heald internal centerless grinders to machine finish the inside of the cylinder barrel, cutting the machining time for this process in half.[226] A specially designed Greenlee multistation machine reduced the machining time on supercharger front sections from 224 to 24 minutes.[227] Using these automatic machine tools not only speeded production but reduced the total number of machine tools required to build the Cyclone 14 and enabled Wright to hire thousands of semi- and unskilled workers in the Cincinnati area who were put through short, intensive training courses on the very machines they would be using in the machine shops.[228]

The planned rate of Cyclone 14 production at Lockland made the Greenlee automatic transfer machine that Wright had developed vital to achieving this goal. The need for 14 cylinders per engine, as well as spare cylinders, meant more cylinder heads had to be manufactured. To reach the scheduled rate of production of 3,000 Cyclone 14 engines a month required 42,000 cylinder heads, roughly 1,350 per day. Automatic transfer machines reduced the number of machine tools and operators and dramatically speeded up cylinder head production. The first Greenlee automatic transfer machine at Lockland consisted of two machines linked together to form one machine 80 ft in length. The first machine had 16 machining stations performing 25 different operations. The cylinder heads, mounted on steel plates, passed via a conveyor to the second machine with 53 stations, which performed 46 operations on the front of the cylinder head. At the end of the machining cycle, a washer automatically cleaned the parts. The machine could turn out a finished cylinder head in 2 minutes, compared to 35 minutes using previous methods.[229]

Wright soon replaced this early model with a more advanced and even longer set of three Greenlee automatic transfer machines that totaled 154 ft in length.[230] The first machine had 16 stations that performed 25 operations on the front of the cylinder head and 20 operations on the rear cylinder head as the cylinder head moved down a conveyor, automatically stopping at each station for machining. After passing through a washing operation, the cylinder heads traveled to the second machine that had 56 stations performing 46 operations on the front of the cylinder head and 37 operations on the rear. The longest operation required 48 seconds to complete, so the cylinder heads moved to the next station in line at this pace, every 48 seconds. While the entire machining operation on a cylinder head took just under an hour, this more advanced automatic transfer machine effectively completed a cylinder head every 48 seconds. Once the cylinder heads had been attached to the cylinder barrels forming a complete unit, they were placed in a third Greenlee automatic transfer machine with 25 stations for the final drilling and boring operations. Collectively, these three machines replaced 75 other types of machine tools, required 202 fewer operators, and saved 1,671 production hours.[231] The Greenlee automatic transfer machines required only 24 operators for three shifts, and of these 20 would be unskilled, whereas using the previous semi-production machines required semiskilled operators and many more skilled workers to set up the machines.[232]

While the Greenlee automatic transfer machines received the most attention in the press, during the war Greenlee Brothers, working with Wright, Studebaker, and other aircraft engine manufacturers, developed other types of high-production machine tools. Studebaker, for example, worked with Greenlee to develop a rotary, six-station, automatic indexing machine that would automatically perform more drilling, tapping, and other operations on the completed cylinder assembly, moving the workpiece automatically from station to station.[233] Greenlee produced a series of these horizontal, rotary automatic indexing machines.[234] Wright and Greenlee developed a more advanced 14-station rotary automatic indexing machine that would drill, bore, and face holes in the front section of the Cyclone supercharger.[235] Two of these 14-station machines replaced seven standard machines, reducing the number of operators from 19 to two and production hours from 139 to 17.4 hours.[236] Later in the war, the Dodge-Chicago factory used an 18-station automatic indexing machine.[237]

Wright worked with Greenlee to develop automatic machine tools specifically for machining crankcases for the Cyclone 14. Each crankcase had 14 flat deck pads to which the cylinders were attached. Each deck pad had to be machined to a smooth surface and then have 20 holes drilled, reamed, counterbored, and tapped so the cylinder flanges could be bolted to the crankcase, totaling 280 holes for all 14 cylinder deck pads. This process normally took 25 machine tools and 571 production hours to complete. Using four Greenlee four and six station machines cut the production time to 65 hours and reduced the number of machine operators from 66 to 9.[238]

Wright did not restrict the new methods of production and process engineering it had developed to the Lockland factory but, in at least one case, passed these methods on to one of Lockland's main subcontractors, the Ohio Crankshaft Company, who manufactured crankshafts for the Cyclone 14. The company had many years of experience in building crankshafts for large gasoline and diesel engines, so was an ideal subcontractor for the Lockland factory. Soon after receiving the crankshaft contract, the Ohio Crankshaft Company received $4.5 million in Government financing to build a new 200,000 ft^2 factory and an additional $3 million to purchase 310 new high-speed machine tools.[239] The Cyclone 14 crankshaft had three sections—center section, rear and counterweight sections, and front section—and contained 30 parts.[240] Manufacturing the crankshaft required 607 separate operations as follows:

(a) Center section:
Rough forging, 65 lb Finished, 26 lb Operations, 135

(b) Rear and counterweight sections:
Rear-rough, 85 lb Finished, 46 lb Operations, 240

(c) Counterweights:
Front counterweight-rough, 30 lb Finished, 16 lb
Rear counterweight-rough, 24 lb Finished, 10 lb
Operations for both weights, 50

(d) Front section:
Rough, 118 lb Finished, 52 lb Operations, 70[241]

At the Lockland factory, the new Ohio Crankshaft Company factory building was laid out according to the principles of line production.[242] The Company divided the factory floor space into three areas corresponding to the main sections of the crankshaft—center section, rear and counterweight sections, and front section—with the machine tools in each section laid out for straight-line production. A system of 96 overhead cranes moved heavy forgings and finished sections along the production line. The latest types of automatic machine tools performed the required machining, grinding, precision boring, and finishing operations, although final finishing was done with handheld tools.[243] Each finished crankshaft passed through three inspections at various stages of production, one from Ohio Crankshaft, one from Wright, and one from AAF inspector before a final inspection by all three representatives and shipment to the Lockland factory (Figure 5.15).[244]

FIGURE 5.15 The new method Wright developed with the Scandia Manufacturing Company to increase the cooling area on Cyclone engine cylinder barrels. The cylinder barrel on the left had 40 fins machined on the barrel. The cylinder on the right had 60 aluminum fins inserted into grooves cut into the side of the cylinder.

Original Wright Aeronautical photo in the author's possession.

In 1944, Wright developed a new means of increasing the cooling area to improve the cooling of higher power Cyclone engines, such as the Cyclone 14BB built at Lockland. This involved increasing the depth of the cooling fins on the cylinder barrel through an entirely new method of production using new types of machine tools. The previous method involved machining the cooling fins on the steel cylinder barrel forging by using an automatic lathe that could cut the 46 fins in one operation, removing 17 lb of steel per cylinder barrel in the process.[245] Increasing the depth of fins on the cylinder barrel would have required a much heavier forging to get the additional fin depth, resulting in a heavier cylinder barrel and complicated machining.

Instead, Wright engineers developed a completely new method invented at the Scandia Manufacturing Company which used thin aluminum fins, in the shape of a "W," inserted in grooves cut into the sides of the cylinder barrel.[246] Using this method, 60 aluminum fins replaced 46 steel fins, increasing the cooling area by 55%, reducing the required thickness of the cylinder barrel, cutting the machining time to less than one-third of the previous method and saving an estimated 12,000 tons of nitralloy steel a year.[247] Wright modified automatic lathes that had cut the cooling fins on the cylinder barrel to cut 30 grooves for the aluminum fins and developed special machine tools to roll thin aluminum sheets into the "W" shape, insert the fins into the cut grooves on the cylinder barrel, and caulk the fins to lock them in place.[248]

One of the most important aspects of line production was production control, defined as "...the planning, scheduling, dispatching, and expediting of machines, materials, and men within the plant."[249] Maintaining quantity production through progressive flow required timely delivery of parts and raw materials to the factory, often ordered months in advance to ensure supply, the correct placement of machine tools in the proper sequence of operations, and a schedule to maintain the pace of production from start to finish, "... feeding parts and subassemblies to the various work areas at the rate and time required."[250] Engineering and design changes, such as the shift to using the "W" aluminum cooling fins, disrupted this flow and had to be planned well in advance. But there were other unplanned interruptions that the Lockland factory had to deal with, particularly changes in aircraft engine production schedules and problems with subcontractors (Figure 5.16).

FIGURE 5.16 Scheduled vs Actual Production at the Wright Lockland Factory in 1944.

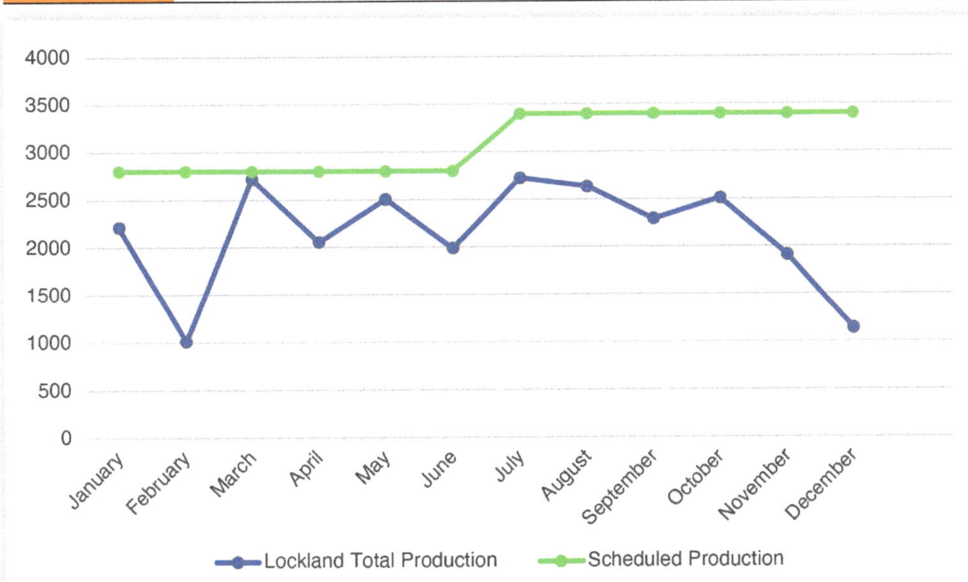

Source: US. Department of Commerce, Civil Aeronautics Administration, Office of Aviation Information, Division of Aircraft Statistics: US Military Acceptances 1940-1945: Aircraft, Engines and Propeller Production, (Washington, DC, 1946); Lilley, Tom, et al: Problems of Accelerating Aircraft Production During World War II, (Cambridge, MA, 1947), Table 20.

The Army Air Force's early 1943 production schedule for the Lockland factory had called for the factory to reach the peak capacity of 4,000 Cyclone 14 engines a month by May 1944, based on aircraft requirements.[251] During that year, the AAF revised the schedule downward, the November 1943 schedule calling for the Lockland factory to reach a peak of 3,400 engines a month by July 1944 and maintaining this rate of production. Even though this would result in a cumulative engine deficit compared to requirements, the AAF could manage the deficit by reducing stocks of spare engines.[252] In January 1944, the AAF decided to cut the production of the Cyclone 14 at the Lockland factory and convert the factory into building the higher-priority Cyclone 18, beginning in November 1944.[253] Even with the AAF's decision to cut back peak Cyclone 14 production to 3,000 engines a month starting in July 1944, this decision played havoc with scheduling Cyclone 14 production for the rest of the year.[254] In April, the Navy requested additional Cyclone 14BB engines that had not been planned for. This led to the fifth revision in Lockland's production schedule since January, causing one AAF officer to comment that "…Wright Aeronautical Corporation are now so confused that they feel it is practically impossible to furnish any kind of production commitment for either the R-2600-B [Cyclone 14BA] or the R-3350 engine [Cyclone 18]."[255]

In early 1944, the Lockland factory experienced another unplanned disruption in production when the Graham-Paige Company produced flawed master rods due to poor management control and machining processes, causing a loss of 1,850 engines from production in February, March, and April, which would result in a production loss of 4,481 engines for the year.[256] By halting or reducing aircraft programs that depended on the Cyclone 14, the AAF restored some semblance of balance to requirements.[257] As part of this rebalancing, the AAF agreed to Wright's proposal to cut back the production for much of 1944 to 2,500 a month.[258] Despite the fact that the Lockland factory built 25,680 Cyclone 14 engines in 1944, these schedule changes caused "serious operating problems."[259] In 1944, the Lockland factory had difficulties meeting its production schedule. Lockland never managed to reach its full production capacity of 4,000 engines a month or the revised target of 3,400 engines a month. It is unclear exactly what problems prevented the Lockland factory from meeting its production target. There may have been problems with the adequate supply of parts from subcontractors, but this failure may also have been due in part to problems with Wright's management, which emerged in an embarrassing fashion during 1943.

The real concern with the quality of management at the Lockland factory stemmed from an investigation by the Senate Special Committee to Investigate the National Defense Program, better known as the Truman Committee from its chairman, Senator Harry S. Truman. Responding to complaints of improper inspection procedures, the Truman Committee reported that the Lockland factory had delivered defective engine parts through lax inspection procedures, faulty record keeping, and management pressure to favor production over inspection.[260] In response to the Truman Committee's investigation, the AAF conducted its own review of the Lockland factory and severely criticized Wright management and its own supervision staff for allowing incompetent management at lower levels, inadequate numbers of insufficiently trained inspectors, improper handling of salvaged and rejected parts, and "excessive machinery, floor space, and manpower," which seems an odd comment given the factory's scheduled monthly production target.[261] The AAF reduced the Lockland factory's Quality Inspection Control rating from A to B.[262]

The AAF's report concluded that the quality control and inspection systems that experienced personnel used at the Paterson factories were poorly documented and insufficiently rigorous for the new and inexperienced inspection personnel at the Lockland factory.[263] While praising the factory manager for the "magnificent job" he had done building up the new organization at Lockland and training the machine operators, the report found that the training of inspection personnel was insufficient.[264] The AAF report admitted that the planned expansion in capacity from 1,000 to 4,000 engines a month by

1944 was an enormous undertaking but criticized Wright's senior management for failing to send a senior executive from Wright's head office in New Jersey to oversee the expansion and to provide competent assistants to take on more responsibility for the full range of activities at the factory.[265] Once the deficiencies came to light, the AAF instituted more stringent inspection procedures, assigned more AAF inspectors, imposed weekly reporting requirements to improve inspection, and increased the engine run-in time from 7 to 10 hours to ensure no defective parts slipped through.[266]

The problem with the production of master rods at the Graham-Paige Company a few months later was another indication of problems with the management at the Lockland factory. The AAF blamed the problems on the failure of the Materials Control System at the factory and Wright's inadequate supervision of its subcontractor.[267] By the summer of 1944, however, the AAF believed that all the problems at Lockland had been resolved. The AAF representative at the Lockland factory was confident that the factory could meet the revised target of 3,000 engines a month and deliveries in July 1944 were short of the objective. The Lockland factory never reached a rate of building 3,000 Cyclone 14 engines a month before converting into the production of the Cyclone 18 in late 1944.[268]

The management and production problems Wright encountered at the Lockland factory raise the question of whether Wright simply ran into diseconomies of scale at Lockland. The concept of economies of scale—how increases in the scale of production can reduce average product costs—is well-known; the concept of diseconomies of scale is perhaps less so.[269] Diseconomies of scale refer to conditions where costs can actually increase as output increases.[270] There can be many sources of diseconomies of scale, such as limits on the supply of a factor of production or declining efficiency in the use of a production factor as its use increases, but management and organizational structure can also be a key contributor to diseconomies of scale.

Management effectiveness may decline as the scale of the organization increases.[271] As C.F. Pratten argues, "…as scale increases, the costs of coordinating and organizing production may rise more than proportionally. The effectiveness of management may decline as the chain of management is extended because of delays in making decisions brought about by the length of the management chain and/or the tendency of those making decisions to get out of touch with events affecting the decisions."[272] In particular, the quality of information may decline as the levels of management expand with the increase in scale, as the "accuracy and amount" of information passing upward declines.[273] Oliver Williamson speaks to the issue of "communication distortion," where lower-level managers tell their superiors what they think their superiors want to hear, thereby distorting the reality of the situation.[274] There may also be, as seems apparent at the Lockland factory, a direct relationship between the scale and the quality of management at middle and lower levels.[275] As firms expand rapidly, as so many did during World War II, they may or may not be successful in recruiting capable managers.

Figure 5.17 shows how employment at the Lockland factory expanded after the AAF's decision to increase the production capacity from 1,000 to 4,000 engines a month. The initial plan for the Lockland factory was to build up to a total of 12,000 employees, but within two years from the start of production, the total had nearly tripled.[276] The situation at Wright's Paterson factories was similar. Between January 1941 and December 1943, the total number of employees at Paterson and the Wood-Ridge factory nearly tripled.[277] Managing this rate of expansion would be challenging at any time, but during wartime with intense pressure to meet the production schedules, it must have been doubly difficult. It is remarkable that in 1939 Wright had a little more than 3,000 employees and four years later had over 70,000. Where and how Wright acquired its middle management and supervisory staff for its seven factories is a question that remains unanswered but warrants further research.

FIGURE 5.17 Increase in Total Number of Employees at Wright Lockland Factory 1941–1943.

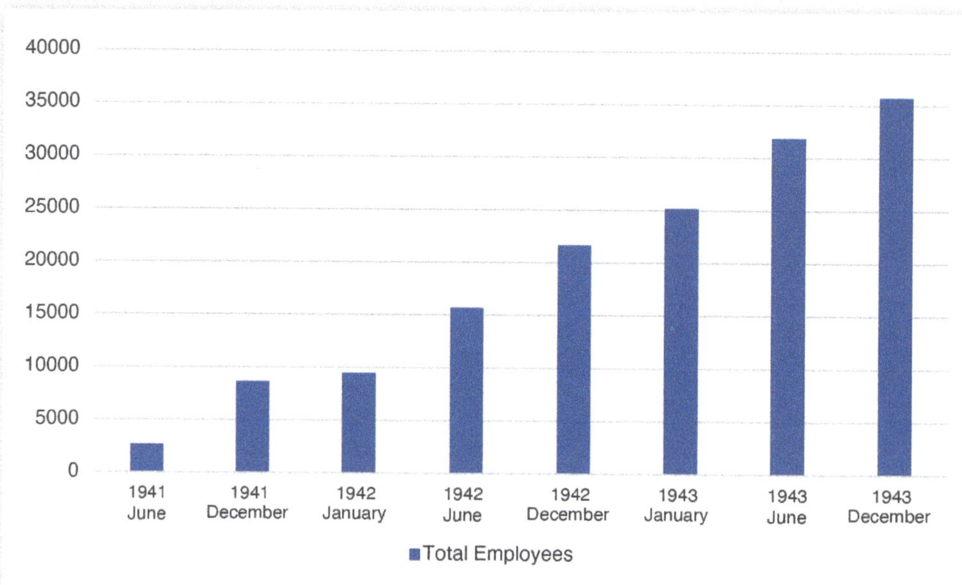

Source: US Department of Commerce, Civil Aeronautics Administration, Office of Aviation Information, Division of Aircraft Statistics: US Military Acceptances 1940-1945: Aircraft, Engines and Propeller Production, (Washington, DC, 1946).

Training was an issue from the very start of planning for the Lockland factory. In the early stages, when Wright expected to have 12,000 employees working at the factory, the company estimated that more than 60% of employees would have limited, if any, experience operating machine tools.[278]

The problems with management, production, and inspection at the Lockland factory did improve, but the AAF remained concerned with the lack of qualified senior-level managers for production, production control, and material controls.[279] The AAF was fortunate that it could meet the shortfall in Cyclone 14 production during 1944 by cutting the production of aircraft that used the Cyclone 14. The Wright Lockland factory's record demonstrates that managing a giant factory was not a simple extrapolation of methods and processes achieved at smaller factories.

Neither Buick nor Chevrolet experienced similar problems with managing production and subcontractors. Production at Buick and Chevrolet was more stable, as indicated in Figure 5.18 which shows the production at the Wright Lockland factory, Buick, and Chevrolet from January until September 1944, when all three factories began converting part of their production into different engine models. This was the year of peak production at all three factories when one might assume that cumulative experience would have translated into maximum efficiency. As the Harvard Business School study of aircraft and aircraft engine production in World War II noted, "probably the greatest contributions of the licensee companies to the engine production program were their existing management organizations that knew how to carry out production engineering in all its phases."[280]

FIGURE 5.18 Monthly Engine Production at Wright Lockland, Buick, and Chevrolet January-September 1944.

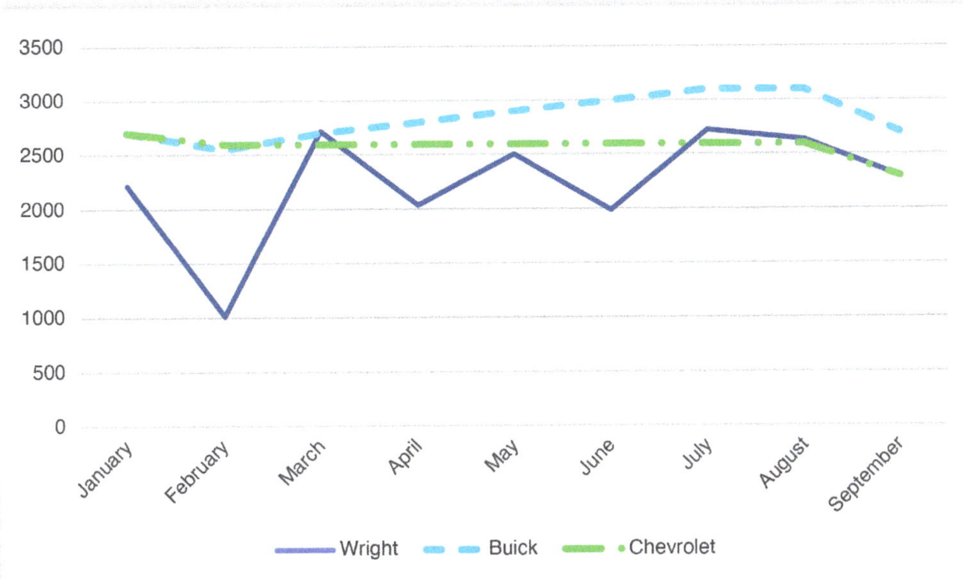

US Department of Commerce, Civil Aeronautics Administration, Office of Aviation Information, Division of Aircraft Statistics: Source: US Military Acceptances 1940-1945: Aircraft, Engines and Propeller Production, (Washington, DC, 1946).

Conclusion

The record of the Buick and Chevrolet divisions and the Wright Lockland family demonstrate again that aircraft engine production in America was on a completely different scale from production in Britain. The Buick, Chevrolet, and Wright Lockland factories each individually built more aircraft engines than the entire No. 1 and No. 2 Shadow Groups building the Bristol Hercules. The total floor space devoted to building the R-1830 Twin Wasp and the R-2600 Cyclone 14 amounted to more than 10 million ft^2, more than double the factory floor space for building the Hercules. At peak production, all three American factories built more engines per month than the combined shadow factories. Greater demand for the R-1830 and the R-2600 engines was clearly a factor; the American military needed tens of thousands of these engines for heavy, medium, and light bombers, including the US Navy's dive and torpedo bombers for its vast carrier fleet. The factories building these engines were generally larger than their British equivalents, in the case of the Wright Lockland factory, bigger than all the shadow factories combined. With no threat of air attack, the American factories could be built as large, single-story structures better suited for line production with no internal obstructions interrupting the flow of work.

As with Studebaker's experience building the R-1820 Cyclone 9, the record of Buick and Chevrolet in building the Pratt & Whitney R-1830 Twin Wasp illustrates the three key components of successful license production the Harvard Business School study enumerated. First, the Pratt & Whitney licensees were assigned an engine that was fully developed and in production; second, they received full cooperation from Pratt & Whitney with training and the transfer of information; and third, the management of the Buick and Chevrolet divisions brought years of experience with mass production to the task of building aircraft engines.[281] The Harvard Business School Study noted that the management of these factories were, however, doing a job "which neither they nor others had ever done before."[282]

In contrast, the experience of the Wright Lockland factory demonstrates the difficulties and challenges of managing a branch factory under wartime pressures. The Harvard Business School study observed that the aircraft and aircraft engine manufacturers who established branch factories had a choice between choosing managers with experience in their own company or men with experience in managing large-scale factories.[283] Invariably, as Wright did, the companies selected men from within their own organizations who by the nature of prewar levels of production had limited experience in managing a large organization. Growing from 8,650 employees to 35,702 employees at the Wright Lockland factory in the space of two years would have challenged the most experienced management. Setting up these branch factories "created a series of extremely complex administrative relationships entirely new to most companies in the industry."[284] The difficulties that the Wright Lockland factory encountered took place at a time when Wright Aeronautical's senior management was struggling with the difficulty of getting its even bigger engine, the Wright R-3350 Cyclone 18, into production.

References

1 These figures come from Ministry of Aircraft Production: *Statistical Review 1939-1945*, (1946), Table 9, p. 26, and U.S. Department of Commerce, Civil Aeronautics Administration, Office of Aviation Information, Division of Aircraft Statistics: *U.S. Military Acceptances 1940-1945: Aircraft, Engines and Propeller Production*, (Washington, DC, 1946).

2 *U.S. Military Acceptances 1940-1945: Aircraft, Engines and Propeller Production.*

3 Ibid., pp. 141-43.

4 Ibid., p. 142.

5 Ibid.

6 Ibid., pp. 141-45.

7 History of Production of Bristol Aircraft engines by the Shadow Industry, Abridged Record of Engines, Spares and Accessory Output from Shadow Industry 1941-1945.

8 Lilley, *Problems Accelerating Aircraft Production in World War II*, p. 33.

9 Ibid., p. 34.

10 Angle, Glenn D.: *Aerosphere 1939*, (New York, 1939), p. 601.

11 Bingham, Victor F.: *Major Piston Aircraft Engines of World War II*, (Shrewsbury, UK, 1998), p. 98.

12 Angle, Glenn D.: *Aerosphere 1939*, (New York, 1939), p. 618.

13 Ibid., p. 620; Angle, Glenn D.: *Aerosphere 1942*, (New York, 1942), p. B-44.

14 This description of the characteristics of the Twin Wasp is drawn from Bridgman, Leonard: *Jane's All the World's Aircraft 1942*, (New York, 1943), pp. 57d-58d.

15 Wright Aeronautical Corporation, *Wright Double-Row Cyclone 14 Aircraft Engines*, (No Date).

16 Ibid.

17 Angle, Glenn D., ed.: *Aerosphere 1939*, (New York, 1939), p. 827.

18 The first version of the Cyclone 14 became the Cyclone 14A when Wright introduced the more powerful 1,700 H.P. model as the Cyclone 14B. This became the Cyclone 14BA when the more powerful Cyclone 14BB came into service.

19 "The Double Row Cyclone 14", *Trade Winds*, Vol. 4, No. 11, (July 1939), p. 7; Wright Aeronautical Corporation, *Wright Double-Row Cyclone 14 Aircraft Engines*, (No Date).

20 Daniel and Cuny: *L'Aviation Française de Bombardement et de Renseignment (1918/1940)*, p. 325.

21 Ibid., p. 239; *French Air Commission: Situation des Contrats au 16 Juin 1940*, AVIA 38/426, NA.

22 *French Air Commission: Situation des Contrats au 16 Juin 1940*, AVIA 38/426, NA.

23 Daniel and Cuny: *L'Aviation Française de Bombardement et de Renseignment (1918/1940)*, p. 239; *French Air Commission: Situation des Contrats au 16 Juin 1940*, AVIA 38/426, NA.

24 *French Air Commission: Situation des Contrats au 16 Juin 1940*, AVIA 38/426, NA.

25 Ibid.

26 Ibid.

27 Report No. 7, Table III-D, p. 12; Aeronautical Section, NDAC, Army, Navy, Airframe and Engine requirements, June 6, 1940, Vol. 269, June 5-6, 1940, pp. 265-67, 282, 285.

28 Report No. 7, Table III-D, p. 12; Butler, Phil, with Dan Hagedorn: *Air Arsenal North America: Aircraft for the Allies 1938-1945 Purchases and Lend-Lease*, (Hinckley, UK, 2004), p. 28.
29 Report No. 7, Table III-D, p. 12.
30 Ibid.
31 Ibid.
32 Ibid., Table I, p. 7.
33 *The Wall Street Journal*, February 23, 1939, p. 13; March 24, 1939, p. 5.
34 *The Wall Street Journal*, March 15, 1940, p. 25; May 28, 1940, p. 39.
35 Schedule A, Total Capital Commitments of the British and French Governments to the Aircraft Industry, May 15, 1940, "Capital Commitments of the British and French Governments to the Aircraft Industry, May 15, 1940", Mr. Young to Secretary Henry Morgenthau, May 16, 1940, Vol. 263, May 16-19, 1940, p. 26; Hall, *North American Supply*, p. 289n.
36 Special Study No. 21, p. 62.
37 Special Study No. 21, p. 38; *The Wall Street Journal*, September 6, 1940, p. 1; United Aircraft Corporation: *The Bee Hive*, Vol. XV, No. 9, 10, (September-October 1940), p. 4.
38 Conference call between William Knudsen and Secretary Henry Morgenthau, October 31, 1940, Vol. 326, October 29-30, 1940, p. 295-99, Diaries of Henry Morgenthau, Jr.
39 *The New York Times*, January 3, 1941, p. 8; *The Wall Street Journal*, January 20, 1941, p. 5.
40 *The Chicago Tribune*, January 27, 1941, p. 1.
41 *The Detroit Free Press*, July 1, 1941, p. 13; *The Chicago Tribune*, July 1, 1941, p. 27.
42 *The Chicago Tribune*, July 1, 1941, p. 27.
43 *The Wall Street Journal*, July 18, 1941, p. 9.
44 *The New York Times*, September 11, 1941, p. 18.
45 Craven and Cate *The Army Air Forces in World War II*, Volume Six: *Men and Planes*, p. 312.
46 Conference on the Air Expansion Program, May 30, 1940, Vol. 267, May 28-31, 1940, p. 363.
47 Ibid., p. 353.
48 Ibid., pp. 354-61.
49 Ibid., p. 356.
50 Assistant Secretary of the Treasury Edward Foley to Secretary Morgenthau, June 13, 1940, Vol. 272, June 12-14, 1940, pp. 124-25, Diaries of Henry Morgenthau.
51 White, *Billions for Defense*, p. 22.
52 Ibid.
53 *The New York Times*, July 27, 1940, p. 1.
54 White, *Billions for Defense*, p. 30.
55 *The Wall Street Journal*, October 24, 1940, p. 11; November 4, 1940, p. 5; Maj Gen George Brett to Assistant Secretary of War, re Contract W535 AC-16288, November 1, 1940, File 452.8 Wright Engines November 21, 1941 to January 14, 1939, Box 817, Entry 290, Army Air Force Central Decimal Files 1939-1942, RG 18, NARA.
56 *Hartford Courant*, May 12, 1929, p. 1.
57 Ibid., p. 2; July 30, 1929, p. 1.
58 *Hartford Courant*, May 15, 1929, p. 24.
59 *Hartford Courant*, August 19, 1929, p. 21; March 22, 1930, p. 6.
60 Ward, J. Carlton, Jr.: "Plant Layout and Production Methods for Modern Aircraft Engines", *S.A.E. Journal (Transactions)*, Vol. 40, No. 5, (May 1937), p. 179.
61 Ibid.
62 Ibid., pp. 179-81.
63 Ibid., p. 180.
64 *The Wall Street Journal*, May 25, 1936, p. 2.
65 *The Wall Street Journal*, February 23, 1939, p. 13; United Aircraft Corporation: *The Bee Hive*, Vol. XIV, Nos. 2 and 3, (February-March 1939), p. 3.
66 Report by Philip Young on Pratt & Whitney, Grumman, Brewster, and Sperry, October 19, 1940, Vol. 323, October 17-20, 1940, p. 323, Diaries of Henry Morgenthau.
67 AuWerter, Jay P.: "Pratt & Whitney Expands", *Aviation*, Vol. 39, No. 7, (July 1940), p. 39.
68 *The Wall Street Journal*, May 28, 1940, p. 2.
69 Report by Philip Young on Pratt & Whitney, Grumman, Brewster, and Sperry, October 19, 1940, p. 25.
70 *The Wall Street Journal*, August 22, 1940, p. 2; Aeronautical Chamber of Commerce: The *Aircraft Yearbook for 1941*, (New York, 1941), p. 310.
71 United Aircraft Corporation: *The Bee Hive*, Vol. XV, Nos. 9 and 10, (September-October 1940), pp. 4-5, 7; "Rapid Expansion of Engine Plant Leaves Time for Improvements", *Engineering News-Record*, (October 24, 1940), p. 70-71.

[72] United Aircraft Corporation: *The Bee Hive*, Vol. XVI, Nos. 1, 2, (January-February 1941, p. 3.

[73] United Aircraft Corporation: *The Bee Hive*, Vol. XVI, Nos. 3 and 4, (March-April 1941), p. 4.

[74] United Aircraft, *The Pratt & Whitney Story*, p. 133.

[75] See Aircraft Engine Report 9-G, Report on Airplane Engine Deliveries (October 1, 1941 through June 30, 1943), November 28, 1941, Office of Production Management, Aircraft Branch Reports November 28, 1941-January 20, 1942, Container 207, and Aircraft Engine Report 9-I, Report on Airplane Engine Deliveries (January 1, 1942 through April 30, 1944), February 26, 1942, War Production Board Reports, Aircraft Production Branch, January 31- April 15, 1942, Container 238, War Production Board Reports, Aircraft Production Branch, Delivery Schedule of Aircraft Engines: Aircraft Report 9-K, Container 239, Papers of Harry L. Hopkins, Special Assistant to the President, 1941-1945, Group 24, Franklin D. Roosevelt Presidential Library and Museum, Hyde Park, NY.

[76] *Hartford Courant*, November 21, 1942, p. 14.

[77] *The New York Times*, February 11, 1943, p. 40; *Hartford Courant*, June 27, 1942, p. 7.

[78] United Aircraft, *The Pratt & Whitney Story*, p. 133; Surplus War Property Division of Defense Plant Corporation: *Advance Listing of Industrial Plants and Plant Sites to be disposed of by Defense Plant Corporation*, (Washington, DC, 1944), p. 20.

[79] Surplus War Property Division of Defense Plant Corporation: *Advance Listing of Industrial Plants and Plant Sites to be disposed of by Defense Plant Corporation*, p. 20.

[80] Crow, Carl: *The City of Flint Grows Up: The Success Story of an American Community*, (New York, 1945), pp. 100-101; *Chicago Tribune*, January 25, 1941, p. 3.

[81] Ibid., p. 101.

[82] *Engineering News-Record*, April 3, 1941, p. 6; *Chicago Tribune*, January 27, 1941, p. 1.

[83] *Chicago Tribune*, January 27, 1941, p. 1.

[84] Defense Plant Corporation: *Manufacturing Plant, Melrose Park, Ill., Plancor 39*, (No date), Entry A176 Pamphlets Relating to Manufacturing Facilities 1940-1945, Container 1, Plancor Numbers 1-499, Record Group 234, Reconstruction Finance Corporation/Defense Plant Corporation, NARA.

[85] *Chicago Tribune*, January 30, 1941, p. 27.

[86] *Manufacturing Plant, Melrose Park, Ill., Plancor 39*.

[87] *Chicago Tribune*, July 1, 1941, p. 27; *Detroit Free Press*, July 1, 1941, p. 13.

[88] *Detroit Free Press*, December 26, 1941, p. 6.

[89] *Detroit Free Press*, July 1, 1942, p. 18.

[90] *Detroit Free Press*, February 7, 1942, p. 15.

[91] Ibid.

[92] Defense Plant Corporation: *Aluminium Foundry, Flint, Michigan, Plancor 504*, (No date), Entry A176 Pamphlets Relating to Manufacturing Facilities 1940-1945, Container 2, Plancor Numbers 500-, RG 234, NARA.

[93] Ibid.

[94] *The Wall Street Journal*, July 2, 1941, p. 3.

[95] Geschelin, Joseph: "Chevrolet's Unique Setup for Airplane Engine Production", *Automotive and Aviation Industries*, Vol. 88, No. 11, (June 1, 1943), p. 42.

[96] *The Wall Street Journal*, July 2, 1941, p. 3; *Detroit Free Press*, June 30, 1943, p. 20.

[97] AAF Historical Office: *Army Air Force Historical Studies: No. 40: Expansion of Industrial Facilities under Army Air Forces Auspices 1940-1945*, (No date), p. 93.

[98] Defense Plant Corporation: *Industrial Plant, Tonawanda, New York, Plancor 155-Plant No. 1*, (No date), Entry A176 Pamphlets Relating to Manufacturing Facilities 1940-1945, Container 1, Plancor Numbers 1-499, RG 234, NARA.

[99] *Steel*, Vol. 115, No. 14, (October 2, 1944), p. 52; *The Iron Age*, Vol. 154, Iss. 14, (October 5, 1944), p. 80.

[100] Defense Plant Corporation: *Industrial Plant, Buffalo, New York, Plancor 155-Plant No. 2*, (No date), Entry A176 Pamphlets Relating to Manufacturing Facilities 1940-1945, Container 1, Plancor Numbers 1-499, RG 234, NARA.

[101] *The Capital Journal*, July 5, 1941, p. 12.

[102] *Detroit Free Press*, December 23, 1942, p. 22.

[103] Ibid.

[104] *Muncie Evening Press*, November 3, 1943, p. 7.

[105] *The Star Press*, December 6, 1943, p. 7.

[106] Freeman, Joshua B.: *Behemoth: A History of the Factory and the Making of the Modern World*, (New York, 2018), pp. xiii-xv.

[107] Holley, *Buying Aircraft*, p. 307.

[108] *Wall Street Journal*, August 22, 1940, p. 1.

109 Civilian Production Administration: *Authorizations of War Industrial Facilities Financed with Public and Private Funds June 1940-July 1945*, (Washington, DC, 1946).
110 See Freeman, *Behemoth*, Chapter 5.
111 Melnikova-Raich, Sonia: "The Soviet Problem of Two 'Unknowns': How an American Architect and a Soviet Negotiator Jump-Started the Industrialization of Russia, Part I: Albert Kahn", *The Journal of the Society for Industrial Archeology*, Vol. 36, No. 2, (2010), pp. 68, 70.
112 Hildebrand, Grant: *Designing for Industry: The Architecture of Albert Kahn*, (Cambridge, MA, 1974). pp. 137, 164-72.
113 Freeman, *Behemoth*, p. 141.
114 "Industrial Buildings...A Buildings Type Study", *Architectural Record*, Vol. 91, No. 1, (January 1942), p. 65.
115 Kahn, Albert: "Design of Plants for Mass Production", *Aero Digest*, Vol. 39, No. 4, (October 1941), p. 156, "What the Manufacturer Wants from His Architect", *Architectural Forum*, Vol. 69, No. 2, (August 1938), p. 96.
116 Kahn, "Design of Plants for Mass Production", p. 156.
117 "Wright Plant Sets New Size Record", *Engineering News-Record*, Vol. 127, No. 1, (July 3, 1941), p. 52.
118 Wright Aeronautical Corporation, Plancor 10, Engineer's Final Report, Part 1, p. 60, Record Group 234, Reconstruction Finance Corporation, Defense Plant Corporation Engineer's Reports, Box 4, Plancor 10, NARA.
119 This description of the Lockland factory is drawn from "Wright Plant Sets New Size Record", Kahn, "Design of Plants for Mass Production", and Engineer's Final Report, Part II, RG 234, Box 4, Plancor 10, NARA.
120 Kahn, "Design of Plants for Mass Production", p. 157.
121 "Producer of Production Lines", *Architectural Record*, Vol. 91, No. 6, (June 1942), p. 43.
122 "Wright Plant Sets New Size Record", p. 52.
123 "The Industry's Largest Engine Factory", *Aero Digest's Aviation Engineering*, Vol. 39, No. 1, (July 1941), p. 156.
124 "Wright Plant Sets New Size Record", p. 52.
125 "The Industry's Largest Engine Factory", p. 159.
126 *U.S. Military Acceptances 1940-1945*, p. 141.
127 See letter Col. S. R. Brentwell, Office of the Assistant Chief of Air Staff, to General B. E. Chidlaw, 21 February 1944, R-2600 Engine Part I, Section I, R-2600 Case Study, Vol. 2, p. 729, File 202.2-13, AFHRA.
128 *Trade Winds*, Vol. 6, No. 11, (April 1943), p. 16.
129 The North Shop was a segregated building, with only African-American workers.
130 *Manufacturing Plant, Lockland, Ohio, Plancor 10*, copy in Record Group 270, War Assets Administration, Real Property Disposal Case Files 1947-1951, Box 684, Wright Aeronautical Corporation, NARA, Chicago office; Modley, *Aviation Facts and Figures 1945*, p. 2; The Defense Plant Corporation's brochure *Advance Listing of Industrial Plants and Plant Sites to be Disposed of by Defense Plant Corporation, Oct. 14, 1944*, (Washington, DC, 1944), lists the total floor area of the Wright Lockland factory as 5,622,171 square feet and Willow Run as 5,022,177. The Dodge-Chicago factory was even bigger.
131 This total is derived from *U.S. Military Acceptances 1940-1945: Aircraft, Engines and Propeller Production*, (Washington, DC, 1946), and Lilley, *Problems Accelerating Aircraft Production in World War II.*
132 *U.S. Military Acceptances 1940-1945: Aircraft, Engines and Propeller Production*; Ministry of Aircraft Production: *Statistical Review 1939-1945.*
133 *Detroit Free Press*, January 23, 1942.
134 *U.S. Military Acceptances 1940-1945: Aircraft, Engines and Propeller Production*, p. 146.
135 *The Iron Age*, Vol. 155, No. 9, (March 1, 1945), p. 95.
136 *U.S. Military Acceptances 1940-1945: Aircraft, Engines and Propeller Production*, pp. 142-48.
137 *U.S. Military Acceptances 1940-1945: Aircraft, Engines and Propeller Production*, pp. 144, 146.
138 *U.S. Military Acceptances 1940-1945: Aircraft, Engines and Propeller Production*, pp. 146, 148.
139 Reck, Franklin M.: "Buick Retools for New Engines", *Aero Digest*, Vol. 45, No. 6, (July 15, 1944), p. 118.
140 Connors, *The Engines of Pratt & Whitney: A Technical History*, Table 12: Applications of the R-1830 Twin Wasp, pp. 90-92.
141 Connors, *The Engines of Pratt & Whitney: A Technical History*, pp. 107-14.
142 Ibid., Table 12: Applications of the R-1830 Twin Wasp, pp. 90-92.
143 Reck, "Buick Retools for New Engines", p. 118.
144 Office of Production Management, Aircraft Branch, Aircraft Engine Report 9-I, February 26, 1942, Report on Airplane Engine Deliveries (January 1, 1942, through April 30, 1944), Papers of Harry L. Hopkins, Special Assistant to the President, 1941-1945, Office of Production Management Reports, Aircraft Branch, November 28, 1941-January 20, 1942, Container 207, Franklin D. Roosevelt Library and Archives, Hyde Park, NY.
145 *U.S. Military Acceptances 1940-1945: Aircraft, Engines and Propeller Production.*
146 Ibid., p. 139.
147 Ward, "Plant Layout and Production Methods for Modern Aircraft Engines", p. 180.
148 Ibid., p. 181.

149 Ibid.
150 AuWerter, "Pratt & Whitney Expands", p. 40.
151 Ibid.
152 "Conveyor Belt System", *Aviation*, Vol. 39, No. 3, (April 1941), p. 45.
153 Employment in the Aircraft Manufacturing Industry, Mr. Haas to Secretary Morgenthau, January 14, 1941, Vol. 347 January 13-14, 1941, p. 356, Diaries of Henry Morgenthau, Jr.
154 *The Wall Street Journal*, September 22, 1941, p. 13.
155 Ibid.
156 Ibid.
157 Wilson, *Slipstream*, p. 240.
158 AuWerter, "Pratt & Whitney Expands", p. 39.
159 Ibid., p. 40.
160 "Operations that Speed up PRATT & WHITNEY Engine Production", *Machinery (New York)*, Vol. 47, No. 11, (July 1941), p. 194.
161 Ibid.
162 Ibid., p. 196.
163 *The Wall Street Journal*, September 22, 1941, p. 13.
164 Ibid.; Hesse, *Engineering Tools and Processes*, p. 501.
165 *The Wall Street Journal*, September 22, 1941, p. 13.
166 Ibid.
167 *U.S. Military Acceptances 1940-1945: Aircraft, Engines and Propeller Production*, pp. 180-89.
168 *The Wall Street Journal*, November 9, 1942, p. 5.
169 Ibid.
170 Crow, *The City of Flint Grows Up*, p. 94.
171 Transcript of conference call between Secretary Morgenthau and Henry Purvis, British Purchasing Commission, October 4, 1940, Vol. 319, October 4-7, 1940, p. 132, Diaries of Henry Morgenthau, Jr.
172 Ibid., p. 133.
173 Walton, *Miracle of World War II*, p. 281.
174 Ibid., pp. 280-281.
175 Wilson, *Slipstream*, p. 239.
176 Ibid.
177 Crow, *The City of Flint Grows Up*, pp. 97-98.
178 United Aircraft: *The Bee Hive*, Vol. XVII, No. 12, (December 1942-January 1943), p. 4.
179 Ibid.
180 Geschelin, Joseph: "Buick's Know How Speeds Production of Airplane Engine Parts', *Automotive and Aviation Industries*, Vol. 87, No. 10, (November 15, 1942), pp. 21-22.
181 Herb, Charles O.: "Buick Now Manufactures Bomber Engines", *Machinery (NY)*, Vol. 48, No. 11, (July 1942), p. 161.
182 Ibid.; Geschelin, Joseph: "Buick's Warplane Engine Plant", *Automotive and Aviation Industries*, Vol. 86, No. 11, (June 1, 1942), p. 20.
183 Herb, Charles O.: "Buick Now Manufactures Bomber Engines", pp. 161-62.
184 Geschelin, "Buick's Warplane Engine Plant", p. 20.
185 Geschelin, "Buick's Know How Speeds Production of Airplane Engine Parts", p. 22.
186 Young, O.W.: "Special-Purpose Machine Tools Step Up War Production", *Steel*, Vol. 112, No. 4, (January 25, 1943), p. 82.
187 Geschelin, "Buick's Warplane Engine Plant", pp. 21-22.
188 Chase, Herbert: "Buick Practice in Machining P&W Cylinder Heads", *Aviation*, Vol. 42, No. 7, (July 1943), pp. 174-76, 354, 357-58, 361.
189 Chase, Herbert: "Mass Machining Features Buick Shortcuts", *Aviation*, Vol. 43, No. 11, (November 1944), p. 141.
190 Ibid.
191 "16-Station Automatic", *Steel*, Vol. 113, No. 23, (December 6, 1943), p. 118.
192 Ibid.
193 Ibid.
194 Ibid.
195 Chase, "Mass Machining Features Buick Shortcuts", p. 142.
196 Ibid., p. 143.
197 "Develops Improved Tooling to Obtain Greater War Output at Lower Cost", *American Machinist*, Vol. 86, No. 12, (June 11, 1942), p. 581.

198 Geschelin, Joseph: "Chevrolet's Unique Setup for Airplane Engine Production", *Automotive and Aviation Industries*, Vol. 88, No. 11, (June 1, 1943), pp. 25-26.
199 Ibid., p. 26.
200 "Conveyors Cut Machining Time On Cylinder Heads", *American Machinist*, Vol. 31, No. 18, (August 31, 1944), p. 90.
201 Ibid.
202 See Geschelin, Joseph: "Chevrolet's Unique Setup for Airplane Engine Production", *Automotive and Aviation Industries*, Vol. 88, No. 11, (June 1, 1943), pp. 24-28 and especially p. 42.
203 "New Ways by Chevrolet", *American Machinist*, Vol. 87, No. 19, (September 16, 1943), p. 120.
204 "Chevrolet Finds New Ways", *American Machinist*, Vol. 87, No. 22, (October 28, 1943), p. 113.
205 Ibid.
206 Ricker, Chester S.: "Continuous Pouring of Magnesium Engine Castings", *Aviation*, Vol. 43, No. 1, (January 1944), p. 159.
207 Ibid., p. 150.
208 "Light-Alloy Forgings", *Aircraft Production*, Vol. 6, No. 63, (January 1944), pp. 13-16.
209 Geschelin, Joseph: "Chevrolet's Unique Setup for Airplane Engine Production", p. 42.
210 *U.S. Military Acceptances 1940-1945: Aircraft, Engines and Propeller Production*, pp. 186, 188, 190.
211 Ibid., pp. 180, 190.
212 Special Studies No. 21, *Aircraft Production Policies under the National Defense Advisory Commission and Office of Production Management, May 1940 to December 1941*, p. 155.
213 See Maj. Frank W. Lavista, AAF Resident Representative, to Commanding General, AAF Material Command, Comments on Bureau of the Budget Report Dated 18 February 1944 Reference: Wright Aeronautical Corporation, Lockland, Ohio, 24 February 1944, R-2600 Case Study, Vol. 2, p. 730, File 202.2-13, A2070, AFHRA. See also Manpower Office, Management Control, Air Technical Service Command, Army Air Forces: *Labor Statistics for the Aircraft Industry*, November 1944-July 1945, File 218.6-7 Labor Statistics, AFHRA.
214 *The Wall Street Journal*, January 7, 1941, p. 7; February 10, 1941, p. 1; "Crankcases by Otis Elevator", *Trade Winds*, Vol. 6, No. 11, (April 1943), pp. 6-7, 18; "Hudson Builds Cyclone Parts', *Trade Winds*, Vol. 6, No. 9, (October 1942), pp. 6-7; "Crankshaft by Ohio", *Trade Winds*, Vol. 6, No. 10, (February 1943), pp. 8-9, 18.
215 *The Wall Street Journal*, January 7, 1941, p. 7; February 10, 1941, p. 1; "Crankcases by Otis Elevator", *Trade Winds*, Vol. 6, No. 11, (April 1943), pp. 6-7, 18.
216 "Cylinder Heads: A Cyclone 14 Begins its Growth with Casting of Cylinder Heads", *Trade Winds*, (May 1941).
217 Geschelin, Joseph: "Wright Ready to Power Bombers", *Automotive and Aviation Industries*, Vol., No., (February 1, 1942), p. 22.
218 "Cornfield to Mass Production in A Single Year-That's the Story of Wright's Cincinnati Division", *Trade Winds*, (November 1941), p. 8.
219 "The Ohio Plant Today", *Trade Winds*, Vol. 6, No. 11, (April 1943), p. 13.
220 *The Wall Street Journal*, June 16, 1941, p. 1.
221 *Wall Street Journal*, (September 22, 1941), p. 18.
222 Geschelin, "Wright Ready to Power Bombers", p. 25; "Manufacturing Plant, Lockland, Ohio, Plancor 10" brochure.
223 Joseph Geschelin's article, "Wright Ready to Power Bombers", contains a list of these machine tools on p. 25.
224 "Wright's Ohio Engine Plant Opens Mass Production Era", *American Aviation*, Vol. 5, No. 3, (July 1, 1941), p. 46.
225 Ibid.
226 Ibid.
227 "Cylinder Heads by the Mile", *Trade Winds*, (November 1941), p. 13.
228 "Cornfield to Mass Production in A Single Year-That's the Story of Wright's Cincinnati Division", *Trade Winds*, (November 1941), pp. 8-9.
229 "High Output Machine Tools, In-Line Production Speed Up Manufacture of Engines at Cincinnati", *Trade Winds*, (July 1941), p. 7.
230 This description is drawn from Linsley, H.E.: "Cyclone Engines Mass Produced", *The Iron Age*, (October 9, 1941), pp. 53-61 and "Mass Production for the Aircraft-Engine Industry", *Mechanical Engineering*, Vol. 64, No. 2, (February 1942), pp. 100-05.
231 Wright Aeronautical Corporation: *Application of High Production Machine Tools*, pp. 5, 8.
232 Linsley, H.E.: "Cyclone Engines Mass Produced", p. 55.
233 *Studebaker Corporation R-2600 Aircraft Engines Construction and Production Analysis*, p. 32.
234 "Wright Aero Produces Engines for the B-29 Superfortress", *Machinery (New York)*, Vol. 50, No. 12, (August 1944), pp. 133-44.
235 *Application of High Production Machine Tools*, p. 15.

[236] Ibid.

[237] "Wright Aero Produces Engines for the B-29 Superfortress", *Machinery (New York)*, Vol. 50, No. 12, (August 1944), pp. 133-44.

[238] *Application of High Production Machine Tools*, pp. 21-23.

[239] McCarthy, J.F.: "Precision Machining Operations on the Cyclone 14 Crankshaft", *Aero Digest's Aviation Engineering*, Vol. 39, No. 6, (December 1941), p. 198.

[240] Geschelin, Joseph: "Crankshafts for the Wright Cyclone", *Automotive and Aviation Industries*, Vol. No. (March 1, 1942), p. 9.

[241] Geschelin, Joseph: "Crankshafts for the Wright Cyclone', p. 9.

[242] Geschelin, Joseph: "Crankshafts for the Wright Cyclone', p. 10.

[243] Geschelin, Joseph: "Crankshafts for the Wright Cyclone', p. 15.

[244] Ibid.

[245] "Aluminum Fins Rolled On Aircraft Engine Cylinders", *The Iron Age*, (February 22, 1945), pp. 66-67.

[246] "Aluminum Fins Rolled On Aircraft Engine Cylinders", p. 66.

[247] "Aluminum Fins Rolled On Aircraft Engine Cylinders", pp. 67-68.

[248] "Aluminum Fins Rolled On Aircraft Engine Cylinders", pp. 68-70.

[249] *Production-Line Technique*, p. 187.

[250] *Production-Line Technique*, p. 195.

[251] Based on Aircraft Production Board Report 9-L of January 1943. See Delivery Schedule of Aircraft Engines Aircraft Report 9-L November 1, 1942 through October 31, 1944, War Production Board Reports, Aircraft Branch, Dec 31, 1942-Jan 15, 1943, Papers of Harry L. Hopkins, Container 240, FDR Library.

[252] See the memo "The November 4th Aircraft Engine Schedule", Morris L. Copeland, Assistant Director, Munitions Branch Statistical Department, to Donald Davis, Vice Chairman, War Production Board, December 7, 1943, War Production Board Select Document File, Aircraft Schedule-Standardization, Box 29, Entry 3A, RG179, NARA.

[253] Summary of the R-2600 Engine Project, p. 6, Case History of the R-2600 Engine Project, Part I, Section I, File 202.2-13, A2070, AFHRA.

[254] Summary of the R-2600 Engine Project, p. 6.

[255] Maj. Gen. C.E. Branshaw, Material Command, to Assistant Chief of Staff, M.M. &D., 11 April 1944, Case History of the R-2600 Engine Project, Part I, Section I, pp. 775-76, File 202.2-13, A2070, AFHRA.

[256] Office of Assistant Chief of Staff, Material, Maintenance, and Distribution, Inter-Desk Memorandum, 17 February 1944, R-2600 Engine Part I, Section I, R-2600 Case Study, Vol. I, p. 724, File 202.2-13, A2070; Maj. Gen. C.E. Branshaw, Material Command, to Assistant Chief of Staff, M.M. &D., 11 April 1944, Case History of the R-2600 Engine Project, Part I, Section I, pp. 775-76, File 202.2-13, A2070, AFHRA.

[257] Aircraft production numbers based on *U.S. Military Acceptances 1940-1945: Aircraft, Engines and Propeller Production*. See also memo to General Meyers, Engine Situation Resulting from Airplane Cuts, 22 May 1944, Case History of the R-2600 Engine Project, Part I, Section I, pp. 794-96, File 202.2-13, A2070, AFHRA.

[258] Brig. Gen. Orval R. Cook, Chief, Production Division, to Deputy Chief of Staff, Wright Field, R-2600-B and R-3350-BA Engine Production at Wright Aeronautical Corporation Plant, Lockland, Ohio, 13 May 1944, Case History of the R-2600 Engine Project, Part I, Section I, pp. 787-93, File 202.2-13, A2070, AFHRA.

[259] Summary of the R-2600 Engine Project, p. 8.

[260] Riddle, Donald H.: *The Truman Committee: A Study in Congressional Responsibility*, (New Brunswick, NJ, 1964), pp. 122-24.

[261] Summary of the R-2600 Engine Project, Part II: Lockland Investigation, R-2600 Case Study, Vol. 3, p. 859, File 202.2-13, AFHRA.

[262] An A rating, the highest quality control rating, indicated that the AAF had complete trust in the organization's quality control methods and had assigned all responsibility for inspection to the rated facility and performed only spot checks. With the B rating, the AAF inspectors carried out inspections on a certain percentage of parts, duplicating and checking on the methods and procedures of the organization. Report of Conditions at Wright Aeronautical Corporation New Jersey Plants, Vol. I., p. 13, File 208,8A, AFHRA.

[263] Report: Truman Sub Committee Investigation of Wright Lockland Engine Plant, 11 April 1943, R-2600 Case Study, Vol. 3, p. 923, File 202.2-13, AFHRA.

[264] Report: Truman Sub Committee Investigation of Wright Lockland Engine Plant, 11 April 1943.

[265] Lt. Gen. William Knudsen, Memorandum for Robert A. Lovett, Assistant Secretary of War for Air, April 17, 1943, R-2600 Case Study, Vol. 3, p. 932, File 202.2-13, AFHRA.

[266] Summary of the R-2600 Engine Project, Part II: Lockland Investigation, R-2600 Case Study, Vol. 3, pp. 859-60, File 202.2-13, AFHRA.

[267] Col. S.R. Brentnall, Office of the Chief of Air Staff, Material, Maintenance, and Distribution, AAF, to Gen. B.W. Chidlaw, 21 February 1944, R-2600 Case Study, Vol. 2, p. 723, File 202.2-13, A2070, AFHRA.

268 Col. S.R. Brentnall, Office of the Chief of Air Staff, Material, Maintenance, and Distribution, AAF, to Gen. B.W. Chidlaw, 21 February 1944.

269 Pratten, C.F.: *Economies of Scale in Manufacturing Industry*, (Cambridge, 1971), p. 3.

270 See Greenwald, Douglas, ed.: *Encyclopedia of Economics*, (New York, 1982), pp. 327-29.

271 *Economies of Scale in Manufacturing Industry*, p. 15.

272 Ibid.

273 *Economies of Scale in Manufacturing Industry*, p. 297.

274 Williamson, Oliver E.: *Markets and Hierarchies: Analysis and Antitrust Implications: A Study in the Economics of Internal Organization*, (New York, 1975), p. 122.

275 Williamson, *Markets and Hierarchies*, p. 298.

276 "30,000 Workers in 6 Wright Plants to Build 3,780,000 H.P. Monthly", *Trade Winds*, (April 1941), p. 4.

277 These numbers are taken from *U.S. Military Acceptances 1940-1945: Aircraft, Engines and Propeller Production.*

278 *The Cincinnati Enquirer*, July 27, 1940, p. 2.

279 Maj. C.J. McKinsey, Acting AAF Resident Representative, Wright Aeronautical Corporation Lockland, Ohio, to District Supervisor, Central Procurement District, AAF Material Command, April 28, 1944, R-2600 Case Study, Vol. 3, p. 1163, File 202.2-13, AFHRA.

280 Lilley, *Problems Accelerating Aircraft Production in World War II*, p. 53.

281 Lilley, *Problems Accelerating Aircraft Production in World War II*, pp. 68-69.

282 Ibid., p. 41.

283 Ibid., p. 64.

284 Ibid.

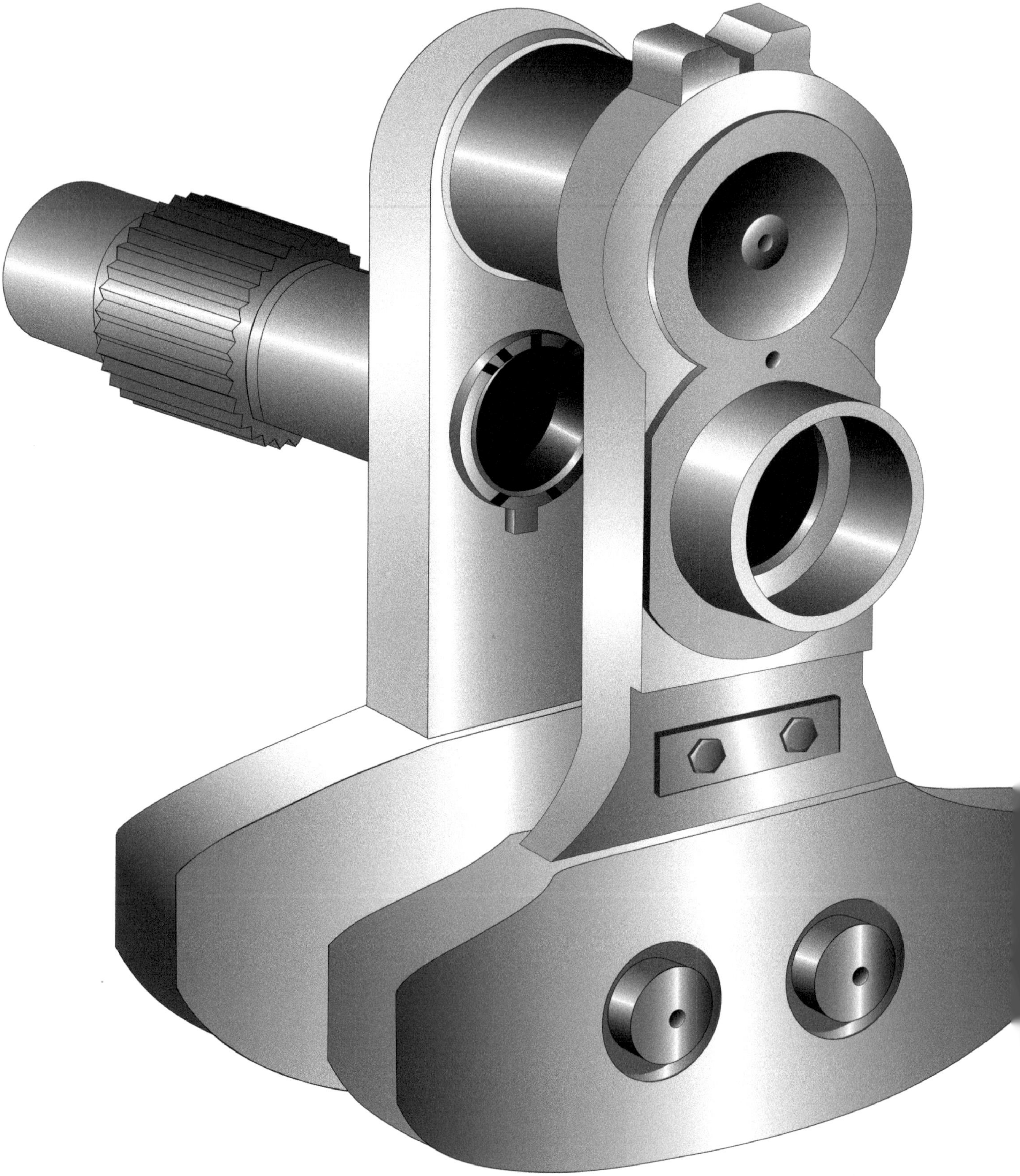

6

Building 18-Cylinder Engines: The Pratt & Whitney R-2800 Double Wasp, Wright R-3350 Duplex Cyclone 18, and Bristol Centaurus

Introduction

During World War II, all the combatants maintained an intensive effort to develop airplanes with higher performance, pushing for more speed, higher altitudes, longer range, greater armament, and, for bombers, heavier bomb loads.[1] Airplane engines were critical to achieving better performance as "the power plant and its accompanying propeller were the keys to aircraft performance, for speed, range, altitude, and rate of climb depended in large measure on the power and efficiency of the propulsion unit."[2] In the race to develop and build aircraft engines with greater horsepower, America pulled well ahead of all other combatants, including Britain. During the war, America put two 18-cylinder air-cooled radial aircraft engines generating over 2,000 hp into quantity production. Pratt & Whitney developed the R-2800 Double Wasp engine, and together with its automotive licensees, Ford Motor Company and Nash-Kelvinator built 114,067 R-2800 engines, while Wright developed the more powerful R-3350 Duplex Cyclone 18 and, with the Dodge Division of the Chrysler Corporation, built 32,240 engines before the end of the war.[3] In comparison, Bristol developed the 2,500 hp Centaurus air-cooled, radial sleeve-valve engine, but built just 2,967 by the end of 1945.[4] The Napier Company developed the liquid-cooled, in-line Napier Sabre engine that had also reached 2,500 hp by the end of the war and built 4,991 engines by the war's end.[5] The Soviet Union developed two 2,000 hp engines during the war, the 14-cylinder Ash-83 and the 18-cylinder M-71, but the production of these engines was limited.[6] Neither Germany nor Japan had much success building engines to match the American R-2800 and R-3350.[7]

The production of the Pratt & Whitney R-2800 and the Wright R-3350 was radically different from the production of the Bristol Hercules. Pratt & Whitney, Wright Aeronautical, and their licensee companies built these engines on a massive scale in what were some of the largest single-story industrial buildings in the world, using the most advanced, automatic, high-production machine tools. The Pratt & Whitney factory in Kansas City, Missouri, built to manufacture the R-2800 and Chrysler's Dodge Division factory in Chicago that built the R-3350 were huge, dwarfing the aircraft engine factories in Britain. In total floor area, the Dodge-Chicago factory was larger than all the factories building Bristol engines combined and was, at the time, the largest single-story industrial building in the entire world. The Pratt & Whitney Kansas City and Dodge-Chicago factories reflected a trend toward what Joshua Freeman has called "industrial giantism," a move to build truly giant factories to take maximum advantage of methods of mass production developed at Ford Motor Company.[8] The Pratt & Whitney Kansas City and Dodge-Chicago factories have received surprisingly little mention, even in works devoted to their designer, the famous industrial architect Albert Kahn. Although the Dodge-Chicago factory was the largest Army Air Force-sponsored project of the war, larger in total area and in total cost than the more famous Ford Willow Run factory, it is Willow Run that is cited as the biggest factory built during the war. The story of the construction of the Pratt & Whitney and Chicago factories, which used new construction methods developed rapidly during the war, has rarely been described.

The key to achieving quantity production at these factories was not size per se, but the adoption of line production to improve flow through the factories. New production methods, developed through production engineering and process engineering at Pratt & Whitney and Wright Aeronautical during 1940–1941, with the added production experience of their automotive licensees, characterized these giant factories. The Pratt & Whitney Kansas City and Dodge-Chicago factories, together with a third large government-financed factory built for Wright Aeronautical at Wood-Ridge, New Jersey, were laid out for line production to achieve continuous flow from beginning to end. The factories relied on new types of advanced high-production machine tools Wright Aeronautical and the automotive companies developed with the American machine tool manufacturers, notably the Greenlee and Foote Burt automatic transfer machines. These advanced machine tools replaced large numbers of standard machine tools, cut the number of operators, and speeded production (Figure 6.1).

FIGURE 6.1 American and British 18-Cylinder Air-cooled, Radial Engine Production, 1939–1945.

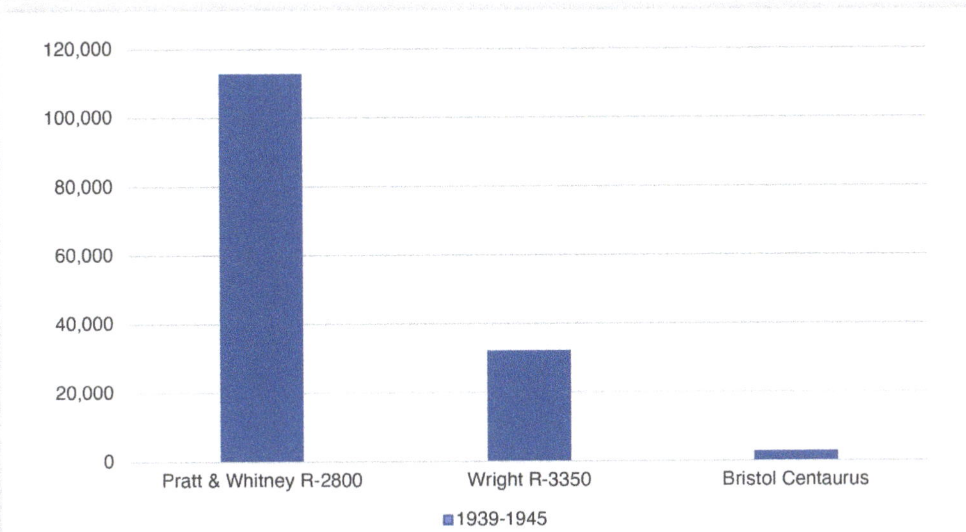

Source: US Department of Commerce, Civil Aeronautics Administration, Office of Aviation Information, Division of Aircraft Statistics: US Military Acceptances, 1940–1945: Aircraft, Engines and Propeller Production (Washington, DC, 1946); Ministry of Aircraft Production: Statistical Review, 1939–1945.

The Engines

In the mid-1930s, Pratt & Whitney had developed the double-row 14-cylinder R-2180 Twin Hornet engine that promised 1,400 hp, only to find that Wright Aeronautical was in the advanced stages of introducing the larger and more powerful 1,600 hp R-2600 Cyclone 14 engine, with the even larger R-3350 Duplex Cyclone in development that promised an initial output of 1,800 hp. Pratt & Whitney turned its initial effort with the R-2180 into a larger, experimental 18-cylinder engine that became the R-2800. Design work on this experimental engine began in March 1937, and by November, a complete engine, designated the R-2800A, had completed 100 hours of running time.[9] In 1939, the R-2800 completed its military type tests and its first flight tests, entering production in 1940.[10] In the R-2800, Pratt & Whitney "provided an outstanding air-cooled engine with a potential great enough eventually to make it at least the equal in performance of any liquid-cooled engine."[11]

The R-2800 Double Wasp had two rows of nine cylinders each, with the cylinder design based on Pratt & Whitney's extensive experience with its earlier poppet-valve engines. The engine was longer than the R-1830 Twin Wasp, but only just over 4½ in. wider in diameter. The Double Wasp cylinder was made of cast aluminum with the same bore as the Pratt & Whitney R-1340 Wasp engine but with a longer stroke. To promote cooling, the cylinder heads had closely spaced cooling fins that stretched casting methods to their limits, with aluminum baffles to direct airflow onto the cylinders.[12] Instead of cutting fins in the cylinder barrel, Pratt & Whitney fitted a cast-finned aluminum muff onto the cylinder barrel.[13] Unlike Wright Aeronautical that switched to a forged steel crankcase in the R-2600, for the R-2800 Pratt & Whitney continued to use a three-section, forged aluminum crankcase developed for the R-1830 Twin Wasp. The company had to put in months of design work and testing to develop a crankshaft that could withstand the intense torsional vibration from two rows of 18 cylinders.[14] The early R-2800 engine series utilized a single-speed, two-stage supercharger. As an 18-cylinder engine, the R-2800 had over 13,000 parts presenting a challenge for manufacturing the engine in quantity (Figure 6.2).

FIGURE 6.2 The Pratt & Whitney 18-cylinder R-2800 Double Wasp engine powered three fighter aircraft vital to the American war effort. The US Navy's Grumman F6F Hellcat, the Navy's premier carrier fighter of the war, and the Vought F4U Corsair, used by Navy and Marine Corps squadrons from carriers and land bases, both used the R-2800. Republic chose the R-2800 for its P-47 Thunderbolt.

80G-235282, RG80G, NARA.

During the war, Pratt & Whitney developed over 50 models of the R-2800 engine. The first production versions in the R-2800 A series developed 1,800 hp for take-off, but Pratt & Whitney soon boosted this to 2,000 hp in the B Series with the R-2800-8 for the US Navy and the R-2800-21 for the Army Air Force. By the end of the war, the C series engines could produce 2,100 hp and up to 2,800 hp with water injection. The Double Wasp was in great demand as an engine for fighters, powering the Republic P-47 Thunderbolt for the Army Air Force with 15,683 built during the war and the Grumman F6F Hellcat (12,275 built) and Chance Vought F4U Corsair (11,426 built) for the US Navy and Marine Corps.[15] The Double Wasp also powered several multi-engine aircraft, including the Martin B-26 Marauder medium bomber and its intended replacement, the Douglas A-26 Invader, and the Lockheed PV-1 and PV-2 Ventura patrol bomber and the Martin PBM-5 flying boat.[16]

The history of the Wright R-3350 Cyclone 18 shows that "engine development is evolutionary—and costly and complex."[17] The Cyclone 18 came out of Wright's experiments to join two Cyclone 9 engines together in the search for greater engine power and the experience the company had gained developing the double-row R-2600 Cyclone 14. Wright abandoned this early twin Cyclone 9 project and instead in 1936 turned to developing a completely new 18-cylinder, two-row engine design with 3,350 in.^{3} capacity.[18] This engine had a three-section steel crankcase, incorporating two crankcase sections from the Cyclone 9, nine cylinders per row, the same cylinder design and master and articulated connecting rods, with the same bore and stroke as the Cyclone 14, and a two-throw crankshaft based on work on the Cyclone 14.[19] The first engine ran in April 1937, producing 1,800 hp, and soon passed a 50-hour endurance test.[20] Rebuilt to incorporate a two-stage supercharger, the experimental Cyclone 18 (designated the R-3350 by the Army Air Corps) failed its first 150-hour test due to problems with the propeller reduction gear.[21] This was just one of three significant problems that would plague the Cyclone 18 until late in the war. Development of the Cyclone 18 continued with the Army Air Corps purchasing small numbers of experimental engines. By 1939, the Cyclone 18 was producing 2,000 hp, but the engine had developed problems with exhaust valve failures and, more seriously, difficulties cooling the cylinders, a problem that continued even after the engine was introduced into combat operations late in the war.[22] When in April 1941 the Army Air Corps placed an order for the Boeing B-29 long-range bomber powered by four R-3350 Cyclone 18 engines, now rated at 2,200 hp, the Cyclone 18 was still very much an experimental engine having flown only a limited number of hours and was far from ready for production. After Pearl Harbor, the Army Air Force gave the B-29 program its highest priority and tripled its order for Cyclone 18s (Figure 6.3).[23]

FIGURE 6.3 The Wright R-3350 Cyclone 18 engine was vital to the success of the Boeing B-29 long-range bomber. The B-29 had double the maximum take-off weight of the Boeing B-17.

B25682, RG342FH, NARA.

The Cyclone 18 had the same diameter as the Cyclone 9, but with two rows of cylinders and accessories, it was nearly 30 in. longer.[24] The front section of the crankcase and the rear supercharger housing were magnesium alloy castings, while the crankcase main section comprised three steel forgings, front, center, and rear. The crankcase main section had 18 mounting pads for the two rows of nine cylinders. The Cyclone 18's two-throw crankshaft, to accommodate two master connecting rods, one for each bank of cylinders, was made up of three sections each forged from a chrome nickel steel alloy and machined for accuracy and finish. As in the Cyclone 14, the crankshaft had two Wright Dynamic Dampers to deal with torsional vibration. The Cyclone 18's cylinders were based on the Cyclone nine-cylinder design, with an aluminum cylinder head and steel cylinder barrel. The two rows of cylinders were staggered to improve the flow of cooling air. Each cylinder had one intake and one exhaust valve. The cylinder head had deep and closely spaced fins for cooling, while the cylinder barrel had the newer "W" aluminum fins inserted in place of the steel fins cut into the cylinder barrel. The initial models of the Cyclone 18 had a carburetor, while later models switched to fuel injection. Like the R-2800, the Cyclone 18 had over 13,000 parts.

Problems with the Cyclone 18 continued, leading the Army Air Force to form the R-3350 Engine Committee composed of representatives of the Army Air Force, Wright Aeronautical, and Pratt & Whitney to coordinate efforts to address these problems. Finding solutions delayed the production of the Cyclone 18. During 1942, Wright completed only 66 Cyclone 18s and 917 in 1943 at a time when the Army Air Force was trying desperately to ramp up B-29 production; production at the Dodge-Chicago factory did not begin until early 1944.[25] Redesigned propeller reduction gears and gear housing helped solve the propeller reduction gear problems, but toward the end of 1943 the Cyclone 18 was still having problems with engine cooling, exhaust valve failures, piston rings, and ignition systems.[26] When the B-29 entered combat in May 1944, operating initially from bases in India, the hot temperatures and high humidity made these problems even worse, continuing when the B-29s moved to bases in the Marianas later in the year. In November 1944, 19% of all sorties (one flight by one aircraft) against Japan were ineffective due to mechanical difficulties.[27] Intensive corrective efforts, a more disciplined approach to operating within limits on engine performance, and a change in tactics to low-level night bombing did much to improve the reliability of the Cyclone 18 in the B-29.

The Bristol Centaurus also emerged during the late 1930s. In 1937, Bristol Aeroplane's Roy Fedden also concluded that an aircraft engine of 2,000 hp or greater would be needed in the years ahead for military airplanes.[28] That year, the Air Ministry had asked the British aircraft engine manufacturers to develop engines of 2,000 hp to power the next generation of heavy bombers for the Royal Air Force.[29] Fedden began designing an 18-cylinder, two-row, air-cooled, radial engine using the sleeve-valve system developed for the Perseus, Hercules, and Taurus engines. The Centaurus' 18 cylinders had the same bore as the Perseus and Hercules, but a longer stroke giving a volume of 3,270 in.^{3}, close to the Wright Cyclone 18's volume of 3,347 in.^{3}, and were expected to start at 2,000 hp.[30] The Centaurus had a three-section forged crankcase, aluminum alloy cylinder barrels with cast aluminum cylinder heads and used a single-stage, two-speed supercharger.[31] The initial tests of the Centaurus began in July 1938, and the engine passed its type test in October 1939 delivering 2,000 hp.[32] Despite this promising beginning, wartime pressures slowed the development of the Centaurus. In 1940, Lord Beaverbrook, heading the new Ministry of Aircraft Production, stopped much development work to concentrate on producing more fighters and a limited number of bomber aircraft.[33] The immediate task of the overworked Bristol Engine department was to concentrate on increasing the production of the Mercury and Pegasus, improving the Hercules, and working out manufacturing techniques to get the

Hercules into production at the shadow factories.[34] Even with the accumulated experience gained from the earlier sleeve-valve engines, Bristol ran into difficulties with the Centaurus. Cooling and engine vibration problems emerged with further testing, but with no immediate need for the Centaurus efforts to address these issues they received a low priority and delayed the engine's entry into service.[35]

Organizing Production

Under President Roosevelt's 50,000 aircraft program, the more immediate need was for the Pratt & Whitney R-2800 engine rather than the Wright R-3350. The NDAC's Aeronautical Section's July 1940 report on aircraft engine manufacturing facilities estimated a need for 8,994 R-2800 engines by July 1941, compared with 1,684 R-3350 engines.[36] Pratt & Whitney had indicated to the government that it could not manufacture the R-2800 engine in the quantities required and at the same time build all the lower-power engines needed for the thousands of training aircraft for the Army Air Corps and the Navy.[37] That August, William Knudsen approached the Ford Motor Company about building the R-2800 engine under license from Pratt & Whitney.[38] Henry Ford had refused to build the Rolls-Royce Merlin under license as he did not want to build engines for a belligerent, but Charles Sorensen and his son, Edsel Ford, persuaded Ford to change his mind and accept Knudsen's request.[39] On August 16, it was announced that Ford had agreed to build 4,000 Pratt & Whitney R-2800 engines.[40] The following month, Knudsen and General Arnold discussed the possibility of substituting the Pratt & Whitney R-2800 for the Wright R-2600, which was in greater demand.[41] Arnold proposed switching the R-2600 engines on the Curtiss C-46 Commando transport for the larger R-2800 engines, which was done, increasing the number of R-2800 engines Ford was to build.[42] Ford was awarded a contract for $130 million in November 1940 and a second contract for $182 million in October 1941 for additional R-2800 engines.[43]

Following America's entry into the war, the demand for the R-2800 accelerated rapidly, more than doubling from 21,723 engines on order in November 1941 to 57,388 engines as of February 1942.[44] The Ford Motor Company received a contract for $255 million in February for more R-2800 engines.[45] The Nash-Kelvinator Corporation had begun building engine parts for Pratt & Whitney in late 1941, but was quickly brought into license production of complete engines.[46] In March 1942, the Navy Department as the coordinating contractor for Pratt & Whitney engines awarded a contract for $47.9 million to Nash-Kelvinator to build the R-2800 engine under license, quickly followed by a second contract for $58 million in July, and a third contract for $142.8 million in December.[47] Ford's production target was to achieve a rate of 1,800 engines a month by February 1943, while Nash-Kelvinator was to reach 700 engines a month by July 1943.[48] With additional production at the Pratt &Whitney Hartford factories, the combined production of the parent company and its two licensees would nearly triple the production of the R-2800 within the space of a year.

A new engine model brought about a change in the structure of R-2800 production. In 1940, Pratt & Whitney started working on a total redesign of the R-2800, which became the C series producing 2,100 hp and even more power with water injection. To cope with higher temperatures and stresses from a greater power, Pratt & Whitney switched from cast to forged cylinder with machined cooling fins to increase the cooling fin area, strengthened the aluminum crankcase, redesigned the crankcase, and strengthened the master rod and connecting rods.[49] For better performance at higher altitudes, Pratt & Whitney developed an innovative two-stage, two-speed supercharger.[50] The changes were so great that the C series engines had few interchangeable parts with the earlier A and B series engines.[51]

The Army Air Force and the Navy wanted the R-2800 C series engines to power more capable versions of the P-47 Thunderbolt and F4U Corsair and the Navy's next series of fighters, the Grumman F7F and F8F, and foresaw a need for more engines than Pratt & Whitney could supply from its Hartford factory, or that Ford and Nash could produce.[52] In April 1942, the Navy decided that a new factory would have to be built to expand R-2800 production and that Pratt & Whitney, though reluctant, would have to manage it. While the Navy intended the factory to build the B series engines, Pratt & Whitney proposed to the Navy that the new factory, to be built in Kansas City, Missouri, be dedicated instead to building the new C series, even though at the time the C series was not ready for production.[53] The Navy decided to take the gamble, and in July contractors broke ground for what would become a giant aircraft engine factory.[54]

The Pratt & Whitney Kansas City factory began building the R-2800 C series engines in 1943. At the end of 1944, the Navy instructed the factory to increase its monthly output by 324 engines by November 1945.[55] To add more production capacity, in early 1944 the Navy contracted with the Chevrolet Division to begin building the C series engine at its factories in Buffalo. Chevrolet completed its first C series engine in August 1944 and commenced production alongside the Pratt & Whitney R-1830 production line.[56] In March 1945, a contract with the Ford Motor Company to build the C series engines was transferred to Chevrolet to avoid disrupting Ford's production of the B series at its engine factory.[57]

In the fall of 1941, the newly redesignated Army Air Force decided to expand heavy bomber production and increase orders for the Boeing B-29 powered by the Wright R-3350 Cyclone 18 engine. This created a demand for more R-3350 engines than Wright could build on its own at its factories in Paterson, New Jersey.[58] With the other major automobile companies already committed to aircraft engine production, the Chrysler Corporation was the most likely candidate to take on the Cyclone 18.[59] Chrysler became Wright's second licensee and the last of the major automobile companies the government brought into aircraft engine production. Chrysler was one of the "Big Three" American automobile companies and in 1940 second only to General Motors, with sales of 1,044,290 vehicles during that year.[60] At the beginning of the war, Chrysler had over 80,000 employees and 19 factories in the US and Canada using more than 20,000 machine tools.[61] On January 30, 1942, the AAF issued a letter of intent to Chrysler covering the production of 10,000 Wright Cyclone 18 engines at a rate of 1,000 engines a month and granted approval for Chrysler to begin construction of a large factory.[62] In April, the Army Air Force changed its contract with Chrysler to increase production to 1,500 R-3350 engines per month, which required an even bigger factory.[63]

Chrysler chose Chicago as the site of the new factory and assigned the production of the R-3350 to its Dodge Division.[64] Dodge set up a new independent organization designated the Dodge-Chicago Plant Division to manage the new factory.[65] This became the biggest Army Air Force-sponsored project of World War II, with a total floor area of 6,793,686 ft^2, substantially bigger than the more famous Ford factory at Willow Run, Michigan, at a cost of more than $173 million.[66] Because of various delays, production of the R-3350 did not begin until January 1944. Wartime production of the R-3350 would be concentrated at the Dodge-Chicago factory.

At the same time, Wright Aeronautical received approval to build a large new factory in New Jersey devoted to building the R-3350 engine to supplement production at the Dodge-Chicago factory. Wright chose a site at Wood-Ridge, New Jersey, just under 7 miles southeast of its main factories in Paterson. Financed through the Defense Plant Corporation, the factory, which became Wright's Plant No. 7, had 1.5 mm ft^2 of total floor space. Construction on the new factory began in July 1942 with engine production beginning in early 1943.[67] In early 1944, the War Department instructed Wright Aeronautical to begin converting its Lockland, Ohio, factory to production of the R-3350, giving the Army Air Force a third source for the engine. The Lockland factory began tooling up for R-3350 production in the early summer and completed its first engine in October 1944 (Figure 6.4).[68]

FIGURE 6.4 Lacking an aircraft in demand for the Royal Air Force, the Bristol Centaurus, the most powerful British air-cooled radial engine built during World War II, saw little operational service during the war. The Centaurus powered the Bristol Buckingham medium bomber, shown here, designed as a replacement for Bristol's own Blenheim light bomber. Other aircraft types, however, proved superior to the Buckingham.

80G-190047, RG80, NARA.

The production of the Bristol Centaurus never approached the numbers of Pratt & Whitney R-2800 and Wright Aeronautical R-3350 engines built during the war. Unlike the R-2800 and the R-3350, the Centaurus never had an aircraft "champion," an aircraft using the Centaurus that was in great demand for the Royal Air Force.[69] American Army Air Force and Navy fighter planes and the Boeing B-29 program created a demand for thousands of R-2800 and R-3350 engines. The need for the Centaurus was minuscule in comparison. The RAF lacked a suitable airframe requiring large numbers of Centaurus engines. Delays in solving the Centaurus' technical problems contributed to the lack of demand for the engine, which was seen mostly as insurance against the failure of the Napier Sabre.[70] The first airplane to use the Centaurus, the Vickers Warwick twin-engine heavy bomber, made its first flight in 1940, but was soon superseded by more capable four-engine bombers.[71] The Warwick was ordered in small numbers primarily as a transport and airborne search and rescue aircraft. In 1941, however, Bristol responded to a request from the Air Ministry for a light bomber to replace the Bristol Blenheim with a twin-engine design using the Centaurus known as the Buckingham, which entered production in 1942.[72] The Buckingham would suffer a similar fate as the Warwick, as the De Havilland Mosquito proved far superior as a bomber. Only

119 Buckinghams and 110 similar Buckmaster trainers were completed.[73] The Centaurus did find more gainful employment in the Hawker Tempest fighter, equipping the Tempest II, but severe vibration problems delayed production until 1944 when the RAF placed contracts with Hawker for 800 fighters and with Bristol Aeroplane for an additional 300.[74]

The production of the Centaurus remained with Bristol Aeroplane throughout the war. The parent factory at Filton built the Centaurus and Bristol added a unique shadow factory, known as the Hawthorn factory, built underground at Corsham, a town near Bath.[75] With uncertainty about the application of the Centaurus continuing into 1942, the Air Supply Council had a difficult time determining how many Centaurus engines would be needed. In January 1943, it was estimated that the Bristol parent factory and the Hawthorn factory could produce 370 Centaurus engines a month, which just matched the level of demand for the Warwick, Buckingham, and the Tempest II, with no margin.[76] From time to time, there was some consideration of converting the Bristol Accrington factory to Centaurus production, but this was never required.[77]

The Factories

The Wright Aeronautical Wood-Ridge Factory

Wright's Wood-Ridge, New Jersey, factory was the first aircraft engine factory to employ a new method of construction, dubbed the "warspeed" method due to the speed of construction and its wartime necessity.[78] In the spring of 1942, as planning for the Wood-Ridge factory was getting underway, the War Production Board announced sharp cutbacks in the availability of steel to conserve steel for warships and other armaments.[79] As a result, structural steel for building construction rapidly became scarce.[80] Wright Aeronautical had hired Albert Kahn's firm to design a large, single-story factory building that would have a span of 38 ft between supporting columns to give adequate room for two rows of machine tools with space for an aisle or conveyor system to move materials and parts.[81] Kahn now had to design a building without using structural steel for columns, beams, and trusses to support the roof. The New Jersey construction firm Mahony-Troast Construction Company, which had previously worked with Wright Aeronautical and Albert Kahn's firm and understood Wright's requirements, came up with an idea for a new structure and a new method of construction.[82]

Mahony-Troast proposed making the entire factory building out of concrete, with steel-reinforced concrete columns and concrete beams supporting a flat slab concrete roof.[83] Kahn and his firm redesigned the building to conform to the new construction method. The construction company built the building using wooden forms that traveled on wheels, one form for each of the 69 20 ft, wide bays that made up the rectangular main manufacturing building.[84] Towers brought the concrete to the top of the traveling forms where hoppers carried the concrete to other forms making the 21 ft high columns, the 38 ft wide beams, and the flat slab roof. When one section was complete, the traveling form moved to the next position. Once the roof, beams, and columns were in place, the construction company poured a concrete floor 10 in. thick and covered it with 2-in. wood blocks.[85] The factory was of the blackout type, with brick walls but no windows, all lighting coming from fluorescent lights. The dimensions of the main manufacturing building were 1,107 ft by 1,387 ft giving a floor area of 1,424,500 ft^2, with a basement of 150,000 ft^2 and an office mezzanine area of 58,600 ft^2.[86] Two test cell buildings with 36 test cells were attached directly to the manufacturing building with 103,000 ft^2 of space. The Defense Plant Corporation provided $21.2 million in financing for the factory, which was designated Plancor 994, and was completed in less than 50 days.[87]

The Wood-Ridge factory followed Albert Kahn's design feature of having all employee services in the basement. The main manufacturing area was divided into separate departments dedicated to manufacturing or assembling a particular component. Raw materials and parts entered the factory at the northeast end, moving progressively down wide corridors to the green assembly area at the southeast end adjacent to the test cells.

The Dodge-Chicago Factory

When Chrysler received the AAF contract for 10,000 Cyclone 18 engines at a rate of 1,000 engines a month, the Dodge-Chicago Plant Division calculated that this could only be accomplished with a greatly expanded factory complex with a total floor area of 5,551,744 ft^2 including a massive machine shop and assembly building that would cover 82 acres under one roof.[88] The AAF directed Chrysler to build the new factory in either Chicago or Milwaukee.[89] After a survey of both cities, Chrysler chose Chicago as the location for the new factory.[90] Chrysler also turned to Albert Kahn, who had designed several of Chrysler's automobile factories and a large government-financed factory where Chrysler was to build tanks for the American Army. While the AAF was concerned that Kahn's firm was overloaded with defense work, Chrysler's management felt confident that the firm could do the job adding that "if this was a job of our own we would not think of employing any other architect."[91]

Albert Kahn's design for the machine and assembly building was a large rectangle, measuring 2,344 ft by 1,525 ft with a floor area of 3,575,633 ft^2.[92] The building had 72 30-ft bays running along the length of the building and 36 38-ft bays along its width, with the concrete arches supported by 3,363 concrete columns.[93] The concrete ribs supporting the roof arches had 5-ton crane attachments for moving heavy parts and machinery. The machine shop, with its long lines of machine tools, took up roughly four/fifths of the building, with the assembly section taking up the remainder. The building was oriented on a north–south axis, with materials from the foundries, forges, and outside suppliers entering at the south end and moving progressively north toward the assembly section and then to the test cells attached to the end of the building. The ten bays in the assembly section were air-conditioned, while the machine shop area had fresh air supplied through filtered supply units. Lighting in the machine shop and assembly building was a combination of natural light through steel sash windows along the sides of the building and artificial lighting. The walls of the building were built from brick rising 8½ ft from the floor, topped by the 12½ ft steel and wood window sash.[94] A contemporary article described the building as "big enough to put the Ford Willow Run plant inside of it and have room for twenty baseball diamonds around the edge."[95]

As he had done at the Wright Lockland factory, Kahn placed all employee services below the main factory production floor. His design incorporated a basement covering 621,000 ft^2. The area comprised two longitudinal tunnels, each 88 ft wide, running the length of the building and connecting to the large employee parking lot at the south of the building. When the Chicago factory was nearing peak employment, 22,000 employees would enter the factory between 0700 and 0730 and leave between 1630 and 1700.[96] A lateral tunnel ran through the center of the basement connecting the two longitudinal tunnels. The basement area had nine cafeterias that could seat 4,000 employees at a time, washrooms and toilets, offices, locker facilities, and stairways to the production floor.[97] As at Lockland, employees would enter and exit the machine shop and assembly building through the basement, resulting in what the *Architectural Forum* called one of the Dodge factory's greatest achievements: traffic planning. "At no point," the *Architectural Forum* noted, "do the two flowing streams—people and products—meet except where they are intended to (Figure 6.5)."[98]

FIGURE 6.5 The Dodge-Chicago factory under construction using the "warspeed" method that Albert Kahn's firm developed to replace steel structures with concrete forms. This innovative method saved substantial amounts of steel.

na051789, Automotive Council for War Production Collection, Detroit Public Library.

While designing the Chicago factory at the same time as the Wood-Ridge factory was under construction, Kahn faced the same challenge of a shortage of structural steel. In May 1942, the government instructed Chrysler to remove almost all structural steel from the plans for the Chicago factory, a building that on paper was the largest single-story industrial building Kahn had ever designed.[99] As with the Wright Wood-Ridge factory, Kahn and his associates decided to build the entire factory out of reinforced concrete, using the "warspeed" method of construction, adapted to the requirements of building a structure covering 80 acres.[100] Kahn designed the roof structure to consist of multiple reinforced concrete arch slabs with a span of 38 ft, supported by concrete ribs and girders resting on concrete columns placed 30 ft apart.[101] The load-bearing capacity of the arch eliminated the need for steel reinforcing rods commonly used in flat concrete roof construction.[102] This replaced structural steel beams and girders. As with the Wright Wood-Ridge factory, the building was built as an open-sided shed structure, with the roof and columns in place but no floor or walls. The original plan called for long 30 ft by 76 ft bays, but by cutting the bays in half to measure 30 ft by 38 ft, there were considerable savings in steel.[103] Using arched slabs as opposed to flat slabs also saved on the amount of concrete required. The supporting girders were 18 in. square and rested on slabs of concrete on the ground.[104] Kahn included a large basement area under the floor of the main manufacturing building with conventional flat slab concrete ceilings. With the "warspeed" method, the steel required was kept to a minimum. The reinforced concrete arch slabs used a steel wire mesh, while the supporting ribs, girders, and columns used steel bars, but the savings in steel over conventional structural steel construction were considerable. Using Kahn's reinforced concrete arched roof structure, the amount of steel needed was only 2.7 lb/ft^2 of floor area compared with 12.7 lb/ft^2 using conventional methods.[105] The savings in steel were apparently enough to build 14 destroyers for the Navy or 600 tanks for the Army (Figure 6.6).[106]

FIGURE 6.6 The George A. Fuller Company, which built the Dodge-Chicago factory, used movable wooden structures to carry the molds for the reinforced concrete roof sections. After a section had cured, the "Trojan Horse" as it was called was moved on to the next station to repeat the process.

na051788, Automotive Council for War Production Collection, Detroit Public Library.

Construction got underway in June 1942 and took a year to complete. Chrysler had hired the George A. Fuller Company as its general contractor. Working with Albert Kahn and his colleagues, Fuller engineers used "warspeed" construction to build the Dodge-Chicago factory buildings, employing 16,000 workers at the peak of construction.[107] The construction company used a new innovation, movable wooden structures on wheels that traveled along rails laid down in front of them. Dubbed "Trojan Horses," these structures carried the forms for molding the reinforced concrete arched roof sections, supporting girders and columns. The forms were long enough to allow for a 120-ft roof section to be poured at one time. The towers were moved into place, and the forms for the arch slabs were raised into position. Movable towers hoisted the wet concrete to the top of the structure to be poured over the molds for the 3-in. roof sections after workers had laid out sections of steel mesh over the form, a process that took 48 hours.[108] After a 72-hour curing period, the forms were stripped from the arch slab, lowered, and moved to the next station to form the next section of the roof.[109] The construction company built 60 traveling Trojan Horses, and to speed up construction of the huge machine shop and assembly building, construction started at both ends at the same time, with the Trojan Horses traveling toward the middle of the building. Completing the machine shop and assembly building took eight months with workers pouring the last concrete for the building on February 26, 1943.[110] Construction of the entire factory complex ended in July 1943.[111] In its final form, with the basement, a second-floor fan room, and small mezzanine, the machine and assembly building had a total area of 4,322,619 ft^2 (Figure 6.7).[112]

FIGURE 6.7 An aerial view of the completed Dodge Chicago factory on September 9, 1945, with the main manufacturing building on the left and the aluminum and magnesium foundries and forges some distance away to the right. At the time, the Dodge factory was the largest single-story industrial building in the world. The main manufacturing building still survives in Chicago.

Chicago Tribune historical photo/TCA.

The Dodge-Chicago had separate aluminum and magnesium foundries. Though the machine and assembly building dwarfed these foundries, these, too, were considerable buildings, each nearly 500,000 ft^2 in area. As at Lockland, they were also rectangular, with open production areas and slightly larger 30 ft by 40 ft bays to accommodate extensive conveyor systems and electrical hoists needed to carry molten metal and castings. The Dodge-Chicago factory complex had its own heat-treating shop and light and heavy forging buildings to forge parts for the Cyclone 18 using huge hammers and presses. The forge buildings, each 66,000 ft^2 in size and made with structural steel to absorb the vibration from forging hammers, were located some 4,000 ft away from the machine shop and assembly building to ensure that the vibrations did not interfere with the precision machining operations in the machine shop.[113] The 179,000 ft^2 heat-treating building sat next to the two forging buildings. An office building of 269,000 ft^2 and the test cells covering 265,000 ft^2 were the other large buildings in the complex. In total, the Dodge-Chicago factory complex had 19 buildings on its nearly 500 acres, with 6,430,000 ft^2 of floor space, nearly 1,400,000 ft^2 larger than the Ford airframe factory at Willow Run.[114]

The Pratt & Whitney Kansas City Factory

Pratt & Whitney's new factory in Kansas City, Missouri, was the third airplane engine factory built during 1942 using the "warspeed" method of construction and the third that Albert Kahn and his firm designed in the space of just a few months. Following the Navy's decision to finance a new Pratt & Whitney factory, the War Production Board recommended Pratt & Whitney consider building the factory in Kansas City, Missouri, as the most promising in terms of availability of labor and services.[115] At the end of April, a team from United Aircraft visited Kansas City and selected a 300-acre site of an old speedway track some 8 miles southeast of the city.[116] United Aircraft selected Albert Kahn and his firm to design a giant new factory building, apparently at the same time that Kahn was working on the design of the Dodge-Chicago factory.[117] With the Kansas City factory, Kahn faced the same challenge, a limitation on the structural steel that he had traditionally used in his industrial factory buildings.[118] He again employed the arch structure to reduce the use of reinforcing steel as he was doing with the Dodge-Chicago factory and adopted the "warspeed" construction method to build the factory in record time with minimal use of strategic materials (Figure 6.8).

FIGURE 6.8 The Pratt & Whitney factory in Kansas City, Missouri, designed by Albert Kahn's firm, had a total floor area of 2.9 million ft^2. This was the third giant aircraft engine factory built in America during World War II, exclusively to build Pratt & Whitney R-2800-C engines.

10005945 P&W factory Kansas City, Kansas City Public Library.

The main manufacturing building Kahn designed was massive, the second-largest American aircraft engine factory built during the war. The factory incorporated Kahn's ideals for industrial buildings, a large rectangular space well suited to line production with minimal disruptions to allow for a continuous flow of materials and parts through the factory space. The overall size of the building was 1,003 ft by 2,026 ft with 2,000,000 ft^2 of floor space, a basement area containing some 500,000 ft^2, and a mezzanine of 241,000 ft^2. An extension at the northeast end of the manufacturing building incorporated test cells and the final assembly area giving a total floor area of 2,940,000 ft^2.[119] This was bigger than the Wright Lockland Machine and Assembly Building No. 2, though not as big as the giant Dodge-Chicago factory's 4,322,619 ft^2. The large basement area followed Kahn's principle of putting all employee facilities—cafeterias, kitchen, locker rooms, toilet facilities, and even a basketball court—underneath the production area.[120] An office building comprising 500,000 ft^2 was attached to the main manufacturing building.

On July 2, 1942, Kahn's firm delivered plans for the new factory to the construction firms bidding for the contract; on July 4, there was a ground-breaking ceremony at the construction site; and on July 12, the Navy awarded contracts to the Long Construction Company of Kansas City, and the Turner Construction Company of New York, which had done work for United Aircraft in the past.[121] When work began excavating the site for the factory, the rain turned the area into a sea of mud, which forced the contractors to build the concrete floor before work started on the roof, the reverse of work done at the Wood-Ridge and Dodge-Chicago factories.[122] Work began on the roof of the factory building using the same "warspeed" method Kahn and the contractors had developed for Wood-Ridge and Dodge-Chicago. For the Kansas City factory, Kahn made the individual bays slightly larger, 40 ft by 40 ft. The wooden forms used to make the arch roof slabs, columns, and supporting ribs and girders were 80 ft long, allowing for the construction of two bays at a time.[123] The wooden frames were placed side by side across the 1,000 ft width of the factory moving on rails placed on the concrete floor. The columns and supporting girders were poured first, and when set, the arch slab roof forms were raised into position and the concrete was poured over the reinforcing steel mesh. When dry, the arch form was lowered and moved forward to make the next section. The entire process took five to seven days to complete two bays.[124] The plant was operational by February 1943, and the construction companies completed work on the entire factory complex on July 3, 1943.[125]

The Ford Motor Company and Nash-Kelvinator Corporation Factories

Where most of the aircraft engine factories built for World War II production followed a large, single-story, rectangular form, the factory the Ford Motor Company built to manufacture the R-2800 engine featured a novel two-story design.[126] After Ford accepted the Government's proposal to build the R-2800 at a rate of 15 engines a day, the company selected one of the last vacant areas at its massive River Rouge auto complex as the site for its aircraft engine factory. The Detroit architectural firms of Giffels & Vallet, Inc., and L. Rossetti designed the factory as a two-story building within a building with the manufacturing area enclosed within a four-story service shell that held offices, lockers, pedestrian walkways, and service rooms.[127] The building was rectangular in shape, 1,008 ft long and

360 ft wide, with the production area initially covering 570,000 ft^2, and the total floor space amounting to 825,039 ft^2.[128] The production area was windowless, with artificial lighting, and was completely air-conditioned. The second story was devoted to machining operations, with five 60 ft wide aisles that were 950 ft long, each with an overhead crane. The first floor contained more machining operations and final engine assembly and was divided into ten 30 ft aisles, also with overhead crane systems. Built before wartime restrictions, the building's construction made extensive use of steel girders and columns. Floors on the first and second stories were concrete slabs surfaced with wood blocks. Construction began in September 1940 and was completed eight months later.[129] The government provided $21 million in financing for the factory.[130]

In July 1941, the government more than doubled its original order, calling on Ford to build 40 R-2800 engines a day.[131] The government provided Ford with an additional $14 million to build a 400 ft long addition to the engine factory and later financing for a larger test facility comprising 28 test cells.[132] By the end of the war, the total area of the Ford aircraft engine factory amounted to 1,573,900 ft^2 and was equipped with nearly 9,000 machine tools.[133] In addition, Ford built an aluminum foundry to cast cylinder heads for the R-2800 and a magnesium smelter and foundry to make magnesium engine parts.[134]

While most of the automotive licensee companies carried out aircraft engine production in new, government-financed wartime factories, Nash-Kelvinator converted its prewar automobile factory in Kenosha, Wisconsin, to build R-2800 engines, with a new government-financed assembly and test facility built next door to the automobile factory. The company had started out making engine parts for Pratt & Whitney at its factories in Lansing, Michigan, but after Pearl Harbor the Navy gave Nash-Kelvinator a contract for 350 engines a month that was soon doubled to 700 engines a month.[135] As Nash-Kelvinator was already expanding production of Hamilton-Standard propellers at its Lansing factories, there were no facilities available for an increase in aircraft engine production. The company decided to convert its prewar automobile factory in Kenosha to building aircraft engines, transferring propeller manufacturing to the company's factories in Lansing.[136] The factories in Lansing that had been intended as facilities for building the R-2800 transferred their machine tools to the Kenosha factory, which covered some 107 acres of land. The main automobile manufacturing building went through a rapid conversion with workmen moving out machine tools, ripping up the automobile assembly lines, and tearing down partitions within the factory to create a huge open area with a main corridor running down the center of the factory building for close to a half a mile.[137] The new assembly building was a single-story, concrete frame structure with brick walls that had 204,800 ft^2 of production space, with a test cell building nearby built with reinforced concrete occupying 68,100 ft^2.[138] The first R-2800 engine built at the Kenosha factory was completed in October 1942 with full production beginning in December.[139]

The Bristol Engine Factory and Hawthorn Shadow Factory

The Bristol engine factory at Filton, also known as the East Works, built most of the Centaurus engines completed during the war. The East Works comprised four machine shop buildings, the Bristol No. 2 Shadow Factory, a small foundry, test cells, and other smaller buildings. In addition, the Bristol Engine Department (re-organized as the Engine Division in October 1942 when Bristol combined the Production and Development Departments) maintained several dispersal facilities around Bristol for storing finished parts and some machining and engine assembly. The machine shop buildings were built to the pattern used in the No. 2 Shadow Group factory buildings, but were smaller in size, with the largest having a floor area of 205,450 ft^2.[140] Several of the buildings contained space for activities other than engine manufacturing. The No. 3 Machine Shop was 179,000 ft^2 in total but had only 100,800 ft^2 devoted to machining operations, with 33,600 ft^2 turned over to engine repair, a major activity of the Filton engine works. In total, the Filton engine factory had 651,000 ft^2 of floor space devoted to engine production, about the same as one of the No. 2 Shadow Group factories. Filton had a smaller number of engine test cells, which were completely separate from the machine shops and located several hundred feet away.

The Hawthorn underground factory came out of the Ministry of Aircraft Production's intention to have a major engine production facility that would be protected from aerial bombing. The Air Supply Board approved the scheme for an underground factory to be built in the Spring Quarry near Bath in December 1940.[141] The initial plan was to build a factory with 2,200,000 ft^2 of space, with 1,500,000 ft^2 allocated to Bristol Aeroplane with 1,000,000 ft^2 for engine production, 200,000 ft^2 to house Bristol's aircraft engine development department, and 150,000 ft^2 for staff space.[142] The plan was to build eight machine shops in this space with storage areas for raw materials and finished parts, two areas for engine assembly, and canteens for the estimated 9,000 workers who would be employed in the factory.[143] Six engine test buildings were to be built above ground, with lifts taking the completed Centaurus engines up for their test runs.[144] The inability to obtain adequate numbers of workers, particularly skilled workers, proved to be an insurmountable obstacle. In July 1943, the Bristol directors had to admit that the engine production program planned for the Hawthorn shadow factory was "lost beyond all hope of recovery."[145] The Hawthorn factory never reached its planned production target of 250 Centaurus engines a month and was substantially over budget.[146]

The Production Record

The Pratt & Whitney R-2800 Double Wasp

From 1940 to 1945, Pratt & Whitney, Ford Motor Company, Nash-Kelvinator, and Chevrolet built 114,067 R-2800 engines.[147] The R-2800 production record shows once again how critical license production was to building the quantities of engines the Army Air Force and Navy needed during the war. Ford, Nash-Kelvinator, and Chevrolet built 70% of all the R-2800 engines completed during the war. Moreover, by continuing the production of the R-2800 B series, Ford and Nash-Kelvinator enabled the Pratt & Whitney Hartford and Kansas City factories and the Chevrolet factory to turn to manufacture the higher performance C series engines. Ford built 57,637 R-2800 engines starting with one model in the R-2800 A series and then shifting to building ten different models in the B series.[148] Three engine models—the R-2800-43, R-2800-59, and R-2800-75—made up 54% of Ford's production; these engines equipped the Curtiss C-46, the Douglas A-26, and the Republic P-47C/D airplanes.[149] Nash-Kelvinator built 17,108 R-2800 engines comprising seven models, but just one model, the R-2800-10W, made up 57.8% of the company's engine production.[150] Starting later than the other licensees, Chevrolet built the smallest number of R-2800 engines, completing 4,282 C series engines in four models.[151] Pratt & Whitney completed 35,040 R-2800 engines at its Hartford and Kansas City factories, with the Kansas City factory building 7,815 of these engines.[152] As the Kansas City factory ramped up production in 1944, the production of the R-2800 at the Hartford factories gradually declined from a peak of 947 engines in February to 445 engines in December as Pratt & Whitney devoted more effort into getting the R-2800 E series and the more powerful R-4360 engine into production.[153] During 1944, the Hartford factories also stopped the production of the R-1340 Wasp and R-985 Wasp Jr. to concentrate on the R-2800, R-2000, R-1830, and the R-4360.[154] Production at the Kansas City factory started in December 1943 when the factory delivered its first engine and built up to 475 engines by December 1944.[155]

An important factor contributing to the Allied air offensive against Germany and Japan was the pace of R-2800 production at the Pratt & Whitney licensee companies, shown in Figure 6.9. While it took 21 months from the start of the Government giving Ford the go-ahead to reach a rate of 500 engines a month, it took Ford only 10 months to reach 1,000 engines a month, compared with 19 months for Pratt & Whitney.[156] In 1943, Ford built nearly twice as many R-2800 engines as Pratt & Whitney and more than double Pratt & Whitney's production in 1944. Ford's peak production of 2,419 engines in June 1944 was more than twice Pratt & Whitney's peak rate of production. That same year, during which the American Army Air Force and Navy flew more combat sorties than in any other year of the war, Ford and Nash-Kelvinator built nearly three times as many R-2800 engines as Pratt & Whitney (Figure 6.10).[157]

FIGURE 6.9 Pratt & Whitney R-2800 Production, 1940-1945.

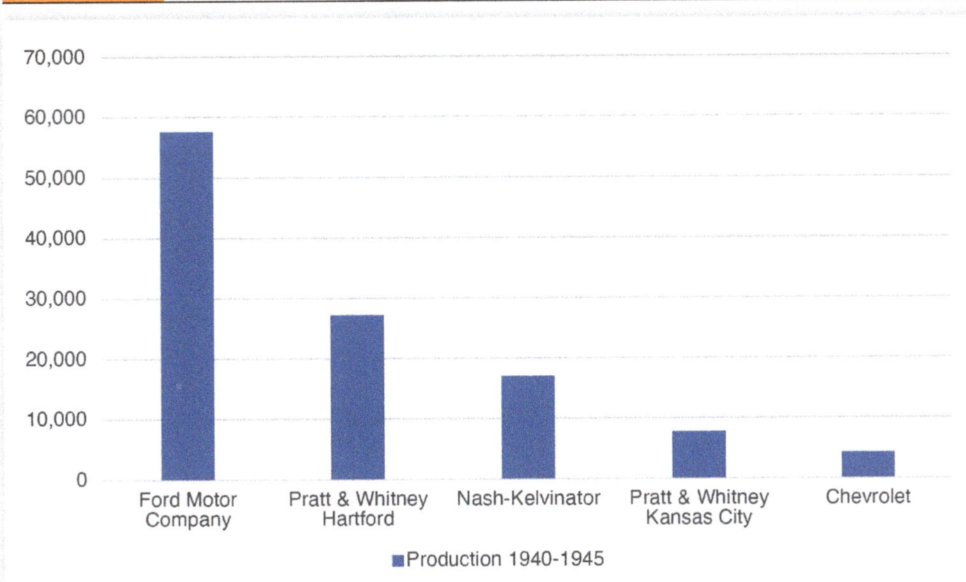

Source: US Department of Commerce, Civil Aeronautics Administration, Office of Aviation Information, Division of Aircraft Statistics: US Military Acceptances, 1940–1945: Aircraft, Engines and Propeller Production (Washington, DC, 1946).

FIGURE 6.10 Pace of R-2800 Production, 1940–1945.

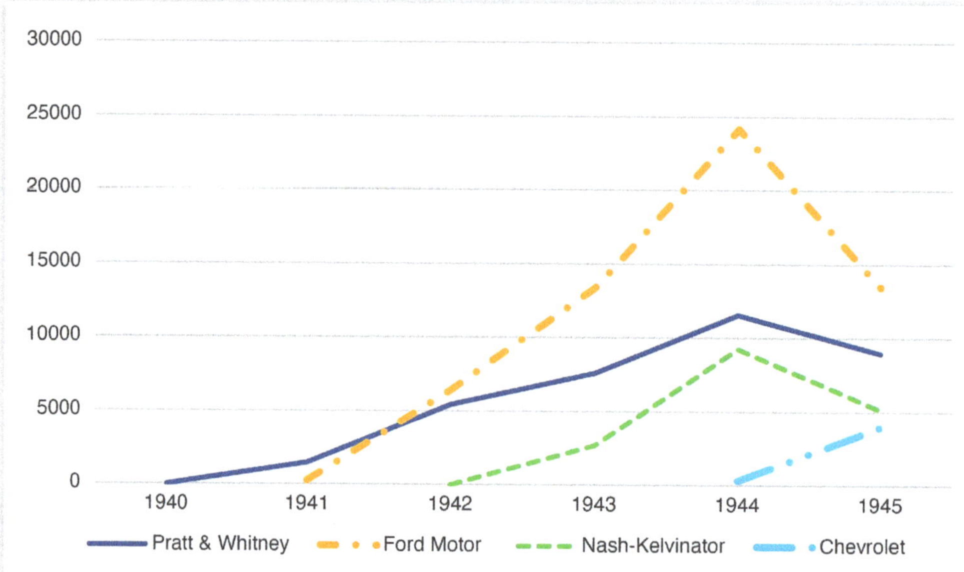

Source: US Department of Commerce, Civil Aeronautics Administration, Office of Aviation Information, Division of Aircraft Statistics: US Military Acceptances, 1940–1945: Aircraft, Engines and Propeller Production (Washington, DC, 1946).

The Wright R-3350 Duplex Cyclone 18

As with the R-2800, Chrysler's Dodge Division, the Wright Aeronautical Company's licensee, built more Wright R-3350 Duplex Cyclone 18 engines than the Wright Company. The Dodge-Chicago factory's record of 18,413 R-3350 engines completed during the war represented 57% of the entire wartime production of the Cyclone 18. While it took nearly two years for the Chicago factory to begin production, once began production increased rapidly and the factory was consistently ahead of its planned production schedule, reaching peak production of 1,600 Cyclone 18 engines a month in March 1945 just 15 months after completing the first production engine. The Chicago factory built only two variants of the Cyclone 18, and like Studebaker and the Cyclone 9, a single variant (the R-3350-23) made up 90% of the Chicago factory's production. The factory began building a fuel injection version of the Cyclone 18 in 1945, completing 1,986 of this version by the end of the war.

To augment Cyclone 18 production at the Chicago factory and relieve Wright's Paterson factories of an additional production burden, the government financed the large new factory at Wood-Ridge, New Jersey, which became Wright Plant No. 7. The production of the Cyclone 18 at Wood-Ridge began in February 1943 with parts supplied from the Paterson factories. As Figure 6.11 shows, it took longer for the Wood-Ridge factory to build up production than it did for the Dodge-Chicago factory. By the war's end, the Wood-Ridge factory had completed 11,793 Cyclone 18 engines, building 17 different models of the Cyclone 18. The Wright Lockland factory built an additional 1,867 R-3350 engines during 1945, taking total wartime production of the R-3350 Cyclone 18 to 32,240.[158]

FIGURE 6.11 Wright Cyclone 18 Production at Wood-Ridge, Dodge-Chicago, and Wright Lockland Factories, 1943–1945.

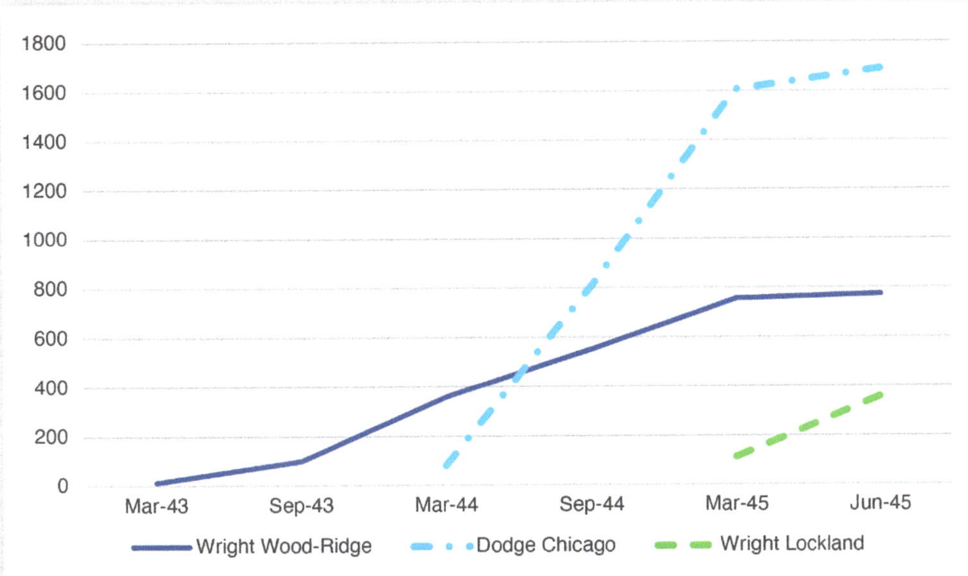

Source: US Department of Commerce, Civil Aeronautics Administration, Office of Aviation Information, Division of Aircraft Statistics: US Military Acceptances, 1940–1945: Aircraft, Engines and Propeller Production (Washington, DC, 1946).

The Bristol Centaurus

Compared to the production of the Pratt & Whitney R-2800 and the Wright R-3350, the number of Bristol Centaurus engines built during World War II was minuscule, amounting to 2,967 engines.[159] Demand for the Centaurus did not emerge until 1944, when modest numbers of the Bristol Buckingham, the Hawker Tempest II, and the Vickers Warwick were in production. The Tempest II might have been built in greater quantities had the Ministry of Aircraft Production, the RAF, and the Hawker Company pushed development work on a Centaurus-powered fighter more enthusiastically.[160] For much of 1942, the Centaurus was still in development. During the year, Bristol's Engine Department's workshops at Filton completed only 29 engines and in early 1943 the Engine Division was still making alterations to production engines and trying to finalize specifications.[161] The production of the Centaurus at the Bristol parent company peaked in the fourth quarter of 1944 when production reached an average of 142 engines a month, with Bristol completing 1,401 Centaurus engines during the year.[162] The Ministry of Aircraft Production had intended that the Bristol Hawthorn underground shadow factory would be the main source of Centaurus engines, with a target of building 250 engines a month, but the factory never lived up to expectations.[163] The Hawthorn factory did not complete its first engine until September 1943 and had built only 36 engines by the end of the year.[164] Production peaked at the Hawthorne factory during the fourth quarter of 1944 when the factory produced an average of 42 Centaurus engines a month.[165] Production at Hawthorn started in 1943, but continuing shortages of labor and machine tools, made it difficult for the Hawthorn factory to meet its production targets, and as the Centaurus was not in great demand, the factory could not be granted a priority for either labor or machine tools.[166] There was regular discussion at Bristol and the Ministry of Aircraft Production about converting the Bristol Accrington shadow factory to Centaurus production, but with fluctuating requirements for the Centaurus and increasing demands for the Hercules until late in the war the Accrington factory never made a full conversion.[167] By the end of the war, the Hawthorn shadow factory had delivered only 525 Centaurus engines, with the Bristol parent engine factory at Filton producing all but 35 of the remaining engines, with the Accrington shadow factory building this small number at the end of the war (Figure 6.12).[168]

FIGURE 6.12 Centaurus Production, 1942–1945.

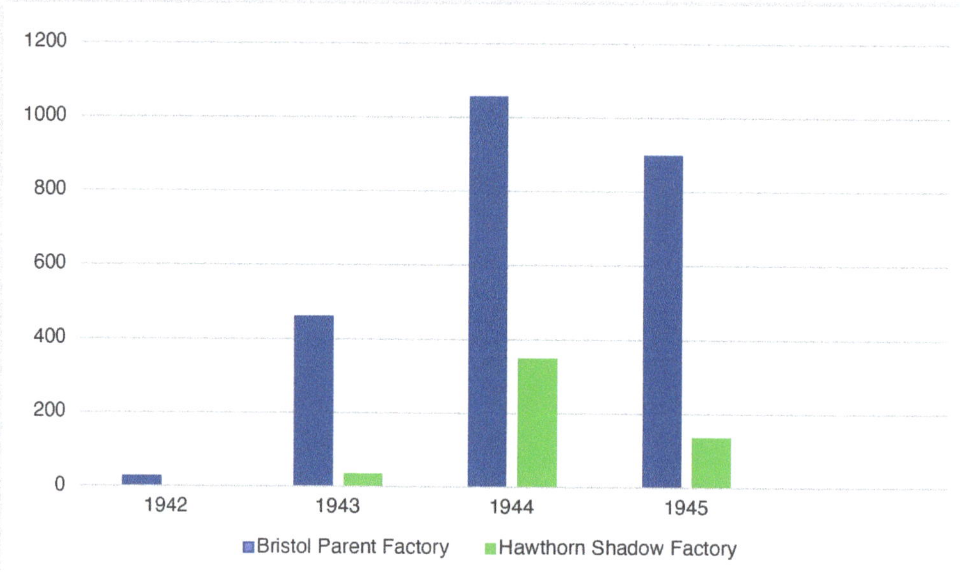

Source: Ministry of Aircraft Production: Statistical Review, 1939–1945, Table 9, p.26.; Bristol Aircraft Company, chart New Engines Delivered-Piston Engines, copy in the Aerospace Bristol Museum archives.

How It Was Done

Building the Pratt & Whitney R-2800

The production of the R-2800 engine at the Pratt & Whitney automotive licensee factories is well covered in the contemporary industry literature, production at the Pratt & Whitney Hartford factories much less so. During 1942, the Hartford factories more than doubled the production of the R-2800, with production reaching 525 engines in August, but increased by just 41% in 1943 with December being the peak month when the factories completed 874 engines.[169] It is highly likely that during this two-year period Pratt & Whitney maintained its methods of flexible production, using standard machine tools with special jigs and fixtures to accommodate changes in engine design.[170] In planning for the changeover to the more powerful R-2800 C series and beginning the production of the 28-cylinder R-4360, Pratt & Whitney's management announced that the program for the new engine would involve instituting "manufacturing processes radically different from those now in use."[171] The company did not elaborate on what these new processes would entail, although one of the new processes related to the new forged cylinder heads which required machined cooling fins using "manufacturing techniques as yet untried in production."[172] This was to be undertaken at the satellite Plant N in Southington, Connecticut. Pratt & Whitney also decided to bring back production of crankshafts for the new engine to the Plant M factory in East Longmeadow, Massachusetts.[173] While management spoke of the need to retool the Hartford factories for building the R-2800 C series, there was no mention of the types of machine tools that would be employed (Figure 6.13).[174]

FIGURE 6.13 The Pratt & Whitney R-2800 Double Wasp engine. During the war, Pratt & Whitney and its automotive licensees built 115,122 R-2800 engines.

Source: © 2023 Raytheon Technologies Corporation, Pratt & Whitney division. Reproduced with permission.

The production of the R-2800 at the new Pratt & Whitney factory in Kansas City used different methods and apparently different types of machine tools. The Kansas City factory gave Pratt & Whitney's management a clean slate, and with intense pressure to build up the production of the R-2800 C series engines, the layout of the factory focused on line production to build flow through the manufacturing process. The factory benefited from the collective experience of Pratt & Whitney's automotive licensees. Production engineers laid out the factory for a continuous flow of raw materials and parts from the east to the west side of the main manufacturing building.[175] Roughly 40% of the parts for the R-2800 engine came from subcontractors.[176] The factory had nearly 5 miles of conveyors to move parts through the machining process.[177] The main engine components—cylinder heads and barrels, crankcase, crankshaft, etc.—each had their own separate department with parts moving progressively from machine to machine without leaving the department.[178]

The continuing challenge for the Kansas City factory was to obtain and train the labor force. By 1943, other war industries in the area had long since swallowed the supply of skilled machinists.[179] Pratt & Whitney set up a training school modeled on the one at the Hartford factory to train workers with no experience in operating machine tools. Instructors trained workers to operate a basic type of machine tool during six weeks of instruction, later compressed to four weeks.[180] To fit the limited experience of the workers, the Kansas City factory shifted from Pratt & Whitney's reliance on standard machine tools, which required a higher level of experience to operate, to more special purpose machine tools with "push-button" operation that required a minimum of experience and thus a shorter period of training.[181] The factory also adopted more high-production, special purpose machine tools to build these engines in quantity.

A complete layout of the machine tools in the Kansas City factory as of June 1944, showing the location and types of machine tools in the entire factory fortunately has survived.[182] The Kansas City factory had 5,000 machine tools.[183] The layout for the Kanas City factory shows these, but also a number of high-production and expensive special purpose machine tools similar to the types found in the Buick and Ford aircraft engine factories.[184] Of particular interest are the Foote Burt automatic transfer machines that Buick developed with the Foote Burt Company.[185] The crankcase department used four Foote Burt machines to machine the front and rear sections of the crankcase, moving the crankcase casting automatically from station to station around a rectangular track during which time the crankcase underwent 21 drilling, boring, and reaming operations.[186] In the cylinder head and cylinder barrel departments, the Kansas City factory used four 13-station, 12-spindle Foote Burt machines, two 18-station, ten-spindle machines, and two large 20-station Foote Burt machines larger than the ones found at the Buick Melrose Park factory.[187] The factory had two more 20-station Foote Burt machines for machining master rods for the R-2800.

Of the three automobile firms that built the R-2800 under license—Ford Motor Company, Nash-Kelvinator, and Chevrolet—the story of production at Ford is the more interesting, and the more important as Ford built more R-2800 engines than any other manufacturer. Shortly after William Knudsen approached Ford with a proposal to build the R-2800 engine, Edsel Ford, president of the company, went to Hartford with Charles Sorensen, Ford's Vice President, and several top Ford engineers, to explore Pratt & Whitney's production methods and determine whether Ford could take on the task.[188] The Ford group "discovered an intricate machine, much more sophisticated than any automobile engine ever built," but told the War Department that Ford would agree to build the engine under license.[189] Over the next several months, while the Ford aircraft engine factory was under construction, Ford sent teams of engineers to Pratt & Whitney's Hartford factories to learn how to build aircraft engines. In June 1941, Ford set up a training school for foremen and workers, but "the scarcity of really competent personnel was painful."[190] There were delays as well in getting the machine tools the Ford factory needed for production.[191] Fortunately, Ford could turn to its own large Tool and Die Department to help modify machine tools and design new machine tools specifically for engine production. During the war, Ford's engineers designed 8,128 machine tools for the engine factory.[192] Still struggling with recruiting workers, training them, and getting machine tools, Ford nonetheless began manufacturing parts for the R-2800 in April 1941 and in August completed its first engine.[193] Production built up rapidly. Ford built 164 engines in December 1941, reached 500 a month by May 1942, and in December 1942 built 798 engines, completing 6,003 engines during 1942, more than Pratt & Whitney built over the same period.[194]

Toward the end of 1942, Joseph Geschelin, who had reported on the automotive industry for many years for the journal *Automotive Industries*, visited the Ford aircraft engine factory. He wrote that the factory was "considered to be one of the finest examples of aircraft engine manufacturing undertaken by the automobile industry."[195] Geschelin noted that the production of the R-2800 at Ford was a self-contained operation, with the company manufacturing all the main components of the engine, the majority of the gears used in the engine, and producing castings and forgings at its ferrous and nonferrous foundries.[196] With the experience accumulated in over a year of production, Ford's methods "have been progressively changed and improved so as to increase productivity and raise the level of quality."[197]

Ford used large, multi-spindle automatic machine tools to manufacture components of the R-2800 such as Bullard Mult-Au-Matics, Heald Borematics, Sundstrand automatic lathes, and Gisholt turret lathes among others, but for certain operations added its own innovations. Manufacturing the finned cylinder muff, which fitted over the cylinder barrel in the R-2800 cylinder assembly, demanded exceptional care to avoid distortion.[198] The inside of the aluminum cylinder muff had to be ground to a fine finish to fit securely over the cylinder barrel, but the thin wall of the cylinder muff could easily become misaligned. To reduce the risk, Ford developed a special water-cooled arbor, a device for holding the cylinder muff, and a means of cooling the grinding wheel during operation to prevent distortion.[199] A Fay automatic lathe with 78 cutting tools set in three tool slides cuts the fins in the cylinder muff to the required depth and spacing.[200] In another example, Ford worked with the Sundstrand Corporation to design a special boring machine for boring, facing, and chamfering (cutting at an angle to a flat surface) the cylinder bores and pads on the crankcase that held the cylinders.[201] The machine had two boring heads that could perform the operations on the upper and lower rows of cylinders in the crankcase at the same time, automatically moving from one cylinder bore to the next.

Ford's most unique and, for a time, controversial innovation was to use centrifugal casting to make cylinder barrels instead of the more traditional method of forging the barrels that Pratt & Whitney and Wright Aeronautical had used for years. Pratt & Whitney disagreed with Ford's decision to take this approach, and the Army Air Force was also skeptical as centrifugal casting had not been used previously in aircraft engines.[202] Ford engineers had experience with centrifugal casting, having done some experimental work for Rolls-Royce in 1940, and believed that using this method would produce cylinder barrels that would be stronger than forged barrels and would use less material.[203] Ford provided more experimental cast steel cylinder barrels for the R-2800 to Army Air Force engineers at Wright Field, which granted authorization to Ford in November 1941 to use this method on Ford-built engines.[204] The process involved mounting a circular steel mold onto a molding machine.[205] The mold rested on supporting rollers inside the mold machine that were driven by an electric motor. The mold machines were mounted on a revolving turntable that could carry eight machines spaced an equal distance apart. After being loaded on the turntable, the mold machine moved to the next station where the inside of the circular steel mold received a spray of graphite. With the circular steel mold spinning at a rate of 1,500 ft a minute, the mold machine moved to the next station where a worker poured in the right amount of molten steel at a temperature of 2,930°F. There were three more cooling stations along the turntable until the mold machine reached a conveyor where the circular steel mold was removed and the still hot cylinder barrel was removed, weighed to ensure the proper amount of steel, and sent on to cooling and rough finishing. After machining inside and out, the centrifugally cast cylinder barrels were indistinguishable from forged cylinder barrels. Ford found that the centrifugal casting method saved 35 lb of steel per cylinder barrel, a savings of 17,500 lb for the 500 cylinder barrels made each day, and drove down the cost of cylinder barrel from $13.38 to $10.50.[206]

In early 1943, Ford was building up to a production rate of 1,800 engines a month when in March the Army Air Force informed the company that its new goal was 3,400 engines a month.[207] The aircraft engine factory did not have the space nor were there enough workers available to reach the new production target. Instead, Ford decided to shift production of many engine components to other Ford factories and have the aircraft engine factory concentrate on engine assembly.[208] Ford shifted machine tools to its famous Highland Park factory in Detroit, factories in Memphis, Tennessee, St. Paul, Minnesota, Kansas City, Missouri, Green Island, New York, factories in Ohio, and factories in the Ford Village Industries.[209] This increased the total floor space devoted to engine production from 2,687,230 to 4,187,831 ft^2 and increased the total number of employees to 32,805 in February 1944, with 15,596 working in the aircraft engine factory in Dearborn.[210]

Ford ordered thousands of new machine tools to meet the new production goal. While four/fifths were standard machine tools that Ford could modify with its own jigs and fixtures, one/fifth of the new tools were special purpose, high-production machine tools with multiple spindles and automatic features designed specifically to speed production.[211] These various special purpose machines combined several operations into one machine, reducing the number of operators required and speeding up machining. One operation on the cylinder sleeve for the R-2800 required 36 single-spindle machines with 36 operators. Ford engineers designed two special machines that replaced the 36 single-spindle machines and their operators.[212] Ford continued to make incremental changes in its machining processes that speeded production. When grinding the aluminum cylinder muff, Ford switched from a 3- to 6-in. grinding wheel, which eliminated the need for the wheel to travel back and forth across the surface to be ground, cutting the operating time by 50%.[213] In another case, Ford put in two infrared drying ovens to dry painted and enameled engine parts, which cut the drying time from 1½ to 2 hours down to 15 minutes.[214] To speed the installation of engines in the test cells, Ford developed a simple coupling device that removed the need to connect and disconnect the separate wires and pipes to the engine.[215] All these small, incremental changes steadily reduced the time needed to build an engine.

From the early stages, Ford had employed various material handling systems in the aircraft engine facture to ensure a smooth and continuous flow of parts using monorail conveyors and tractors.[216] Ford had also adapted automobile assembly line operations to aircraft engine assembly, dividing the engine into subassemblies that workers assembled on benches as in other aircraft engine factories, moving these subassemblies to large assembly stands that moved on rails from station to station.[217] The final assembly department occupied an area of 1,400 ft by 60 ft using overhead monorails, cranes, and hoists to move parts and subassemblies as needed.[218] When told to build production of up to 3,400 engines a month, Ford added two more assembly lines to the aircraft engine factory.[219] Ford's productivity improvements reduced the manpower needed to assemble 60 engines per shift from 650 to 231.[220]

Chevrolet's challenge in building the R-2800 while continuing to manufacture the R-1830 engine was to integrate the assembly of two completely different engines, with no interchangeable parts, on a single assembly line. By 1943, there were 17 factories in the Chevrolet Division system contributing to Pratt & Whitney engine production, producing parts for the main factories at Buffalo and Tonawanda.[221] When asked in early 1944 to take on the production of the R-2800 C series while continuing but gradually reducing the production of the R-1830, Chevrolet had to undertake a sizeable reorganization.[222] Announcing the new contract in March 1944, M.E. Coyle, general manager of the Chevrolet Division, said, "the urgency of this program makes it imperative that Chevrolet begin production at the earliest possible date. We have scheduled the start of production in approximately seven months. A complex and involved program of plant rearrangement, retooling and installation of new equipment that would normally require a year of determined effort, lies ahead."[223]

One of Chevrolet's immediate tasks was to secure a source of forged aluminum cylinder heads and cylinder muffs for the new R-2800 C series engines. By 1943, Chevrolet had four forging factories: two prewar factories, a factory in Saginaw, Michigan, that forged and cast iron parts for Chevrolet auto-engines, and the second in Muncie, Indiana, which had been converted to forging aluminum aircraft and aircraft engine parts.[224] The Muncie factory forged crankcase sections for Pratt & Whitney engines and started forging cylinder heads for the R-2800.[225] With financing from the Defence Plant Corporation, Chevrolet expanded its wartime forging factory at Anderson, Indiana, to produce forged aluminum cylinder heads, modernizing its two existing buildings, adding 55,000 ft^2 of floor space, and 350 new machine tools and presses.[226] At the same time, Chevrolet undertook an expansion of the Tonawanda factory. This third phase of expansion added more floor space and a rearrangement and modification of all the engine test cells to accommodate the larger R-2800 engine.[227] Chevrolet set up the production of the new R-2800 engine with remarkable speed. The first Chevrolet-built R-2800 C series engine completed its test run in August 1944, five months and 18 days after the Army Air Force requested Chevrolet to undertake R-2800 production.[228]

Assembling the larger 18-cylinder R-2800 on the same assembly line with two models of the 14-cylinder R-1830 engine with no interchangeable parts between the two types of engines required carefully organized scheduling, a number control system for all parts and subassemblies, and double checking of all parts all along the assembly line.[229] A production schedule, determined in advance, set the number and types of engines to be assembled during each shift. Each part and subassembly received a number that corresponded to the model number of the engine being assembled. Specially trained personnel checked parts to ensure that the correct numbered part reached the right station on the assembly line at the right time. Like other aircraft engine factories, the Chevrolet engine assembly factory used benches running at right angles to the assembly line, feeding parts toward the line where engines on stands moved on tracks down the line, the assemblies being added as the engine advanced. Eighteen subassemblies and many individual parts made up a complete engine. After the green test in the engine test cells, the engine went back to the teardown line where the assembly process went into reverse, following the same scheduling and number control system to ensure that all parts of the same engine model remained together during and after inspection.[230] During March 1945, a month before Chevrolet ended the production of the R-1830, the Tonawanda and Buffalo factories completed 525 R-2800 engines and 605 R-1830 engines, approximately 22 R-2800 engines and 25 R-1830 engines a day on the same assembly line.[231]

Building the Cyclone 18 at the Dodge-Chicago Factory

The Dodge-Chicago factory was designed for integrated production of the Cyclone 18, relying less on subcontracting. The Dodge-Chicago factory's facilities were more comprehensive than at either the Wright Lockland or Wood-Ridge factories. As previously mentioned, the Chicago factory had large aluminum and magnesium foundries, the most modern in America at the time, with two separate forges for heavy and light forging. The initial plan was to have the Chicago factory make 305 separate parts for the Cyclone 18 as follows:

Machined parts	113
Assemblies	107
Forgings	75
Magnesium castings	8
Aluminum castings	2
Total	**305**[232]

The number of machined parts was increased from 113 to 158 when suitable subcontractors could not be found and when engineering changes added parts to the Cyclone 18. This increase proved more than the Chicago factory could accommodate.[233] The AAF agreed that the needed parts could be manufactured in other Chrysler and Dodge Division factories.[234] By the end of 1944, the level of subcontracting at the Chicago factory was about half that of the Lockland factory.[235] This was partly the result of timing. By late 1942, Dodge found it difficult to find subcontractors as most of the suitable firms were already heavily committed to other munitions manufacturers.[236] Fortunately, the Chicago factory did not depend on subcontractors for any of the main parts of the Cyclone 18 and, unlike the Lockland factory, never suffered a delay in production from the failure of a subcontractor to deliver parts.[237]

The Dodge-Chicago factory was the only aircraft engine factory in America that made its own forgings. The factory had two foundry buildings, one for light forgings and one for heavy forgings. The two foundries made crankcases, crankshafts, cylinder barrels, and other smaller steel parts for the Cyclone 18. The foundry used 35,000 lb steam hammers to make the three sections of the main crankcase. These came to the foundry as steel billets 24 in. in diameter.[238] A rotary hearth furnace heated the billets to 2300°F. The nine-man furnace crew removed the heated billet and placed it on the lower half of the die on the steam hammer, which hammered the billet against the die with 15 to 20 hard blows.[239] The forging next went to a trimming press, which removed the excess metal and punched a 10-in. hole in the center of the piece. When cooled, the forged crankcase sections went to the machine shop and assembly building for finishing and then to final assembly. The steel crankshafts went through a similar process, arriving as steel billets, and then heated and placed in less powerful steam hammers for forging, and excess metal was trimmed off and sent to the machine shop for finishing. Cylinder barrels were formed from solid steel billets using an upsetter, a forging machine where the workpiece was placed between two dies and pressure was exerted horizontally, as opposed to vertically. A furnace heated the steel billet until the front section was red hot. A team of three or four workmen removed the billet from the furnace and placed it in the upset forge, where the dies and hammers forged a hollow cylindrical barrel of the right size and shape and cut it off from the billet automatically, dropping the finished workpiece onto a conveyor to go to a cooling bin.[240]

The aluminum and magnesium foundries at the Chicago factory were the largest of their type in America.[241] The aluminum foundry concentrated solely on casting aluminum cylinder heads for the Cyclone 18. Completing 1,600 Cyclone 18 engines required 28,800 cylinder heads a month, 18 per engine, which works out to casting about 930 a day (as well as 930 cylinder barrels). In addition, the Chicago aluminum factory made cast cylinder heads for the Wright Wood-Ridge factory, sending 209,924 Cyclone 18 cylinder heads to Wood-Ridge in 1944.[242] The Chicago factory used the same casting process that Wright did, creating sand molds hardened in an oven before receiving the molten aluminum. The Chicago factory developed a process of reinforcing the fins in the sand mold that required only 240 steel pins per mold, substantially less than the 1,400 that Wright used, allowing for faster mold assembly.[243] The magnesium foundry made castings for the front section of the crankcase, the front and rear housings of the supercharger, and several other smaller parts. Both foundries were designed for line production and made extensive use of mechanical overhead and roller conveyors to speed production. Production flowed through "molding, baking, core assembly, pouring, shaking out, cleaning and inspection."[244]

Machining parts for 1,600 Cyclone 18 engines a month ultimately required over 10,000 machine tools, nearly three times as many as were in the Wood-Ridge factory.[245] The Chicago factory benefited from the work Wright had done developing high-production machine tools. The Chicago factory used many of the same types of machine tools that were used to build the Cyclone 18 at Wright's Wood-Ridge factory, such as Bullard Mult-Au-Matic multi-spindle drilling and boring machines, Cincinnati Hydro-Tel vertical milling machines, Ex-Cell-O automatic thread grinders, Greenlee 3-, 5-, 7-, 9- and 18-station horizontal machines, Heald Boromatics, and Rehnberg-Jacobson multi-station automatic indexing machines.[246] Two 18-station Greenlee horizontal automatic indexing machines could drill, ream, and counterbore 100 holes in the supercharger front section, replacing eight radial drilling machines and two radial tapping machines, reducing the handling operations from ten to two.[247] To round the edges of the many gears used in the Cyclone 18, Wright replaced a time-consuming process using portable tools with a Packer-Matic machine with automatic indexing, which reduced the process from 150 to 3 minutes per gear.[248]

Given the number of cylinder assemblies required for 1,600 Cyclone 18 engines every month, the Greenlee automatic transfer machine was arguably even more important to production at the Dodge-Chicago factory than at Lockland. Plans for the machine shop and assembly building incorporated stronger foundations for three Greenlee Automatic Transfer Machines, the largest of which weighed 275,000 lb.[249] In the event, the Chicago factory set up only two Greenlee automatic transfer machine lines. These were even bigger than the transfer machines at Lockland. Each transfer line had three sets of machines that were 209 ft long, 55 ft longer than the Lockland machines. These machines performed 269 operations, more than the Lockland machines, and could turn out a completed cylinder head every 45 seconds.[250] Using the automatic transfer machines the Chicago factory could complete 70% of all the machining operations the cylinder head required in a fraction of the time required with previous methods.

The massive machine shop and assembly building were ideal for line production with nothing to interfere with the placement of rows of machine tools except the heat treatment department on the east side of the building. Materials and parts entered at the south end of the building and moved progressively north toward the final assembly section, a 22-acre space that was fully air-conditioned, before the fully assembled engines moved to the engine test cells. Like the Lockland factory, production at the Chicago factory was organized around components, with each component having its own section of the building. The work flowed past rows of machine tools via conveyors. Since Dodge had been unable to find enough subcontractors to manufacture the many gears needed for the 18 cylinders in the Cyclone 18, Dodge had to manufacture these itself at the Chicago factory and other Chrysler and Dodge factories.

Manufacturing gears for the Cyclone 18 provides another example of how the aircraft engine companies borrowed knowledge and experience from the automotive industry to develop new methods of large-scale production. Within the machine shop and assembly building at the Dodge-Chicago factory, the gear department took up more space than any other component.[251] The big Wright Cyclone 14 and 18 double-row engines required many gears of different sizes and shapes for the propeller reduction gear, to operate the cams controlling the intake and exhaust valves, supercharger, generator, and other accessories. Making gears was laborious.[252] Gears started as a steel alloy forging. A gear cutter cut the rough outline of the gear teeth, and then the gear went through a hardening process. Wright had developed a method of hardening the gear first and then grinding the gear teeth to the proper shape using a special two-wheel gear shaper that would remove 0.004 to 0.006 in. of metal from each side of the gear teeth using abrasive wheels. The final process, finish grinding, was to round off sharp corners of the gear teeth by hand, a delicate task, and one not suited to large-scale production. However, it was the accepted practice in the aircraft engine industry.[253]

Before the war, the automotive industry, with its high demand for gears for automobile transmissions, developed a different method of manufacturing gears better suited to large-scale production. This method, called gear shaving, used a cutting edge to shave off excess metal instead of a grinding machine.[254] Shaving instead of grinding gears saved considerable time allowing a higher rate of production with fewer machine tools. It was more accurate and less dependent on the skill of the operator.[255] Faced with a looming shortage of gear-finishing machine tools, the War Department instructed the aircraft engine companies and their licensees to switch to gear shaving where it was practical to do so.[256] Wright undertook an experimental development program to work out a method of using gear shaving for aircraft engine gears, which would take into account the difference between automobile and aircraft engine gears in design and performance requirements.[257] This program was successful and the factories building Wright engines shifted to this new method of manufacturing gears (Figure 6.14).

FIGURE 6.14 Final assembly of a Wright R-3350 Cyclone 18 engine. Two workers are lowering the front section onto the nearly fully assembled engine. Once assembled, the engine would go to the test cells for test running before final inspection and shipment.

Wright Aeronautical photo, author's collection.

The Chicago factory had to cope with changes in the number of parts to be machined at the factory and the need to produce parts for other Wright factories building the Cyclone 18. In January 1943, even before completing construction of the factory buildings, the Chicago factory began machining work on 113 machined parts for the Cyclone 18, increasing this to machining 158 parts later in the year. From May through September 1944, Dodge brought the total number of parts to be machined at the factory to 184.[258] The Chicago factory transferred some production to other Chrysler Dodge factories, but continued to produce about 130 machined parts until the end of the war.[259] The Chicago factory also machined 135 different parts for the Wright Wood-Ridge factory, sending 75,674 machined parts to Wood-Ridge in 1944.[260] The Chicago factory also provided Cyclone 18 parts to the Lockland factory as it transitioned from building the Cyclone 14 to the Cyclone 18.[261] In addition to finished engines and parts for other Wright factories, the Chicago factory produced over 34 million spare parts for the Cyclone 18 engine from the fall of 1943 to the end of the war.[262] To get other Dodge factories into production, the Dodge-Chicago factory provided skilled staff and machine tools. Over the summer of 1944, the Dodge Main factory took over the production of oil pumps for the Cyclone 18 from Dodge-Chicago. The Chicago factory built up a 30-day supply of oil pumps and then shipped 479 machine tools to the Dodge Main factory where they were added to newly ordered machine tools and other tools shipped from other Dodge factories. With support from Dodge-Chicago, the Dodge Main factory began manufacturing oil pumps before the 30 days had run out.[263] In total, the Chicago factory transferred the production of 34 separate parts and 13 assemblies to other Dodge factories.[264]

Production at the Dodge-Chicago factory was not without its challenges. During the war, the factory received 6,274 design changes for the R-3350 engine from Wright Aeronautical.[265] These design changes resulted in 48,500 engineering releases and change notices to the Planning and Production divisions for changes in tools, manufacturing methods, and materials.[266] New processes had to be integrated into the existing production lines, and the production of a new fuel injection Cyclone 18 engine had to be ramped up as quickly as possible. Shifting to the new aluminum "W"-shaped fins on the cylinder barrel resulted in scrapping many automatic lathes, acquiring new machine tools to make the cuts in the cylinder barrel and insert the aluminum fins, and a major revision of the cylinder barrel production line.[267]

Getting the fuel injection Cyclone 18 into production was fraught with problems. At several points along the way, Chrysler had to persuade the AAF to allow the Dodge factory to continue the production of the carburetor engine so as not to disrupt production and cause labor and machine tools to be idle.[268] General Arnold had called for 120 B-29s with fuel injection engines to be built by October 1944, which again proved hopelessly unrealistic.[269] When the AAF approached Chrysler about building the fuel injection Cyclone 18 at the Dodge-Chicago factory, Chrysler stated that the company had no experience with fuel injection engines.[270] Chrysler was completely dependent on Wright for engineering drawings and guidance on production methods and the Bendix Corporation for fuel injectors. The fuel injector system consisted of two pumps, one for each row of cylinders, with nine plungers in each row that forced a set mixture of fuel and air into the combustion chamber of each of the Cyclone 18's eighteen cylinders.[271]

The fuel injection engine had 97 new parts that were different from the standard Cyclone 18 that the Dodge factory was building and was in many ways a new engine. The fuel injection engine would require new materials, new machine tools, fixtures, and gages and new subcontractors as Dodge decided it could build only eight or nine of the new parts required.[272] Despite not having all the necessary machine tools and fixtures, the Dodge-Chicago factory completed its first fuel injection engine for test purposes in September. However, with continued engineering problems, the need for new and different machine tools, and delays in getting fuel injection systems from Bendix, the Chicago factory did not begin manufacturing fuel injection engines until January 1945.[273]

In the interim, Chrysler engineers used their extensive experience to redesign, at the AAF's request, the fuel lines leading from the fuel injection pumps to the cylinders, which in Wright's original design were "routed tortuously and inaccessibly; hard to install, and harder to service: if a line broke, it could not be replaced without removing the engine from the nacelle, a job of as much as 400 man-hours."[274] Dodge engineers redesigned the fuel lines so that a mechanic could completely remove the lines and reinstall them in 2 hours. As the production of the fuel injection engine built up over the next few months, the AAF ordered the Chicago factory to plan to stop the production of the standard carburetor engine in November 1945 and to progressively switch to the fuel injection engine, increasing production to a level of 1,600 engines by that same month.[275]

The Chrysler Corporation and the Dodge Division's extensive experience with mass production enabled the Dodge-Chicago factory's Engineering Department to improve manufacturing methods for the Cyclone 18 that speeded production and cut costs, mostly by eliminating unnecessary features on the engine.[276] The Engineering Department introduced 25 major changes to the Cyclone 18 engine. These changes cut 174 machining operations, eliminated 62 assembly operations, and canceled 58 unnecessary parts from the engine.[277] Dodge engineers developed a special test stand fixture that cut the time needed to install an engine in the test cell from 2 hours to 10 minutes. In another innovation, Dodge connected the engines under test directly to generators to generate electricity for the factory around the clock.[278] Chrysler estimated that these improved manufacturing techniques not only reduced the cost per engine, saving the government upward of $100 million.[279] The Chicago factory cut the cost of building the Cyclone 18 engine dramatically, from an estimated cost of $25,314 per engine to an actual cost of $11,357 for the standard version and $12,954 for the fuel injection version.[280]

Another innovation that the Chicago factory introduced, again borrowed from the automobile industry, was to adopt shot peening to strengthen connecting rods. Shot peening was a process of bombarding the metal surface of the connecting rod with tiny metal or ceramic pellets to strengthen the surface layer where fatigue failures often started.[281] Shot peening also eliminated the need to polish the connecting rod, saving 36 man-hours of production time per engine.[282] Developing this method was a joint effort of the aircraft engine manufacturers and their licensees through the Aircraft Engine Division of the Automotive Council for War Production, a grouping that combined representatives of the two industries to work out new production methods that would speed production and reduce labor.[283]

Production Challenges at the Chicago Factory

The Chicago factory's experience with production control provides another, if different, example of how challenging organizing large-scale production was during the war. The Dodge factory never experienced the disruptions and management control problems of the Lockland factory. The biggest challenge the Dodge management faced was simply getting into production. Shortages of machine tools persisted for months. The problems that had to be overcome to build the Cyclone 18 in quantity covered a broad spectrum of challenges in engineering, production, availability of materials and labor, scheduling, choice of engine model, and coordination that at times must have seemed insurmountable. The breadth of these problems and the urgency of the Cyclone 18 and the B-29 program taxed the management of Wright, Chrysler, and its Dodge Division and the AAF to the limit. Despite the urgency, it took the Dodge-Chicago factory 29 months from the contract date to reach a production rate of 500 engines a month, longer than the average of 15 to 20 months for other aircraft engine factories.[284]

Setting up the production line, however, is dependent on having a product that is ready for production. Only then can the production engineers determine the machine tools needed to manufacture the product, the machining operations required, the correct sequence of operations, the labor that will be needed, and the parts and raw materials that will be necessary. Once the factory has put all the components of the production line in place, production control is relatively straightforward.[285] When the Chrysler Corporation received the contract to build the Cyclone 18, the engine was far from ready for production. In 1942, the Cyclone 18 was still very much an experimental engine. Between the time Chrysler received the contract to build the Cyclone 18 in February 1942 and November 1943, Wright put through more than 2,000 engineering changes on the Cyclone 18, 500 of which required changes in tooling.[286] As the Harvard Business School study on the problems of accelerating aircraft production noted, "the early months of the R-3350 [Cyclone 18] history indicate the futility of commencing to tool a shop for an engine too far in advance of the time when its design is reasonably well established."[287]

Delays in getting machine tools affected the Dodge-Chicago factory's ability to set up a production line. In March 1942, Dodge received from Wright a list of 4,999 machine tools required to build the Cyclone 18, with the less than helpful note that the list "...had been prepared hurriedly and would undoubtedly be subject to considerable revision."[288] The AAF then decided to give priority to accelerating machine tool deliveries to the Wright Wood-Ridge factory in the hopes of getting Wood-Ridge into production quickly.[289] By April 1943, the Dodge factory had placed orders for 7,050 machine tools but had received only 32% of the tools on order.[290] More importantly, the Dodge-Chicago factory had received only 50% of the machine tools needed to set up a pilot production line, where Dodge engineers could actually test manufacturing methods and confirm the sequence of operations.[291]

The AAF did not help matters by changing its requirements for the Cyclone 18. In April 1942, the AAF ordered Chrysler to plan for the production of 1,500 Cyclone engines a month instead of 1,000 a month and instructed Chrysler to plan to build 750 Cyclone 18BA engines and 750 Cyclone 18BB a month.[292] As these engines had important differences in their construction, they required separate production lines. Chrysler began ordering machine tools for the two different engine models and setting up production lines for the BA and BB models, only to have the AAF decide in October 1943, once the process was well underway, to drop plans for the Dodge-Chicago factory to build the Cyclone 18BB and instead to build 1,500 Cyclone 18BA engines a month, quickly raised to 1,600 a month to be achieved by May or June 1945.[293]

When the AAF canceled the order for the Cyclone 18BB, Dodge found that only 752 of 2,537 machine tools the factory had ordered for the Cyclone 18BB production line could be transferred to the Cyclone 18BA line.[294] By the end of 1943, the factory was still short of machine tools.[295] The start of Cyclone 18 production at the Chicago factory had to be pushed back from March 1943 to July 1943 and finally to January 1944, nearly two years from the date of Chrysler receiving the contract. The rapid increase in production at the Dodge factory during 1944 demonstrates how important production control was to quantity production. With a standardized engine ready for production, production processes could be clarified and sequenced, parts and raw materials could be obtained from vendors to a set schedule, adequate numbers of machine tools could be acquired and placed in production lines, and production could accelerate rapidly.

Building the R-3350 at the Wright Wood-Ridge Factory

The key characteristic of the Wood-Ridge factory was the factory's reliance on the high-production machine tools that Wright Aeronautical had developed before the factory initiated production.[296] The Wood-Ridge factory built the entire R-3350 engine, relying on other Wright factories in the Paterson area and the Wright Lockland factory for certain parts. Wood-Ridge received studs, nuts, bolts, and reduction gears from the Wright Plant 4, magnesium engine sections from the magnesium factory at Plant 3, cylinder heads and other castings from the aluminum factory at Plant 2, and a large number of gears from the Lockland factory.[297] In return, Wood-Ridge made steel crankcase forgings and forged cylinder heads for the R-1820 Cyclone 9 built in the Paterson factories.[298]

The Wood-Ridge factory received an early version of the Greenlee automatic transfer machine, with one 18-station and one 20-station machine connected together for machining operations on cylinder heads.[299] This may have been enlarged later in the war. The factory acquired several smaller Greenlee automatic indexing machines for milling magnesium supercharge sections. These included three-way nine-station and five-way 18-station horizontal indexing machines that could perform multiple operations in sequence, automatically moving the workpiece at the end of each stage.[300] In addition, the factory had several special Greenlee drilling machines that cost between $43,000 and $53,000 each.[301] Other high-production machine tools included Barnes special drills, Bullard Mult-Au-Matic four-spindle machines, Cincinnati Milling four-spindle Hyrdro-Tel milling machines, Ex-Cell-O special drilling machines, and different Heald, Landis, and Norton boring and grinding machines.[302] The emphasis at the Wood-Ridge factory, as at the Lockland factory, was on having machine tools perform multiple operations to speed production and reduce the number of operators, a critical factor at Wood-Ridge where recruiting workers became an issue.

Building the Bristol Centaurus

There is little information available on how Bristol built the 18-cylinder Centaurus engine. Contemporary British aviation and industry magazines, principally *Aircraft Production* and *Machinery (London)*, do not contain detailed descriptions of Centaurus production as they do for the Bristol Hercules. This is probably because Bristol built comparatively few Centaurus engines during the war and to wartime security restrictions. It is likely that in many respects, building a Centaurus engine was similar to building a Hercules. As the Centaurus was a development of the Hercules, the two engines had much in common in terms of the design and composition of their components. Apparently, however, while machine tools designed for manufacturing the Centaurus could be used to manufacture the Hercules, the machine tools in the Bristol factories used to manufacture the Hercules could not be used to manufacture components for the Centaurus.[303] For this reason, Bristol recommended purchasing new machine tools for Centaurus production and specifically requested the Ministry of Aircraft Production to order American machine tools. In June 1941, Bristol prepared a proposal for the Ministry to build the Centaurus at the Bristol parent engine factory at Filton and the Bristol No. 3 Shadow Factory at Accrington and requested £450,000 (approximately $1.8 million) to purchase 132 machine tools and other specialized equipment in America.[304] On July 4, the MAP's Air Supply Board approved an expenditure of £200,000 ($800,000) to purchase machine tools and other equipment for the parent factory to build the Centaurus, with a planned output of 80 engines a month, rising to 100 a month when Bristol stopped the production of the Taurus engine.[305] A year later, the Air Supply Board approved an additional £200,000 for machine tools at the parent factory (Figure 6.15).[306]

FIGURE 6.15 The Bristol Centaurus IV 18-cylinder sleeve-valve engine. The Bristol Engine Department at Filton and the Hawthorn shadow factory build 716 Centaurus IV engines before the end of the war. This model powered the Bristol Buckingham, Hawker Tempest II, and the Vickers Warwick twin engine bomber.

Rolls-Royce Heritage Trust-Bristol Branch.

While the production of the Centaurus began at the parent engine factory in Filton toward the end of 1942, after the Centaurus IV passed its type test in July, debate continued over the rate of production to be achieved and how to allocate production between the parent factory, the Hawthorn factory, and the Accrington shadow factory.[307] A May 1943 proposal to enable the Accrington shadow factory to build 150 Centaurus engines a month gives an indication of the types of special purpose machine tools Bristol wanted to order from America for Centaurus production.[308] On the list of equipment that Bristol wanted were several automatic machine tools that were also found in the American aircraft engine factories such as Bosch multi-spindle drilling machines, Cincinnati Milling Hydromatic machines, and Heald Borematics, but none of the larger automatic indexing machines that were employed in the American factories.

At its Filton engine factory, Bristol built the Hercules and, until late 1943, the sleeve-valve Taurus engine, in addition to the Centaurus. While the exact layout of the machine shops at the engine factory at Filton, where Bristol built most of the Centaurus engines during the war, is unknown, Bristol's board of directors noted the limited ability to make improvements in production methods at the factory due to the layout of machine tools compared with the more modern layouts in the shadow factories.[309] Based on this concern, it is likely that the machine shops at the Filton engine factory were not set up for line production. The legacy of the dispersal of facilities was another constraint on production efficiency. Bristol had shifted certain operations and departments to facilities around the Bristol area. Bristol had moved engine assembly to a building in Paulton, some 20 miles from Filton that had 36,000 ft^2 of floor space.[310] Parts and components for the Centaurus would have been trucked from Filton and other dispersal sites to Paulton, and completed engines trucked back to the engine test cells at the Filton factory, adding time to the production process.

The production of the Centaurus at the Filton factory took months to build up. As of May 1943, the Centaurus was still in the process of production development. The number of engines built that month was below target, but the Engine Division was slowly eliminating production problems.[311] The problems persisted into July, with a lack of testing facilities and a shortage of boost controls from America slowing production.[312] Over the summer of 1943, the Bristol Engine Division transferred Hercules assembly from the facility at Paulton to allow Paulton to concentrate on the Centaurus.[313] The rate of production improved during 1944, although shortages of certain components sourced from America continued as did some technical issues. Had demand for the Centaurus been greater, it is likely that the MAP and Bristol would have put more effort into improving production.

Conclusion

Achieving quantity production of the big, 18-cylinder Pratt & Whitney R-2800 Double Wasp and the Wright R-3350 Duplex Cyclone 18 engines, as well as the pace of production, was a remarkable achievement. Nearly 80% of these engines were built during the final 18 months of the war. Once again, the contribution of the automotive companies to achieving the quantities needed was vital. Ford, Nash-Kelvinator, Chevrolet, and the Dodge Division built 54% of all the R-2800 and R-3350 engines made during the war, and 61% of the engines built over the last 18 months.

Ford benefited from the fact that during 1941 as the company was gearing up to build the R-2800, Pratt & Whitney was accelerating the production of the engine and sharing this knowledge with Ford's engineers. When Ford began working on the R-2800, the engine was ready for production. Ford went on to share its experience and production knowledge with Nash-Kelvinator and Chevrolet.[314] Pratt & Whitney's licensing of the R-2800 B series to Ford and Nash-Kelvinator allowed Pratt & Whitney to devote its efforts to developing the more advanced C series and getting this engine into production at the Pratt & Whitney Kansas City factory and Chevrolet.[315] In the case of these automotive companies, their experience met the three conditions for effective licensing arrangements as outlined in the Harvard Business School study.[316]

The Dodge Division's experience getting the Wright R-3350 into production illustrates the problems that did arise in some licensing arrangements. In early 1942, when Chrysler received the War Department contract to build the R-3350, the engine was not ready for production. That year Wright's Paterson factories completed only 67 R-3350 engines.[317]

Wright engineers were perfectly willing to share information with Dodge engineers, but were themselves struggling to work out problems with the engine, and until the engine reached a production standard, it could not be built. Delays in reaching production status caused delays in acquiring needed machine tools, setting back production by a year, from January 1943 to January 1944. It took the Dodge Division 29 months from getting the go-ahead to reach a rate of 500 engines a month, the longest period of any of the automobile licensees. Yet once begun, production accelerated and the Dodge-Chicago regularly surpassed its production targets, reflecting the Dodge Division's long experience with production engineering.

All the branch and licensee factories building the R-2800 and R-3350 engines adopted line production methods, including extensive use of mechanical conveyors, and simpler manufacturing processes to increase production volumes and cope with the influx of semi- and unskilled workers. Several factories employed the latest designs in high-production machine tools such as the Greenlee and Foote Burt automatic transfer machines and other advanced automatic machine tools. The systems of line production that the factories developed retained the flexibility to adjust to the frequent changes in design that Pratt & Whitney and Wright put through to improve engine performance and reliability. Designing successful production systems and manufacturing processes for engines that required some 13,000 component parts, as well as mobilizing and coordinating the work of thousands of subcontractors, and training tens of thousands of employees was an achievement that has received little recognition. The successful production of these 18-cylinder engines helped put the American aviation industry in a dominant position in the postwar market for commercial airliners.

The story of the Bristol Centaurus is one of what might have been. Had there been an aircraft "champion" for the Centaurus that the RAF required in quantity, if Bristol Aeroplane had had the resources to devote more effort to Centaurus' development earlier in the war, the Centaurus production record might well have been different. By the end of the war, the Centaurus had reached 2,500 hp, equal to the Wright R-3350. Post-World War II, the Centaurus had a long and successful career as an engine in military transport aircraft, but did not fare well as an engine for commercial airliners, a market where the American air-cooled radial piston engines dominated until the advent of the jet engine. The attempt to manufacture the Centaurus in the underground Hawthorn factory failed to meet expectations, but the Bristol parent factory proved able to meet the limited demand for the Centaurus in the final years of the war.

References

1 Loftin, Laurence K., Jr.: *Quest for Performance: The Evolution of Modern Aircraft*, (Washington, DC, 1985), p. 103.

2 Craven and Cate, Volume Six, *Men and Planes*, p. 194.

3 Production data taken from *U.S. Military Acceptances 1940-1945: Aircraft, Engines and Propeller Production*.

4 Ministry of Aircraft Production: *Statistical Review 1939-1945*, Table 9, p. 26.

5 Ibid., Table 9, p. 27.

6 Kotelnikov, Vladimir: *Russian Piston Aircraft Engines*, (Ramsbury, Wiltshire, UK, 2005), pp. 128-129.

7 To reach 2,000 hp the German BMW Flugmotorenbau and Daimler Benz companies resorted to coupling two standard engines together, but the 28-cylinder BMW 803 engine never went into production and the Daimler-Benz DB606 and DB610, combining two DB601 liquid-cooled, inline engines in the first case and two DB605 engines for the DB610, went into limited production for the ill-fated Heinkel He 177 heavy bomber. Late war German engines did, however, produce more than 2,000 hp by using water injection to increase the density of the fuel-air mixture and lower the risk of detonation. In Japan, the 18-cylinder Nakajima Homare Ha-45 was

intended to reach 2,000 hp, but production engines only reached 1,900 hp. The late-war 18-cylinder Mitsubishi MK9/Ha-211 had a rated power of 2,200 hp, but was not put into production. For German engines see Douglas, Calum E.: *The Secret Horsepower Race: Western Front Fighter Engine Development*, (Horncastle, UK, 2020) and for Japanese engines see Goodwin, Mike and Peter Starkings: *Japanese Aircraft engines: 1910-1945*, (Poland, 2017).

8 Freeman, Joshua B.: *Behemoth: A History of the Factory and the Making of the Modern World*, (New York, 2018), pp. xiii-xv.

9 White, Graham: *R-2800 Pratt & Whitney's Dependable Masterpiece*, (Shrewsbury, UK, 2001), p. 100.

10 Ibid., pp. 100-01.

11 Craven and Cate, Volume Six, *Men and Planes*, p. 195.

12 White, Graham: *R-2800 Pratt & Whitney's Dependable Masterpiece*, p. 26.

13 Ibid., pp. 25-26.

14 Ibid., pp. 34-60.

15 Dean, Francis H.: *America's Hundred-Thousand: U.S. Production Fighters of World War Two*, (Atglen, PA, 1997), pp. 277, 507, 590.

16 White, *R-2800 Pratt & Whitney's Dependable Masterpiece*, pp. 375-78.

17 "From 1700 to 2700 Horsepower in 10 Years", *Aviation Week*, (May 24, 1948), p. 26.

18 Ibid.

19 Ibid.

20 "Summary of the R-3350 Engine Project", p. 1.

21 Ibid.

22 Ibid., p. 2.

23 Ibid.; McCutcheon, Kimble D.: "Wright R-3350 'Cyclone 18'", Aviation Engine Historical Society, https://www.enginehistory.org/Piston/Wright/WrightR-3350.pdf, accessed 13 December 2019.

24 This description of the Cyclone 18 comes from Wright Aeronautical Corporation: *Wright Cyclone 18 Aircraft Engine, Series C18BA, Preliminary Instruction Book*, (November 1942).

25 *U.S. Military Acceptances 1940-1945: Aircraft, Engines and Propeller Production.*

26 Case History of the R-3350 Engine, p. 126.

27 Kent, S.R.: "Cyclone 18 Performance in Combat Areas", Society of Automotive Engineers Summer Meeting, June 2-7, 1946, p. 1.

28 Heaven, *Bristol Piston Engines Since 1914*, HS5/33.

29 Bingham, *Major Piston Aircraft Engines of World War II*, pp. 58-59.

30 Heaven, *Bristol Piston Engines Since 1914*, HS5/33.

31 Bingham, *Major Piston Aircraft Engines of World War II*, pp. 58-60.

32 Heaven, *Bristol Piston Engines Since 1914*, HS5/33.

33 Furse, *Wilfird Freeman*, pp. 136, 334.

34 Ministry of Aviation, *Narrative: Reciprocating Aircraft engines & Engine Accessories Production Programmes*, para. 42.

35 Furse, *Wilfird Freeman*, pp. 173, 334.

36 Airplane Division Report No. 7, Table D, p. 12.

37 Mr. Young to Secretary Morgenthau, October 19, 1940, Volume 323, October 17-20, 1940, Diaries of Henry Morgenthau, Jr.

38 Nevins, Allan, and Frank Ernest Hill: *Ford: Decline and Rebirth 1933-1962*, (New York, 1962), pp. 177-78.

39 Ibid., p. 178.

40 Ibid., p. 178; *The Wall Street Journal*, August 17, 1940, p. 1, September 18, 1940, p. 7.

41 Special Study No. 21, p. 141.

42 Ibid.

43 *The Wall Street Journal*, October 9, 1941, p. 11.

44 See Aircraft Branch Aircraft Engine Report 9-G, November 28, 1941, and Aircraft Branch Aircraft Engine Report No. 9-I, February 26, 1942, Office of Production Management Reports, Box 207, Papers of Harry Hopkins.

45 War Production Board: Statistics Division: *Alphabetical Listing of Major War Supply Contracts Cumulative Through February 1943*, Vol. 2, DEL-KEN, (Washington, DC, 1943), p. 308.

46 Hyde, Charles K.: *Storied Independent Automakers: Nash, Hudson, and American Motors*, (Detroit, MI, 2009), p. 71.

47 War Production Board: Statistics Division: *Alphabetical Listing of Major War Supply Contracts Cumulative Through February 1943*, Vol. 3, KEN-Rie, (Washington, DC, 1943), p. 567; *The Wall Street Journal*, June 30, 1942, p. 12.

48 Aircraft Branch Aircraft Engine Report No. 9-K, September 9, 1942, Office of Production Management Reports, Box 239, Papers of Harry Hopkins.
49 White, *R-2800 Pratt & Whitney's Dependable Masterpiece*, pp. 144-71.
50 Ibid., pp. 185-87.
51 Ibid., p. 144.
52 United Aircraft, *The Pratt & Whitney Story*, p. 136.
53 Ibid., pp. 136-137.
54 *The Kansas City Times*, July 4, 1942, p. 1.
55 *The Wall Street Journal*, December 23, 1944, p. 5.
56 *The Iron Age*, Vol. 154, No. 7, (August 17, 1944), p. 134.
57 *The Wall Street Journal*, March 9, 1945, p. 3.
58 *Aircraft Production Policies under the National Defense Advisory Commission*, p. 100.
59 Ibid.
60 Hyde, Charles K.: *Riding the Roller Coaster: A History of the Chrysler Corporation*, (Detroit, MI, 2003), p. 120.
61 "Motor Cars to Munitions: Chrysler: Didn't Wait for New Machines; Revamped the Ones They Had", *American Machinist*, Vol. 86, No. 12, (June 1942), p. 549.
62 Hyde, *Riding the Roller Coaster: A History of the Chrysler Corporation*, p. 141; United States Air Force: United States Air Force Numbered Historical Studies, No. 40: *Expansion of Industrial Facilities under AAF Auspices 1940-1945*, (No Date), p. 113.
63 Hyde, *Riding the Roller Coaster: A History of the Chrysler Corporation*, p. 141.
64 Ibid.
65 Ibid.
66 United States Air Force Numbered Historical Studies, No. 40: *Expansion of Industrial Facilities under AAF Auspices 1940-1945*, p. 113.
67 *The Wall Street Journal*, June 10, 1942, p. 5.
68 *The Cincinnati Enquirer*, June 16, 1944, p. 16, October 10, 1944, p. 17.
69 I owe this idea of an aircraft "champion" to my good friend and fellow Museum of Flight volunteer Steve Ellis.
70 Ministry of Aviation, *Narrative: Reciprocating Aircraft Engines & Engine Accessories Production Programmes*, para. 216.
71 Andrews, C.F. and E.B. Morgan: *Vickers Aircraft Since 1908*, (London, 1988), pp. 364-71; Furse, *Wilfird Freeman*, p. 334.
72 Barnes, C.H.: *Bristol Aircraft Since 1910*, 3rd Ed., (London, 1988), pp. 312-14.
73 Ibid., pp. 315-16.
74 Mason, Francis K.: *The Hawker Typhoon and Tempest*, (Bourne End, Bucks., UK, 1988), pp. 117-23.
75 Ministry of Aviation, *Narrative: Reciprocating Aircraft Engines & Engine Accessories Production Programmes*, para. 41.
76 Ministry of Aviation, *Narrative: Reciprocating Aircraft Engines & Engine Accessories Production Programmes*, para. 222.
77 Ministry of Aviation, *Narrative: Reciprocating Aircraft Engines & Engine Accessories Production Programmes*, para. 222, 232.
78 "Single-Story Concrete Factory-A War-inspired Innovation", *Engineering News-Record*, Vol. 129, No. 17, (October 22, 1942), p. 117.
79 *The Wall Street Journal*, April 8, 1942, p. 3.
80 *The Wall Street Journal*, April 28, 1942, p. 1, June 10, 1942, p. 5.
81 "Single-Story Concrete Factory-A War-inspired Innovation", p. 117.
82 Ibid.
83 Ibid.
84 Ibid., p. 121.
85 Ibid., p. 122.
86 Defense Plant Corporation: *Industrial Plant: Wood-Ridge, New Jersey, Plancor 994*, (No date), Entry A176 Pamphlets Relating to Manufacturing Facilities 1940-1945, Container 1, Plancor Numbers 500-, Record Group 234, Reconstruction Finance Corporation/Defense Plant Corporation, NARA.
87 Surplus Property Administration: *Aircraft Plants and Facilities: Report of the Surplus Property Administration to the Congress, January 14, 1946*, (Washington, DC, 1946), Exhibit B, Section 3; "Single-Story Concrete Factory-A War-inspired Innovation", p. 117.
88 Dodge Chicago Plant Chronology 1942, p. 7 -p. 50, p. 35, Note of April 11, 1942.
89 Dodge Chicago Plant Chronology 1942, p. 7 -p. 50, p. 7, Note of January 5, 1942.

90 Dodge Chicago Plant Chronology 1942, p. 7 -p. 50, p. 10.

91 Dodge Chicago Plant Chronology 1942, p. 7 -p. 50, p. 25.

92 This description is drawn from the Defense Plant Corporation pamphlet "Airplane Engine Plant, Chicago, Illinois Plancor 792-Brochure A", and "Aluminum & Magnesium Castings Airplane Engine Plant, Chicago Illinois, Plancor 792-Brochure B", copy in Record Group 270, War Assets Administration, Real Property Disposal Case Files 1947-1951, Box 243, Chrysler Corporation, National Archives and Records Administration, Chicago office.

93 "Dodge Chicago Plant Division of Chrysler Corp.", *Architectural Forum*, Vol. 73, No. 6, (December 1943), p 54.

94 "Multiple-Arch Concrete Roof Saves Steel", p. 100.

95 *The Chicago Tribune*, December 15, 1942, p. 29.

96 Management Control, Central District, AAF Air Technical Service Command: *History of Dodge Chicago*, September 1945, p. 240, File 204.11-2, AFHRA.

97 "Hitler's 500 Acre Headache", Popular Mechanics Magazine, (November 1943).

98 "Dodge Chicago Plant Division of Chrysler Corp.", p. 52.

99 Dodge Chicago Plant Chronology 1942, pp. 7-50, p. 48.

100 "Multiple-Arch Concrete Roof Saves Steel", *Engineering News-Record*, Vol. 130, No. 2, (January 14, 1943), p. 99.

101 Ibid., p. 99.

102 *The Kansas City Star*, July 4, 1943, p. 42.

103 "Multiple-Arch Concrete Roof Saves Steel", p. 100.

104 Ibid., p. 99.

105 Ibid., p. 99.

106 *The Chicago Tribune*, December 15, 1942, p. 29.

107 Ibid.

108 "Building an 82-acre Concrete Roof with Traveling Retractable Forms", *Engineering News-Record*, Vol. 130, No. 24, (June 17, 1943), pp. 63-64.

109 Ibid., p. 64.

110 Dodge Chicago Plant Chronology 1942, pp. 51-102, p. 65, note of July 1, 1942; *The Wall Street Journal*, February 23, 1943, p. 8.

111 Dodge Chicago Plant Chronology 1943, pp. 103-153, p. 135, memo of July 1, 1943. Ironically, although the Dodge Chicago factory was intended to be a "temporary" building, to last only seven to ten years, it is still in operation today as the main factory for the Tootsie Roll Company making a variety of candies.

112 See "Airplane Engine Plant, Chicago, Illinois, Plancor 792-Brochure A", copy in Record Group 270, War Assets Administration, Real Property Disposal Case Files 1947-1951, Box 243, Chrysler Corporation, NARA, Chicago office.

113 See "Heat Treating and Forging Facilities, Airplane Engine Plant, Chicago, Illinois, Plancor 792-Brochure C", copy in Record Group 270, War Assets Administration, Real Property Disposal Case Files 1947-1951, Box 243, Chrysler Corporation, NARA, Chicago office; "Dodge Chicago Division of Chrysler Corporation", *Factory Management and Maintenance*, Industrial Plant Section, Vol. 102, (April 1944), pp. B-72-76.

114 Square footages taken from Surplus Property Administration: *Aircraft Plants and Facilities: Report of the Surplus Property Administration to the Congress, January 14, 1946*, Exhibit B.

115 Pratt & Whitney Aircraft Corporation of Missouri: *Announcing Production of the Pratt &Whitney Aircraft's R-2800-C Double Wasp Engine*, (Kansas City, MO, 1944), p. 3.

116 Ibid., pp. 3-4.

117 This may account for the AAF's concern with Kahn's workload at the time.

118 *The Kansas City Star*, July 4, 1943, p. 42.

119 Defense Plant Corporation: *Manufacturing Plant, Kansas City, Missouri, Plancor 1213*, (No date), Entry A176 Pamphlets Relating to Manufacturing Facilities 1940-1945, Container 1, Plancor Numbers 500-, Record Group 234, Reconstruction Finance Corporation/Defense Plant Corporation, NARA.

120 *The Kansas City Star*, February 21, 1943, p. 52.

121 *The Kansas City Star*, July 2, 1942, p. 3; *The Kansas City Times*, July 4, 1942, p. 1; *The New York Times*, July 12, 1942, p. RE3.

122 Pratt & Whitney Aircraft Corporation of Missouri: *Announcing Production of the Pratt &Whitney Aircraft's R-2800-C Double Wasp Engine*, p. 10.

123 *The Kansas City Star*, December 20, 1942, p. 58; February 21, 1943, p. 52.

124 *The Kansas City Star*, July 4, 1943, p. 45.

125 HAER Mo-118: Pratt & Whitney Plant (Bannister Federal Complex) (Kansas City Plant), Written Historical and Descriptive Data, Field Records, Historic American Engineering Record, Midwest Regional Office, National Park Service, Department of the Interior, (2012).

126 "Plane Engine Plant Sets New Precedents", *Engineering News-Record*, Vol. 127, No. 5, (July 31, 1941), p. 48.

127 Ibid.
128 Ibid., pp. 48-49.
129 O'Callaghan, Timothy J.: *Ford in Service to America: Mass Production for the Military in World War II*, (Jefferson, NC, 2009), pp 91-92.
130 *Detroit Free Press*, December 8, 1940, p. 3, July 3, 1941, p. 4.
131 *Detroit Free Press*, July 3, 1941, p. 4.
132 Ibid.
133 *Aircraft Plants and Facilities: Report of the Surplus Property Administration to Congress, January 14, 1946*, Exhibit B; O'Callaghan, *Ford in Service to America: Mass Production for the Military in World War II*, p. 90.
134 O'Callaghan, *Ford in Service to America: Mass Production for the Military in World War II*, pp. 107-09.
135 Hyde, *Storied Independent Automakers*, pp. 71-72.
136 *Kenosha Evening News*, March 3, 1942, p. 3.
137 *Kenosha Evening News*, October 31, 1942, p. 1.
138 Surplus War Property Division: *Advance Listing of Industrial Plants and Plant Sites*, p. 141.
139 *Kenosha Evening News*, October 31, 1942, p. 1; Hyde, *Storied Independent Automakers*, p. 73.
140 Data on the Bristol East Works buildings is taken from a Ministry of Supply report on Manufacturing Sites/Factories for the Bristol Aeroplane Company, Ltd. In the J.V. Connolly Collection, National Aerospace Library, Farnborough, United Kingdom.
141 Air Supply Board Meeting No. 56, 24 December 1940, AVIA 10/179, NA.
142 Minutes of Directors' Committee Meeting, 18 February 1941, BAC Directors' Committee Minutes Vol. 9, (8 October '40 to 24 June '41).
143 See the illustration on pp. 152-53 in McCamley, N.J.: *Secret Underground Cities*, (Barnsley, South Yorkshire, UK, 2007).
144 Ibid., pp. 172-73.
145 Minutes of Directors' Committee Meeting, 13 July 1943, BAC Directors' Committee Minutes Vol. 13, (13 July '43 to 27 June '44).
146 Hornby, *Factories and Plant*, pp. 207-08.
147 This total comes from company production data in *U.S. Military Acceptances 1940-1945: Aircraft, Engines and Propeller Production*.
148 Lilley, *Problems Accelerating Aircraft Production in World War II*, Table 24, p. 110; *U.S. Military Acceptances 1940-1945: Aircraft, Engines and Propeller Production*.
149 White, Graham: *R-2800 Pratt & Whitney's Dependable Masterpiece*, pp. 108-09.
150 Lilley, *Problems Accelerating Aircraft Production in World War II*, Table 25, p. 110; *U.S. Military Acceptances 1940-1945: Aircraft, Engines and Propeller Production*; White, Graham: *R-2800 Pratt & Whitney's Dependable Masterpiece*, pp. 117-18.
151 Lilley, *Problems Accelerating Aircraft Production in World War II*, Table 23, p. 109; *U.S. Military Acceptances 1940-1945: Aircraft, Engines and Propeller Production*.
152 HAER Mo-118: Pratt & Whitney Plant (Bannister Federal Complex) (Kansas City Plant).
153 Lilley, *Problems Accelerating Aircraft Production in World War II*, Table 17, p. 104, Table 19, p. 106.
154 *U.S. Military Acceptances 1940-1945: Aircraft, Engines and Propeller Production*, p. 147.
155 Lilley, *Problems Accelerating Aircraft Production in World War II*, Table 17, p. 104, Table 19, p. 106.
156 Ibid., Table 2, p. 85.
157 All production data comes from *U.S. Military Acceptances 1940-1945: Aircraft, Engines and Propeller Production*. Data on Army Air Force sorties from *Army Air Forces Statistical Digest 1945*, (Washington, DC, 1946), Table 118, p. 220, Navy sorties from *Naval Aviation Combat Statistics World War II*, (Washington, DC, 1946), Table 3, p. 16.
158 All production figures from *U.S. Military Acceptances 1940-1945: Aircraft, Engines and Propeller Production* and Lilley, *Problems Accelerating Aircraft Production in World War II*, Table 18, p. 105, Table 19, p. 106.
159 Ministry of Aircraft Production: *Statistical Review 1939-1945*, Table 9, p. 26.
160 Mason, Francis K.: *The Hawker Typhoon and Tempest*, (Bourne End, Bucks, 1988), pp. 117-18.
161 Minutes of Directors' Committee Meeting, 9 February 1942, p. 1, BAC Directors Committees Minutes, Vol. 12, (5 January 1942 to 29 June 1943).
162 Ministry of Aircraft Production: *Statistical Review 1939-1945*, Table 9, p. 26.
163 *Narrative: Reciprocating Aircraft engines and Engine Accessories Production and Programmes*, para. 222.
164 Ibid. para. 233; Ministry of Aircraft Production: *Statistical Review 1939-1945*, Table 9, p. 26.
165 Ministry of Aircraft Production: *Statistical Review 1939-1945*, Table 10, p. 28.
166 *Narrative: Reciprocating Aircraft engines and Engine Accessories Production and Programmes*, para. 222-223.
167 Ibid., para. 222, 227-34.
168 Ministry of Aircraft Production: *Statistical Review 1939-1945*, Table 12, p. 31.

[169] Lilley, *Problems Accelerating Aircraft Production in World War II*, Table 17, p. 104.
[170] *Steel*, Vol. 111, No. 3, (July 20, 1942), pp. 62-63.
[171] United Aircraft Corporation: *The Bee Hive*, Vol. XIX, No. 4, (September-October-November 1943), p. 15.
[172] Ibid.
[173] Ibid.
[174] *The Wall Street Journal*, October 20, 1943, p. 8.
[175] *The Kansas City Star*, July 4, 1943, p. 48; Reck, Franklin M.: "'Double Wasp' Engine Now Mass Produced at Kansas City", *Aero Digest*, Vol. 45, No. 4, (May 15, 1944), p. 128.
[176] *The Kansas City Star*, July 4, 1943, p. 48.
[177] Reck, "'Double Wasp' Engine Now Mass Produced at Kansas City", p. 129.
[178] *The Kansas City Star*, July 4, 1943, p. 48.
[179] *Announcing Production of the Pratt &Whitney Aircraft's R-2800-C Double Wasp Engine*, p. 8.
[180] Ibid., p. 9; *The Kansas City Times*, December 9, 1943, p. 1.
[181] Reck, "'Double Wasp' Engine Now Mass Produced at Kansas City", p. 129.
[182] See Defense Plant Corporation Plancor 1213, Pratt & Whitney Aircraft Corp. of Mo. Kansas City, Mo, MACHINERY LAYOUT FACTUAL APPENDIX "A", Pratt & Whitney Aircraft Kansas City, Missouri, Factual Appendix "A" Schedules, Defense Plant Corporation, Engineers' Reports and Appendices Plancors 1212-1213, Box 462, RG 234, NARA.
[183] Reck, "'Double Wasp' Engine Now Mass Produced at Kansas City", p. 128.
[184] Geschelin, Joseph: "Ford War Plant Facilitates Straight-Line Production", *Automotive and Aviation Industries*, Vol. 86, No. 11, (June 1, 1942), pp. 18-24, 102, 104. See also the list of machine tools in the Buick Flint and Melrose Park factories, in General Motors Corporation, Buick Motor Division, Flint, Michigan-Melrose Park, Illinois, Factual Appendix "A", Plancor 39, 1-1-40 through 5-31-42, Defense Plant Corporation, Engineers' Reports and Appendices, Plancor 39, Box 29, RG 234, NARA.
[185] The Foote Burt Company also marketed its products under the brand name Footburt. Data from http://vintagemachinery.org/mfgindex/detail.aspx?id=4016.
[186] Reck, "'Double Wasp' Engine Now Mass Produced at Kansas City", p. 129.
[187] This information is drawn from the machine tool layout for the Kansas City factory.
[188] Nevins, Allan, and Frank Ernest Hill: *Ford: Decline and Rebirth 1933-1962*, (New York, 1962), p. 178.
[189] O'Callaghan, Timothy J.: *Ford in Service to America: Mass Production for the Military in World War II*, (Jefferson, NC, 2009), p. 89.
[190] Nevins and Hill, *Ford: Decline and Rebirth 1933-1962*, p. 180.
[191] Ibid., pp. 180, 201.
[192] Hyde, *Arsenal of Democracy*, p. 55.
[193] Ibid., p. 53.
[194] *U.S. Military Acceptances 1940-1945: Aircraft, Engines and Propeller Production*, pp. 140-43.
[195] Geschelin, Joseph: "Ford War Plant Facilitates Straight-Line Production", *Automotive and Aviation Industries*, Vol. 86, No. 11, (June 1, 1942), p. 18.
[196] Ibid., pp. 18, 24.
[197] Ibid., p. 19.
[198] Ibid., p. 23.
[199] Ibid.
[200] Ibid., p. 22.
[201] Ibid., pp. 22-23.
[202] Nevins and Hill, *Ford: Decline and Rebirth 1933-1962*, p. 179.
[203] Ibid.; Ricker, Chester S.: "Ford's Centrifugal Cast Cylinders", *Aviation*, Vol. 42, No. 6, (June 1943), p. 134.
[204] Ricker, "Ford's Centrifugal Cast Cylinders", p. 134.
[205] This description of the process is taken from Ricker, "Ford's Centrifugal Cast Cylinders", pp. 135-37, 331.
[206] O'Callaghan, *Ford in Service to America: Mass Production for the Military in World War II*, p. 93.
[207] Hyde, *Arsenal of Democracy*, p. 56.
[208] Ibid.
[209] O'Callaghan, *Ford in Service to America: Mass Production for the Military in World War II*, p. 93.
[210] O'Callaghan, *Ford in Service to America: Mass Production for the Military in World War II*, p. 93; Hyde, *Arsenal of Democracy*, pp. 56-57; *U.S. Military Acceptances 1940-1945: Aircraft, Engines and Propeller Production*, p. 196.
[211] Reck, Franklin M.: "Many Production Innovations in Ford P & W Engine Plant", *Aero Digest*, Vol. 43, No. 3, (September 1943), pp. 256, 259-60.
[212] O'Callaghan, *Ford in Service to America: Mass Production for the Military in World War II*, p. 93.
[213] Reck, "Many Production Innovations in Ford P & W Engine Plant", p. 257.

214 Ibid., p. 259.
215 O'Callaghan, *Ford in Service to America: Mass Production for the Military in World War II*, p. 94.
216 Geschelin, "Ford War Plant Facilitates Straight-Line Production", p. 19.
217 Ibid., p. 102.
218 Hyde, *Arsenal of Democracy*, pp. 55-56.
219 Reck, "Many Production Innovations in Ford P & W Engine Plant", pp. 256-57.
220 Hyde, *Arsenal of Democracy*, p. 56.
221 *Muncie Evening Press*, September 9, 1943, p. 12.
222 *The Boston Globe*, August 11, 1944, p. 13.
223 *Washington County News*, March 30, 1944, p. 1.
224 *The Star Press*, September 10, 1944, p. 9; *The Iron Age*, Vol. 154, No. 4, (July 17, 1944), p. 138.
225 Ibid.
226 *The Iron Age*, Vol. 154, No. 14, (October 5, 1944), p. 80; *The Indianapolis News*, July 22, 1944, p. 11.
227 *Mill & Factory*, Vol. 35, No. 5, (November 1944), p. 188.
228 *The Boston Globe*, August 11, 1944, p. 13.
229 "Two Aircraft Engines on One Assembly Line", *American Machinist*, Vol. 89, No. 5, (March 1, 1945), p. 106.
230 Ibid., p. 107.
231 *U.S. Military Acceptances 1940-1945: Aircraft, Engines and Propeller Production*, p. 196.
232 Report of C.E. Dalton, January 31, 1943, Dodge Chicago Plant Chronology 1943, pp. 103-153.
233 Memo of C.E. Dalton, February 29, 1944, p. 15, Dodge Chicago Plant Chronology 1944, pp. 1-50.
234 Memo of F.J. Damborn, March 2, 1944, p. 16, Dodge Chicago Plant Chronology 1944, pp. 1-50.
235 Memo of William Butler, Chief, Aircraft Section, Munitions Branch, Bureau of Planning and Statistics, War Production Board, May 15, 1944, File 315.3043 Aircraft Engines-Deliveries and Shipments, War Production Board Policy Documentation File 315.3043-315.3143, Box 1126, Entry 1, RG 179, NARA.
236 Management Control, Central District, AAF Air Technical Service Command: *History of Dodge Chicago*, September 1945, p. 156, File 204.11-2, AFHRA.
237 *History of Dodge Chicago*, p. 161.
238 "Dodge Chicago Plant Makes Own Forgings for Army B-29 Super-Fortress Plane Engines", *Industrial Heating*, Vol. XI, No. 10, (October 1944), p. 1588.
239 Humphrey, William V.: "Aircraft Engine Forging Operations", *Industrial Aviation*, (July 1944), p. 58.
240 A Chrysler documentary from 1945, "Engines for Superbombers", shows this procedure. The documentary was available on YouTube on 7 October 2019.
241 Gude, William G.: "Aluminum Castings by Dodge Chicago Plant", *The Foundry*, Vol. 72, No. 9, (September 1944), p. 80.
242 Letter of K.T. Keller, President, Chrysler Corporation, January 9, 1945, p. 1, Dodge Chicago Plant Chronology 1945, pp. 1-50.
243 Fuller, Curtis: "Superfort Power Plant", *Flying*, Vol. 35, No. 1, (July 1944), p. 72.
244 Gude, William G.: "Magnesium Castings by Dodge Chicago Plant", *The Foundry*, Vol. 72, No. 8, (August 1944), p. 79.
245 Report of E.G. Nelson, Machine Records, February 24, 1945, p. 25, Dodge Chicago Plant Chronology 1945, pp. 1-50.
246 "Wright Aero Produces Engines for the B-29 Superfortress", *Machinery (New York)*, Vol. 50, No. 12, (August 1944), pp. 133-44.
247 "Wright Aero Produces Engines for the B-29 Superfortress", pp. 141-42.
248 "Wright Aero Produces Engines for the B-29 Superfortress", pp. 139-40.
249 The outline for these Greenlee machines appears on a set of original plans for the Dodge Chicago factory from October 1942 in the possession of Albert Kahn Associates in Detroit, Michigan. I would like to thank Donald Bauman, Director of Architecture and Historic Preservation for giving me access to these plans.
250 *Steel*, Vol. 115, (September 4, 1944), p. 74; Mansfield, J.H.: "Production Robots Cut Machining Costs", *Society of Automobile Engineers Journal*, Vol. 55, (February 1947), p. 44.
251 See the documentary "Engines for Superbombers."
252 See "Wright, Leader in Gear Design, Created Profile, Now Standard", *Trade Winds*, (May 1941); Brown, P.W. and Earle V. Farrar: "Design and Manufacture of Aircraft Gearing", *Machinery (NY)*, Vol. 49, No. 3, (November 1942), pp. 219-221.
253 Pfeffer, Charles G.: "Shaved Aircraft Engine Gears", *The Iron Age*, Vol. 155, No. 15, (April 12, 1945), p. 54.
254 "Shaved Aircraft Engine Gears", p. 54.
255 Ibid.
256 Harris, A.W.: "Shaved Gears for Aircraft Engines", *The Iron Age*, Vol. 152, No. 6, (August 5, 1943), p. 59.
257 Pheffer, "Shaved Aircraft Engine Gears", p. 54.

[258] Report by the Engineering Special Division, Dodge-Chicago Plant, August 21, 1945, p. 99, Dodge Chicago Plant Chronology 1945, pp. 51-100.
[259] Ibid.
[260] Letter of K.T. Keller, President, Chrysler Corporation, January 9, 1945, p. 1, Dodge Chicago Plant Chronology 1945, pp. 1-50.
[261] Memo of C.E. Dalton, Chief Liaison Executive, Dodge Chicago Plant, January 12, 1945, p. 2, Dodge Chicago Plant Chronology 1945, pp. 1-50.
[262] Chrysler Corporation press release of March 23, 1945, p. 38, Dodge Chicago Plant Chronology 1945, pp. 1-50.
[263] Memo of J.B. Morrow and D.E. Newsted, July 12, 1944, p. 57, Dodge Chicago Plant Chronology 1944, pp. 51-118.
[264] Report by W.L. McNeil, Central Planning, October 3, 1944, p. 88, Dodge Chicago Plant Chronology 1944, pp. 51-118.
[265] Stout, *Great Engines and Great Planes*, p. 23.
[266] Ibid.
[267] Stout, *Great Engines and Great Planes*, p. 27.
[268] Telegram of L.A. Moehring, Comptroller, Chrysler Corporation, December 11, 1944, p. 109, Dodge Chicago Plant Chronology 1944, pp. 51-118.
[269] Memo of C.E. Dalton, Dodge-Chicago, covering meeting at Wright Field, July 5, 1944, pp. 54-55, Dodge Chicago Plant Chronology 1944, pp. 51-118.
[270] Memo of C.E. Dalton, Dodge Chicago Factory, July 4, 1944, p. 53, Dodge Chicago Plant Chronology 1944, pp. 51-118.
[271] Stout, *Great Engines and Great Planes*, p. 34.
[272] Manpower Office, Management Control, Air Technical Service Command, Army Air Forces: *Labor Statistics for the Aircraft Industry*, November 1944-July 1945, File 218.6-7 Labor Statistics, AFHRA.
[273] C.E. Dalton, Dodge-Chicago, January 17, 1945, p. 5, Dodge Chicago Plant Chronology 1945, pp. 1-50.
[274] Stout, *Great Engines and Great Planes*, p. 35.
[275] Letter of Brig. Gen. E.W. Rawlings, Administrator Aircraft Production Board, June 11, 1945, p. 68, Dodge Chicago Plant Chronology 1945, pp. 51-100.
[276] Engineering Report #PE-6-1, Dodge Chicago Plant, pp. 75-76, Dodge Chicago Plant Chronology 1945, pp. 51-100.
[277] Engineering Report #PE-6-1, Dodge Chicago Plant.
[278] These two examples are drawn from the Chrysler documentary "Engines for Superbombers."
[279] Address by K.T. Keller, President, April 17, 1945, p. 46, Dodge Chicago Plant Chronology 1945, pp. 1-50.
[280] Stout, *Great Engines and Great Planes*, pp. 45-6.
[281] Horger, O.J.: "Mechanical and Metallurgical Advantages of Shot Peening", *The Iron Age*, Vol. 155, No. 13, March 1945, p. 40.
[282] Stout, *Great Engines and Great Planes*, p. 36; "Shotblasting Aircraft Engine Parts", *American Machinist*, Vol. 87, No. 13, (June 24, 1943), p. 88.
[283] "Shotblasting Aircraft Engine Parts", p. 88.
[284] Lilley, *Problems of Accelerating Aircraft Production During World War II*, pp. 84-85.
[285] Muther, *Production-Line Technique*, p. 23.
[286] Summary of the R-3350 Engine Project, p. 3, Case Study of the R-3350 Engine Project, p. 5, File 202.2-12, A2070. AFHRA.
[287] Summary of the R-3350 Engine Project, p. 87.
[288] R.H. Hetrick, Comptroller's Office, Memo of March 20, 1942, p. 29, Dodge Chicago Plant Chronology 1942, pp. 7-50.
[289] Report of C.E. Dalton, May 4 and May 5, 1943, p. 125, Dodge Chicago Plant Chronology 1943, pp. 103-153.
[290] Report of K.H. Kingsley, April 5, 1943, p. 121, Dodge Chicago Plant Chronology 1943, pp. 103-153.
[291] Progress Report File, March 25, 1943, Dodge Chicago Plant Chronology 1943, pp. 103-153.
[292] Dodge Chicago Plant Chronology, 1942, pp. 35, 38, 39.
[293] Dodge Chicago Plant Chronology 1943, pp. 130, 139, 161, 164, 166, 169.
[294] Memo of Meeting, C.E. Dalton, October 15, 1943, p. 166, Dodge Chicago Plant Chronology 1943, pp. 154-179.
[295] Report of K.H. Kingsley, December 31, 1943, p. 179, Dodge Chicago Plant Chronology 1943, pp. 154-179.
[296] Sutton, Kenneth E.: "Wright Aero Produces Engines for the B-29 Superfortress", *Machinery (New York)*, Vol. 50, No. 12, (August 1944), p. 134; *The Morning Call*, September 2, 1943, p. 1.
[297] Wright Aeronautical Corporation: *Wright at the Moment*, Vol. 5, No. 18, (August 17, 1944), p. 2.
[298] Ibid.

299 Factual Appendix "A", Plancor 994, Wright Aeronautical Corporation, Wood-Ridge, New Jersey, Defence Plant Corporation Engineers' Reports and Appendices, Plancors 993-994, Box 403, Reconstruction Finance Corporation, RG 234, NARA.

300 Sutton, "Wright Aero Produces Engines for the B-29 Superfortress", pp. 140-42.

301 Factual Appendix "A", Plancor 994, Wright Aeronautical Corporation.

302 Ibid.; also see Sutton, "Wright Aero Produces Engines for the B-29 Superfortress" for descriptions of other advanced machine tools.

303 Air Supply Board Meeting No. 145, 31 October 1941, AVIA 10/188, NA.

304 Minutes of Directors' Committee Meeting, 17 June 1941, BAC Directors' Committee Minutes Vol. 9, (8 October '40 to 24 June '41).

305 Air Supply Board Meeting No. 205, 2 July 1942, AVIA 10/195, NA.

306 Ibid.

307 See Air Supply Board Meeting No. 201, 4 June 1942, Meeting No. 203, 16 June 1942, AVIA 10/195; Air Supply Board Meeting No. 250, 20 May 1943, AVIA 10/199, NA.

308 Air Supply Board Meeting No. 250, 20 May 1943, AVIA 10/199, NA.

309 Minutes of Directors' Committee Meeting, 8 December 1942, BAC Directors' Committee Minutes Vol. 11, (6 January 1942 to 22 December 1942).

310 Minutes of Directors' Committee Meeting, 25 March 1941, BAC Directors' Committee Minutes Vol. 9, (8 October '40 to 24 June '41).

311 Minutes of Directors' Committee Meeting, 11 May 1943, BAC Directors' Committee Minutes Vol. 12, (5 January '43 to 29 June '43).

312 Minutes of Directors' Committee Meeting, 13 July 1943, BAC Directors' Committee Minutes Vol. 13, (13 July '43 to 27 June '44).

313 Minutes of Directors' Committee Meeting, 24 September 1943, BAC Directors' Committee Minutes Vol. 13, (13 July '43 to 27 June '44).

314 O'Callaghan, *Ford in Service to America: Mass Production for the Military in World War II*, p. 93.

315 Lilley, *Problems Accelerating Aircraft Production in World War II*, p. 68.

316 Ibid., pp. 68-69.

317 Ibid., Table 18, p. 105.

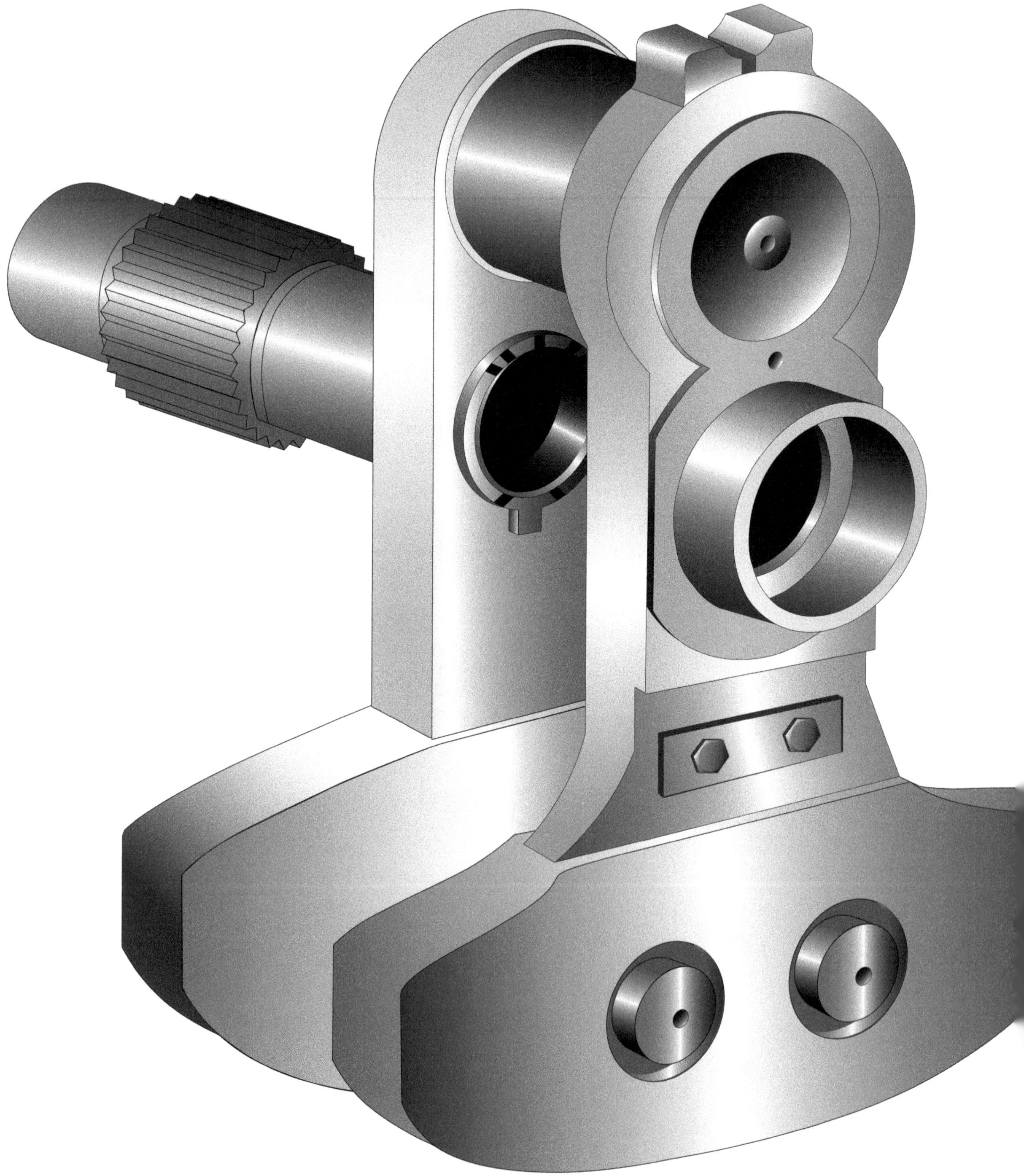

7

Comparing American and British Aircraft Engine Production

Introduction

As the chapters on Pratt & Whitney and Wright Aeronautical engine production have shown, wartime aircraft engine production in America was on a completely different scale from production in Britain. During World War II, the American aircraft engine industry built three times the number of engines of all sizes as the British aircraft engine industry and in a shorter period of time. Figure 7.1 shows that in 1944, the peak year of American production, the American industry built four and half times the number of aircraft engines built in Britain that same year. Pratt & Whitney and its licensees alone built more engines than the entire British industry. There are three factors that can explain the difference in volume of production between America and Britain: first, the disparity of resources available; second, differences in labor productivity; and third, differences in the organization of production, specifically factory size and composition of machine tools.

FIGURE 7.1 Comparative American and British aircraft engine production, 1940–1945.

Source: US Department of Commerce, Civil Aeronautics Administration, Office of Aviation Information, Division of Aircraft Statistics: US Military Acceptances, 1940–1945: Aircraft, Engines and Propeller Production (Washington, DC); Ministry of Aircraft Production: Statistical Review, 1939–1945.

Disparity of Resources

Given the disparity in resources that could be brought to wartime production, it is not surprising that America built substantially more aircraft engines for more aircraft than Britain. Using GDP as a measure, America had substantially greater resources available than Britain, even when adding in resources of the British Commonwealth and colonies. At the start of the war in 1939, though still recovering from the effects of the Depression, American GDP was three times greater than that of the UK.[1] By 1944, the peak year of American wartime production, American GDP was four times greater.[2] America had a population nearly three times bigger than the population of Britain.[3] The pool of potential workers in America, adults in the age group from 20 to 64, was at 77.1 million in 1939 greater than the entire population of the UK.[4]

The disparity in resources is evident in the relative expenditures on aircraft engine production during the war. In Britain, government expenditure on aircraft engine factories between 1936 and 1945, both shadow factories and extensions to factories of the aircraft engine manufacturers amounted to approximately $439 million.[5] From July 1940 to August 1945, the American government provided $1.3 billion to the aircraft engine manufacturers and their automobile licensees, including the General Motors Allison Division building the Allison V-1710 liquid-cooled engine and Packard Motor Company building the Rolls-Royce Merlin V-1650 under license.[6] Within this total, the automobile companies building Pratt & Whitney and Wright Aeronautical engines under license received $701.8 million in financing for factories and equipment, nearly double the entire British government expenditure on aircraft engine facilities (Table 7.1).[7]

TABLE 7.1 Government financing for aircraft engine factories.

Factory	Government financing
Pratt & Whitney Hartford Area	$90.3 million
Pratt & Whitney Kansas City	$85.5 million
Buick Melrose Park	$125.6 million
Chevrolet Tonawanda	$134.1 million
Ford Dearborn	$128 million
Nash-Kelvinator	$36.7 million
Wright Paterson area	$39.7 million
Wright Wood-Ridge	$70.2 million
Wright Lockland	$141.1 million
Dodge-Chicago	$181 million
Studebaker Chicago, Ft. Wayne, South Bend	$93.3 million

Source: Civilian Production Administration: Industrial Statistics Division: *War Industrial Facilities Authorized July 1940–August 1945* (Washington, DC, 1946).

The more important factor for aircraft engine production in wartime was that the American aircraft engine companies had a much larger industrial base to draw on than their British counterparts. The American motor vehicle industry, the source of the licensee companies during the war, was far larger. In 1937, the year of peak production before the outbreak of war, the British motor vehicle industry built 493,256 cars and commercial vehicles.[8] In contrast, in 1941 after the American motor vehicle industry had recovered from the Depression, the industry built 3,779,682 automobiles and 1,060,948 commercial vehicles, nearly ten times the number of British motor vehicles built in 1937.[9] Figure 7.2 shows the disparity in production between individual American companies that built aircraft engines under license and the British Standard Motor Company, one of the automotive companies in the Shadow Factory Scheme. The American automobile industry's extensive experience in production engineering and production management contributed greatly to the quantity production of aircraft engines by the automotive licensee companies during World War II.[10] The licensees were companies that "understood the importance of production engineering and that knew how to use it."[11]

FIGURE 7.2 Automobile production in 1939.

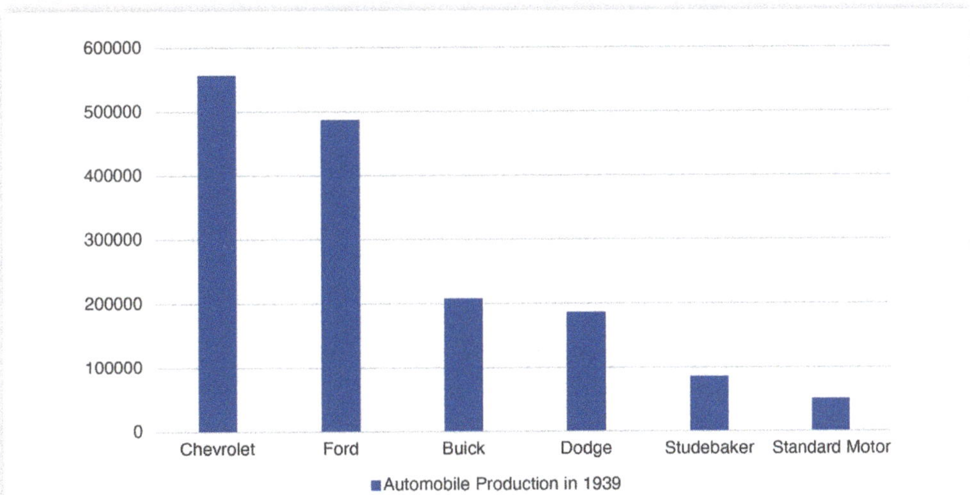

Source: https://en.wikipedia.org/wiki/U.S._Automobile_Production_Figures; Robson, Graham: The Book of the Standard Motor Company (Poundbury, Dorchester, UK, 2011), pp. 63-64.

The size and continued expansion of the American automobile industry were also important to the development of new types of machine tools and the growth of the American machine tool industry.[12] Demand from the automotive industry for improved production processes, faster production, and improvements in automobile components stimulated progress in the design and construction of machine tools.[13] The automotive industry served as a "vector of innovation" for the industry, driving the industry to introduce machine tools with greater automaticity, speeds, precision, and accuracy.[14] Like the automotive industry, the American machine tool industry was larger than the British industry. In 1935, the British industry produced machine tools with a value of approximately $24 million, while the American industry's production amounted to $85 million.[15] During 1942, the year of peak machine tool production, the American industry built 307,000 machine tools of all types compared with 95,800 machine tools built in Britain.[16] With a deep understanding of the requirements of the automobile industry and other metal cutting industries that had certain characteristics in common with the aircraft engine industry, the machine tool company engineers and tool and die makers made an equally significant contribution in the shift to quantity production of aircraft engines.[17]

Comparing Labor Productivity at American and British Aircraft Engine Factories

A comparative study of British and American aircraft engine production during World War II raises the issue of comparative labor productivity. Little data have surfaced on the man-hours required to build aircraft engines at these wartime factories, particularly over time, but some data are available on total employment during the war.[18] One simple measure of productivity is to look at the output per worker, in this case the number of workers per engine built. Using this albeit simplistic measure of productivity, it can be demonstrated that the American wartime aircraft engine factories were more productive than their British counterparts.

There is historical precedence for the difference in productivity between America and Britain. Stephen Broadberry has argued that since the late nineteenth century America has maintained an advantage in productivity over Britain, particularly in manufacturing.[19] Even during the Depression years, American productivity in manufacturing was double that of Britain.[20] Broadberry attributes the difference to "a distinctive American machine-intensive and resource-using technology that economized on skilled shop floor labor."[21] This was particularly true for certain sectors of manufacturing producing large quantities of standardized products, such as the automobile industry, where unskilled labor could be combined with machines to greatly expand production.[22] Broadberry contrasts the American mass production approach with the European "flexible production technology" that relied on more readily available skilled labor than machine tools and confronted markets demanding greater product differentiation.[23] It can also be argued that the two different approaches utilized different types of machine tools, which directly affected productivity. American mass production methods required more special purpose, high-production machine tools to obtain large-scale production, while standard machine tools were better suited to the European flexible production approach.

In looking at comparative labor productivity, it is necessary to bear in mind that there were obvious differences between American and British factories in the type of aircraft engine produced—single-row versus double-row, 9, 14, or 18 cylinders—how many models of a particular engine were built, the number of shifts and hours worked per week, the number and types of machine tools available, the size and layout of the factory space, the percentage of subcontracting, and the demand for engines and size of contracts with the military, among other variables. The aircraft engine manufacturing companies such as Bristol, Pratt & Whitney, and Wright Aeronautical had many more responsibilities than their licensees. The number of employees at the parent firms included many working in design, development, and administration and in experimental machine shops. Coming up with a meaningful measure of comparative productivity is thus challenging.

Comparing productivity at American and British aircraft engine companies is challenging, not simply because of the multiple variables between factories, but more so because of a lack of data on employment at the British aircraft engine factories. With minor exceptions, data are available on employment and production on a monthly basis for nearly every airframe and engine factory in America from 1940 to 1945.[24] The equivalent publication from the Ministry of Aircraft Production provides only averages of engine production by quarter for most major factories, but combines production at No. 1 and No. 2 Shadow Groups into one figure, making it impossible to determine a breakdown of production at the different factories.[25] Estimating total workers is also difficult due to the relative paucity of information on employment during the war particularly how factories calculated the number of workers. The Ministry of Aircraft Production's *Statistical Bulletin* lists the employment of "operatives" at the different airframe and aircraft engine factories, but only for selected periods from March 1942 to September 1945, and does not define what constituted an "operative." The more comprehensive lists of labor at the different factories that were used to produce these data do not seem to have been preserved in the records of the Ministry of Aircraft Production.

Because of constraints on the supply of labor in Britain during the war, calculations of productivity were important to the Ministry of Aircraft Production as a basis for calculating labor requirements. Calculating productivity at the factories was never easy. MAP's own calculations often underestimated increases in productivity.[26] Part of the problem was that the firms themselves had "little experience of measuring labor requirements under conditions of large scale production, with a much higher degree of tooling, etc., than they were accustomed to."[27] Reports on monitoring and reporting on productivity improvements have yet to surface in the archives. From what information is available, it is clear that there were variations in levels of productivity and improvements in productivity over time as production increased.

It is also important to point out that at the time productivity was less important than production. There was a far greater emphasis on meeting monthly production schedules than on reducing man-hours per engine.[28] As Alexander Field has observed, at a time when government contracts were on a cost-plus basis, "success is measured by one's ability to produce large quantities of ordnance quickly."[29] The difficulty, as Field notes, is that there are few detailed studies of wartime productivity, particularly on an individual industry basis.[30] From the records surveyed, there does not appear to have been many efforts at the time to determine relative productivity in the aircraft engine factories in either Britain or America, though there is much information on monthly production targets.[31] In the British Ministry of Aircraft Production's planning process, the more important questions for program planning for an aircraft engine factory were the date when production would start and how fast it would take to reach the level of peak production.[32] In the British system of production, with frequent modifications and changes in engine models, the expectation was that a few aircraft or aircraft engine models would reach peak production before modifications would require changes in production processes.[33] The more important calculation was to estimate labor requirements at the factories for a given rate of production in order to determine the manpower allocation among the Admiralty, the Ministry of Aircraft Production, and the Ministry of Supply.[34]

Looking at American aircraft engine production, it can be shown that productivity varied considerably among the American aircraft engine factories. Using the number of workers per engine built, Figure 7.3 shows that the three most productive American aircraft engine factories, those factories with the fewest workers per engine built, were those managed by three automobile companies—Studebaker, Buick, and Chevrolet—building engines under license. Two of the biggest aircraft engine factories—the Dodge-Chicago factory and the Pratt & Whitney factory in Kansas City—required more workers per engine built. The Wright Lockland factory was the only giant factory to approach the productivity levels of Studebaker, Buick, and Chevrolet, although sustaining productivity at Lockland proved difficult. This finding contradicts the assumption that giant factories were, or should be, more productive. The function of the giant factory was to increase output and improve efficiency through economies of scale, yet these productivity figures imply that in fact this was not the case.[35]

FIGURE 7.3 Total workers per engine built during peak production at the Principal American air-cooled radial aircraft engine factories, 1943–1945.

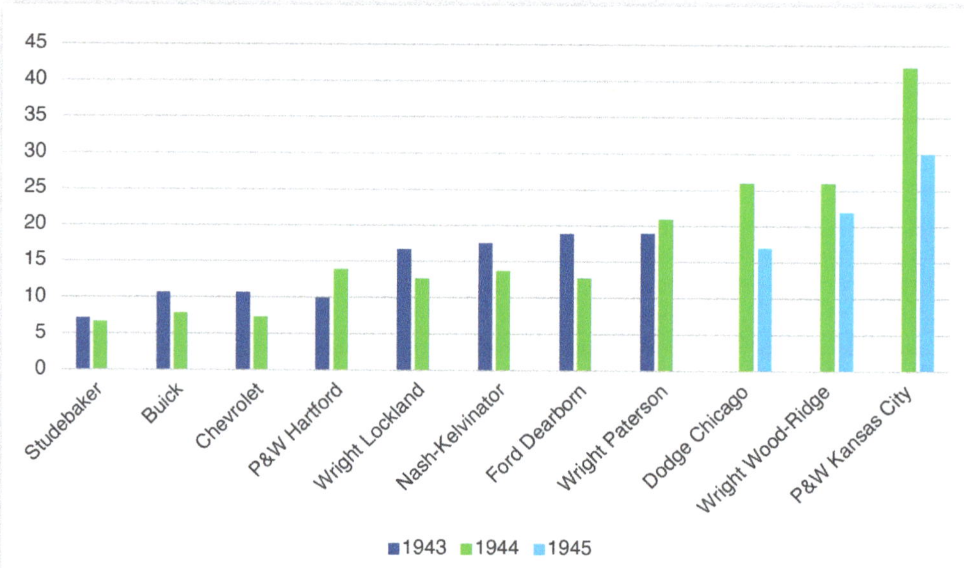

Source: US Department of Commerce, Civil Aeronautics Administration, Office of Aviation Information, Division of Aircraft Statistics: US Military Acceptances, 1940–1945: Aircraft, Engines and Propeller Production (Washington, DC).

Explaining these variations in productivity is difficult as there are multiple variables to consider and comparable data on these variables are not always available. Three factors that can be considered are the type of engine in production, the reliance on subcontracting, and production experience. Regarding the type of engine, the Harvard Business School study makes clear that building established engine models in production before America's entry into the war contributed to better productivity.[36] Building proven engines with fewer variants, where the licensee companies could draw on the technical and production knowledge of the parent firms, clearly improved productivity. The complexity of the engine was also a critical factor in productivity. Building an 18-cylinder engine with nearly three times the number of parts was, not surprisingly, more demanding than building a 9-cylinder engine. As Figure 7.4 shows, the Ford Dearborn factory building the 18-cylinder Pratt & Whitney Double Wasp engine required nearly twice as many total workers per engine as Studebaker did building the 9-cylinder Cyclone 9.

FIGURE 7.4 Best record of total workers per engine for American factories building 9-, 14-, and 18-cylinder air-cooled radial aircraft engines.

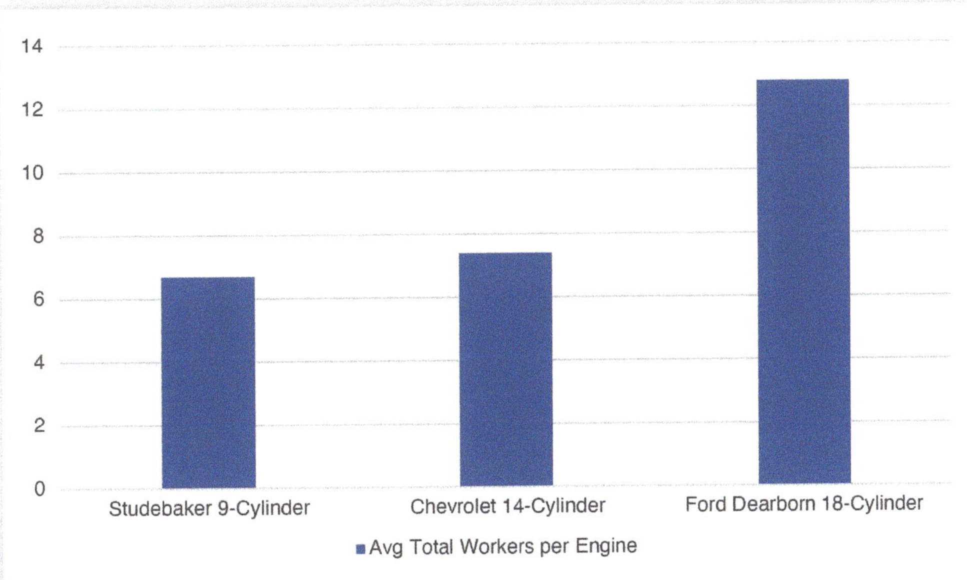

Source: US Department of Commerce, Civil Aeronautics Administration, Office of Aviation Information, Division of Aircraft Statistics: US Military Acceptances, 1940–1945: Aircraft, Engines and Propeller Production (Washington, DC, 1946); Holley, Irving Brinton, Jr.: Buying Aircraft: Material Procurement for the Army Air Forces, United States Army in World War II: Special Studies (Washington, DC, 1964), pp. 580-81; Manpower Office, Management Control, Air Technical Service Command, Army Air Forces: Labor Statistics for the Aircraft Industry, November 1944–July 1945, File 218.6-7 Labor Statistics, AFHRA.

A possible correlation between the level of subcontracting and productivity is not clear-cut. There are data for the level of subcontracting at the aircraft engine factories from November 1944 through February 1945.[37] Charts 7–5 show the reliance on subcontracting at the different factories. The percentages given for this period may not, however, be an accurate reflection of the true reliance on subcontracting at earlier periods in the war. The Studebaker factory, for example, relied on subcontractors for 50% of the value of the Cyclone 9 engine, while in early 1943 the Wright Lockland factory depended on subcontracting for 65% of the parts in the Cyclone 14.[38] It would seem that a higher reliance on subcontractors would result in a higher output of engines per worker at the aircraft engine factories. The Ford Dearborn and Nash-Kelvinator figures do not support this assumption. For 1944, Ford and Nash-Kelvinator had almost identical figures for the number of direct workers per engine, yet Ford's reliance on subcontracting was one or five times that of

Nash-Kelvinator. Similarly, the Wright Lockland factory had a greater reliance on subcontracting than either the Buick or Chevrolet factories (nearly double the percentage at Chevrolet) but was not notably more productive (Figure 7.5).

FIGURE 7.5 Percentage of subcontracting, November 1944–February 1945.

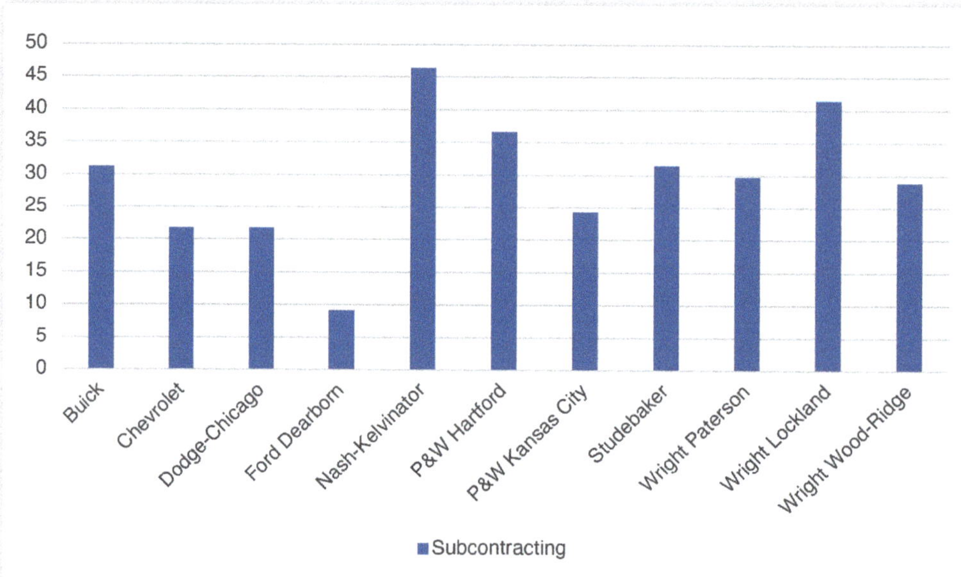

Source: Manpower Office, Management Control, Air Technical Service Command, Army Air Forces: Labor Statistics for the Aircraft Industry, November 1944–July 1945, File 218.6-7 Labor Statistics, AFHRA.

The implication is that it was not necessarily the type of engine nor the level of reliance on subcontractors that determined productivity, but the organization of production within the factory; that is, how well the management deployed methods of production and process engineering. It is noteworthy that for each type of engine—9-cylinder, 14-cylinder, and 18-cylinder—the factories that were the most productive in terms of the fewest number of workers per engine built and that had the most sustained record of productivity were the factories managed by automobile licensees. As the Harvard Business School study notes, these firms brought to aircraft engine production "management organizations that knew how to carry out production engineering in all its phases."[39] This is perhaps most evident in the record of the three factories building the biggest and newest air-cooled radial engines. Pratt & Whitney Kansas City undertook the production of the redesigned Pratt & Whitney R-2800 C series, while the Dodge-Chicago factory and the Wright Wood-Ridge factory built the new Wright Aeronautical

R-3350 Cyclone 18. Figure 7.6 shows that the Dodge-Chicago factory continued to improve productivity in the final year of the war to achieve a level of productivity close to that of Ford and Nash-Kelvinator. The Pratt & Whitney and Wright factories were less successful in this effort.

FIGURE 7.6 Average total workers per engine built for factories building 18-cylinder air-cooled radial aircraft engines, 1944–1945.

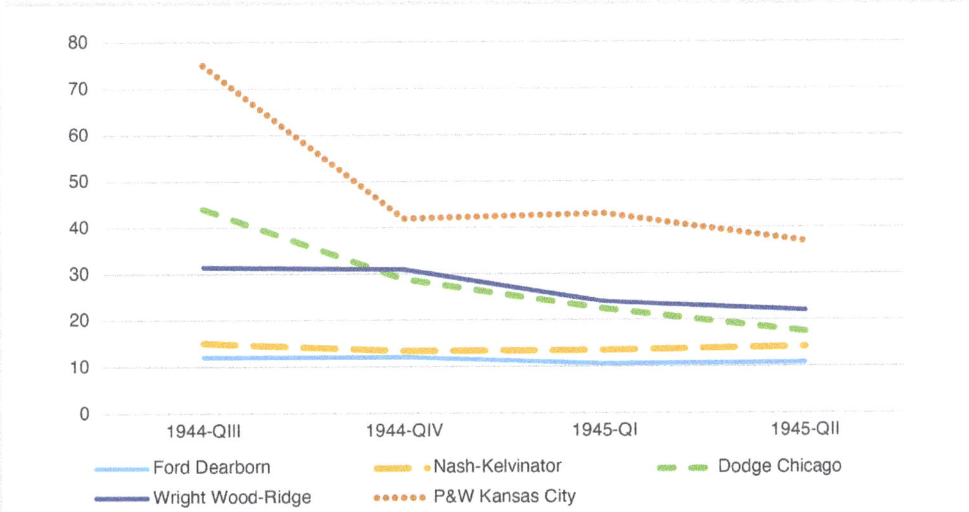

Source: US Department of Commerce, Civil Aeronautics Administration, Office of Aviation Information, Division of Aircraft Statistics: US Military Acceptances, 1940–1945: Aircraft, Engines and Propeller Production (Washington, DC, 1946); Holley, Irving Brinton, Jr.: Buying Aircraft: Material Procurement for the Army Air Forces, United States Army in World War II: Special Studies (Washington, DC, 1964), pp. 580-81; Manpower Office, Management Control, Air Technical Service Command, Army Air Forces: Labor Statistics for the Aircraft Industry, November 1944–July 1945, File 218.6-7 Labor Statistics, AFHRA.

There is some evidence to indicate that even before the start of the war Pratt & Whitney and Wright Aeronautical employed manufacturing methods that were more productive than Bristol. Pratt & Whitney and Wright appear to have required half the number of workers per engine built as shown in Figure 7.7. Data for the war years show a similar relationship, particularly when comparing factories building similar types of engines as shown in Figure 7.8. No. 1 Shadow Group and Studebaker both built 9-cylinder engines, while the most appropriate comparison for the Bristol Accrington factory and the combined No. 1 and No. 2 Shadow Group factories, which built the 14-cylinder Hercules, would be the Buick, Chevrolet, and Wright Lockland factories that also built 14-cylinder engines. Studebaker was twice as productive as the factories in No. 1 Shadow Group, with an average of five workers per engine built compared with ten workers at No. 1 Shadow Group factories. Buick, Chevrolet, and Wright Lockland were twice as productive as the factories in No. 1 and No. 2 Shadow Groups and the Bristol No. 3 Shadow Factory at Accrington, with half the number of workers per engine built.

FIGURE 7.7 Prewar average total workers per engine built per month.

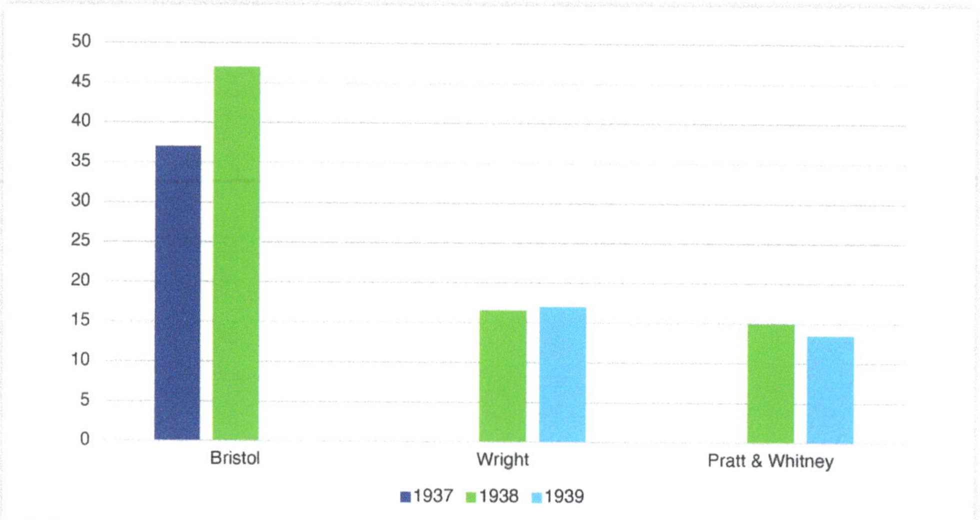

Source: US Department of Commerce, Civil Aeronautics Administration, Office of Aviation Information, Division of Aircraft Statistics: US Military Acceptances, 1940–1945: Aircraft, Engines and Propeller Production (Washington, DC, 1946); Lilley, Tom, et al: Problems of Accelerating Aircraft Production During World War II, P. 12; Manpower Office, Management Control, Air Technical Service Command, Army Air Forces: Labor Statistics for the Aircraft Industry, November 1944–July 1945, File 218.6-7 Labor Statistics, AFHRA; Mr. Haag to Secretary Morgenthau, "Employment in the Aviation Manufacturing Industry," January 14, 1941, Diaries of Henry Morgenthau, Vol. 347-January 13–14, 1941, FDR Library Digital Collections; Bristol Aeroplane Company, Summary of All Employees for 1937 and 1938, copy in the archives of Aerospace Bristol Museum.

FIGURE 7.8 Average number of total workers per engine for American and British factories for peak month of production, 1941–1945.

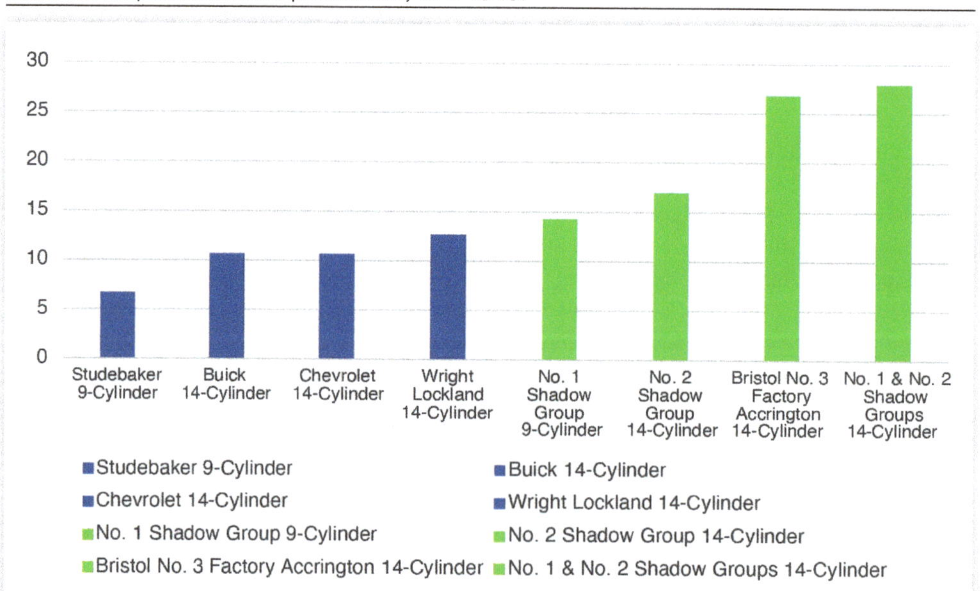

Source: US Department of Commerce, Civil Aeronautics Administration, Office of Aviation Information, Division of Aircraft Statistics: US Military Acceptances, 1940–1945: Aircraft, Engines and Propeller Production (Washington, DC, 1946); Ministry of Aircraft Production: Statistical Review, 1939–45.

Production at the American factories building 9- and 14-cylinder aircraft engines was on a different scale to the production of Bristol engines, as Figure 7.9 indicates. Since American airframe production was also greater than British airframe production, monthly production targets at the American aircraft engine factories were significantly higher than those at the British aircraft engine factories. Only the combined factories in No. 1 and No. 2 Shadow Groups building the Bristol Hercules approached the American factories in terms of peak monthly production. As Figure 7.10 shows, the American factories employed more workers, but not substantially more. Studebaker and the No. 1 Shadow Group factories had approximately the same number of employees, while the combined No. 1 and No. 2 Shadow Group factories had more employees than the Buick, Chevrolet, or Wright Lockland factories.

FIGURE 7.9 Peak monthly production at American and British factories building 9- and 14-cylinder air-cooled radial engines, 1941–1945.

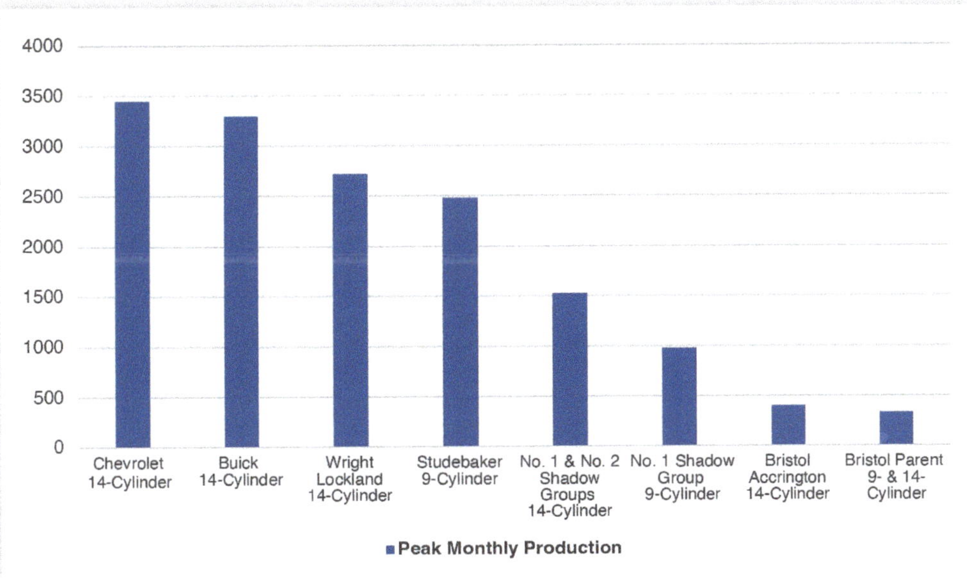

Source: US Department of Commerce, Civil Aeronautics Administration, Office of Aviation Information, Division of Aircraft Statistics: US Military Acceptances, 1940–1945: Aircraft, Engines and Propeller Production (Washington, DC, 1946); Ministry of Aircraft Production: Statistical Review, 1939–45; History of Production of Bristol Aeroengines by the Shadow Industry, Abridged Record of Engines, Spares and Accessory Output from Shadow Industry for 1938–1944.

FIGURE 7.10 Peak monthly employment at American and British factories building 9- and 14-cylinder air-cooled radial aircraft engines, 1941–1945.

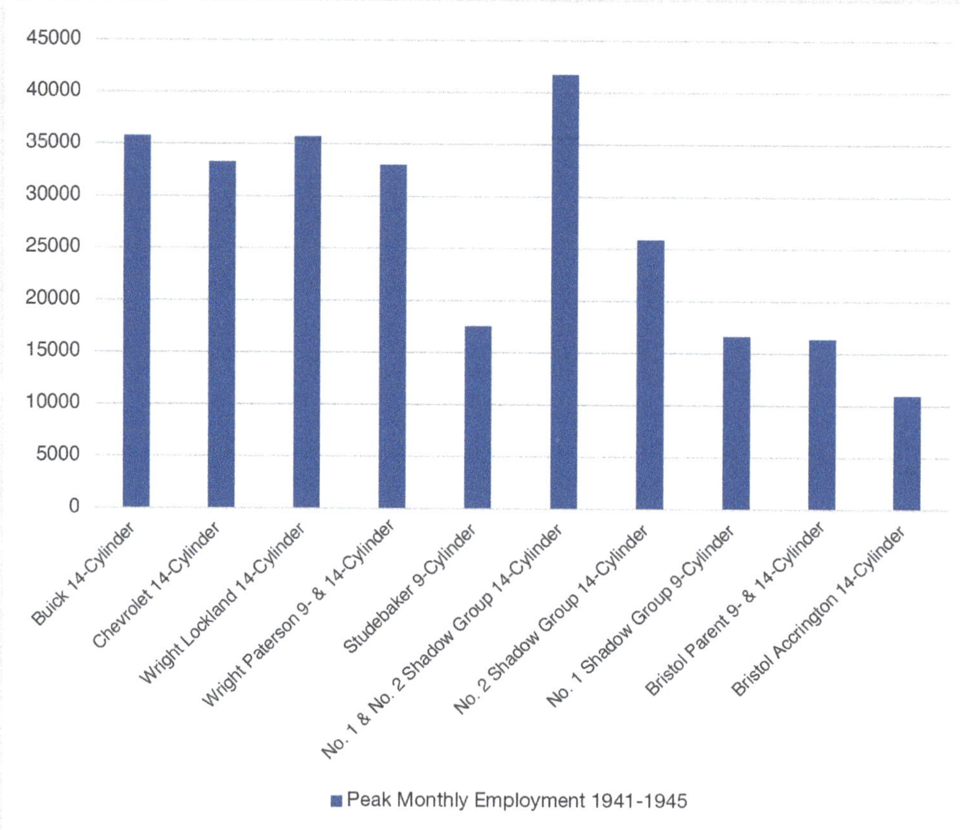

Source: US Department of Commerce, Civil Aeronautics Administration, Office of Aviation Information, Division of Aircraft Statistics: US Military Acceptances, 1940–1945: Aircraft, Engines and Propeller Production (Washington, DC, 1946); Ministry of Aircraft Production: Statistical Review, 1939–1945.

Hornby notes that there was a wide variation in productivity at the different British factories and asserts that it was related to the degree of subcontracting.[40] Hornby describes the difficulties in expanding capacity at a specialist subcontractor for the aircraft engine manufacturers after the war began, but states that while some of the new aircraft engine factories operated with a lower-than-normal reliance on subcontracting, many others achieved a higher level of subcontracting due to limitations on civilian production that allowed other manufacturers to switch to manufacturing components and subassemblies for the aircraft engine factories.[41] In these cases, which did vary from factory to factory, Hornby says the level of subcontracting at the aircraft engine factories approached that of the airframe factories, where subcontracting reached a level of 50% and often higher, compared with an average of 40% at the American aircraft engine factories.[42]

Unfortunately, precise data on the level of subcontracting at the factories building Bristol engines do not seem to have survived. However, anecdotal evidence suggests that these factories were heavily reliant on subcontracting, at a level that was at least equal to or higher than the levels at the Wright Lockland and Studebaker factories. For example, none of the factories in No. 1 or No. 2 Shadow Groups, nor the Bristol Accrington factory, had their own foundries for casting aluminum or magnesium engine parts nor facilities for forging engine parts. They were therefore entirely dependent on outside suppliers for cylinder heads, cylinder barrels, crankshafts, and crankcases, as well as specialized bearings, piston rings, carburetors, and magnetos.

Factories and Machine Tools

Why the large difference in productivity between American and British factories? The answer lies in factory design and composition of machine tools. The difference in the scale of intended production was critical. As shown in Figure 7.9, there was a vast difference between monthly aircraft engine production targets in Britain and America.[43] Where the target for the Bristol Accrington factory was 400 Hercules engines per month, the target for the Wright Lockland factory was 3,000 Cyclone 14 engines a month. The answer lies in factory design and composition of machine tools. When faced with unprecedented demands for engines but with full government financing available, Pratt & Whitney and Wright Aeronautical and their automotive licensees had the ability to design, build, and equip factories that could meet this demand. Monthly production targets could be broken down into the number of parts and components required to build the specified number of engines. From the number of parts per engine, production and process engineers could determine the number of machining and other operations needed to build the required number of parts and, from this, determine the number and type of machine tools that would be needed for a given number of daily shifts. Plant layout engineers could then determine the size of the factory needed to accommodate all the machine tools.

The size and design of the factories that built these engines in America and Britain were in sharp contrast. In America, wartime aircraft engine factories were designed from the beginning to facilitate flow production. Critical to flow production was "the passage of the part from operation to operation in a direct and uninterrupted sequence."[44] With no threat of air attack, American factories could be built as large, single-story buildings accommodating all phases of production and with no internal obstructions restricting the flow. The large American factories could accommodate conveyor systems to speed the flow of parts. In contrast, the shadow factories in Britain had to incorporate measures to reduce their vulnerability to air attack. Instead of large factory buildings, the government recommended several smaller buildings placed apart to reduce the risk of damage to all and protective blast walls built within buildings to contain an explosion in one area of the building.[45] The Bristol Accrington factory is an example of how the constraints of air raid precautions imposed on factory design resulted in a factory layout that was less than ideal for flow production. The Accrington factory, it will be recalled, had five separate machine shops with engine test cells separated from the final assembly shop, an arrangement that was not conducive to efficient flow production. The other factories in No. 2 Shadow Group also had multiple buildings. These factories were, too, smaller than almost all their American counterparts. At around 800,000 ft^2 of floor area, the Accrington factory was some 200,000 ft^2 smaller than the Studebaker factory at South Bend, Indiana, and even smaller than the large Pratt & Whitney Kansas City, Wright Lockland, and Dodge-Chicago factories. To contrast the size of British and American aircraft engine factories, Figure 7.11 shows the outline of the Accrington factory superimposed on the outline of the Wright Lockland factory's main machine shop and assembly building. The entire Accrington factory could fit within the main machine shop and assembly building with room to spare. This was the difference between the second and third generations of aircraft engine factories.

FIGURE 7.11 Factory comparison.

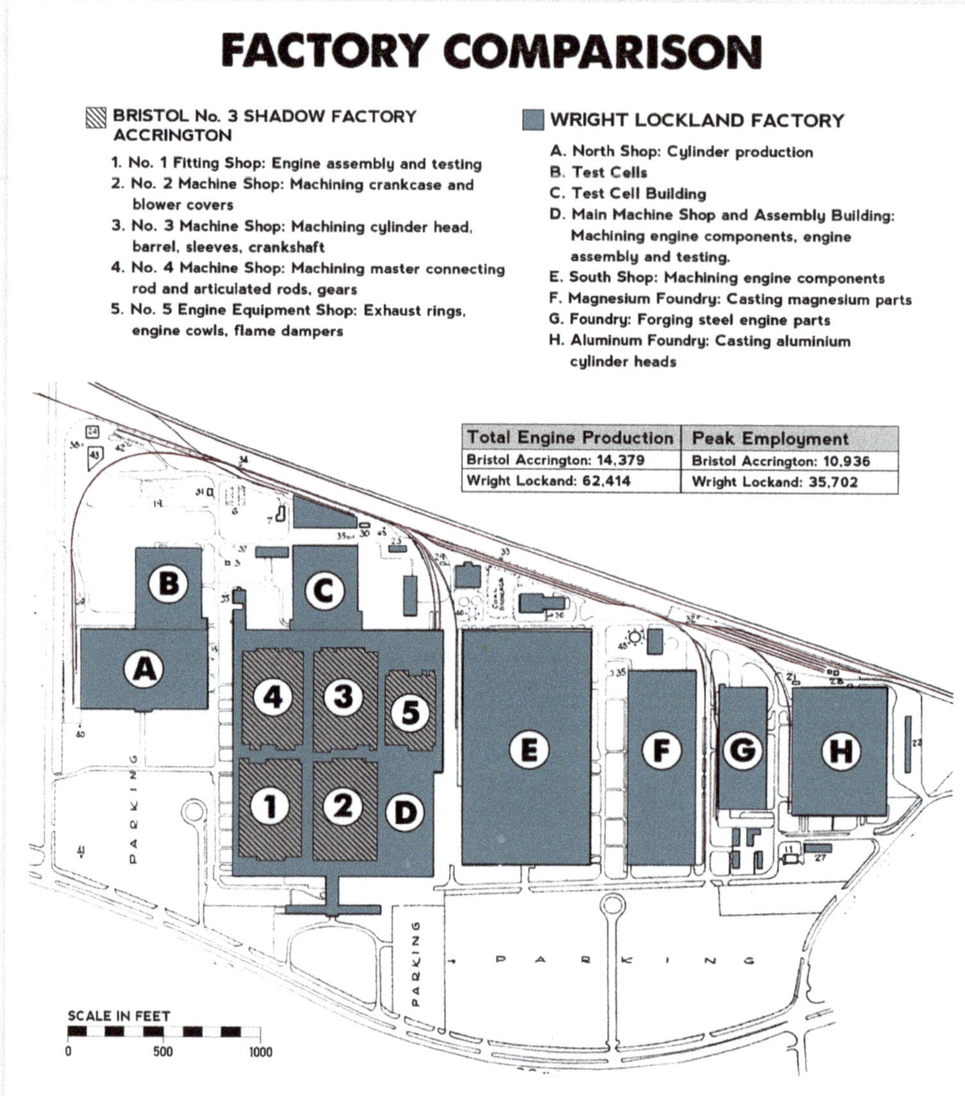

Source: Plancor 10, Manufacturing Plant Lockland, Ohio. Property Defense Plant Corporation, Record Group 234, NARA; AIR 2/3684, National Archives, Kew, UK.

Similarly, when faced with unprecedented demands for engines, Pratt & Whitney and Wright turned to the mass production methods and machine tools developed for large-scale production in the automobile industry and adapted them to large-scale production of Pratt & Whitney and Wright engines, passing on these methods and machines to their automotive licensees, learning and adopting methods and machine tools from their licensees as well. While Wright in particular focused on developing advanced, high-production machine tools that would provide for large-scale production and replace many standard machine tools, Bristol, while incorporating a certain number of high-production machine tools as a complement to standard machine tools, preferred to rely on these standard machine tools, mirroring the contrasting characteristics of American and European production methods. While Bristol and the shadow factories did make use of special purpose machine tools, many of which were designed by Bristol and manufactured in America, most of the machine tools at these factories were standard types. The British aircraft engine factories do not appear to have adopted the more advanced machine tools that could be found in many of the American aircraft engine factories.

The Missions to Study Production

The reports of British and American government missions to study American and British production methods in the fall of 1942 support the argument that the superior productivity of the American factories stemmed from advantages in factory size, greater attention to production engineering in factory layout, and greater use of special purpose, high-production machine tools. In September 1942, the Ministry of Aircraft Production sent a small group of executives from the Society of British Aircraft Constructors, A.V. Roe, Boulton-Paul, Vickers, Vickers-Armstrong, Ford, and High Duty Alloys to visit airframe, aircraft engine, and component and material manufacturers in the US.[46] The mission spent two weeks in America and visited the major airframe and aircraft engine manufacturers, including Packard, Ford's Dearborn aircraft engine factory, the Wright factories in Paterson, and the Pratt & Whitney factories at Hartford. The scale of American airframe and aircraft engine production greatly impressed the mission, with some factories "...four to five times the size of our largest factories."[47] In their report, the members of the mission focused on three key elements of the American system of production: the characteristics of the factories, quality of the engineering departments, and methods of production.

The mission stressed the size of American factories as their most important characteristic. Size determined not only the manufacturing methods that could be used, but, more importantly, the layout of the factories. The American factories were designed for large-scale production from the start. The mission concluded that "the fundamental difference in the average size of the plants—large units in the US and dispersal into small units forced upon us by conditions in the UK—vitally affects the methods of production."[48] Vulnerability to air attack was not an issue, and with greater space available, American factories could be built on a scale surpassing what could be done in Britain. The mission was impressed that many of the American factories were built without windows and used artificial lighting and air conditioning to control temperatures. The lighting was superior to conditions in British factories, and the combination of artificial lighting and temperature control meant that day and night shifts could operate under the same conditions, a definite improvement over British factories.[49] The mission believed that better lighting and the light-colored walls of the factories were "valuable aids to production."[50]

In general, the factories the mission visited had much larger engineering departments than equivalent British factories and a greater number of engineering staff relative to direct production workers.[51] The mission's members gained the impression that in the American engineering departments there was less of a "watertight division" between design and manufacture than in Britain and that American engineers were as concerned with problems of manufacturing as with problems of design.[52] They noted, as a result, the larger number of engineering staff devoted to production planning. As one example of this orientation, the mission found that as an aid to productivity engineering departments prepared considerable numbers of perspective drawings of parts, which were easier for unskilled workers to interpret than standard engineering drawings.[53]

The two aspects of American production methods that impressed the mission the most were the extensive use of moving assembly lines (for components) and special purpose machine tools. Moving assembly lines, mostly using powered conveyors, were "standard practice" in the factories the mission visited.[54] The mission found that regardless of the extent of moving assembly lines, "all plants go to considerable trouble to lay out their shops to avoid unnecessary movement of material, and to arrange their final assembly lines to this end."[55] The extensive conveyor systems the mission saw in the American factories they visited were clearly better suited to these large factories than to the smaller, dispersed factories in Britain, but where the level of production could justify the cost, the mission concluded that "a substantial savings in man hours can be achieved" and recommended that more British factories adopt these systems.[56] The mission recognized clearly that the benefit of moving assembly lines was "to control and stimulate the production of the component parts which the assembly line demands in order that progress be maintained," one of the most critical features of mass production.[57]

The mission observed that "the experience of the American automobile industry is probably responsible for the practice in aircraft manufacture of using special purpose machinery to save handling time to a greater extent than in the case in the UK."[58] Not surprisingly, the mission stated that the "best example of this outlook" was the Greenlee automatic transfer machine the mission saw in operation at the Wright factory in Paterson, New Jersey. The mission gave a detailed description of how the automatic transfer machine could produce one completed cylinder head every 50 seconds, noting that this large machine required only nine male operatives and "three girls" to perform all loading, unloading, and gauging duties.[59] However, the mission realized that introducing these systems into British factories where, as at the Bristol Accrington factory, blast walls interfered with the flow of work in the shop, and where not all components were manufactured in the same building would be a difficult undertaking.[60]

Critical to the argument this chapter is making on comparative productivity, the mission's report stated clearly that the American aircraft engine factories the mission visited made much greater use of special purpose machine tools than comparable British factories.[61] This reflected, the mission believed, the underlying principle of American manufacturing that "economy of labor is regarded as more important than economy of equipment," the same point that Stephen Broadberry would make some 50 years later.[62] The mission noted that "where complicated and expensive multipurpose tools can be employed there appears to be little difficulty in obtaining suitable machines specially designed for the job."[63] The mission attributed this tendency to use special purpose machinery in aircraft engine manufacturing to the experience of the American automobile industry and its development of high-production machine tools.

At the end of the year, a second British mission arrived in America at the invitation of the American government.[64] Sir Roy Fedden, Bristol's former head of engine development but serving as a special advisor to the Minister of Aircraft Production, led a group of 11 British aeronautical engineers and production specialists. The purpose of the mission was to investigate American airframe and aircraft engine production and coordination of British and American technical and design efforts.[65] The Fedden Mission to America, as it was called, spent three months visiting the major airframe, aircraft engine, and component manufacturing plants across the country. With more time available than the Ministry of Aircraft Production's mission, Fedden and his team conducted a more thorough review of American production practices. Among the aircraft engine factories they visited were the Packard and Ford plants in Detroit building the Rolls-Royce Merlin under license, the Allison factory in Indianapolis that built the liquid-cooled V-1710 aircraft engine, the Wright factories in Lockland and Paterson, and the Pratt & Whitney factories at Hartford.

The Fedden Mission's conclusions provide additional support to the argument that factory size, layout, and greater use of special purpose machine tools were the key factors behind America's greater productivity. The Mission concluded that the layout of the American aircraft engine factories was superior to the layout of British factories. The Mission's report noted that while manufacturing practices had much in common, "the actual layout of shops and the use of line production with or without conveyors is quite different...and it is felt that we have been slower to take advantage of this than the Americans."[66] The liaison between design and production was closer in American factories, with a larger staff devoted to working out the details of shop layout and machine placement for a given rate of engine production. The Mission was impressed with the layout for line production at the parent aircraft engine factories, but commented that "in some of the 'shadow' plants [the automobile licensee factories they visited] the layout is really excellent."[67] While acknowledging the difficulties that air raid precautions imposed in Britain that were absent in America, the Mission believed that "considerable improvements in the layout of British shops are still possible."[68]

The Mission came to these conclusions after making a careful study of the production organization at the aircraft engine factories they visited. The Mission's report noted that "the most important lessons are to be learnt from the general planning and laying out of the American engine factories and the administrative details."[69] The most important point was that the American factories had devoted time and effort toward efficient planning from the start, by implication more than comparable British factories had done. The departments responsible for production engineering derived the plant layout from a thorough study of the delivery program and standard times for machine tool operations, working out the placement of machine tools with the goal of obtaining balanced line production and eliminating unnecessary movement between machining operations. As with the MAP Mission, the Fedden Mission was impressed with how American factories made extensive use of conveyor systems to speed the flow of work from manufacture to final assembly. While recognizing the difficulties British factories faced from dispersal and the need for blast walls and shelters within factories, the Mission believed that British factories "could benefit by more careful study when laying out both machine and assembly shops."[70] By implication, British factories devoted less attention to speeding up the flow of parts within the factory.

The members of the Fedden Mission were equally impressed with the special purpose machine tools they saw at the American aircraft engine factories, concluding that "more American special machine tools for use in Britain would assist our engine factories to greater production."[71] Visiting the aircraft engine factories, the Mission saw that "first class, new, and in many cases specialized machine tool equipment is used throughout all the engine manufacturing plants."[72] The Mission's report made note of several types of special purpose machine tools that apparently were not used in British factories but were common in the American factories they visited, among them multi-head tapping machines, special multi-head horizontal drilling machines, and multi-spindle boring machines.[73] They were less enthusiastic about the Greenlee automatic transfer machines they saw, arguing that a stoppage at one point along the machine would disrupt the entire operation and noting that in British factories such a large machine had a greater risk of damage from air attack.[74]

In return for the American government's mission sponsoring the Ministry of Aircraft Production's mission to review the American aircraft industry, the Ministry invited a group of representatives of the American aircraft industry to visit Britain.[75] None other than the discoverer of the learning curve effect, Theodore P. Wright, the aeronautical engineer and executive with the Curtiss-Wright Corporation and then serving as Deputy Director of Aircraft Production at the War Production Board, led a group of eight senior executives from different companies to Britain from October 6 to November 11, 1942. During their time in England, the group visited more than 22 factories building airframes, aircraft engines, propellers, and metal components.[76] Included in the tour were visits to the Standard Motor No. 2 Shadow Factory building the Bristol Hercules and the Ford factory at Trafford Park building the Rolls-Royce Merlin.[77] On his return to America, Wright wrote up a report on the group's observations of British production.

Wright observed that the greatest difference between British and American factories was the need for dispersal in Britain to avoid the risk of air raids. "Naturally," his report commented, "it is difficult to carry out much line production under these conditions. As this situation is quite general, it naturally prohibits the efficiencies which we hope to attain in this country."[78] Wright and the other American representatives noted that in the British factories they visited "the extent of the use of machine tools in England is far less than here [America] while the use of conveyor systems was almost nil."[79] Wright found that there were not many single purpose tools in evidence at the factories, leading him to conclude that "the conception of true line production was far less" than in America. Overall, Wright did not believe that output per man or per man-hour was as high as in America, a condition he attributed to the differences in factory facilities and the types of machine tools in use at the factories.[80] Wright estimated that British output was probably 20% lower due to the dispersal of factories and bombing precautions in the factories, probably referring to blast walls.[81] Although it is noted that the working hours were some 15% greater than in America, the more important factor was the use of less-productive machine tools. "The use of older tools and general-purpose tools as against single purpose tools," he said, "is a factor which naturally decreases production."[82]

These reports from the three missions were all from the end of 1942, when the demand for capital equipment in both Britain and America had, for the most part, peaked.[83] Thereafter, the gap almost certainly increased. Although declining, the Ministry of Aircraft Production's requirements for machine tools continued into 1944.[84] There may, however, have been constraints on the number of high-production machine tools Britain could import from America. In November 1943, machine tools were removed from Lend-Lease, which means Britain would have had to purchase additional machine tools using its foreign exchange reserves.[85] Indeed, British imports of machine tools from America dropped from $69 million in value in 1943 to $19 million in value in 1944.[86] American high-production machine tools were expensive. Most appear to have cost between $8,000 and $16,000, four to five times the cost of standard machine tools, but some advanced types ranged from $25,000 to $53,000 for the larger Bullard multi-spindle Multi-Au-Matics and Greenlee multi-station machines used in the factories building Wright engines.[87] We do not know what was imported later, but these types of high-production machine tools also do not appear on the list of machine tools Britain acquired through Lend-Lease in 1943.[88] We also know from contemporary late-war periodicals describing Bristol shadow factories and the Wright Lockland and Wood-Ridge factories, and lists of machine tools at the Wright and Pratt & Whitney factories, that British factories were not using high-production machine tools.

Holley's Dilemma of Mass Production: Quantity versus Quality

Finally, there is the question of which production system, American or British, was better suited to responding to the quantity versus quality dilemma. As Irving Holley noted, obtaining quantity in production with continuous improvement in quality is "an ideal combination that is hard to obtain."[89] During the war, the constant demand for improvements in weaponry called for continual modifications and changes in design, but changes in design are the antithesis of large-scale production. The trade-off, as David Edgerton says, is "between scale efficiency and flexibility: large-scale production was more efficient, but flexibility in production was essential to accommodate new weapons and modifications."[90] In his history of British war production, M.M. Postan argues that "when modifications were so frequent, methods of mass production in elaborate production lines, equipped with special purpose tools, lost much of their value."[91] Comparing British factories with what he described as the more specialized American factories, Postan said that "it remains true that the less specialized factories with their simpler equipment found it easier to introduce modifications and to change over from one type or mark of weapon to another than the factories elaborately equipped for mass production."[92]

Postan addresses his argument principally to airframe production, contrasting the British approach of piecemeal modification on the production line with the American approach of freezing a design to obtain larger production runs and using separate modification centers to incorporate modifications.[93] Jonathan Zeitlin takes the same line of argument on airframe production in favor of the British flexible system of production, arguing that "the American system of temporarily frozen designs and retrospective modifications rather than continuous improvement during the course of production likewise exerted a negative impact on both the quality and quantity of combat-ready aircraft."[94] David Edgerton comes to a similar conclusion on airframe production, arguing that while American airframe factories may have been more productive than their British counterparts, when the additional labor required at modification centers is factored into American production, "the overall productivity was the same."[95]

Whether this argument applies equally as well to aircraft engine production is debatable. Postan extends his argument about airframe production to aircraft engine production, giving as his example the fact that during the war the more specialized, production-oriented Ford Trafford Park factory could produce the Merlin engine with a single-stage, two-speed supercharger more rapidly than the Rolls-Royce factories, but was unable to change over to the Merlin with a two-stage supercharger "without a complete re-equipment of the machining and assembly shops."[96] The Ford factory was more efficient, but less flexible in its ability to introduce new models. Postan's argument implies that the British aircraft engine factories were more effective at modifying their engines than the American factories, but were they? None of the major Allied aircraft engines went through the war without some modification. The question is where and how was this done and how efficiently and timely did the factories introduce modifications into production.

It is noteworthy that in contrast to American airframe production during World War II, there were no modification centers in America for aircraft engines. Modifications were done at the factories, just as in Britain. As in Britain, almost all the modifications to an engine came from the parent engine manufacturer and were passed on to the licensee or branch factories. Engines were modified. Between 1931 and 1941, for example, Wright built 41 variants of the Cyclone 9 for commercial airlines and military services.

The Army Air Force and Navy used dash numbers to designate major changes in an engine model, for example, the -65 and -97 to designate different variants of the Cyclone 9 and the -9, -20, -23, and -29 for variants of the Cyclone 14. It is uncertain whether a change in the dash number for an American aircraft engine is the same as a change in the Mark number for a British aircraft engine. If they are roughly comparable, then in comparison with the 41 Marks of the Merlin engine that Rolls-Royce produced during the war, Wright produced 31 variants of the Cyclone 9 and 22 variants of the Cyclone 14. From another perspective, of the 41 Marks of the Merlin, ten marks made up approximately 75% of total Merlin engine production, while eight variants of the Wright Cyclone 9 made up 91% of production, seven variants of the Wright Cyclone 14 made up 86% of production, and six marks of the Bristol Hercules made up 98% of production. It is difficult to argue that in taking the Cyclone 18 from its initial power rating of 1,800 to 2,500 hp in the Cyclone 18BB model Wright was less capable of modifying its engines than Rolls-Royce. In other words, the patterns were very similar, but it is possible that America did not have to make so many design changes, because it had many more high-powered engines to choose from.

Rolls-Royce's progressive development of the Merlin, taking the engine from 1,030 hp in 1939 to 1,710 hp by 1944, was a remarkable achievement, but arguably Rolls-Royce had no alternative.[97] The 1,750 hp Vulture, Rolls-Royce's first attempt at an engine more powerful than the Merlin, was a failure, while the Griffon, which produced 2,000 hp, was delayed until late in the war.[98] Pratt & Whitney and Wright did not introduce the same number of modifications to their 1939 model engines, the Pratt & Whitney Twin Wasp, and the Wright Cyclone 9, because they had no need to, as they both produced high-power engines as well—they were more successful than Rolls-Royce in making the jump from 1,000 to 2,000 hp. Rolls-Royce achieved a 66% increase in power over the four years of development of the Merlin. Wright achieved the same by moving from the Cyclone 9 at 1,200 hp, to the Cyclone 14, an initial increase of 41% in power, and on to the Cyclone 18, which gave an 83% increase in power over the Cyclone 9. Pratt & Whitney achieved something similar by moving from the 1,200 hp Twin Wasp to the 2,000 hp Double Wasp engine. America, fortunately, had the industrial, technical, and financial resources to accomplish this.

There are in case other points to be made against Postan's critique of the Ford Trafford Park plant's supposed inability to switch to another model of the Merlin. Postan implies that Rolls-Royce or some government authority wanted Ford to switch to the two-stage

Merlin but decided against this due to the inability of Ford to build the newer engine with its existing machine tools. It is not clear, however, that this was the case. Ford continued building the single-stage Merlin until the end of the war because this was the engine that powered the AVRO Lancaster, which also remained in production until the end of the war. There was also pressure from the Ministry of Aircraft Production to keep older models of the Merlin in production to avoid the risk of any shortages.[99]

Secondly, Postan implies, but gives no evidence to support this, that the Rolls-Royce factory at Derby could switch to the two-stage Merlin *without* reequipping their machine shops. Given the extent of changes between the two versions of the Merlin, this seems questionable. Even standard machine tools would have required changes in tooling and fixtures to manufacture new parts. So, the issue is surely about the relative costs of changing, and no information is provided. The mass production methods in use at the Ford Trafford Park factory and in the American aircraft engine factories that Postan criticizes may well have incorporated more flexibility than he assumed. As a contemporary textbook on production commented, "line production offers opportunity for the introduction of far more flexibility than is generally assumed."[100] Although the conversion of a factory to a new layout or new equipment was not without costs and problems, it was not unusual. The "complete re-equipment of the machining and assembly shops" that Postan decries was an expected element, and cost, of the manufacturing process in time of peace or war. The reality was that companies had to cope with changes in the design of products, fluctuations in demand, the discontinuation of a product, and new production processes as part of their normal business.[101]

Given the record of the factories building Bristol and Wright engines, it is difficult to agree with Postan's conclusion that "factories elaborately equipped for mass production" were less capable of coping with frequent design changes, modifications to engines in production, shifting to the production of different variants of an engine, and converting to the production of completely new engine models than the less-specialized factories he praises.

References

1 Harrison, Mark: "The Economics of World War II: An Overview", in Harrison, Mark, ed.: *The Economics of World War II: Six Great Powers in International Comparison*, (Cambridge, 1998), p. 10.

2 Ibid., p. 10.

3 Ibid., pp. 3, 7.

4 U.S. Department of Commerce, Bureau of the Census: *Statistical Abstract of the United States 1941*, (Washington, DC, 1942), Table 10, p. 20.

5 Hornby, *Factory and Plant*, p. 253.

6 This approximate total is drawn from Civilian Production Administration: Industrial Statistics Division: *War Industrial Facilities Authorized July 1940-August 1945*, (Washington, DC, 1946).

7 Civilian Production Administration: Industrial Statistics Division: *War Industrial Facilities Authorized July 1940-August 1945*.

8 Society of Motor Manufacturers and Traders Limited: *The Motor Industry of Britain Centenary Book 1896-1996*, (London, 1996), p. 119.

9 U.S. Department of Commerce: *Historical Statistics of the United States 1789-1945*, (Washington, DC, 1949), Series K 225-235, p. 223.

10 Lilley, *Problems Accelerating Aircraft Production*, p. 53.

11 Ibid.

12 Wagoner, Harless D.: *The U.S. Machine Tool Industry from 1900-1950*, (Cambridge, MA, 1968), p. 22.

13 Ibid.

14 Scranton, "From Depression to Globalization: Reconfiguring 20th Century American Machinery and Machine Tool Building", pp. 11-14.
15 Central Statistical Office: *Fighting with Figures: A Statistical Digest of the Second World War*, (London, 1995), Table 7.29, p. 176; Wagoner, *The U.S. Machine Tool Industry from 1900-1950*, Table 6, p. 363.
16 Ibid.
17 Lilley, *Problems Accelerating Aircraft Production*, p. 53.
18 Manpower Office, Management Control, Air Technical Service Command, Army Air Forces: *Labor Statistics for the Aircraft Industry*, November 1944-July 1945, does provide figures for man-hours per engine in the July 1945 issue, but it is unclear if these figures are just for the month of July or for a longer period of time.
19 Broadberry, Stephen N.: *The Productivity Race: British Manufacturing in International Perspective, 1850-1990*, (Cambridge, 1997), p. 3; Broadberry, Stephen and Mary O'Mahony: "Britain's Productivity Gap with the United States and Europe: A Historical Perspective", *National Institute Economic Review*, No. 189, (July 2004), p. 76.
20 *The Productivity Race: British Manufacturing in International Perspective, 1850-1990*, Table A3.1(c), p. 48.
21 *The Productivity Race*, p. 89.
22 *The Productivity Race*, pp. 79-80.
23 Broadberry and O'Mahony: "Britain's Productivity Gap with the United States and Europe: A Historical Perspective", p. 76. This is also Jonathan Zeitlin's argument. See, for example, Zeitlin, Jonathan: "The Historical Alternatives Approach", in Jones, Geoffrey G. and Jonathan Zeitlin, eds.: *The Oxford Handbook of Business History*, (Oxford, 2008), pp. 120-40.
24 This compiled in U.S. Department of Commerce, Civil Aeronautics Administration, Office of Aviation Information, Division of Aircraft Statistics: *U.S. Military Acceptances 1940-1945: Aircraft, Engines and Propeller Production*, (Washington, DC, 1946). *Problems of Accelerating Aircraft Production During World War II* also has data on monthly production going back to 1938 for Pratt & Whitney and Wright Aeronautical.
25 See Ministry of Aircraft Production: *Statistical Review 1939-45*.
26 *Labour in the Munitions Industry*, p. 205.
27 *Labour in the Munitions Industry*, p. 206.
28 This observation comes from Dr. Cristiano Ristuccia, Faculty of Economics, Cambridge University.
29 Field, Alexander J.: "The Impact of the Second World War on US Productivity Growth", *The Economic History Review*, New Series, Vol. 61, No. 3 (August 2008), p. 676.
30 "The Impact of the Second World War on US Productivity Growth", p. 672.
31 See for example "The November 4th Aircraft Engine Schedule", Morris L. Copeland, Assistant Director, Munitions Branch Statistical Department, to Donald Davis, Vice Chairman, War Production Board, December 7, 1943, War Production Board Select Document File, Aircraft Schedule-Standardization, Box 29, Entry 3A, RG179, NARA. This memo focuses exclusively on aircraft engine requirements compared to planned production schedules to determine if there will be surpluses or deficits in production of engines.
32 Devons, Ely: *Planning in Practice: Essays in Aircraft Planning in War-Time*, (Cambridge, 1950), p. 44.
33 *Planning in Practice*, p. 45.
34 Inman, P.: *Labour in the Munitions Industry*, (London 1957), pp. 201-05.
35 *Behemoth*, p. xiv.
36 Ibid.
37 Manpower Office, Management Control, Air Technical Service Command, Army Air Forces: *Labor Statistics for the Aircraft Industry*, November 1944-July 1945, File 218.6-7 Labor Statistics, AFHRA.
38 Air Technical Service Command Plans (T-5), Logistics Planning Division: *Studebaker Corporation R-2600 Aircraft Engines Construction and Production Analysis*; NA, AVIA 10/99, The Fedden Mission to America December 1942-March 1943, Final Report, Section 4: Engines, Power Plants and Propellers, Air Ministry and Ministry of Aircraft Production: Miscellaneous Unregistered Papers. Miscellaneous. Visit to America: report by Sir Roy Fedden.
39 *Problems of Accelerating Aircraft Production During World War II*, p. 55.
40 *Factories and Plant*, p. 263.
41 *Factories and Plant*, pp. 264-65.
42 *Factories and Plant*, pp. 227, 265.
43 *Industry and Air Power*, p. 262.
44 Wollard, *Principles of Mass and Flow Production*, p. 48.
45 *Wartime Building Construction*, (New York, 1942), pp. 124, 125-28.
46 NA, AVIA 10/104, Report of the British Mission to the United States of America to Study Production Methods (September-October 1942), Air Ministry and Ministry of Aircraft Production Miscellaneous Unregistered Papers. Missions and Visits. British Mission to America to Study Production Methods: report of High Duty Alloys Ltd.

47 Report of the British Mission to the United States of America, p. 2.
48 Report of the British Mission to the United States of America, p. 3.
49 Ibid.
50 Ibid.
51 Report of the British Mission to the United States of America, p. 5.
52 Ibid.
53 Report of the British Mission to the United States of America, p. 6.
54 Report of the British Mission to the United States of America, p. 9.
55 Ibid.
56 Report of the British Mission to the United States of America, p. 10.
57 Ibid.
58 Report of the British Mission to the United States of America, p. 32.
59 Report of the British Mission to the United States of America, p. 33.
60 Ibid.
61 Report of the British Mission to the United States of America, Section 2. p. 32.
62 Report of the British Mission to the United States of America, p. 10.
63 Report of the British Mission to the United States of America, p. 11.
64 *The New York Times*, January 3, 1943, p. 31.
65 Ibid.
66 NA, AVIA 10/99, The Fedden Mission to America December 1942-March 1943, Final Report, Section 4: Engines, Power Plants and Propellers, 4A-1.01-02, Air Ministry and Ministry of Aircraft Production: Miscellaneous Unregistered Papers. Miscellaneous. Visit to America: report by Sir Roy Fedden.
67 The Fedden Mission to America Final Report, 4A-1.02.
68 Ibid.
69 The Fedden Mission to America Final Report, 4A-4.01.
70 The Fedden Mission to America Final Report, 4A-3.01.
71 The Fedden Mission to America Final Report, 4A-1.02.
72 The Fedden Mission to America Final Report, 4A-3.01.
73 The Fedden Mission to America Final Report, 4A-4.01.
74 The Fedden Mission to America Final Report, 4A-4.02.
75 *Flight*, (December 3, 1942), p. 600-01.
76 *Flight*, (December 3, 1942), p. 601.
77 Minutes of Meeting on 20 November 1942, Standard Motor Company Board Minutes 1939-1945. The minutes reported that the layout of the factory had received "the highest compliments" from the members of the American mission.
78 Wright, T.P.: "U.S. Aircraft Production Mission to England, October 6 to November 11, 1942: Report on Personal Observations of the Chairman", *Articles and Addresses of Theodore P. Wright*, Vol. II: Aircraft Production, Uses of Aviation, (Buffalo, NY, 1961), p. 212.
79 "U.S. Aircraft Production Mission to England", p. 212.
80 "U.S. Aircraft Production Mission to England", p. 213.
81 Ibid.
82 Ibid.
83 Hall, Hessel D.: *History of the Second World War, United Kingdom Civil Series, War Production Series: North American Supply*, (London, 1955), p. 406.
84 By the end of 1943 Britain's of labour were "almost exhausted". See Inman, *Labour in the Munitions Industry*, p. 195. The comment on the Ministry of Aircraft Production's need for high production machine tools is in Hornby, *Factories and Plant*, pp. 316-17.
85 *North American Supply*, p. 406.
86 Stoughton, Bradley: *History of the Tools Division War Production Board*, (New York, 1949), p. 85.
87 Wright Aeronautical Corporation, Cincinnati, Ohio, Factual Appendix "A", Plancor 10, and Wood-Ridge, New Jersey, Factual Appendix "A", Plancor 994, Defense Plant Corporation Engineers" Reports and Appendices, Plancors 993-994, Boxes 4 and 403, RG 234, NARA.
88 Forester Weekly List of M/T Requirements, Operating Offices Files 1940-45, Bureau of Supplies and Requirements and Supplies Branch, Box 1600, Entry 323, Foreign Economic Administration, RG 169, NARA.
89 *Buying Aircraft*, p. 512.
90 Edgerton, David: *Britain's War Machine: Weapons, Resources and Experts in the Second World War*, (London, 2011), pp. 207-08.
91 *British War Production*, p. 411.

[92] Ibid.

[93] *British War Production*, pp. 342-43.

[94] Zeitlin, *Flexibility and Mass Production at War*, p. 60.

[95] *British War Machine*, p. 208.

[96] *British War Production*, p. 411.

[97] Although there was also a great deal to be gained by keeping the Merlin in production to use existing factories, plant and experienced workers.

[98] None of the British engines of 2,000 hp or greater were built in quantity during the war. The 2,000 hp Napier Sabre suffered prolonged development problems, while the Bristol Centaurus also suffered delays in getting into production. Rolls-Royce built 4,748 Griffon engines, Napier built 4,991 Sabres and Bristol completed 2,767 Centaurus engines by the end of the war.

[99] Lloyd, Ian: *Rolls-Royce: The Merlin at War*, (London, 1987), p. 122.

[100] *Production-Line Technique*, p. 253.

[101] Ibid.

Bibliography

Primary Sources

Archives

- Aerospace Bristol Museum
  - Records of the Bristol Aeroplane Company
- Albert Kahn Associates, Detroit, MI
- Air Force Historical Research Agency, Maxwell AFB, AL
  - A2070: Case Study of the R-2600 Aircraft Engine Project
  - N0875: Case Study of the R-3350 Aircraft Engine Project
  - A2095: History of Dodge-Chicago
  - A2171: Report of Conditions at Wright Aeronautical New Jersey Plants
- Aviation Hall of Fame & Museum of New Jersey, Teteboro, NJ
  - Papers relating to the Wright Aeronautical Corporation
- Baker Library Special Collections, Harvard Business School, Cambridge, MA
- Chicago History Museum, Chicago, IL
- Detroit Public Library, Rose & Robert Skillman Branch Library, Detroit, MI
  - National Automotive History Collection: Records of the Automotive Council for War Production
- Franklin D. Roosevelt Presidential Library and Museum
  - Harry Hopkins Papers
  - Diaries of Henry Morgenthau, Jr., April 27, 1933–July 27, 1945
- Library of Congress, Washington, DC
  - Papers of General Henry H. Arnold
  - Papers of Robert P. Patterson

- Modern Records Centre, University of Warwick
  - Records of the Rover Company, Ltd. (Aero Engines) and the Standard Motor Company.
- National Archives, Kew, London UK
  - AIR 2. Air Ministry registered files:
    - Piece numbers 629, 1738, 1821, 1822, 1823, 1842, 2015, 2194, 2324, 2325, 2326, 2366, 2577, 2873, 3355, 3683, 3684, 3685.
  - AIR 6. Records of meetings of the Air Board and Air Council:
    - Piece numbers 43, 44, 45, 46, 47, 48, 49, 50, 51, 52, 53, 54, 55, 56, 57, 58, 59, 60.
  - AIR 10. Air Ministry Air Publications and Reports:
    - Piece numbers 310, 1544, 1545, 1546, 1547, 1548, 1549, 1550, 1551.
  - AIR 19. Air Ministry Private Office papers:
    - Piece numbers 3, 4, 5, 6, 7, 9, 73, 74, 282.
  - AIR 20. Air Ministry unregistered papers:
    - Piece numbers 3065, 3066, 3554.
  - AIR 41. Air Ministry and Ministry of Defence: Air Historical Branch: Narratives and Monographs: Piece number 8.
  - AVIA 9. Ministry of Aircraft Production Private Office papers.
    - Piece numbers 5, 9, 21, 44, 45, 70.
  - AVIA 10. Ministry of Aircraft Production unregistered papers:
    - Piece numbers 5, 6, 7, 14, 15, 18, 20, 32, 41, 94, 99, 104, 151, 152, 153, 154, 155, 156, 157, 158, 159, 160, 161, 162, 163, 164, 165, 166, 167, 168, 169, 170, 171, 174, 175, 176, 177, 178, 179, 180 181, 182, 183, 184, 185, 186, 187, 188, 189, 190, 191, 192, 193, 194, 195, 196, 197, 198 199, 200, 201, 202, 203, 204, 205, 206, 207, 214, 230, 231, 232, 262, 264, 267, 268, 269, 276, 284, 311, 315, 316, 317, 318, 327, 332, 378, 382, 383, 384, 393, 616, 617, 621.
  - AVIA 15. Ministry of Aircraft Production files:
    - Piece numbers 414, 423, 724, 907, 1344, 1390, 1583, 1852, 2102, 2346, 2349, 2485, 2537, 2538, 2576, 2910, 3746, 3765, 3825.
  - AVIA 22. Ministry of Supply registered files:
    - Piece numbers 2661, 2684, 2685.
  - AVIA 38. Ministry of Supply and Ministry of Aircraft Production: North American Supply Missions, Second World War:
    - Piece numbers 81, 314, 421, 423, 426, 442, 450, 455, 485, 489, 532, 611, 623, 639, 664, 738, 781, 790, 803, 868.
  - AVIA 46. Ministry of Supply files series I (Establishments):
    - Piece numbers 35, 65, 70, 71, 72, 73, 82, 88, 91, 95, 97, 98, 106, 112, 118, 121, 159, 160, 179, 261, 306, 307.

- AVIA 53. Ministry of Supply: Contracts, registered files:
  - Piece number 554.
- BT 87. Ministry of Aircraft Production: Ministers" and Officials" papers:
  - Piece numbers 93, 177.
- CAB 64. Minister for the Co-ordination of Defence: Registered Files (CD series):
  - Piece number 17.
- CAB 68. War Cabinet Memoranda (WP(R) series):
  - Piece number 2.
- CAB 102. Cabinet Office historical section: official war histories (Second World War), civil:
  - Piece numbers 45, 53, 274, 275, 332.
- CAB 140. War Cabinet and Cabinet Office: Historical Section:
  - Piece number 32.
- CAB 141. War Cabinet and Cabinet Office: Central Statistical Section: Working Papers:
  - Piece number 87.
- FO 115. Foreign Office: Embassies and Consulates, United States of America. General Correspondence. North American Supplies:
  - Piece number 3771.
- SUPP 3. Principal Supply Officers Committee:
  - Piece number 44.

- National Archives and Records Administration, Chicago, IL
  - Record Group 270: War Assets Administration: Real Property Disposal Case Files 1947-1951, Dodge-Chrysler, Studebaker, Wright Aeronautical.
- National Archives and Records Administration, College Park, MD
  - Record Group 18: Records of the Army Air Force, Entry 219, Material Command, Col. A.J. Lyons Project Files 1939-1941; Entry 290, Central Decimal File, General Correspondence 1939-1942; Entry 292, Central Decimal File, General Correspondence 1942-44.
  - Record Group 107: Office of the Under Secretary of War.
  - Record Group 169: Entry 322, Foreign Economic Administration, Bureau of Supplies.
  - Record Group 179: War Production Board, Policy Documentation Files, Entry 1; Select Document Files, Entry 3.
  - Record Group 234: Reconstruction Finance Corporation: Defense Plant Corporation, Entry 146, Engineers" Reports and Appendices.
  - Record Group 342: Records of the United States Air Force Commands, Activities, and Organizations, Entry P-26, "Sarah Clark" Correspondence Files.

- Royal Air Force Museum, Hendon, London, UK
  - X003-8876/066: History of Production of Bristol Aero Engines by the Shadow Industry Operating Under the Joint Aero Engine Committee in Conjunction with the Bristol Aeroplane Company Limited.
- Rolls-Royce Heritage Trust-Bristol Branch, Filton, UK
  - Records of the Bristol Aeroplane Company
- Studebaker Museum, South Bend, IN

Magazines, Newspapers, and Yearbooks

- Aero Digest
- Aeronautical Engineering Review
- Aeroplane
- Aerosphere
- Air Annual of the British Empire
- Aircraft Engineering
- Aircraft Production
- Aircraft Yearbook
- American Aviation
- American Machinist
- Architectural Forum
- Architectural Record
- Automotive Industries
- Automotive and Aviation Industries
- Aviation
- Chicago Tribune
- Detroit Free Press
- Engineering News-Record
- Factory Management and Maintenance
- Flight
- Foundry
- Hartford Courant
- Journal of the Society of Automotive Engineers
- Machinery (Lon)
- Machinery (NY)
- Mechanical Engineering
- Mill & Factory

- New York Times
- Steel
- Studebaker Spotlight
- The "Bristol" Review
- The Iron Age
- The Tool Engineer
- Wall Street Journal
- Wright Trade Winds
- Wright at the Moment

Government Documents

Air Service Command, Army Air Force: Technical Order No. 02-35JA-2: *Handbook of Service Instructions: R-3350-13, -21, -23, -35 Aircraft Engines*, (February 15, 1943).

------: AN 02-35JA-2: Service Instructions: R-3350-13, -18, -19, -21, -23, -35, -41 Aircraft Engines, (25 September 1944, Revised November 1, 1944).

Army Air Forces: Army Air Forces Statistical Digest World War II, 2nd Printing, (Washington, D.C., 1945).

Civilian Production Administration: Industrial Mobilization for War: History of the War Production Board and Predecessor Agencies, 1940-1945, (Washington, D.C., 1946).

Civilian Production Administration: *Minutes of the Advisory Commission to the Council of National Defense, June 12, 1940 to October 22, 1941*, Historical Reports on War Administration, War Production Board, Documentary Publication No. 1, (Washington, D.C., 1946).

Civilian Production Administration: *Minutes of the Council of the Office of Production Management, December 21, 1941 to January 14, 1942*, Historical Reports on War Administration, War Production Board, Documentary Publication No. 2, (Washington, D.C., 1946).

Civilian Production Administration: *Minutes of the War Production Board, January 20, 1942 to October 9, 1945*, Historical Reports on War Administration, War Production Board, Documentary Publication No. 4, (Washington, D.C., 1946).

Civilian Production Administration: Industrial Statistics Division: *Alphabetic Listing of Major War Supply Contracts, Cumulative June 1940 through September 1945*, Vol. 1 A-C, Vol. 2 D-j, Vol. 3 K-Rex, Vol. 4 Rey-Z, (Washington, D.C., 1946).

Civilian Production Administration: Industrial Statistics Division: *War Industrial Facilities Authorized July 1940-August 1945*, (Washington, D.C., 1946).

Ministry of Aircraft Production: *Statistical Review 1939-1945*, (London, 1946).

Surplus War Property Division of Defense Plant Corporation: *Advance Listing of Industrial Plants and Plant Sites to be Disposed of by Defense Plant Corporation, Oct. 14, 1944*, (Washington, D.C., 1944).

Surplus Property Administration: Aircraft Plants and Facilities: Report of the Surplus Property Administration to the Congress, January 14, 1946, (Washington, D.C., 1946).

U.S. Department of Commerce, Bureau of the Census: *Statistical Abstract of the United States 1941*, (Washington, D.C., 1942).

U.S. Department of Commerce, Civil Aeronautics Administration, Office of Aviation Information, Division of Aircraft Statistics: *U.S. Military Acceptances 1940-1945: Aircraft, Engines and Propeller Production*, (Washington, D.C. 1946).

War Production Board: Historical Reports on War Administration: Special Studies No. 21, Aircraft Production Policies under the National Defense Advisory Commission and Office of Production Management, May 1940 to December 1941, (Washington, D.C., 1946).

PRIMARY SOURCES

Articles and Books

"16-Station Automatic", *Steel*, Vol. 113, No. 23, (December 6, 1943), p. 118.

"40 Years of Machine Tool Progress", *Automotive Industries*, Vol. 81, No. 7, (October 1, 1939), pp. 52-53, 352, 354, 356, 358, 360, 362, 364.

"30,000 Workers in 6 Wright Plants to Build 3,780,000 h.p. Monthly", *Trade Winds*, (April 1941), pp. 3-5.

"A Huge Plant for Engine Assembly", *Aircraft Production*, (December 1941), pp. 111-113.

"Aero Engine Manufacture II: Methods Employed by the Bristol Aeroplane Co., Ltd., Filton, Bristol, on the Mark VI. Jupiter Engine", *Machinery (London)*, Vol. 28, No. 725, (September 2, 1926), pp. 637-641.

"Aero Engine Manufacture III: Methods Employed by the Bristol Aeroplane Co., Ltd., Filton, Bristol, on the Mark VI. Jupiter Engine", *Machinery (London)*, Vol. 28, No. 726, (September 9, 1926), pp. 669-673.

"Aero Engine Manufacture IV: Methods Employed by the Bristol Aeroplane Co., Ltd., Filton, Bristol, on the Mark VI. Jupiter Engine", *Machinery (London)*, Vol. 28, No. 727, (September 16, 1926), pp. 711-715.

"Aero Engine Manufacture V: Methods Employed by the Bristol Aeroplane Co., Ltd., Filton, Bristol, on the Mark VI. Jupiter Engine", *Machinery (London)*, Vol. 28, No. 728, (September 23, 1926), pp. 733-738.

"Aero Engine Manufacture VI: Methods Employed by the Bristol Aeroplane Co., Ltd., Filton, Bristol, on the Mark VI. Jupiter Engine", *Machinery (London)*, Vol. 28, No. 729, (September 30, 1926), pp. 780-783.

"Aero Engine Manufacture VII: Methods Employed by the Bristol Aeroplane Co., Ltd., Filton, Bristol, on the Mark VI. Jupiter Engine", *Machinery (London)*, Vol. 29, No. 730, (October 7, 1926), pp. 1-5.

Aircraft Division, National Defense Advisory Council: "Report on a Study of Aircraft Engine Manufacturing Capacity", July 22, 1940, (Report 7), Record Group 179, Records of the War Production Board, Entry 1, War Production Board Policy Documentation File, Box 1125, File 315.30421, National Archives and Records Administration (NARA), Washington, D.C.

"Aircraft Engines by Buick", *American Machinist*, Vol. 86, No. 9, (April 30, 1942), pp. 394-403.

Aircraft Industries Association of America, Inc.: *Aviation Facts and Figures 1945*, (New York: McGraw-Hill Book Company, 1945).

"Albert Kahn", *Architectural Forum*, Vol. 69, No. 2, (August 1938), pp. 87-101.

Alford, L.P.: *Cost and Production Handbook*, (New York: The Roland Press Company, 1940).

"Aluminum Fins Rolled on Aircraft Engine Cylinders", *The Iron Age*, (February 22, 1945), pp. 66-70.

Anderson, A.G.: *Industrial Engineering and Factory Management*, (New York: The Ronald Press Company, 1928).

Angle, G.D.: *Aerosphere 1939*, (New York: Aircraft Publications, 1939).

------: *Aerosphere 1942*, (New York: Aircraft Publications, 1942).

"Austin Production Methods: An Improved Layout of the Engine Shop", *The Automobile Engineer*, Vol. XXV, No. 329, (February 1935), pp. 43-47.

"Automatic Bore-Finishing Machines", *Aircraft Engineering*, Vol. 8, Issue 7, (July 1936), pp. 201-202.

AuWerter, J.P.: "Pratt & Whitney Expands", *Aviation*, Vol. 139, No. 7, (July 1940), pp. 38-41, 116, 118.

"Aviation Engine Plant", *Mill & Factory*, Vol. 35, No. 5, (November 1944), pp. 128-129, 188, 190.

Banks, F.R.: "The Art of the Aviation Engine", *Journal of the Royal Aeronautical Society*, Vol. 52, Issue 453, (September 1948), pp. 527-550.

Bastian, P.J.: "Production Control in Aircraft-Engine Manufacture", *Machinery (NY)*, (March 1944), pp. 205-207.

Berghell, A.B.: *Production Engineering in the Aircraft Industry*, (New York: McGraw-Hill Book Company, 1944).

Berlin, D.R. and P.F. Rossmann: "Applying Automotive Methods to Aircraft Production", *Aviation*, Vol. 40, No. 1, (January 1941), pp. 42-43, 122, 144; Vol. 40, No. 3, (March 1941), pp. 46-47, 122.

Biermann, A.E.: "The Design of Metal Fins for Air-Cooled Engines", *Society of Automotive Engineers Journal (Transactions)*, Vol. 41, No. 3, (September 1937), pp. 388-392.

Blank, H.E. Jr.: "Precision Is the Watchword in the Pratt & Whitney Aircraft Engine Plant. Exacting Requirements Prompt a Technique Unique in Production Methods", *Automotive Industries*, Vol. 77, No. 1, (July 3, 1937), pp. 10-19, 37.

Bristol Aeroplane Co. Ltd.: *"Bristol" Hercules VI-XI-XVI-XVII-XVIII Operator's Handbook*, (Filton, UK, August 1945).

Bristol Aeroplane Co. Ltd.: *The "Bristol" Pegasus Aircooled Radial Engines*, Vol. III, Types XVII and XVIII, (Filton, UK, March 1940).

Bristol Aeroplane Co. Ltd.: "History of the Production of Bristol Aero Engines by the Shadow Industry Operating under the Joint Aero Engine Committee in Conjunction with the Bristol Aeroplane Company Limited", (January 1945), X003-887/066, Royal Air Force Museum.

Bristol Aeroplane Co. Ltd.: *The Power Behind Their Wings: An account of the part played by the Bristol Aeroplane Company in the development of the air-cooled radial aero-engine in Great Britain*, (No Date).

"Bristol Built: The Evolution of Bristol Aero-Motors", *The Aeroplane*, (March 13, 1942), pp. 296-300.

"The Bristol Centaurus", *The Aeroplane*, (July 5, 1945), pp. 11-17.

"Bristol Fashion II: The Rise of the Radial", *Flight*, (March 9, 1939), pp. 236-244.

"Bristol Pegasus Engine", *The Automobile Engineer*, Vol. XXVI, No. 346, (June 1936), pp. 221-224.

"Bristol Sleeve Valve Engines", *Aircraft Engineering*, Vol. 9, No. 102, (August 1937), pp. 218-222.

Brown, P.W.: "Wright Cyclone Production Speeded up by Special Methods", *Machinery (NY)*, Vol. 49, No. 12, (August 1943), pp. 172-176.

------: "Wright Turns to Line Production", *American Machinist*, Vol. 85, No. 13, (June 25, 1941), pp. 595-607.

Brown, P.W. and H.E. Linsley: "Engines from the Ground Up", *American Machinist*, Vol. 84, No. 11, (May 29, 1940), pp. 371-374.

------: "More Horsepower for National Defense", *Aero Digest*, Vol. 39, No. 4, (October 1941), pp. 176-182, 305.

Buckingham, E.: *Production Engineering*, (New York: Wiley & Sons, Inc., 1942).

"Buick's Cylinder Head Foundry", *Aero Digest*, Vol. 42, No. 2, (January 1943), pp. 192-193, 195-96, 199.

"Build the Plant around the Process", *Factory Management and Maintenance*, Vol. 97, No. 4, (April 1939), p. B-35.

"Building an 82-acre Concrete Roof with Traveling Retractable Forms", *Engineering News-Record*, Vol. 130, No. 24, (June 17, 1943), pp. 63-65.

"Building Types: Factories", *Architectural Record*, Vol. 82, No. 2, (February 1938), pp. 99-139.

"Building Types: Factory Design", *Architectural Record*, Vol. 89, No. 1, (January 1941), pp. 138-154.

Burghardt, H.D.: *Machine Tool Operation: Part I: Lathe, Bench, Forge*, (New York: McGraw-Hill Book Company, 1936); *Part II: Drilling, Planning, Milling, Grinding*, (New York: McGraw-Hill Book Company, 1937).

Campbell, K.: "Cylinder Cooling and Drag of Radial Engine Installations", *Society of Automotive Engineers Journal (Transactions)*, Vol. 43, No. 6, (December 1938), pp. 515-527.

"Canley Works Extensions: Improved Methods and New Machines at the Standard Motor Co., Ltd.", *The Automobile Engineer*, Vol. XXVII, No. 354, (January 1937), pp. 2-9.

Chase, H.: "Buick Practice in Machining P&W Cylinder Heads", *Aviation*, Vol. 42, No. 7, (July 1943), pp. 174-176, 354, 357-58, 361.

-------"Mass Machining Features Buick Shortcuts", *Aviation*, Vol. 43, No. 11, (November 1944), pp. 141-143.

Chatfield, C.H. and C.F. Taylor: *The Airplane and Its Engine*, (New York: McGraw-Hill Book Company, 1928).

Chatfield, C.H., C.F. Taylor, and S. Ober: *The Airplane and Its Engine*, 5th Ed., (New York: McGraw-Hill Book Company, 1949).

"Chevrolet Finds New Ways", *American Machinist*, Vol. 87, No. 22, (October 28, 1943), pp. 112-113.

Clinton, A.C.: "Machining Operations on the Bristol 'Mercury' Engine", in Nelson, W/C H.: *Aero Engineering*, Vol. II, Part 1: Production, (London, 1939), pp. 342-390.

------: "Mercury and Pegasus Engines", in Nelson, W/C H.: *Aero Engineering*, Vol. I, Part 2: Principles and Construction, (London, 1939), pp. 573-638.

------: "Short Survey of Aero-engine Production", in Nelson, W/C H.: *Aero Engineering*, Vol. II, Part 1: Production, (London, 1939), pp. 287-324.

Collison, T.: *Flying Fortress: The Story of the Boeing Bomber*, (New York: Charles Scribner's & Sons, 1943).

------: *The Superfortress Is Born: The Story of the Boeing B-29*, (New York: Duell, Sloan, & Pearce, 1945).

Connery, R.H.: *The Navy and Industrial Mobilization in World War II*, (Princeton, NJ: Princeton University Press, 1951).

"Cornfield to Mass Production in a Single Year-That's the Story of Wright's Cincinnati Division", *Trade Winds*, (November 1941), pp. 7-9.

"Conveyor Belt System", *Aviation*, Vol. 39, No. 3, (April 1941), pp. 44-45, 124.

"Conveyors Cut Machining Time on Cylinder Heads", *American Machinist*, Vol. 31, No. 18, (August 31, 1944), pp. 90-93.

"Crankcases by Otis Elevator", *Trade Winds*, Vol. 6, No. 11, (April 1943), pp. 6-7, 18.

"Crankshaft by Ohio", *Trade Winds*, Vol. 6, No. 10, (February 1943), pp. 8-9, 18.

Darnton, R.R.: "Buick Forges Aircraft Engine Parts", *American Machinist*, Vol. 87, No. 15, (July 22, 1943), pp. 98-100.

De Leeuw, A.L.: "Machine Tools of the Present and the Future", *The Iron Age*, Vol. 136, No. 10, (September 5, 1935), pp. 50-54, 215-20.

"Develops Improved Tooling to Obtain Greater War Output at Lower Cost", *American Machinist*, Vol. 86, No. 12, (June 11, 1942), pp. 581-585.

Devons, E.: *Planning in Practice: Essays in Aircraft Planning in War-Time*, (Cambridge: Cambridge University Press, 1950).

Diemer, H.: *Factory Organization and Administration*, 2nd Ed., (New York: McGraw-Hill Book Company, 1914).

"Dodge Chicago Division of Chrysler Corporation", *Factory Management and Maintenance*, Industrial Plant Section, Vol. 102, (April 1944), pp. B-72-B-76.

"Dodge Chicago Plant Division of Chrysler Corp.", *Architectural Forum*, Vol. 73, No. 6, (December 1943), pp. 50-58.

"Dodge Chicago Plant Makes Own Forgings for Army B-29 Super-Fortress Plane Engines", *Industrial Heating*, Vol. XI, No. 10, (October 1944), pp. 1581-1594.

"Drilling Operations on the Vauxhall Cylinder Block", *Machinery (London)*, Vol. 43, No. 1120, (March 29, 1934), pp. 769-771.

Duston, F.C.: "Airplane Engine Plant Uses Carbide Tools in Small-lot Production", *Machinery (NY)*, Vol. 40, No. 11, (July 1934), pp. 673-674.

Engelbach, C.F.R.: "Some Notes on Re-organizing a Works to Increase Production", *Proceedings of the Institute of Automobile Engineers*, Vol. XXII, (1927-28), pp. 496-544.

"Engine-Testing Equipment: New Electrical Installation at Longbridge Works", *The Automobile Engineer*, Vol. XXV, No. 332, (May 1935), pp. 177-178.

"Engines for the Air Force", *American Machinist*, Vol. 83, No. 11, (May 31, 1939), pp. 369-376.

"Engines from Line Production", *Aviation*, (July 1941), pp. 54-55.

Evans, W.D.: "Recent Productivity Trends and Their Implications", *Journal of the American Statistical Association*, Vol. 42, No. 238 (June 1947), pp. 211-223.

Fairbrother, E.: "Aero-Engine Production", *Aircraft Engineering*, Vol. 16, Issue 7, (July 1944) pp. 209-210.

Faurote, F.L.: "Aeronautical Aluminum Foundry", *The Iron Age*, Vol. 124, No. 14, (December 12, 1929), pp. 1587-1592.

------: "Assembling an Airplane Engine", *The Iron Age*, Vol. 124, No. 10, (September 5, 1929), p. 597.

------: "Layout of a Manufacturing Plant", *The Iron Age*, Vol. 124, No. 16, (December 26, 1929), pp. 1730-1731.

------: "Machining Airplane Cylinders", *The Iron Age*, Vol. 124, No. 8, (August 22, 1929), pp. 462-464.

------: "Machining the Wright Crankshaft", *The Iron Age*, Vol. 124, No. 7, (August 15, 1929), pp. 402-403.

------: "Machining Two Connecting Rods", *The Iron Age*, Vol. 124, No. 6, (August 8, 1929), pp. 337-340.

------: "'Making Aluminum Aircraft Castings', 'Layout of a Manufacturing Plant'", *The Iron Age*, Vol. 124, No. 15, (December 19, 1929), pp. 1650-1654.

------: "Making a Whirlwind Crankcase", *The Iron Age*, Vol. 124, No. 9, (August 29, 1929), pp. 525-528.

------: "New Wright Aeronautical Engines", *The Iron Age*, Vol. 124, No. 3, (July 18, 1929), pp. 148-152.

Fedden, A.H.R.: "Aircraft Power Plant Past and Future", Thirty-Second Wilbur Wright Memorial Lecture, *Royal Aeronautical Society Journal*, Vol. XLVIII, (1944), pp. 337-459.

------: "Next Decade's Aero Engines Will Be Advanced but Not Radical", *Society of Automotive Engineers Journal (Transactions)*, Vol. 33, No. 6, (December 1933), pp. 377-400.

------: "Trend of Air-Cooled Aero-Engines-The Next Five Years", *Society of Automotive Engineers Journal (Transactions)*, Vol. 41, No. 4, (October 1937), pp. 437-467.

"Fifteen Curtiss-Wright Factories Speed Defense Output", *Aero Digest*, Vol. 39, No. 4, (October 1941), pp. 160, 164, 166, 168.

Finney, B.: *Arsenal of Democracy: How Industry Builds Our Defense*, (New York: McGraw-Hill Book Company, 1941).

"Ford Airplane Engine Plant Has Many Distinctive Features", *Mill & Factory*, Vol. XXIX, No. 4, (October 1941), pp. 59-66.

"Ford Methods in Rolls-Royce Merlin Engine Production", *Machinery (London)*, Vol. 62, No. 1579, (January 14, 1943), pp. 29-33.

"Forged Cylinder Heads Require New Technique", *Aviation*, Vol. 43, No. 6, (June 1944), pp. 142-145, 248, 251.

Frass, A.P.: *Aircraft Power Plants*, (New York: McGraw-Hill Book Company, 1943).

"From 1700 to 2700 Horsepower in 10 Years", *Aviation Week*, (May 24, 1948), pp. 26-27, 31.

Fuller, C.: "Superfort Power Plant", *Flying*, Vol. 35, No. 1, (July 1944), pp. 70-73, 170.

"G-200 Cyclone", *Aviation*, Vol. 38, No. 7, (July 1939), pp. 56, 90, 95.

Gadd, E.R.: "Metallurgical Problems of Aero Engine Manufacture", *Royal Aeronautical Society Journal*, Vol. XXL, (1936), pp. 622-630.

Gagg, R.F.: "Current Problems in Airline Engines", *Society of Automotive Engineers Journal (Transactions)*, Vol. 36, No. 2, (February 1935), pp. 50-57.

------: "New Production Lines for Aircraft Engines", *Mechanical Engineering*, Vol. 63, No. 3, (March 1941), pp. 177-179.

Geschelin, J.: "Advanced Tooling Produces Newest Studebakers", *Automotive Industries*, Vol. 80, No. 13, (April 1, 1939).

------: "Buick's Know How Speeds Production of Airplane Engine Parts", *Automotive and Aviation Industries*, Vol. 87, No. 10, (November 15, 1942), pp. 20-25, 80, 82.

------: "Buick's Warplane Engine Plant", *Automotive and Aviation Industries*, Vol. 86, No. 11, (June 1, 1942), pp. 20-27, 82.

------: "Chevrolet's Unique Setup for Airplane Engine Production", *Automotive and Aviation Industries*, Vol. 88, No. 11, (June 1, 1943), pp. 24–28, 42.

------: "Conversion Almost Complete", *Automotive and Aviation Industries*, Vol. 87, No. 7, (October 1, 1942), pp. 52-55, 266, 228.

------: "Conversion for War: Influence of Automotive Mass-Production Methods", *SAE Transactions*, Vol. 50, No. 7, (July 1942), pp. 278-283.

------: "Dodge Spent $6,000,000 and Got a Model Truck Plant", *Automotive Industries*, Vol. 80, No. 5, (February 4, 1939), pp. 130-141.

------: "Ford War Plant Facilitates Straight-Line Production", *Automotive and Aviation Industries*, Vol. 86, No. 11, (June 1, 1942), pp. 18-24, 102, 104.

------: "Production at Chrysler Plant, Windsor, Ontario Multi-Model Engine Assembly Adapted to Mass Production at Chrysler Plant, Windsor, Ontario", *Automotive Industries*, Vol. 79, No. 7, (August 6, 1938), pp. 164-172, 180-81.

------: "Sixty-three Parts and Assemblies from Studebaker to Cyclones", *Automotive and Aviation Industries*, Vol. 88, No. 12, (June 15, 1943), pp. 32-35, 67-68, 70.

------: "Studebaker Know-how Speeds Flying Fortress Engine Output", *Automotive and Aviation Industries*, Vol. 88, No. 7, (April 1, 1943), pp. 18-22, 70, 72, 74.

------: "Wright Ready to Power Bombers", *Automotive and Aviation Industries*, Vol. 86, No. 3, (February 1, 1942), pp. 22-29, 74, 76.

Gough, H.J.: "Materials of Aircraft Construction", Twenty-Sixth Wilbur Wright Memorial Lecture, *Royal Aeronautical Society Journal*, Vol. XLII, (October 1938), pp. 921-1032.

"The Government Aircraft 'Shadow' Scheme", *Machinery (London)*, Vol. 51, No. 1308, (November 4, 1937), pp. 133-136.

Gude, W.G.: "Aluminum Castings by Dodge Chicago Plant", *The Foundry*, Vol. 72, No. 9, (September 1944), pp. 80-83, 198-200.

------: "Magnesium Castings by Dodge Chicago Plant", *The Foundry*, Vol. 72, No. 8, (August 1944), pp. 78-81, 178-80.

Hall, H.D.F.: *The Machine Tool Primer*, (Newark, NJ, 1943).

Heldt, P.M.: *High-Speed Combustion Engines: Design: Production: Tests*, 13th Ed., (Nyack, NY: P. M. Heldt, 1946).

Herb, C.O.: "Buick Now Manufactures Bomber Engines", *Machinery (NY)*, Vol. 48, No. 11, (July 1942), pp. 161-166.

------: "Buick Manufactures Bomber Engine", *Machinery (NY)*, Vol. 49, No. 5, (January 1943), pp. 149-153.

------: *Machine Tools at Work*, (New York: The Industrial Press, 1942).

"Herculean Development", *Aeronautical Engineering, The Aeroplane*, (November 28, 1941), pp. 599-601.

"Hercules Engine Components: Machining Operations on the Crankcase, Impeller, Sleeve, Cylinder Barrel and Connecting Rods", *Aircraft Production*, (July 1944), pp. 309-319.

"Hercules Progress", *Flight*, (November 11, 1943), pp. 528-530.

"Hercules Sleeve Valve Engine Production", *Machinery (London)*, Vol. 65, No. 1678, (December 7, 1944), pp. 617-625.

Hesse, H.C.: *Engineering Tools and Processes: A Study of Production Technique*, (New York: D. Van Nostrand Company, Inc., 1941).

Hill, H.C.: "Design Problems in the Quantity Production of Aircraft Engines", *SAE Journal (Transactions)*, Vol. 46, No. 1, (January 1940), pp. 18-24.

------: "Engineering Aspects of the Single-Row Cyclone G-200 Series Engine", Wright Aeronautical Corporation, (1940), pp. 1-11.

------: "More Aircraft Engines for National Defense", Paper presented at the National Aeronautic Meeting of the Society of Automotive Engineers, March 13-14, 1941, pp. 1-7.

Hirsch, W.Z.: "Manufacturing Progress Functions", *The Review of Economics and Statistics*, Vol. 34, No. 2 (May 1952), pp. 143-155.

Holben, M.M.: "Some Aircraft-Engine Production Methods", *SAE Journal (Transactions)*, Vol. 52, No. 10, (October 1944), pp. 492-500.

"Hudson Builds Cyclone Parts", *Trade Winds*, Vol. 6, No. 9, (October 1942), pp. 6-7.

Humphrey, W.V.: "Aircraft Engine Forging Operations", *Industrial Aviation*, Vol. 1, No. 2, (July 1944), pp. 58-59.

Immer, J.R.: *Layout Planning Techniques*, (New York: McGraw-Hill Book Company, 1950).

"Increase in Production Capacity of Machine Tools in Ten Years", *Machinery (NY)*, Vol. 43, No. 5, (July 1937), pp. 325-327.

"Increasing Productivity and Technological Improvements in Defense Industries", *Monthly Labor Review*, No. 34, (January 1942), pp. 34-48.

"Inspection: Routine Employed by the Bristol Aeroplane Co., Ltd., to Ensure Quality and Accuracy", *The Automobile Engineer*, Vol. XXIII, No. 306, (May 1933), pp. 169-170.

James, T.: "Getting Going at the Shadow Factories", *The Aeroplane*, (November 3, 1937), pp. 529-539.

Johnson, J.B.: "Aircraft-Engine Materials", *Society of Automotive Engineers Journal (Transactions)*, Vol. 40, No. 4, (April 1937), pp. 153-164.

Jones, E.J.H.: *Production Engineering: Jig and Tool Design*, (New York: Chemical Publishing, Inc., 1941).

Jones, H.A.: *The War in the Air*, Vol. VI, (Oxford: Oxford University Press, 1937).

Judge, A.W.: *Aircraft Engines*, Vol. One, (London: Chapman and Hall, 1941), Vol. Two, (London: Chapman and Hall, 1942).

------: *Motor Manuals: Volume I: Automobile Engines in Theory, Design, Construction, Operation, Testing & Maintenance*, (London: Chapman and Hall, 1942).

Kahn, A.: "Design of Plants for Mass Production", *Aero Digest*, Vol. 39, No. 4, (October 1941), pp. 156-159, 305.

"Keeping the Record Straight", *Automotive and Aviation Industries*, Vol. 86, No. 4, (February 15, 1942), pp. 17-19, 60.

Kimball, D.S.: *Principles of Industrial Organization*, 4th Ed., (New York: McGraw-Hill Book Company, 1933).

Klemin, A.: "A Miracle of Production", *Scientific American*, (December 1944), pp. 256-258.

Kudlich, D.U.: "Engineering Liaison and Production Control", *SAE Journal (Transactions)*, Vol. 50, No. 5, (May 1942), pp. 169-176.

Lansburgh, R.H.: *Industrial Management*, (New York: Wiley & Sons, Inc., 1940).

Leak, A.H.: "Coordinating Aircraft-Engine Design and Production", *SAE Journal (Transactions)*, Vol. 44, No. 2, (February 1939), pp. 85-92.

------: "Problems Co-ordinating Engine Design and Production", (Wright Aeronautical Corporation, 1940), pp. 1-19.

LeGrand, H.R.: "Right Gears for Wright Engines", *American Machinist*, Vol. 80, No. 1, (January 1, 1936), pp. 4-6.

"Light-Alloy Forgings", *Aircraft Production*, Vol. 6, No. 63, (January 1944), pp. 13-16.

Linsley, H.E.: "A Preview of Plant Layout", *American Machinist*, Vol. 85, No. 10, (May 14, 1941), pp. 419-420.

------: "Aircraft Engines on the Production Line", *Mechanical Engineering*, Vol. 65, No. 7, (July 1943), pp. 493-496.

------: "Automatic Tooling Speeds Production in the New Wright Aero Plant", *Machinery (NY)*, Vol. 48, No. 6, (February 1942), pp. 109-118.

------: "Cyclone Engines Mass Produced", *The Iron Age*, (October 9, 1941), pp. 53-61.

------: "Forged Cylinder Heads Increase Power of Cyclone Engine", *Aero Digest's Aviation Engineering*, Vol. 40, No. 6, (June 1942), pp. 141-142.

------: "Machining the Wright Cyclone Forged Cylinder Head", *The Iron Age*, (January 13, 1944), pp. 47-53.

------: "Mass Production for the Aircraft-Engine Industry", *Mechanical Engineering*, Vol. 64, No. 2, (February 1942), pp. 100-105.

"Line Production of the Rolls-Royce Merlin Engine", *Machinery (London)*, Vol. 61, No. 1567, (October 22, 1942), pp. 449-456.

Liston, J.: *Aircraft Engine Design*, (New York: McGraw-Hill Book Company, 1942).

Lucato, C.R.: "At Cowley To-day", *The Morris Owner*, (August 1934), pp. 556-558.

"Machine Tool Developments", *Machinery (London)*, Vol. 49, No. 1261, (December 10, 1936), pp. 301-317.

"Machine Tool Developments in 1935", *Machinery (London)*, Vol. 47, No. 1209, (December 12, 1935), pp. 312-333.

"Machining Aero-engine Crankcase Castings", *Machinery (London)*, Vol. 61, No. 1471, (November 19, 1942), pp. 561-565.

"Machining Aero-engine Crankshafts and Airscrew Shafts", *Machinery (London)*, Vol. 49, No. 1252, (October 8, 1936), pp. 33-38.

"Machining Aero-engine Crankshafts and Camshafts", *Machinery (London)*, Vol. 61, No. 1572, (November 26, 1942), pp. 589-596.

"The Manufacture of Aero-engine Connecting Rods", *Machinery (London)*, Vol. 61, No. 1573, (December 3, 1942), pp. 617-623.

"The Manufacture of Aero-engine Pistons", *Machinery (London)*, Vol. 61, No. 1574, (December 10, 1942), pp. 645-649.

"Machine Work Saved by Buick Change in Heat Treating", *American Machinist*, Vol. 87, No. 7, (April 1, 1943), pp. 75-77.

"Machining Aero-Engine Cylinder Barrels", *Machinery (London)*, Vol. 65, No. 1679, (December 14, 1944), pp. 645-653.

"Machining Aero-Engine Cylinder Blocks", *Machinery (London)*, Vol. 61, No. 1570, (November 12, 1942), pp. 533-538.

"Machining Aero Engine Cylinders", *Machinery (London)*, Vol. 41, No. 1057, (January 12, 1933), pp. 425.

"Machining Airplane Engine Cylinder Heads", *Machinery (London)*, Vol. 65, No. 1681, (December 28, 1944), pp. 701-705.

"Machining Airplane Engine Propeller Shafts", *Machinery (London)*, Vol. 65, No. 1682, (January 4, 1944), pp. 1-6.

"Machining Bristol Engine Parts", *Aircraft Engineering*, Vol. 6, No. 2, (February 1934), pp. 40-43.

"Machining 'Bristol' Hercules Connecting Rods", *Machinery (London)*, Vol. 65, No. 1686, (February 1, 1944), pp. 109-114.

"Machining the Bristol Mercury and Pegasus Cylinder Heads", *Machinery (London)*, Vol. 41, No. 1058, (January 19, 1933), pp. 453-458.

"Machining Cast Crankshafts and Camshafts", *Machinery (London)*, Vol. 47, No. 1208, (December 5, 1935), pp. 269-274.

"Machining Cylinder Heads for the Merlin Engine", *Machinery (London)*, Vol. 65, No. 1672, (October 26, 1944), pp. 449-456.

"Machining Cylinder Fins", *Aircraft Production*, Vol. VI, No. 66, (April 1944), pp. 155-156.

"Machining Cylinder Skirts and Liners for the Merlin Engine", *Machinery (London)*, Vol. 65, No. 1673, (November 2, 1944), pp. 477-481.

"Machining Ford 8-h.p. Cylinder Blocks", *Machinery (London)*, Vol. 41, No. 1045, (October 20, 1932), pp. 69-71.

"Machining Light-Alloy Castings for the Merlin Engine", *Machinery (London)*, Vol. 65, No. 1676, (November 23, 1944), pp. 561-565.

"Machining Methods on Wright Aero-Engines", *Machinery (London)*, Vol. 59, No. 1516, (October 30, 1940), pp. 113-119.

"Machining a Gear Bank", *Aircraft Production*, Vol. III, No. 27, (January 1941), pp. 5-6.

"Machining Master Connecting Rods for Radial Aero Engines", *Machinery (London)*, Vol. 49, No. 1253, (October 15, 1936), pp. 61-65.

Mansfield, J.H.: "Production Robots Cut Machining Costs", *Society of Automobile Engineers Journal*, Vol. 55, (February 1947), pp. 42-44.

"Manufacturing the Cyclone, Part I", *Aircraft Production*, Vol. II, No. 4, (April 1940), pp. 109-115.

"Manufacturing the Cyclone, Part II", *Aircraft Production*, Vol. II, No. 5, (May 1940), pp. 151-158.

"Manufacturing the Ford V-Eight", *The Automobile Engineer*, Vol. XXVI, No. 343, (March 1936), pp. 95-101.

"Manufacturing the Rolls-Royce Merlin XX, Part I", *Aircraft Production*, Vol. IV, No. 45, (July 1942), pp. 433-444.

"Manufacturing the Rolls-Royce Merlin XX, Part II", *Aircraft Production*, Vol. IV, No. 46, (August 1942), pp. 486-497.

"Manufacturing the Rolls-Royce Merlin XX, Part III", *Aircraft Production*, Vol. IV, No. 47, (September 1942), pp. 548-558.

"Manufacturing the Rolls-Royce Merlin XX, Part IV", *Aircraft Production*, Vol. IV, No. 48, (October 1942), pp. 592-601.

"Mass Production Brings Mass Inspection", *Western Flying*, Vol. XXII, No. 8, (August 1942), pp. 64-65.

"Mass Production Methods for Building Wright Cyclones", *Aeronautical Engineering Review*, (August 1942), pp. 13-17.

Mead, G.J.: "Aircraft Powerplant Trends", *Society of Automotive Engineers Journal (Transactions)*, Vol. 41, No. 4, (October 1937), pp. 455-467.

Miller, R.E.: "Machine Tools Developed to Meet Lower-Cost Production", *The Iron Age*, Vol. 133, No. 1, (January 4, 1934), pp. 119-128.

"Milling Operations on the Wright 'Cyclone'", *American Machinist*, Vol. 78, No. 7, (March 28, 1934), pp. 245-248.

Ministry of Labour and National Service: *Women in Engineering*, (London: No Date).

------: *Women in Industry*, (London, 1945).

"Modern Plant Insures Clouds of Planes", *Mill & Factory*, Vol. XXVII, No. 1, (July 1940), pp. 55-59.

Modley, R., ed.: *Aviation Facts and Figures 1945*, (New York: McGraw-Hill Book Company, 1945).

Molloy, E.: The "Complete Engineer" Series, Volume 2: *Aircraft Production*, (London: George Newnes Ltd., 1941).

"Motor Cars to Munitions: Chrysler: Didn't Wait for New Machines; Revamped the Ones They Had", *American Machinist*, Vol. 86, No. 12, (June 1942), pp. 549-561.

"Multi-Tooling for Aero-Engine Production", *Aircraft Engineering*, Vol. 11, Issue 5, (May 1939), pp. 217-219.

"Multiple-Arch Concrete Roof Saves Steel", *Engineering News-Record*, Vol. 130, No. 2, (January 14, 1943), pp. 99-100, 121-22.

Muther, R.: *Production Line Technique*, (New York: McGraw-Hill Book Company, 1944).

"New Heat-treatment Department at Bristol Aeroplane Co., Ltd.", *Machinery (London)*, Vol. 52, No. 1350, (August 25, 1938), pp. 637-639.

"New Ways by Chevrolet", *American Machinist*, Vol. 87, No. 19, (September 16, 1943), pp. 119-120.

"New Wright Line-production Foundry Features Company's Plant No. 3 in Fair Lawn, New Jersey", *Trade Winds*, Vol. 5, No. 7, (November 1940), pp. 6-7.

"Nickel Steels and Alloys in Aircraft Engines", *Aero Digest*, Vol. 29, No. 1, (July 1936), pp. 38-39.

Nixon, F.: "Design of Aero-Engines for Production", *Aircraft Engineering*, Vol. 11, Issue 3, (March 1939), pp. 127-131.

Nutt, A.: "A Survey of Engine Development", *Journal of the Aeronautical Sciences*, Vol. 3, No. 4, (February 1936), pp. 137-141.

------: "Aviation Faces Biggest Production Problem", *Automotive Industries*, Vol. 83, No. 7, (October 1, 1940), pp. 321-342, 413.

Oates, J.A.: "The Bristol Hercules: Part I. Sleeve Valve Design and Operation: Manufacturing the Cylinder and Sleeve", *Aircraft Production*, Vol. IV, No. 40, (February 1942), pp. 176-186.

------: "The Bristol Hercules: Part II. Production of the Master Connecting Rod, Airscrew Shaft and Cylinder Head: the Sleeve Ball and Housing", *Aircraft Production*, Vol. IV, No. 41, (March 1942), pp. 176-186.

------: "The Bristol Hercules: Part III. The Gear Department: Engine Assembly and Testing: Heat Treatment and Process Sections: Laboratories and Standards Room", *Aircraft Production*, Vol. IV, No. 42, (April 1942), pp. 303-309.

------: "The Hercules Crankshaft: Design Features: Machining and Assembly Operations at a Shadow Factory", *Aircraft Production*, Vol. VI, No. 64, (February 1944), pp. 59-66.

------: "Hercules Engine Components: Machining Operations on the Crankcase, Impellor, Sleeve, Cylinder Barrel and Connecting Rods", *Aircraft Production*, Vol. VI, No. 69, (July 1944), pp. 309-319.

------: "Manufacturing the Hercules Piston: Part I. Design Details: Press Forging Methods Employed by Specialloid Ltd.: Machining Sequence; Milling the Pockets", *Aircraft Production*, Vol. XI, No. 77, (March 1945), pp. 128-139.

------: "Manufacturing the Hercules Piston: Part II. Fettling and Heat Treatment: Later Stages of Machining: Diamond Turning and Boring: Final Inspection", *Aircraft Production*, Vol. XI, No. 78, (April 1945), pp. 176-183.

"Only the Productive Can Be Strong-Only the Strong Can Be Free: A Report of the 1940 SAE National Aircraft Production Meeting, held in Los Angeles", *Aviation*, Vol. 39, No. 12, (December 1940), pp. 64-65, 144, 148.

"Operations in the Production of the 'Bristol' Hercules Engine", *Machinery (London)*, Vol. 65, No. 1687, (February 8, 1944), pp. 137-140.

"Operations that Speed up PRATT & WHITNEY Engine Production", *Machinery (New York)*, Vol. 47, No. 11, (July 1941), pp. 194-197.

"Over 900 Wright Employees Are Cited for Long Service; 52 Receive 15-year Awards", *Trade Winds*, Vol. Two, No. Seven, (October-November 1935), p. 2.

"Over 36,500 Operations Are Required in the Production of the Wright Cyclone", *Trade Winds*, Vol. Two, No. Six, (September 1935), p. 2.

"P&W Engine Production", *Aero Digest*, Vol. 34, No. 1, (January 1939), pp. 57-59.

"Packard Production of the Merlin Engine", *Machinery (London)*, Vol. 65, No. 1676, (August 17, 1944), pp. 177-182.

Pagé, V.W.: *The Ford Models V8, B and A Cars: Construction-Operation-Repair*, (New York: Norman W. Henley Publishing, 1933).

Phair, W.A.: "In-line Machining of Aircraft Cylinder Heads", *The Iron Age*, (August 12, 1943), pp. 66-73.

"Plane Engine Plant Sets New Precedents", *Engineering News-Record*, Vol. 127, No. 5, (July 31, 1941), pp. 48-53.

"Power at the Salon", *Flight*, (November 26, 1936), pp. 575-579.

"Pratt & Whitney's Advanced Methods of Engine Building", *Machinery (New York)*, Vol. 46, No. 11, (July 1940), pp. 174-177.

Pratt & Whitney Aircraft Corporation of Missouri: *Announcing Production of the Pratt &Whitney Aircraft's R-2800-C Double Wasp Engine*, (Kansas City, MO, 1944).

"Precision Is the Watchword in the Pratt & Whitney Aircraft Engine Plant. Exacting Requirements Prompt a Technique Unique in Production Methods", *Automotive Industries*, Vol. 77, No. 1, (July 3, 1937), pp. 10-19, 37.

"Producer of Production Lines", *Architectural Record*, Vol. 91, No. 6, (June 1942), pp. 39-43.

"Producing Aero-Engine Parts on Automatic Machines", *Machinery (London)*, Vol. 49, No. 1251, (October 1, 1936), pp. 1-4.

"Producing Morris Eight Engines; The Foundry and Machine Methods at the Coventry Works", *The Automobile Engineer*, Vol. XXV, No. 335, (August 1935), pp. 291-296.

"Production of the Merlin Engine", *Machinery (London)*, Vol. 65, No. 1675, (November 16, 1944), pp. 533-537.

"Progressive Assembly of Wright Engines", *Aero Digest*, Vol. 39, No. 4, (October 1941), pp. 192, 194, 305.

Pye, D.R.: *The Internal Combustion Engine*, Volume II: *The Aero-Engine*, (Oxford: Oxford University Press, 1934).

"Rapid Expansion of Engine Plant Leaves Time for Improvements", *Engineering News-Record*, (October 24, 1940), pp. 70-71.

Rautenstrauch, W.: *The Design of Manufacturing Enterprises: A Study in Applied Industrial Economics*, (New York: Pittman Publishing Corporation, 1941).

Ray, S.: "Quality Control of Wright Engine Parts", *Automotive Industries*, Vol. 85, No. 4, (August 15, 1941), pp. 14-17, 60-62.

Reck, F.M.: "Buick Retools for New Engines", *Aero Digest*, Vol. 45, No. 6, (July 15, 1944), p. 118.

------: "'Double Wasp' Engine Now Mass Produced at Kansas City", *Aero Digest*, Vol. 45, No. 4, (May 15, 1944), pp. 128-129.

------: "Many Production Innovations in Ford P & W Engine Plant", *Aero Digest*, Vol. 43, No. 3, (September 1943), pp. 256-260.

Ricker, C.S.: "Continuous Pouring of Magnesium Engine Castings", *Aviation*, Vol. 43, No. 1, (January 1944), pp. 150-153, 157-59.

------: "Ford's Centrifugal Cast Cylinders", *Aviation*, Vol. 42, No. 6, (June 1943), pp. 134-137, 331.

Ross, D.: *An Appraisal of Prospects for the Aircraft Manufacturing Industry, Aviation: Manufacturing-Transportation*, (New York: White, Weld & Co., May 1940).

Rothbarth, E.: "Causes of the Superior Efficiency of U.S.A. Industry as Compared with British Industry", *The Economic Journal*, Vol. 56, No. 223 (September 1946), pp. 383-390.

"SAE Talks Production", *Aviation*, Vol. 38, No. 11, (November 1939), pp. 32-33, 84, 86, 108.

"Shadow Factories in Being", *Flight*, (November 4, 1937), pp. 445-449.

"Single-Story Concrete Factory – A War-Inspired Innovation", *Engineering News-Record*, Vol. 129, No. 17, (October 22, 1942), pp. 117-212.

Smith, G.G.: "The Bristol Hercules", *Flight*, (November 27, 1941), pp. 379-383.

"Some Machining Operations in the Manufacture of Bristol Air-Cooled Aero Engines", *The Machine Tool Review*, Vol. 22, No. 139, (March-April 1934), pp. 33-37.

"Sometimes It's 'The Little Things That Count' Even in Aircraft Engines; Studs, For Example", *Trade Winds*, Vol. Three, No. Four, (January 1937), pp. 7, 18.

"Special Equipment in the Production of the Merlin Engine", *Machinery (London)*, Vol. 62, No. 1582, (February 4, 1943), pp. 113-117.

"Special Machine Tools Boost Engine Production", *The Tool Engineer*, (November 1943), pp. 76-78, 81.

Stout, W.W.: *Great Engines and Great Planes*, (Detroit, MI: Chrysler Corporation, 1947).

"Surface Hardening of Engine Parts by the Nitriding Process", *Aero Digest*, Vol. 38, No. 1, (January 1941), pp. 114, 235.

Sutton, K.E.: *Production and Quality*, (Wright Aeronautical Corporation, 1941).

------: "Progressive Assembly of Wright Engines", *Aero Digest*, Vol. 39, No. 4, (October 1941), pp. 192, 194, 305.

------: "Wright Aero Produces Engines for the B-29 Superfortress", *Machinery (New York)*, Vol. 50, No. 12, (August 1944), pp. 134-144.

Swan, A.: *Handbook of Aeronautics, Volume II: Aero-engine Design and Practice*, (London: Sir Isaac Pitman and Sons, 1938).

"The Bristol Centaurus", *The Aeroplane*, (July 6, 1945), pp. 11-17.

"The 'Bristol' Engine Inspection Department", *The "Bristol" Review*, Engine Issue No. 11, (November 1936), pp. 30-35.

"The Bristol Hercules", *Flight*, (November 27, 1941), pp. 379-383.

"The Coming of the Sleeve Valve", *Aircraft Engineering*, Vol. 9, No. 102, (August 1937), pp. 203-204.

"The Design of Industrial Buildings", *Architectural Record*, Vol. 81, No. 2, (February 1937), pp. BT3-B40.

"The Engine Fitting, Stripping and Re-Building Department", *The "Bristol" Review*, Engine Issue No. 10, (June 1936), pp. 35-38.

"The Government Aircraft 'Shadow' Scheme: Progress at the Factories", *Machinery (London)*, Vol. 51, No., (November 4, 1937), pp. 133-136.

"The Government Shadow Factories", *Engineering*, (November 5, 1937), pp. 511-514.

"The Hercules Crankshaft: Design Features; Machining and Assembly Operations at a Shadow Factory", *Aircraft Production*, Vol. VI, No. 64, (February 1944), pp. 59-66.

"The Industry's Largest Engine Factory", *Aero Digest's Aviation Engineering*, Vol. 39, No. 1, (July 1941), pp. 156, 159-60, 205.

"The Making of Reduction Gears for Wright Aircraft Engines", *Automotive Industries*, Vol. 84, No. 4, (February 15, 1941), pp. 151-154.

"The Manufacture of 'Bristol' Aero Engine Components", *The "Bristol" Review*, Engine Issue No. 8, (November 1934), pp. 23-26.

"The Manufacture of High Performance Aero Engines", *Machinery (London)*, Vol. 48, No. 1249, (September 17, 1936), pp. 737-756.

"The Manufacture of Rolls-Royce Aero-engines", *Machinery (London)*, Vol. 55, No. 1427, (February 15, 1940), pp. 525-544.

"The Manufacture of Motor-car Engine Components", *Machinery (London)*, Vol. 47, No. 1215, (January 23, 1936), pp. 489-492.

"The New Engine Building Department", *The "Bristol" Review, Engine* Issue No. 13, (June 1938), p. 35.

"The New Double-Bank Sleeve-Valve Engine", *The "Bristol" Review*, Engine Issue No. 12, (June 1937), pp. 15-16.

"The Ohio Plant Today", *Trade Winds*, Vol. 6, No. 11, (April 1943), pp. 12-13, 16.

"The Packard-Built Merlin", *Aircraft Production*, Vol. VI, No. 70, (August 1944), pp. 389-394.

"The Production of Aero-engine Crankcases and Superchargers", *Machinery (London)*, Vol. 48, No. 1250, (September 24, 1936), pp. 774-778.

"The Production of Aero-engine Crankshafts: Methods Employed at a Government 'Shadow' Factory in the Manufacture of the Bristol Hercules Engine", *Machinery (London)*, Vol. 60, No. 1534, (March 5, 1942), pp. 145-147, 149-51.

"The Production of the Merlin Engine", *Machinery (London)*, Vol. 65, No. 1671, (October 19, 1944), pp. 421-427.

"The Production of the Rolls-Royce Merlin Engine", *Machinery (London)*, Vol. 62, No. 1580, (January 21, 1943), pp. 57-60.

"The Production of Supercharger Impellers", *Machinery (London)*, Vol. 60, No. 1535, (March 12, 1942), pp. 169-173.

"The Shadow Factories: A Preliminary Survey of Plant and Methods", *The Automobile Engineer*, Vol. XXVII, No. 3645, (November 1937), pp. 444-446.

"The Standard Shadow Factory: Specialized Production Methods for Bristol Mercury Cylinder Units", *The Automobile Engineer*, Vol. XXVIII, No. 370, (April 1938), pp. 123-130.

"The Standard Shadow Factory Specialized Production Methods for Bristol Mercury Cylinder Units", Part II, *The Automobile Engineer*, Vol. XXVIII, No. 371, (May 1938), pp. 154-156.

"The Vauxhall Works", *The Automobile Engineer*, Vol. XXVI, No. 348, (August 1936), pp. 313-317.

The Times: British War Production 1939-1945: A Record, (London, 1945).

"The Wright Aeronautical Corporation Plant at Paterson, N.J.", *Mill & Factory*, Vol. XXIV, No. 6, (June 1939), pp. 143-150, 300-01.

"Transfer Machines Speed Engine Output", *The Tool Engineer*, (November 1943), pp. 82-84.

"Traveling Form System as Used for Longer Spans", *Architectural Record*, Vol. 92, No. 6, (December 1942), pp. 54-55.

Tumser, A.A.: "Quality Control: Checking and Inspection Procedure in the Production of Wright Engines", *Aero Digest's Aviation Engineering*, Vol. 39, No. 5, (November 1941), pp. 152, 155-56, 159-60.

Turner, A.A.: "Quality Control", *Aero Digest's Aviation Engineering*, Vol. 39, No. 5, (November 1941), pp. 152-155.

"Turret Lathe Operations on the Hercules Crank Case", *Machine Tool Review*, Vol. 31, No. 184, (March-April 1943), pp. 26-28.

Tuttle, W.G.: "Production Changes Necessary for Women Workers", *Automotive and Aviation Industries*, Vol. 88, No. 8, (April 15, 1943), pp. 24-26, 74, 76.

"Two Aircraft Engines on One Assembly Line", *American Machinist*, Vol. 89, No. 5, (March 1, 1945), pp. 106-107.

United Aircraft Corporation: *The Pratt & Whitney Story*, (Hartford, CT: United Aircraft Corporation, 1950).

United States Air Force: United States Air Force Numbered Historical Studies, No. 40: *Expansion of Industrial Facilities under Army Air Force Auspices*, 1940-1945 (No Date).

------: United States Air Force Numbered Historical Studies, No. 106: *Distribution of Air Material to the Allies*, 1940-1944 (No Date).

"Valve Spring Manufacture", *Machinery (London)*, Vol. 62, No. 1581, (January 28, 1943), pp. 85-89.

Waddington, J.L.: "American Methods and Modes-II", *Flight*, Vol. XXXVI, No. 1605, (September 28, 1939), pp. 269-271.

Ward, J.C., Jr.: "Plant Layout and Production Methods for Modern Aircraft Engines", *Society of Automotive Engineers Journal (Transactions)*, Vol. 40, No. 5, (May 1937), pp. 178-189.

------: "Machine Tools in the Aero-engine Industry", *Machinery (London)*, Vol. 55, No. 1418, (December 14, 1939), p. 275.

Ware, B.T.: "Production Flow at the Austin Works", *Machinery*, Vol. 44, (June 7, 1934), pp. 178-189.

Warner, E.L.: "Production for Victory", *Automobile and Aviation Industries*, Vol. 87, No. 7, (October 1, 1942), pp. 56-59, 216-18.

"'Warspeed' System of Construction in Concrete", *Architectural Record*, Vol. 92, No. 6, (December 1942), p. 52.

"Wartime Construction of Factory Buildings...Building Type Study No. 72", *Architectural Record*, Vol. 92, No. 6, (December 1942), pp. 51-57.

"What the Manufacturer Wants from His Architect", *Architectural Forum*, Vol. 69, No. 2, (August 1938), pp. 96-97.

"What's Happening and What's Ahead in Industrial Plant Development", *Factory Management and Maintenance*, Vol. 99, No. 4, (April 1941), pp. B-37-B-52, 76.

Wilkinson, P.H.: *Aircraft Engines of the World 1941*, (New York: P.H. Wilkinson, 1941).

------: *Aircraft Engines of the World 1945*, (New York: P.H. Wilkinson, 1945).

Wilson, E.E.: *Slipstream: The Autobiography of an Air Craftsman*, (New York: McGraw-Hill Book Company, 1950).

Winter, O.W.: "Unskilled Labor Creates New Tooling Problems", *American Machinist*, Vol. 87, No. 6, (March 18, 1943), pp. 95-96.

Woollard, F.G.: "Some Notes on British Methods of Continuous Production", *Proceedings of the Institute of Automobile Engineers*, Vol. XIX, (1924-25), pp. 419-474, 885-90.

Wright, I.H.: "Developments in Machine Tools", *Machinery (London)*, Vol. 49, (February 11, 18, 1937), pp. 607-611, 625-26.

"Wright Aero Will Expend $1,350,000 for Extensions to Factory and New Equipment", *Trade Winds*, Vol. 3, No. 7, (May 1937), pp. 2-3.

"Wright Aero's Electro-plating Equipment among Most Modern and Versatile in Industry", *Trade Winds*, Vol. 5, No. 7, (November 1940), pp. 10-11.

"Wright Aeronautical's Line Conveyor Technique Speeds Painting of Cyclone and Whirlwind Parts", *Trade Winds*, Vol. 5, No. 7, (November 1940), pp. 14-15.

"Wright G-200 Cyclone, Rated at 1200 H.P., Released for Both Domestic and Export Sale", *Trade Winds*, Vol. 4, No. 10, (Navy Number 1939), pp. 6-8.

"Wright G-200 Series Cyclone Engines", *Aero Digest*, Vol. 35, No. 1, (July 1939), pp. 113-114.

"Wright Plant Sets New Size Record", *Engineering News-Record*, Vol. 127, No. 1, (July 3, 1941), pp. 52-54.

Wright Aeronautical Corporation: *Application of High Production Machine Tools*, (No Date).

------: *The G-100 Series 1100 H.P. Wright Cyclone*, (No Date).

------: *G-200 Series Wright Cyclone 1200 Horsepower*, (No Date).

------: *Wright Double-Row Cyclone 14 Aircraft Engines*, (No Date).

------: *Preliminary Instruction Book: Wright Cyclone 14 Aircraft Engine: Series C14-BB*, (Paterson, NJ, 1943).

------: *Preliminary Instruction Book: Wright Cyclone 18 Aircraft Engine: Series C18BA*, (Paterson, NJ, 1942).

------: *Service Manual: Wright Cyclone 9GC Aircraft Engines*, 3rd Ed., (Paterson, NJ, 1944).

------: *Service Manual: Wright Cyclone 14BA Aircraft Engines*, 3rd Ed., Paterson, NJ, 1944).

------: *Service Manual: Wright Cyclone 14BB Aircraft Engines*, 2nd Ed., (Paterson, NJ, 1944).

------: *Service Manual: Wright Cyclone 18BA Aircraft Engines*, 2nd Ed., (Paterson, NJ, 1944).

------: *Wright Cyclone Engines: Series 18BA Service Manual*, 4th Ed., (Paterson, NJ, 1947).

------: *Wright Cyclone Engines: Series 18BD Service Manual*, 1st Ed., (Paterson, NJ, 1946).

------: *Wright Double-Row Cyclone 14 Aircraft Engines*, (No Date).

Wright, T.P.: "America's Answer: Gearing Our Aviation Industry to the National Defense", *Aviation*, Vol. 38, No. 6, (June 1939), pp. 26-29, 78, 80, 82, 84.

------: "Factors Affecting the Cost of Airplanes", *Journal of Aeronautical Sciences*, Vol. 3, No. 4, (February 1936), pp. 122-128.

------: "U.S. Aircraft Production Mission to England, October 6 to November 11, 1942: Report on Personal Observations of the Chairman", *Articles and Addresses of Theodore P. Wright*, Vol. II: Aircraft Production, Uses of Aviation, (Buffalo, NY, 1961).

CONTEMPORARY FILMS

Chevrolet Motor Company: "Master Hands", (1936), available on YouTube at https://www.youtube.com/watch?v=Pr8rCNaASlc, accessed 21 February 2020.

Ford Motor Company: "Trip through the River Rouge", (circa late 1930s), available on YouTube at https://www.youtube.com/watch?v=Xa0PAg7FfMk, accessed 21 February 2020.

Rolls-Royce: "Contribution to Victory", (circa 1946), available on YouTube at https://www.youtube.com/watch?v=-fo7SmNuUU4, accessed 25 November 2019.

War Department Official Training Film 1-135: "Aircraft Engines", (1940), available on YouTube at https://www.youtube.com/watch?v=4vBPlcquAD8 accessed 3 March 2020.

Wilding Picture Productions, Inc.: "Engines for Super Bombers", (1945), available on YouTube at https://www.youtube.com/watch?v=_t3akMEm9bI, accessed 25 November 2019.

Wright Aeronautical Corporation: "Wright Builds for Air Supremacy", (1941), available on YouTube at https://www.youtube.com/watch?v=sBfFpcdyd5Q, accessed 25 November 2019.

SECONDARY SOURCES

Articles and Books

Albrecht, A.B.: *The American Machine Tool Industry: Its History, Growth, Restructuring & Recovery: A Personal Perspective*, 3rd Ed. (Ashland, OH: Bookmasters, 2016. ISBN 9-785967-20336-7).

Alchian, A.: "Reliability and Progress Curves in Airframe Production", *Econometrica*, Vol. 31, No. 4, (October 1963), pp. 679-693.

Aldcroft, D.H.: "The Performance of the British Machine Tool Industry in the Interwar Years", *Business History Review*, Vol. 40, No. 3, (Autumn 1966), pp. 281-296.

Anderson, J.D., Jr.: *The Aeroplane: A History of Its Technology*, (Reston, VA: American Institute of Aeronautics and Astronautics, 2002. ISBN 1-56347-525-1).

Andrews, C.F. and E.B. Morgan: *Vickers Aircraft Since 1908*, New Edition, (London: Putnam, 1988. ISBN 0-85177-815-1).

Bailey, G.J.: *The Arsenal of Democracy: Aircraft Supply and the Anglo-American Alliance, 1938-1942*, (Edinburgh: Edinburgh University Press, 2013. ISBN 978-0-4747-7).

Baloff, N.: "The Learning Curve-Some Controversial Issues", *The Journal of Industrial Economics*, Vol. 14, No. 3, (June 1966), pp. 275-282.

Barnett, C.: *The Audit of War: The Illusion and Reality of Britain as a Great Nation*, (London: Macmillan, 1986. ISBN 0333353765).

Barnes, C.H.: *Bristol Aircraft Since 1910*, 3rd Ed., (London: Putnam, 1988. ISBN 0-85177-823-2).

------: *Handley Page Aircraft Since 1907*, (London: Putnam, 1976. ISBN 0-85177-803-8).

------: *Shorts Aircraft Since 1900*, James, D.N., Rev. ed. (London: Putnam, 1989a. ISBN 0-85177-819-4).

Bergeron, L. and M.T. Maicellori-Pontois: *Industry, Architecture, and Engineering*, (New York: Harry N. Abrams, 2000. ISBN 0-81093-473-6).

Bilstein, R.E.: *The American Aerospace Industry: From Workshop to Global Enterprise*, (New York: Twayne Publishers, 1996. ISBN 0-8057-9838-2).

Bingham, V.: *Major Piston Aero Engines of World War II*, (Shrewsbury, UK: Airlife Publishing, 1998. ISBN 1-84037-012-2).

Bitran, G.R. and L. Chang: "Productivity Measurement at the Firm Level", *Interfaces*, Vol. 14, No. 3, (May-June 1984), p. 29.

Baime, A.J.: *The Arsenal of Democracy: FDR. Detroit, and an Epic Quest to Arm an America at War*, (Boston, MA: Houghton Mifflin Harcourt, 2014. ISBN 0-54771-928-0).

Blaszczyk, R.L.: "'Reign of the Robots': The Homer Laughlin China Company and Flexible Mass Production", *Technology and Culture*, Vol. 36, No. 4, (October 1995), pp. 863-911.

Botticelli, P.: "The British Engineering Press during the Second Industrial Revolution: Responses to Corporate Capitalism", *Business History Review*, Vol. 71, (Summer 1997).

Bowden, S. and D.M. Higgins: "British Industry in the Interwar Years", in *Cambridge Economic History of Modern Britain, Volume 2: Economic Maturity 1860-1939*, (Cambridge, 2004), pp. 374-402.

Bowyer, M.J.F.: *The Stirling Bomber*, (London: Faber and Faber, 1980. ISBN 0-571-11101-7).

Bradley, B.H.: *The Works: The Industrial Architecture of the United States*, (New York: Oxford University Press, 1999. ISBN 0-19509-000-4).

Breer, C.: *The Birth of the Chrysler Corporation and Its Engineering Legacy*, Edited by A.J. Yanik, (Warrendale, PA: Society of Automotive Engineers, 1995. ISBN 1-56091-524-2).

Bresnahan, T.F. and D.M.G. Raff: "Intra-Industry Heterogeneity and the Great Depression: The American Motor Vehicles Industry, 1919-1935", *The Journal of Economic History*, Vol. 51, No. 2 (June 1991), pp. 317-331.

Broadberry, S.N.: "How Did the United States and Germany Overtake Britain? A Sectorial Analysis of Comparative Productivity Levels, 1870-1990", *The Journal of Economic History*, Vol. 58, No. 2, (June 1998), pp. 375-407.

------: "Technological Leadership and Productivity Leadership since the Industrial Revolution: Implications for the Convergence Debate", *The Economic Journal*, Vol. 104, No. 453 (March 1994a), pp. 291-302.

------: *The Productivity Race: British Manufacturing in International Perspective, 1850-1990*, (Cambridge: Cambridge University Press, 1997a. ISBN 978-0521584401).

Broadberry, S. and C. Burhop: "Comparative Productivity in British and German Manufacturing before World War II: Reconciling Direct Benchmark Estimates and Time Series Projections", *The Journal of Economic History*, Vol. 67, No. 2, (June 2007), pp. 315-349.

Broadberry, S. and M. O'Mahony: "Britain's Productivity Gap with the United States and Europe: A Historical Perspective", *National Institute Economic Review*, No. 189, (July 2004), pp. 72-85.

Brooks, P.W.: *The Modern Airliner: Its Origin and Development*, (London: Putnam, 1961).

Bryan, F.R.: *Rouge Pictured in Its Prime*, (Detroit, MI: Wayne State University Press, 2003. ISBN 978-0-9727843-0-6).

Bucci, F.: *Albert Kahn: Architect of Ford*, (New York: Princeton Architectural Press, 1993. ISBN 978-18278271846).

Burnham, R.: *A Concrete Atlantis: U.S. Industrial Building and European Modern Architecture 1900-1925*, (Cambridge, MA: MIT Press, 1989. ISBN 978-0262022446).

Butler, P. and D. Hagedorn: *North American Arsenal: Aircraft for the Allies 1938-1945 Purchases and Lend-Lease*, (Hinckley, UK: Midland Publishing, 2004. ISBN 1-85780-163-6).

Cairncross, S.A.: *Planning in Wartime: Aircraft Production in Britain, Germany and the USA*, (Basingstoke, UK: Macmillan Academic and Professional, 1990. ISBN 0-333-53840-4).

Carew, M.G.: *Becoming the Arsenal: The American Industrial Mobilization for World War II, 1938-1942*, (New York: University Press of America, 2010. ISBN 978-0761846680).

Carlsson, B.: "The Development and Use of Machine Tools in Historical Perspective", *Journal of Economic Behaviour and Organization*, Vol. 5, (1984), pp. 91-114.

Carstensen, F.V.: Review of *From the American System to Mass Production, 1800-1932: The Development of Manufacturing Technology in the United States*, *The Business History Review*, Vol. 59, No. 2 (Summer, 1985).

Central Statistical Office: *Fighting with Figures: A Statistical Digest of the Second World War*, (London: Central Statistical Office, 1995. ISBN 0-11-620719-1).

Chandler, A.D., Jr.: *Scale and Scope: The Dynamics of Industrial Capitalism*, (Cambridge, MA: The Belknap Press, 1990. ISBN 0-674-78994-6).

Church, R.: *Herbert Austin: The British Motor Car Industry to 1941*, (London: Europa, 1979. ISBN 0905118294).

------: *The Rise and Decline of the British Motor Industry*, (Cambridge: Cambridge University Press, 1994b. ISBN 978-0521552837).

Church, R. and M. Miller: "The Big Three: Competition, Management and Marketing in the British Motor Industry, 1922-1939", in Supple, B., ed.: *Essays in British Business History*, (Oxford: Clarendon Press, 1979. ISBN 0198770871).

Collins, P. and M. Stratton: *British Car Factories from 1896: A Complete Historical, Geographic, Architectural & Technological Survey*, (Goodmanstone, Dorset: Veloce, 1993. ISBN 1874105049).

Connors, J.: *The Engines of Pratt & Whitney: A Technical History*, (Reston, VA: American Institute of Aeronautics and Astronautics, 2010. ISBN 978-1600867118).

Constant, E.W., II: *The Origins of the Turbojet Revolution*, (Baltimore, MD: The Johns Hopkins University Press, 1980. ISBN 978-0801822223).

Craven, W.F. and J.L. Cate, eds.: *The Army Air Forces in World War II*, Volume Six: *Men and Planes*, (Chicago, IL: University of Chicago Press, 1955).

Cunningham, W.G.: *The Aircraft Industry: A Study in Industrial Location*, (Los Angeles, CA: Lorrin L. Morrison, 1951).

Crouch, T.: *Wings: A History of Aviation from Kites to the Space Age*, (New York: W.W. Norton & Co., 2003. ISBN 0-393-05767-4).

Darley, G.: *Factory*, (London: Reaktion Books, 2003. ISBN 1-86189-155-5).

Dean, F.H.: *America's Hundred-Thousand: U.S. Production Fighters of World War Two*, (Atglen, PA: Schiffer Publishing, 1997).

DeGarmo, P., J.T. Black, R.A. Kosher, and B.E. Klamecki: *Materials and Processes in Manufacturing*, 9th Ed., (New York: Wiley, 2003. ISBN 9780471033066).

De Jong, H. and P. Woltjer: "Depression Dynamics: A New Estimate of the Anglo-American Manufacturing Productivity Gap in the Interwar Period", *The Economic History Review*, Vol. 64, No. 2 (May 2011), pp. 472-492.

de Syon, G.: "What the Wrights Wrought: The Centennial of Flight in Recent Literature", *Technology and Culture*, Vol. 45, No. 2, (April 2004), pp. 350-357.

Douglas, C.E.: *The Secret Horsepower Race: Western Front Fighter Engine Development*, (Horncastle, UK: Tempest Books, 2020. ISBN 978-1-911658-50-4).

Dunbabin, J.P.D.: "British Rearmament in the 1930s: A Chronology and Review", *The Historical Journal*, Vol. XVIII, No. 3, (1975), pp. 587-609.

Edgerton, D.: *England and the Aeroplane: Militarism, Modernity and Machines*, Revised Edition, (London: Penguin, 2013. ISBN 978-0-141-97516-0).

------: *Britain's War Machine: Weapons, Resources and Experts in the Second World War*, (London: Allen Lane, 2011. ISBN 978-0-713-99918-1).

------: "Eléments pour une nouvelle histoire global de la production du XXème siècle", in *Le monde du génie industriel au XXe siècle: Autour de Pierre Bézier et des machines-outils*, (2016a).

------: "From Innovation to Use: Ten Eclectic Theses on the Historiography of Technology", *History and Technology*, Vol. 16, (1999a), pp. 111-136.

------: *Warfare State: Britain, 1920-1970*, (Cambridge: Cambridge University Press, 2006a. ISBN 978-0-521-85636-2).

Elphick, P.: *Liberty: The Ships that Won the War*, (Annapolis, MD: Naval Institute Press, 2001. ISBN 1-59114-451-5).

Eltscher, L.R. and E.M. Young: *Curtiss-Wright: Greatness and Decline*, (New York: Twayne Publishers, 1998. ISBN 0-8057-9829-3).

Emerson, H.P. and D.C.E. Naebring: *Origins of Industrial Engineering: The Early Years of a Profession*, (Norcross, GA: Industrial Engineering and Management Press, 1988. ISBN 9780898060973).

Emiliani, M. and P.J. Seymour: "Frank George Woollard: Forgotten Pioneer of Flow Production", *Journal of Management History*, Vol. 17, Issue 1, pp. 66-87.

Fear, J.: "War of the Factories", in Geyer, M. and A. Tooze, eds.: *The Second World War*, Volume III, *Total War: Economy, Society and Culture*, (Cambridge: Cambridge University Press, 2015. ISBN 978-1-107-03995-7).

Fearon, P.: "Aircraft Manufacturing", in Buxton, N.K. and D.H. Aldcroft, eds.: *British Industry Between the Wars: Instability and Industrial Development 1919-1939*, (London: Scolar Press, 1979. ISBN 0-85967-383-9), pp. 216-240.

------: "The British Airframe Industry and the State, 1918-1935", *The Economic History Review*, Vol. 27, No. 2, (May 1974), pp. 236-251.

------: "The Formative Years of the British Aircraft Industry, 1913-1924", *The Business History Review*, Vol. 43, No. 4, (Winter, 1969), pp. 476-495.

Fenton, J., ed.: *Gasoline Engine Analysis for Computer Aided Design*, (London: Mechanical Engineering Publications, 1986. ISBN 0852986343).

Ferguson, R.G.: "One Thousand Planes a Day: Ford, Grumman, General Motors and the Arsenal of Democracy", *History and Technology*, Vol. 21, No. 2, (June 2005), pp. 149-175.

Ferry, W.H.: *The Legacy of Albert Kahn*, (Detroit, MI: Wayne State University Press, 1987. ISBN 9780814318881).

Field, A.J.: *A Great Leap Forward: 1930s Depression and U.S. Economic Growth*, (New Haven, CT: Yale University Press, 2011. ISBN 978-0300151091).

------: "The Impact of the Second World War on US Productivity Growth", *The Economic History Review, New Series*, Vol. 61, No. 3 (August 2008), pp. 672-694.

Flamming, J.M.: *Chrysler Chronicle: An Illustrated History of Desoto-Dodge-Imperial-Plymouth*, (Lincolnwood, IL: Publications International, 1995. ISBN 978-0785329015).

Foreman-Peck, J., S. Bowden, and A. McKinlay: *The British Motor Industry*, (Manchester, UK: Manchester University Press, 1995. ISBN 0719026121).

Freeman, J.B.: *Behemoth: A History of the Factory and the Making of the Modern World*, (New York: W.W. Norton & Co., 2018. ISBN 978-0-393-24631-5).

Furse, A.: *Wilfrid Freeman: The Genius Behind Allied Survival and Air Supremacy 1939 to 1945*, (Staplehurst, Kent: Spellmount Limited, 2000. ISBN 1-86227-079-1).

Gibbs-Smith, C.H.: *Aviation: An Historical Survey from Its Origins to the End of World War II*, (London: Her Majesty's Stationary Office, 1970).

Giffard, H.: "A Hard Sell: Factors Influencing the Interwar Adoption of Tungsten Carbide Cutting Tools in Germany, Britain, and the United States", *Technology and Culture*, Vol. 56, No. 4, (October 2015), pp. 895-927.

------: *Making Jet Engines in World War II: Britain, Germany, and the United States*, (Chicago, IL: University of Chicago Press, 2016b. ISBN 978-0-226-38859-5).

Goodwin, M. and P. Starkings: *Japanese Aero-Engines: 1910-1945*, (Poland: MMP Books, 2017).

Gowing, M.: "The Organization of Manpower in Britain during the Second World War", *Journal of Contemporary History*, Vol. 7, No. ½, (January-April 1972), pp. 147-167.

Gregor, N.: *Daimler-Benz in the Third Reich*, (New Haven, CT: Yale University Press, 1998. ISBN 978-0-300-07243-3).

Gropman, A.L.: *Mobilizing U.S. Industry in World War II*, McNair Paper 50, (Washington, D.C.: Institute for National Strategic Studies, 1996).

Gunston, B.: *The Development of Piston Aero Engines*, Reprinted, (Sparkford, Somerset, UK: Patrick Stephens Ltd, 1994. ISBN 1-85260-385-2).

------: *Fedden: The Life of Sir Roy Fedden*, Rolls-Royce Heritage Trust Historical Series No. 26, (Derby, UK: Rolls-Royce Heritage Trust, 1998a. ISBN 978-872922-13-3).

------: *Rolls-Royce Aero Engines*, (Sparkford, Somerset, UK: Patrick Stephens Ltd, 1989b. ISBN 1-85260-037-3).

Haight, J.M.V.: *American Aid to France 1938-1940*, (New York: Athenium, 1970).

Hall, H.D.: *History of the Second World War, United Kingdom Civil Series, War Production Series: North American Supply*, (London: Her Majesty's Stationary Office, 1955).

Hallion, R.P.: *Taking Flight: Inventing the Aerial Age from Antiquity through the First World War*, (Oxford: Oxford University Press, 2003. ISBN 0-19-516035-5).

Hannah, L.: "Strategic Games, Scale, and Efficiency, or Chandler Goes to Hollywood", in Coopey, R. and P. Lyth, eds.: *Business in Britain in the Twentieth Century: Decline and Renaissance?* (Oxford: Oxford University Press, 2009. ISBN 9780199226009).

Hansell, H.S., Jr.: *The Strategic Air War against Germany and Japan: A Memoir*, (Washington, D.C.: Office of Air Force History, 1986. ISBN 0-912799-39-0).

Hansen, J.R.: "Aviation History in the Wider View", *Technology and Culture*, Vol. 30, No. 3 (July 1989), pp. 643-656.

Harrison, M., ed.: *The Economics of World War II: Six Great Powers in International Comparison*, (Cambridge: Cambridge University Press, 1998. ISBN 0-521-62046-5).

Hartley, K.: "The Learning Curve and Its Application to the Aircraft Industry", *The Journal of Industrial Economics*, Vol. 13, No. 2, (March 1965), pp. 122-128.

Hassell, P.: "The Bristol Sleeve Valve Aero Engines", in Starr, F., E.L. Marshall, and B. Lawton, eds.: *The Piston Engine Revolution: Papers from a Conference on the History of the Reciprocating Internal Combustion Engines Held at the Museum of Science and Industry, Manchester, 14-17 April 2011*, (Manchester, UK: The Newcomen Society, 2012. ISBN 978-0-904685-15-2).

Heaven, J.: *Bristol Piston Engines Since* 1914, unpublished manuscript, copy in the possession of the Rolls-Royce Heritage Trust-Bristol Branch.

Heinrich, T.B.: "Jack of All Trades: Cramp Shipbuilding, Mixed Production, and the Limits of Flexible Specialization in American Warship Construction, 1940-1945", *Enterprise and Society*, Vol. 11, No. 2, (June 2010), pp. 275-315.

------: *Warship Builders: An Industrial History of U.S. Naval Shipbuilding, 1922-1945*, (Annapolis, MD: Naval Institute Press, 2020. ISBN 9781682475379).

Herman, A.: *Freedom's Forge: How American Business Produced Victory in World War II*, (New York: Random House, 2012. ISBN 978-1-4000-6964-4).

Heron, S.D.: *History of Aircraft Piston Engines*, (Detroit, MI: Ethyl Corporation, 1961).

Heywood, J.B.: *Internal Combustion Engine Fundamentals*, (New York: McGraw Hill, 1988. ISBN 9780070286375).

Hildebrand, G.: *Designing for Industry: The Architecture of Albert Kahn*, (Cambridge, MA: The MIT Press, 1974. ISBN 9780262080545).

Hirst, P. and J. Zeitlin: "Flexible Specialization versus Post-Fordism: Theory, Evidence and Policy Implications", *Economics and Society*, Vol. 10, No. 1, (February 1991), pp. 1-56.

Hoke, D.R.: *Ingenious Yankees: The Rise of the American System of Manufacture in the Private Sector*, (New York: Columbia University Press, 1990. ISBN 0-231-06756-9).

Holley, I.B., Jr.: *Buying Aircraft: Material Procurement for the Army Air Forces, United States Army in World War II: Special Studies* (Washington, D.C.: U.S. Government Printing Office, 1964).

------: "A Detroit Dream of Mass-Produced Fighter Aircraft: The XP-75 Fiasco", *Technology and Culture*, Vol. 28, No. 3, (July 1987), pp. 578-593.

Hooks, G.: *Forging the Military Industrial Complex: World War II's Battle of the Potomac*, (Urbana, IL: University of Illinois Press, 1991. ISBN 9780252017506).

Hornby, W.: *History of the Second World War, United Kingdom Civil Series, War Production Series: Factories and Plant*, (London: Her Majesty's Stationary Office, 1958).

Hounshell, D.A.: "Automation, Transfer Machinery and Mass Production in the U.S. Automobile Industry in the Post-World War II Era", *Enterprise & Society*, Vol. 1, No. 1, (2000), pp. 100-138.

------: "Ford Eagle Boats and Mass Production during World War I", in Smith, M.R., ed.: *Military Enterprise and Technological Change: Perspectives on the American Experience*, (Cambridge, MA, 1985a), pp. 175-201.

------: *From the American System to Mass Production: The Development of Manufacturing Technology in the United States*, Baltimore, MD: The Johns Hopkins University Press, 1984a. ISBN 0-8018-2975-5).

Hughes, T.P.: *American Genesis: A Century of Invention and Technological Enthusiasm, 1870-1970*, Chicago, IL: The University of Chicago Press, 2004. ISBN 0-226-35927-1).

Hyde, C.K.: *Arsenal of Democracy: The American Automobile Industry in World War II*, Detroit, MI: Wayne State University Press, 2013. ISBN 978-0-8143-3951-0).

------: "Assembly-Line Architecture: Albert Kahn and the Evolution of the U.S. Auto Factory, 1905-1940", *The Journal of the Society for Industrial Archeology*, Vol. 22, No. 2 (1996), pp. 5-24.

------: *Riding the Roller Coaster: A History of the Chrysler Corporation*, (Detroit, MI: Wayne State University Press, 2003a. ISBN 0-8143-3091-6).

Hyde, H.M.: *British Air Policy between the Wars 1918-1939*, (London: William Heinemann Ltd., 1976. ISBN 434-47983-7).

Immer, J.R.: *Layout Planning Techniques*, (New York: McGraw Hill, 1950).

Inman, P.: *History of the Second World War, United Kingdom Civil Series, War Production Series: Labour in the Munitions Industry*, (London: Her Majesty's Stationary Office, 1957).

Jarret, P., ed.: *Aircraft of the Second World War: The Development of the Warplane 1939-1945*, (London: Putnam Aeronautical Books, 1997. ISBN 0-85177-875-5).

------: *Biplane to Monoplane: Aircraft Development 1919-1939*, (London: Putnam Aeronautical Books, 1997b. ISBN 0-85177-874-7).

Jolly, M.: "Employment Variability and Mass Production Technology in the British Automobile Industry during the Interwar Period", *Journal of European Economic History*, Vol. 25, No. 2, (Fall 1996), pp. 425-440.

Kennedy, D.M.: *The Oxford History of the United States*, Volume IX: *Freedom from Fear: The American People in Depression and War, 1929-1945*, (Oxford: Oxford University Press, 1999. ISBN 978-0195038347).

Klein, M.: *A Call to Arms: Mobilizing America for World War II* (New York: Bloomsbury Press, 2013. ISBN 978-1-59691-607-4).

Koham, C.M.: *Works and Buildings*, History of the Second World War, United Kingdom Civil Series, (London: Her Majesty's Stationary Office, 1952).

Koistinen, P.A.C.: *Arsenal of World War II: The Political Economy of American Warfare 1940-1945*, (Lawrence, KS: University Press of Kansas, 2004. ISBN 0-7006-1308-0).

------: *Planning War, Pursuing Peace: The Political Economy of American Warfare 1920-1939*, (Lawrence, KS: University Press of Kansas, 1998b. ISBN 0-7006-0890-7).

Kotelnikov, V.: *Russian Piston Aero Engines*, (Ramsbury, Wiltshire, UK: The Crowood Press Ltd, 2005. ISBN 1-86126-702-9).

Lacey, J.: *Keep from All Thoughtful Men: How U.S. Economists Won World War II*, (Annapolis, MD: Naval Institute Press, 2011. ISBN 978-1-59114-491-5).

Lane, F.C., with B.D. Coll, G.J. Fischer, and D.B. Tyler: *Ships for Victory: A History of Shipbuilding under the U.S. Maritime Commission in World War II*, (Baltimore, MD: The Johns Hopkins University Press, 2001. ISBN 0-8018-6752-5).

Launius, R.D., ed.: *Innovation and the Development of Flight*, (College Station, TX: Texas A&M University Press, 1999. ISBN 978-0890968765).

------: "Introduction: Patterns of Innovation in Aeronautical Technology", in Launius, R.D., ed.: *Innovation and the Development of Flight*, (College Station, TX, 1999b).

Leviton, S.A. and D. Werneke: *Productivity: Problems, Prospects, and Policies*, (Baltimore, MD: The Johns Hopkins University Press, 1984. ISBN 9780801830228).

Lewchuk, W.: *American Technology and the British Vehicle Industry*, (Cambridge: Cambridge University Press, 1987. ISBN 0521302692).

------: "Fordist Technology and Britain: The Diffusion of Labour Speed-Up", in Jeremy, D., ed.: *The International Diffusion of Technology*, (London, 1992a).

Lewis, R.: "Redesigning the Workplace: The North American Factory in the Interwar Period", *Technology and Culture*, Vol. 42, No. 4, (October 2001), pp. 665-684.

Little, S.E. and M. Grieco: "Shadow Factories, Shallow Skills? An Analysis of Work Organization in the Aircraft Industry in the Second World War", *Labour History*, Vol. 52, No. 2, (May 2011), pp. 193-216.

Lloyd, I.: *Rolls-Royce: The Merlin at War*, (London: The Macmillan Press, 1978. ISBN 0-333-24016-2).

Lloyd-Jones, R. and M.L. Lewis: *Alfred Herbert Ltd. and the British Machine Tool Industry, 1887-1983*, (Aldershot, UK: Routledge, 2006. ISBN 978-0754605232).

Lockyer, K., A. Muhlemann, and J. Oakland: *Production and Operations Management*, 5th Ed., (London: Pitman, 1988. ISBN 0273028731).

Loften, L.K., Jr.: *Quest for Performance: The Evolution of Modern Aircraft*, (Washington, D.C.: National Aeronautics and Space Administration, 1985).

Lombardi, P.: *Short Stirling: The First of the RAF Heavy Bombers*, (Croydon, UK: Fonthill Media Ltd., 2015. ISBN 978-1-78155-473-9).

Lumsden, A.: *British Piston Aero-Engines and Their Aircraft*, reprint of 1994 edition, (Ramsbury, Marlborough, UK: Airlife Publishing, 2003. ISBN 1 85310 294 6).

Lund, E.: "The Industrial History of Strategy: Reevaluating the Wartime Record of the British Aviation Industry in Comparative Perspective, 1919-1945", *The Journal of Military History*, Vol. 62, No. 1, (January 1998).

Lyddon, D.: "The Myth of Mass Production and the Mass Production of Myth", *Historical Studies of Industrial Relations*, Vol. 1, (March 1996), pp. 77-105.

Mandel, E.: *Late Capitalism*, Verso Edition, (Oxfordshire: Verso, 1980. ISBN 978-1859842027).

Marshal, E.L.: "A Lifelong Love Affair—Sir Harry Ricardo and the Sleeve Valve", *The International Journal for the History of Engineering and Technology*, Vol. 83, No. 1, (January 2013), pp. 62-95.

Marshall, E. and B. Lawton, eds.: *The Piston Engine Revolution: Papers from a Conference on the History of Reciprocating Internal Combustion Engines held at the Museum of Science and Industry, Manchester, 14-17 April 2011*, (Manchester, UK: The Newcomen Society, 2012. ISBN 978-0-904685-15-2), pp. 401-420.

Mason, F.K.: *The British Bomber Since 1914*, (Annapolis, MD: Naval Institute Press, 1994. ISBN 1-55750-085-1).

------: *The British Fighter Since 1912*, (London: Putnam, 1992b. ISBN 0-85177-852-6).

------: *The Hawker Typhoon and Tempest*, (Bourne End, Bucks, UK: Aston Publications Ltd., 1988. ISBN 0-946627-19-3).

Matloff, M. and E.M. Snell: *United States Army in World War II: The War Department: Strategic Planning for Coalition Warfare 1941-1942*, (Washington, D.C.: U.S. Government Printing Office, 1953).

McCamley, N.J.: *Secret Underground Cities*, (Barnsley, South Yorkshire, UK: Leo Cooper, 2001. ISBN 0850525853).

McFarland, K.D. and D.L. Roll: *Louis Johnson and the Arming of America*, (Bloomington, IN: Indiana University Press, 2005. ISBN 0-253-34626-6).

McPherson, T.A.: *The Dodge Story*, (Glen Ellyn, IL: Crestline Publications, 1975. ISBN 091261207X).

Melnikova-Raich, S.: "The Soviet Problem of Two 'Unknowns': How an American Architect and a Soviet Negotiator Jump-Started the Industrialization of Russia, Part I: Albert Kahn", *The Journal of the Society for Industrial Archeology*, Vol. 36, No. 2, (2010), pp. 57-80.

Meredith, R.: *Phoenix: A Complete History of the Luftwaffe 1918-1945*, Volume 2: *The Genesis of Air Power*, (Solihull, UK: Helion & Company Ltd., 2017. ISBN 978-1-910777-27-5).

Meyer, S.: "The Persistence of Fordism: Workers and Technology in the American Automobile Industry, 1900-1960", in Lichtenstein, N. and S. Meyer, eds.: *On the Line: Essays in the History of Autowork*, (Urbana, IL: University of Illinois Press, 1989. ISBN 0252015398), pp. 73-99.

Midwest Regional Office, National Park Service, U.S. Department of Interior: Historic American Engineering Record: Pratt & Whitney Plant (Kansas City Plant), 1500 & 2000 East Bannister Road, Kansas City, MO, Written Historical and Descriptive Data, Field Records, HAER MO-118, (2012).

Miller, M. and R.A. Church: "Motor Manufacturing", in Buxton, N.K. and D.H. Aldcroft, eds.: *British Industry between the Wars: Instability and Industrial Development 1919-1939*, (London: Scolar Press, 1979. ISBN 0-85967-383-9), pp. 179-215.

Miller, R. and D. Sawers: *The Technical Development of Modern Aviation*, (New York: Praeger Publishers, 1970).

Mills, E.D.: *The Modern Factory*, (London: Architectural Press, 1951).

Neale, M.C., ed.: *An Account of a Partnership-Industry, Government and the Aero Engine: The Memoirs of George Purvis Bulman*, Rolls-Royce Heritage Trust Historical Series No. 31, (Darby, UK: Rolls-Royce Heritage Trust, 2002. ISBN 1-872922-21X).

Nelson, D.: "Mass Production and the U.S. Tire Industry", *Journal of Economic History*, Vol. XLVII, No. 2, (June 1987), pp. 329-339.

Nevins, A. and F.E. Hill: *Ford: Decline and Rebirth 1933-1962*, (New York: Charles Scribner's Sons, 1962).

Nixon, F.: "Aircraft Engine Developments during the Past Half Century", *Journal of the Royal Aeronautical Society*, Vol. 70, (January 1966), pp. 150-167.

Nuwer, M.: "From Batch to Flow: Production Technology and Work-Force Skills in the Steel Industry, 1880-1920", *Technology and Culture*, Vol. 29, No. 4, (October 1988), pp. 808-838.

Nye, D.E.: *America's Assembly Line*, (Cambridge, MA: The MIT Press, 2013. ISBN 978-0-262-01871-5).

O'Brien, P.P.: *How the War Was Won: Allied Air-Sea Power and Allied Victory in World War II*, (Cambridge: Cambridge University Press, 2015. ISBN 978-1-107-01475-6).

O'Callaghan, T.J.: *Ford in Service to America: Mass Production for the Military in World War II*, (Jefferson, NC: McFarland & Company, Inc., 2009. ISBN 978-0-7864-4485-4).

Olsen, B. and J. Cabadas: *The American Auto Factory*, (St. Paul, MN: Motorbooks International, 2002. ISBN 978-0760310595).

Overy, R.: *William Morris, Viscount Nuffield*, (London: Europa, 1976. ISBN 0900362847).

Pattillo, D.M.: *Pushing the Envelope: The American Aircraft Industry*, (Ann Arbor, MI: University of Michigan Press, 1998. ISBN 0-472-10869-7).

Pedden, G.C.: *Arms, Economics and British Strategy: From Dreadnaughts to Hydrogen Bombs*, (Cambridge: Cambridge University Press, 2007. ISBN 978-0-521-10838-6).

Piore, M.J. and C.F. Sabel: *The Second Industrial Divide: Possibilities for Prosperity*, (New York: Basic Books, 1984. ISBN 9780465075621).

Postan, M.M.: *History of the Second World War, United Kingdom Civil Series, War Production Series: British War Production*, (London: His Majesty's Stationary Office, 1952).

Postan, M.M., D. Hay, and J.D. Scott: *History of the Second World War, United Kingdom Civil Series, War Production Series: Design and Development of Weapons: Studies in Government and Industrial Organization*, (London: Her Majesty's Stationary Office, 1964).

Pratten, C.F.: *Economies of Scale in Manufacturing Industry*, (Cambridge: Cambridge University Press, 1971. ISBN 0521096693).

Price, B.: "Developments of Aero-Engine in War 1915 to 1950: Impacts of Knowledge Management and Heuristics in the Configuration of Aero-Engine Architecture", in Starr, F., E. Marshall, and B. Lawton, eds.: *The Piston Engine Revolution: Papers from a Conference on the History of Reciprocating Internal Combustion Engines held at the Museum of Science and Industry, Manchester, 14-17 April 2011*, (Manchester, UK: The Newcomen Society, 2012. ISBN 978-0-904685-15-2), pp. 401-420.

Pugh, P.: *The Magic of a Name: The Rolls-Royce Story: The First Forty Years*, (Duxford, Cambridge: Icon Books, 2000. ISBN 1-84046-151-9).

Rae, J.B.: *American Automobile Manufacturers: The First Forty Years*, (Philadelphia, PA: Chilton Company, 1969).

------: *The American Automobile Industry*, (Boston, MA: Twayne Publishers, 1984b. ISBN 9780805798036).

------: *Climb to Greatness: The American Aircraft Industry, 1920-1960*, (Cambridge, MA: The MIT Press, 1968).

------: "Financial Problems of the American Aviation Industry, 1906-1940", *Business History Review*, Vol. 39, No. 1, (Spring, 1965), pp. 99-114.

Raff, D.M.G.: "Making Cars and Making Money in the Interwar Automobile Industry: Economies of Scale and Scope and the Manufacturing behind the Marketing", *Business History Review*, Vol. 65, No. 4 (Winter, 1991), pp. 721-753.

------: *What Happened at Highland Park?*, (Washington, D.C.: National Bureau of Economic Research, 2003b).

Ricardo, H.R.: *The High Speed Internal Combustion Engine*, 5th Ed., (London: Blackie & Sons Ltd., 1968).

Richardson, K.: *The British Motor Industry*, (London: Macmillan Press, 1977. ISBN 9780208016973).

Riddle, D.H.: *The Truman Committee: A Study in Congressional Responsibility*, (New Brunswick, NJ: Rutgers University Press, 1964).

Ristuccia, C.A. and J. Adam Tooze: "Machine Tools and Mass Production in the Armaments Boom: Germany and the United States, 1929-44", *The Economic History Review*, Vol. 66, No. 4, (2013), pp. 953-974.

------: "The Cutting Edge of Modernity: Machine Tools in the United States and Germany 1930-1945", *Cambridge Working Papers in Economics*, CWPE 0342, (September 2003c).

Ritchie, S.: "A New Audit of War: The Productivity of Britain's Wartime Aircraft Industry", *War and Society*, Vol. 12, No. 1, (May 1994), pp. 125-147.

------: *Industry and Air Power: The Expansion of British Aircraft Production, 1935-1941*, (London: Routledge, 1997c. ISBN 0-7146-4343-2).

Robertson, A.J.: "The British Airframe Industry and the State in the Interwar Period: A Comment", *The Economic History Review, New Series*, Vol. 28, No. 4, (November 1975), pp. 648-657.

Robson, G.: *The Book of the Standard Motor Company*, (Poundberry, Dorchester, UK: Veloce Publishing, 2011. ISBN 978-1-845843-43-4).

Rosenberg, N.: "Technological Change in the Machine Tool Industry, 1840-1910", *Journal of Economic History*, Vol. 23, No. 4, (December 1963), pp. 414-443.

------: *Inside the Black Box: Technology and Economics*, (Cambridge: Cambridge University Press, 1982. ISBN 9780521248082).

Rubenstein, J.M.: *Making and Selling Cars: Innovation and Change in the U.S. Automotive Industry*, (Baltimore, MD: The Johns Hopkins University Press, 2001. ISBN 9780801867149).

Ryden, A.J.: "Flow Production", *Work Study*, Vol. 2, Issue 12, (December 1953), pp. 26-31.

Sabel, C.F. and J. Zeitlin: "Historical Alternatives to Mass Production: Politics, Markets, and Technology in Nineteenth-Century Industrialization", *Past and Present*, Vol. 108 (August 1985), pp. 133-176.

Sabel, C.F. and J. Zeitlin, eds.: *World of Possibilities: Flexibility and Mass Production in Western Industrialization*, (Cambridge: Cambridge University Press, 1997. ISBN 978-0521495554).

Schlaifer, R.: *Development of Aircraft Engines*, Heron, S.D.: *Development of Aviation Fuels: Two Studies of Relations Between Government and Business*, (Cambridge, MA: Graduate School of Business Administration, Harvard University, 1950).

Scott, J. and R. Hughes: History of the Second World War: United Kingdom Civil Series, War Production Series: *The Administration of War Production*, (London: Her Majesty's Stationary Office, 1955).

Scranton, P.: *Endless Novelty: Specialty Production and American Industrialization, 1865-1925*, (Princeton, NJ: Princeton University Press, 1997. ISBN 0-691-02973-3).

------: "Diversity in Diversity: Flexible Production and American Industrialization, 1880-1930", *The Business History Review*, Vol. 65, No. 1, (Spring, 1991), pp. 27-90.

------: "From Depression to Globalization: Reconfiguring 20th Century American Machinery and Machine Tool Building", Practical Machinist: 06-14-2003.

Seymour, P.: *Wolseley Radial Aero Engines: Lord Nuffield's Thwarted Venture*, (Stroud, UK: Tempus Publishing Ltd., 2006. ISBN 978-0-7524-3915-0).

Sherrard, P.: *Rolls-Royce Hillington: Portrait of a Shadow Factory*, Rolls-Royce Heritage Trust Historical Series No. 44, (Derby, UK: Rolls-Royce Heritage Trust, 2011).

Simonson, G.R.: "The Demand for Aircraft and the Aircraft Industry", *Journal of Economic History*, Vol. XX, No. 3, (September 1960).

"Single-Story Concrete Factory - A War-Inspired Innovation", *Engineering News-Record*, Vol. 129, No. 17, pp. 117-122.

Smith, H.: *Aircraft Piston Engines: From the Manly Baltzer to the Continental Tiara*, (New York: McGraw-Hill Book Company, 1981. ISBN 0-07-058472-9).

Smith, M.R.: *Harpers Ferry Armory and the New Technology: The Challenge of Change*, (Ithaca, NY: Cornell University Press, 1980. ISBN 9780801409844).

Smith, M.R., ed.: *Military Enterprise and Technological Change: Perspectives on the American Experience*, (Cambridge, MA: The MIT Press, 1985. ISBN 9780262192392).

Smith, R.E.: *The Army and Economic Mobilization, United States Army in World War II: The War Department*, (Washington, D.C.: U.S. Government Printing Office, 1959).

Smith, R.K.: "The Intercontinental Airliner and the Essence of Airplane Performance, 1929-1939", *Technology and Culture*, Vol. 24, No. 3, (July 1983), pp. 428-449.

Society of Motor Manufacturers and Traders Limited: *The Motor Industry of Britain Centenary Book 1896-1996*, (London, 1996).

Stoughton, B.: *History of the Tools Division War Production Board*, (New York: McGraw-Hill Book Company, 1949).

Stratton, M.: "Skating Rinks to Shadow Factories: The Evolution of British Aircraft Manufacturing Complexes", *Industrial Archeology Review*, Vol. 18, No. 2, (Spring 1996), pp. 223-244.

Sullivan, M.P.: *Dependable Engines: The Story of Pratt & Whitney*, (Reston, VA: American Institute of Aeronautics and Astronautics, 2008. ISBN 978-1-56347-957-1).

Tapper, O.: *Armstrong Whitworth Aircraft since 1913*, (London: Putnam, 1988. ISBN 0-85177-826-7).

Taylor, C.F.: "Smithsonian Annals of Flight", Vol. 1, No. 4: *Aircraft Propulsion: A Review of the Evolution of Aircraft Piston Engines*, (Washington, D.C.: U.S. Government Printing Office, 1971).

------: "Engine Design in 1936", *Aviation*, Vol. 36, No. 1, (January 1937), pp. 21, 67.

------: *The Internal Combustion Engine in Theory and Practice*, Volume 1: Thermodynamics, Fluid Flow, Performance, 2nd Ed. Revised, (Cambridge, MA: The MIT Press, 1985b. ISBN 978-0262700269).

------: *The Internal Combustion Engine in Theory and Practice*, Volume 2: Combustion, Fuels, Materials, Design, 2nd Ed. Revised, (Cambridge, MA: The MIT Press, 1985c. ISBN 978-0262700276).

Thoms, D.: *War, Industry and Society: The Midlands 1939-45*, (London: Routledge, 1989. ISBN 0-415-02272-X).

Thoms, D. and T. Donnelly: *The Motor Car Industry in Coventry Since the 1890's*, (Abingdon, Oxon, UK: Routledge, 2018. ISBN 978-1-138-06013-5).

Thomson, R.: *Eras of Technological Convergence: Machine Tools and Mechanization in the United States, 1820-1929*, Paper Prepared for the Economic History Association Meetings, September 2010.

Tolliday, S.: *The Rise and Fall of Mass Production*, Volume I, The International Library of Critical Writings in Business History, (Cheltenham, UK: Edward Elgar Publishing Ltd., 1998. ISBN 1-85898-042-9).

------: "Transferring Fordism: The First Phase of the Overseas Diffusion and Adaptation of Ford Methods, 1911-1939", Actes du GERPISA No. 11, 2000x.

Tooze, J.A.: "Endless Possibilities? Historical Alternatives to Mass Production Fifteen Years On", available at www.adamtooze.com/wp-content/uploads/2016/03/Tooze-Alternatives-to-Mass-Production-Review-Sabel-Zeitlin-Herrigel-and-Scranton-2000.pdf, accessed 30 June 2017.

------: "Review of World of Possibilities. Flexibility and Mass Production in Western Industrialization by Charles Sabel and Jonathan Zeitlin; Industrial Constructions. The Sources of German Industrial Power by Gary Herrigel; Endless Novelty. Specialty Production and American Industrialization 1865-1925 by Philip Scranton", *Social History*, Vol. 25, No. 2 (May 2000), pp. 247-250.

------: *The Wages of Destruction: The Making and Breaking of the Nazi War Economy*, (London: Viking, 2006b. ISBN 978-0670038268).

Underwood, J.S.: *The Wings of Democracy: The Influence of Air Power on the Roosevelt Administration*, (College Station, TX: Texas A&M University Press, 1991. ISBN 978-0890963883).

Vander Meulen, J.: *The Politics of Aircraft: Building an American Military Industry*, (Lawrence, KS: University Press of Kansas, 1991. ISBN 0-7006-0505-3).

------: *Building the B-29*, (Washington, D.C.: Smithsonian Institution Press, 1995. ISBN 1-56098-609-3).

Vetter, H.G.: *The U.S. Economy in World War II*, (New York: Columbia University Press, 1985. ISBN 978-0231057684).

Vincenti, W.G.: "Technological Knowledge without Science: The Innovation of Flush Riveting in American Airplanes, ca. 1930-ca.1950", *Technology and Culture*, Vol. 25, No. 3, (July 1984), pp. 540-576.

------: *What Engineers Know and How They Know It: Analytical Studies from Aeronautical History*, (Baltimore, MD: The Johns Hopkins University Press, 1990. ISBN 0-8018-4588-2).

Wagner, R.: *American Combat Planes of the 20th Century: A Comprehensive Reference*, (Reno, NV: Jack Bacon & Company, 2004. ISBN 0-930083-17-2).

Wagoner, H.D.: *The U.S. Machine Tool Industry from 1900-1950*, (Cambridge, MA: The MIT Press, 1968).

Walton, F.: *Miracle of World War II: How American Industry Made Victory Possible*, (New York: Macmillan Company, 1956).

Warner, G.: *The Bristol Blenheim: A Complete History*, (Manchester, UK: Crécy Publishing, 2002, ISBN 0 947554 92 0).

White, G.T.: *Billions for Defense: Government Financing by the Defense Plant Corporation during World War II*, (Birmingham, AL: University of Alabama Press, 1980. ISBN 0-8173-5177-9).

White, G.: *Allied Piston Engines of World War II*, (Shrewsbury, UK: Airlife, 1995. ISBN 1-85310-734-4).

------: *R-2800: Pratt & Whitney's Dependable Masterpiece*, (Shrewsbury, UK: Airlife, 2001. ISBN 1-84037-335-0).

Whitney, D.D.: *V for Victory: The Story of the Allison V-1710 Aircraft Engine 1929-1948*, (Atglen, PA: Schiffer Publishing, 1998. ISBN 0-7643-0561-1).

Wild, R.: *Mass-Production Management*, (London: Wiley, 1972. ISBN 047194405X).

Williams, K., T. Cutler, J. Williams, and C. Haslem: "The End of Mass Production?", *Economy and Society*, Vol. 16, No. 3, (August 1987), pp. 405-439.

Williams, K., C. Haslem, and J. Williams: "Ford versus 'Fordism': The Beginning of Mass Production?", *Work, Employment & Society*, Vol. 6, No. 4, (December 1992), pp. 517-555.

Williams, K., C. Haslem, J. Williams, A. Adcroft, and S. Johal: "The Myth of the Line: Ford's Production of the Model T at Highland Park, 1909-16", *Business History*, Vol. 35, No. 3, (1993), pp. 66-87.

Williams, R.: *Keywords: A Vocabulary of Culture and Society*, New Edition, (London: Fontana Press, 1988. ISBN 0006861504).

Williamson, O.E.: *Markets and Hierarchies: Analysis and Antitrust Implications: A Study in the Economics of Internal Organization*, (New York: Free Press, 1975. ISBN 0029353602).

Wilson, J.M. and A. McKinlay: "Rethinking the Assembly Line: Organization, Performance and Productivity in Ford Motor Company, c. 1908-27", *Business History*, Vol. 52, No. 5 (August 2010), pp. 760-778.

Wilson, M.R.: *Destructive Creation: American Business and the Winning of World War II*, (Philadelphia, PA: University of Pennsylvania Press, 2016. ISBN 978-0-8122-4833-3).

Winter, J.: *Industrial Architecture: A Survey of Factory Building*, (London, 1970).

Woltjer, P.: *The Great Escape: Technological Lock-in vs Appropriate Technology in Early Twentieth Century British Manufacturing*, GGDC Research Memorandum 141, (University of Groningen, 2013).

Womack, J.P, D.T. Jones, and D. Roos: *The Machine that Changed the World*, (New York: HarperPerennial, 1991. ISBN 0060974176).

Woodbury, R.S.: *Studies in the History of Machine Tools*, (Cambridge, MA: The MIT Press, 1972. ISBN 0262730332).

Woollard, F.G.: *Principles of Mass and Flow Production*, 55th Anniversary Special Reprint Edition, (The CLBM LLC, 2009. ISBN 978-0-9722591-8-7).

Zeitlin, J.: "Flexibility and Mass Production at War: Aircraft Manufacture in Britain, the United States, and Germany, 1939-1945", *Technology and Culture*, Vol. 36, No. 1, (January 1995), pp. 46-79.

------: "The Historical Alternatives Approach", in Jones, G.G. and J. Zeitlin, eds.: *The Oxford Handbook of Business History*, (Oxford: Oxford University Press, 2009. ISBN 978-0-19-957395-0), pp. 120-140.

Index

A

Aircraft Engine Committee, 141
Aircraft training, 89
Allison V-1710 liquid-cooled engine, 272
Aluminum, 33
American aircraft engine factories, 276
  vs. British aircraft engine production
    British and American government missions, production methods, 285–289
    18-cylinder air-cooled, radial engine production, 226, 227
    disparity of resources, 272–274
    factories and machine tools, 283–285
    government financing, 272, 273
    Holley's dilemma of mass production, 289–291
    labor productivity, 274–285
    for peak month of production, 1941–1945, 279, 280
    Prewar average total workers per engine built per month, 279, 280
    production 1940–1945, 271, 272
  high-production machine tools, automobile industry, 286
  reliance on subcontracting and production experience, 276
  total workers per engine built during peak production, 276
American automobile licensees, total wartime engine production, 122
American Packard Motor Company, 179
Anglo-French Purchasing Commission, 185
Anglo-French purchasing program, 184
Armstrong Whitworth Albemarle, 150, 151
Army Air Corps and Navy, 187
  factory deliveries, 187
Articulated rods, 159, 160
Austin Motor Company, 13
AVRO Lancaster, 153, 154

B

Baker vertical, 40-spindle, five-station machine, 115, 116
Batch production, 4–5, 7, 33
Baush multiple back-counterbore machine, 116
Boeing B-29 program, 232, 233
Boeing B-29 Superfortress, 1
Brewster SB2A dive bomber, 185
Bristol Accrington factory, 148, 149, 162, 286
Bristol Aeroplane Company, 6–7, 62, 234
Bristol Aquila, 49
Bristol Centaurus, 230, 231, 233
  automatic machine tools, 259
  at Filton factory, 260
  Hercules, development, 258
  IV 18-cylinder sleeve-valve engine, 259
  production record, 245, 246
Bristol Engine Factory, 241–242
Bristol Hercules engine
  Air Ministry, 144
  air-raid protection, 146, 148
  Bristol Accrington factory, 148, 149
  Bristol No. 2 Shadow Factory, 145, 146, 150
  Bristol No. 3 Shadow Factory, 145–147
  components, 159–160
  crankcase, 160
  crankshaft, 157, 158, 160
  enormous scale, 137
  large-scale production, 138
  machine tools, 137–138
  master rod and articulated rods, 160
  origins of, 139
  poppet-valve engine, 139
  production organization
    capital expenditure, 144
    development, 143
    distinctive characteristics, 139
    Mercury engine, 140–142
    motor firms, 143
    Pegasus engine, 140–142
    RAF, 140
    requirements, 141
  production record
    Armstrong Whitworth Albemarle, 150, 151
    AVRO Lancaster, 153, 154
    dilemma of mass production, 157
    factors, 152
    Handley Page Halifax, 153, 154
    machine tools and shortages, 154, 155
    RAF, 152
    Saro Lerwick, 150, 151
    shadow factories, 155, 156
    Vickers Wellington, 152, 153
  propeller shaft, 160
  quantity production
    Accrington factory, 162–164
    Bristol Blenheim aircraft, 164
    flow production, 165
    Pratt & Whitney factory, 162–163
    principle, 161–162
  role of No. 1 shadow group, 171
  sleeve-valve cylinder barrel, 158–159
  sound-proof testing, 161
  supercharger, 160
  tools
    automatic machine tools, 167
    components, 170–171
    deskilling, 170
    development, 165, 166
    machine operations, 170
    multi-spindle drilling machines, 167–168
    shortage of skilled labor, 167, 169
    special-purpose, high-production machines, 168, 169
    work force, 170
Bristol Jupiter engines, 48
Bristol Mercury and Pegasus engine
  Daimler Motors, 59–60, 62
  delays and disruptions, 73–74

design improvements, 45
design principles, 60–61
drawings, specifications, and schedules, 72
expenditures, 62
labor and management access, 58–59
larger-scale production transition, 74–76
machine tools, 61–62, 72–73
production organization
Aero Engine Committee, 54–56
Air Ministry, 53, 56–57
air parity, 47
allocation of components, 57
application, 49
Aquila, 49
engine capacity, 48
engine testing, 49
expansion capital, 52–53
inspection procedures, 49
machine tools quality, 49
motor firms, 51, 52
octane fuels, 49
"1-7" arrangement, 55, 56
orders, 51–52
production capacity, 51
representatives, 53–54
Scheme A, 47
Scheme C, 48–50
Scheme F, 50–51
supercharging levels, 48
production record
Aero Engine Committee, 67–68
airframes, 69
demands, 64–65
endurance test, 64
expenditures, 65, 68, 70–71
Government financing, 67, 68
labor shortage, 66
monthly output, 71
Munich Crisis, 68–69
orders, 69–70
production problems, 64
production switching, 67
representatives, 66
requirement, 65–66
recommendations, 73
Rootes shadow factory, 62–63
Rover shadow factory, 62
Standard Motor Company, 58, 60, 62
technical information and assistance, 72
Bristol Mercury VIS of 1933, 21
articulated rods, 25
assembly and engine testing, 27
blower (supercharger) unit, 26
component and subassemblies, 22
connecting rods, 25
cooling fins, 23
crankshaft, 24–25
cylinder barrel, 22–24
cylinder head, 24
inlet and exhaust valves, 24
inspection, 22
milling machines, 24
R.R. 56 aluminum alloy, 26
Bristol No. 2 Shadow Factory, 145, 146, 150
Bristol No. 3 Shadow Factory, 143, 145–147, 149, 153
Bristol Pegasus
accessories, 15
cost, 15
engines 1939–1945, peak monthly production, 121
requirements, 15
Bristol shadow factories, 289
British aircraft engine production
vs. American aircraft engine production
British and American government missions, production methods, 285–289
disparity of resources, 272–274
factories and machine tools, 283–285
government financing, 272, 273
Holley's dilemma of mass production, 289–291
labor productivity, 274–285
for peak month of production, 1941–1945, 279, 280
Prewar average total workers per engine built per month, 279, 280
production 1940–1945, 271, 272
British Air Ministry, 89
British Air Ministry, 89
British and American government missions
extensive conveyor systems, 286
production methods, 285–289
watertight division, 285
Buckingham, 233, 234
Buick Division's factory in Melrose Park, 190–192
Bulk production, 4–5
Bullard Mult-Au-Matic multispindle drilling and boring machines, 252–253

C

Carbon manganese steel, 12
Chevrolet Division system, 250
Chevrolet Factories, 191–192
Chicago aluminum factory, 252
Chrysler Corporation, 232
and Dodge Division, 256
Cincinnati Hydro-Tel vertical milling machines, 252–253
Cincinnati Milling Hydromatic machines, 259
Cincinnati Milling Machine Company, 118
Committee of Defence Policy and Requirements, 51
Communication distortion, 213
Crankcase, 159, 160
Crankshaft, 157–160
Curtiss CW-20 transport, 185
Curtiss Hawk 75, 184
Curtiss-Wright Corporation, 288
Custom production, 4–5
Cyclone 14BB engines at Lockland, 210
Cylinder block, 13

D

Daimler Motor Company, 48, 51
Defence Plant Corporation, 250
Defence Requirements Committee, 47
Deskilling, 7
Dodge-Chicago factory, 1, 226, 230, 235–238
aluminum and magnesium foundries, 252
fuel injector system, 255, 256
gear department, 253
heavy forgings, 252
high-production machine tools, 252
integrated production of the Cyclone 18, 251
light forgings, 252
machining work, Cyclone 18, 255
management control problems, 256
productivity level, 255, 279
challenges, 256–258
R-3350 engine to supplement production, 232
rotary hearth furnace, 252
Dodge Main factory, 255
Douglas A-26 Invader, 228
Douglas DB-7, 184
Drop-forging process, 12

E

Eaton Manufacturing Company for propeller shafts, 205
e Bristol No. 3 Shadow Factor, 194
Electro-plating operations in Plant No. 1 and Plant No. 2, 110
Employee skill levels, 35
Employment, 28, 32, 33
Engine power transformation, 36–37
Ex-cell-o automatic multiple drilling machine, 116
Ex-Cell-O automatic thread grinders, 252–253

F

Fairey Battle, 50
Fedden Mission, 287
Foote Burt automatic transfer machines, 249
Ford Dearborn factory, 277
Ford Flathead V-8 engine
accessories, 15
connecting rods and valves, 13
cost of, 15

crankshaft, camshaft, pistons, 13
cylinder blocks, 12–13
gray cast iron, 12, 17
performance, 13
production volume, 20
quality control, inspection, and testing practices, 13
Ford Motor Company, 231, 248
automatic machine tools, 248
centrifugal casting, 249
C series engines, 232
experimental cast steel cylinder barrels, 249
high-production machine tools, multiple spindles and automatic features, 249–250
machining processes, incremental changes, 250
material handling systems, 250
and Nash-Kelvinator Corporation Factories, 240–241
production target, 231
productivity level, 279
reliance on subcontracting, 277–278
French Air Force, 88
French aviation industry, 88

G

General Motors factories, 192, 232
Gloster Gauntlet fighter, 49
Government financing for aircraft engine factories, 272, 273
Graham-Paige Motor Company, 212, 213
for Cyclone 14, 205
Gray cast iron, 12
Greenlee automatic transfer machines, 114, 115, 118, 253, 288
3-, 5-, 7-, 9- and 18-station horizontal machines, 252–253
and Foote-Burt automatic transfer machines, 226
at Lockland, 208, 209
massive machine shop and assembly building, 253
Greenlee multistation machine tool at Lockland factory, 207

H

Hamilton Standard Propeller Company, 31
Handley Page Halifax, 153, 154
Handley Page Hampden, 50, 51, 65–66
Hardening process, gears, 253
Hartford factory, 248
Harvard Business School study, 215, 256
Hawker Tempest fighter, 234
Hawthorn Shadow Factory, 241–242
Heald Borematics, 259
and Rehnberg-Jacobson multi-station automatic indexing machines, 252–253
Hiduminium, 24
High Duty Alloys Ltd., 171
Holley's dilemma of mass production, 289–291
Horizontal milling machine, 200
Hudson Motors Car Company, 48
for Cyclone 14 pistons and rocker arms, 205

I

Industrial plant design, 194
International Harvester Company, 115
Interwar period, 27
American aviation industry, 28
aviation publications, 30
Bristol Aeroplane Company, 30, 31, 34–36
British aircraft industry, 28
British automobile industry, 28
capital requirements, 29
financial risks, 28
Ford Motor Company, 28, 31–32
funding, 28, 29
Pratt & Whitney aircraft engine factory, 31, 33–35
sales fluctuations, 28
Wright Aeronautical Corporation, 30–32, 35

K

Kansas City factory, 248, 249

L

Labor-intensive process, 34
Labor productivity, 34–35
Land-based maritime reconnaissance, 89
Learning curve effect, 288
Light bomber squadrons, 50, 51
Line balance, 4
Line production, 109
Lockheed PV-1 and PV-2 Ventura patrol bomber, 228
Lockland factory, 194, 205
quality inspection control rating, 212, 214

M

Machine tools, 7–8, 35
Daimler Motors, 62, 73
Rootes shadow factory, 62, 73
Rover shadow factory, 62
Standard Motor Company, 61–62, 73
Magna-Flux test machine, 19
Martin 167F, 184
Martin PBM-5 flying boat, 228
Martin PBM patrol bomber, 185
Mass production
continuous flow production, 3
definition, 2
design freezing, 4
elements, 2–3
Ford factory, 3
limitations, 5–6
line production, 4
Master rod, 160
Mercury engine, 140–142
Ministry of Aircraft Production, 275
Morris Motors factory, 13
Multi-engine flying boats, 183

N

Nash-Kelvinator Corporation Factories, 115, 231, 240–241, 277–278
productivity level, 279
National Defence Advisory Commission (NDAC), 93
National Defense Advisory Council, 205
Nickel–chromium–molybdenum steel alloys, 17
Nitriding, 36
North Shop building, 194

O

Ohio Crankshaft Company
for Cyclone 14 crankshafts, 205
factory building, 210

P

Packard Motor Company, 272
Paterson factories, 118, 197
Pegasus engine, 140–142
Piston, 159
Pratt & Whitney Aircraft Division, 6–7
Pratt & Whitney 18-cylinder R-2800 Double Wasp engine, 226, 228
automotive licensees, 248
14-cylinder R-1830 engine with no interchangeable parts, 251
machine tools, types, 247, 248
manufacturing processes, 246
Plant M factory, 246–247
production record, 242–244
R-2800 C series, 232
R-2800 Double Wasp, 227
R-2800 engine, 231
Pratt & Whitney 14-cylinder R-1830 Twin Wasp engine factories, 179, 181, 197, 198
automotive production, 205
at Buick factories, 201–205
at Chevrolet factories, 201–205
design features, 182
drilling and tapping operations, 200
history of, 180
licensing production, 185
machining operation, 200
prewar batch production methods, 197, 198
process engineering, 180
production engineering, 180
quantity production, 200, 205
Pratt & Whitney factories, 188–190
and Chicago factories, 226
construction methods, 226
productivity level, 279
Pratt & Whitney R-2800 C series production, 278

Pratt & Whitney Kansas City factory, 226, 232
Pratt & Whitney Twin Wasp
  accessories, 15
  requirements, 15
Precision engineering, 36
Principal Supply Officers' Committee, 48
Production tolerance, 13
Productive capacity, 48

Q

Quality control, 13

R

Radial drill press, 200
R-3350 Duplex Cyclone, 227
Reconstruction Finance Corporation, 187
Reliability, 13
Repetitive manufacturing (ReM), 3
Republic P-47 Thunderbolt for the Army Air Force, 228
Rolls-Royce Merlin, 137, 140, 153, 154, 171, 172
Rootes Ryton factory, 148
Rootes shadow factory, 62–63
  production problems, 64
Rover Motor Company, 48
  machine tools, 62
  production record, 64
Royal Air Force (RAF), 28, 139, 140, 152, 230
R.R. 59 alloy, 24
R-1830 Twin Wasp, 195–196

S

Saro Lerwick, 150, 151
Scandia Manufacturing Company, 210, 211
Senate Special Committee, 212
Shot peening process of bombarding metal surface, 256
Single-purpose machine tools, 199
Single-spindle vertical milling machine, 200
Snyder milling machines, 123
Snyder Tool & Engineering Company, 118
Soviet Government's plans for rapid industrialization, 193
"Spot assembly" system, 109–110
Standard Motor Company, 48, 58, 60, 62
Standard Motor No. 2 Shadow Factory, 194
Steel, 33, 34
Stellite, 17
Studebaker's wartime production
  Studebaker-built Cyclone engines, 119, 120
  Wright R-1820 Cyclone 9, 121–126
    green test, 123
    labor productivity, 125
    manufacturing methods, 124
    mechanized and gravity conveyors and overhead monorails, 124
    methods and processes, 123
    peak monthly production, 121
    technical drawings and engineering information, 126

T

Technical assistance, 55
Tensile strength, 36
Testing practices, 13
Tonawanda factory, 250
Truman Committee, 212
Two-spindle vertical milling machines, 200

U

United Aircraft Corporation, 31, 185

V

Vauxhall Motors, 13
Versailles Treaty, 47
Vickers Wellesley, 50, 51
Vickers Wellington, 65–66, 152, 153
Vought and Sikorsky airplane companies, 31

W

Wartime production of Twin Wasp, 186
Wellington bombers, 50, 51
Wood-Ridge factories, 251
Wright Aeronautical Corporation, 6–7
Wright Aeronautical R-1820 Cyclone 9, 187
Wright Aeronautical R-2600 Cyclone 14, 187
Wright Aeronautical Wood-Ridge Factory, 234–235
Wright factories
  at Lockland, Ohio, 192–195, 215
  productivity level, 279
  R-2600 engine, 184
  R-3350 Engine Committee, 230
  R-1830 engines, 185
Wright Lockland factories, 192–195, 215, 251
  reliance on subcontracting, 278
Wright R-1820 Cyclone 9, 14
  accessories, 15
  air-conditioning, 98
  artificial lighting, 98
  automatic milling machines, 113
  capital assistance, 100
  characteristics, 98–99
  comparative data, 87
  conveyor systems, 110
  drilling 180-cylinder pad holes, 112
  factory building, 99–101
  Ford River Rouge complex, 98
  forged cylinder head, 117
  gear finishing machines, 113
  high-production machine tools, 111
  inspection and quality control, 119–121
  internal and external grinding machines, 113
  machining techniques, 87
  manufacturing methods
    components, 16–17
    high-tensile strength steel alloys, 17
    inspection, 18, 19
    machining operations, 19–20
    minimum weight, 17
    nitriding and heat treatment, 17
    production volume, 20–21
    quality control, 18, 19
    reliability, 17
    standardization of designs, 19
  mechanical stress, 16
  minimum weight, 15–16
  multiple work pieces using a single machine tool, 113
  performance, 16
  power output, 86
  precision boring machines using cemented carbide tools, 113
  process engineering, 110–118
  production engineering, 108–110
  production organization
    Aeronautical Section's program planning, 95
    Allied financing, 93
    Anglo-French Purchasing Commission, 91–92
    April Congress, 89
    building capacity, 89
    depreciation policy, 91
    driving force, 88
    granting licenses, 95–96
    Munich Crisis, 88–89
    Neutrality Act, 90
    plant and personnel expansion, 94–95
    procurement programs, 93
    production planning, 93
    Studebaker Corporation, 96–98
    subcontractors report, 94
    tax code, 90
    working in shifts, 90–91
  production record, 104–106
  requirements, 15
  Studebaker Corporation, 108
    access to labor, 101
    Chicago factory, 102–103
    flow production, 103
    foreign and domestic orders, 107
    Fort Wayne factory, 101–102
    machining operations, 106–107
    raw materials and shipment, 103
    rough cast/forged parts, 107
    South Bend factory, 101–102
  Studebaker's production, 121–126
  tools of process engineering, 111
  weight reduction, 87

Wright R-2600 Cyclone 14, 179, 196–197
  Boeing Model 314 Clipper for Pan-American Airways, 182
  casting of aluminum alloy cylinder head, 206
  double-row air-cooled radial engine, 183
  high-production machine tools, 207, 208
  history of, 180
  line production assembly methods, 207
  line production methods, 206
  machining operations, 206–207
  management and organizational structure, 213
  management effectiveness, 213
  planned rate of, 208
  scale diseconomies, 213
  scheduled vs actual production, 211
  serious operating problems, 212
  unplanned disruption in production, 212
Wright R-3350 Cyclone 18 engine, 226, 228, 229
  assembly of, 254
  experimental development program, 254
  production record, 244, 245
  propeller reduction gear, 253
Wright Wood-Ridge factory, 252
  R-3350 at, 258

About the Author

Edward M. Young received a BA in Political Science from Harvard University and an MA from the University of Washington. He served two years as a Peace Corps volunteer in Thailand. He spent eight years as a commercial banker before joining Moody's Investors Service, the bond rating agency. Joining Moody's as a Senior Analyst, he later became a Managing Director and Group Managing Director, spending twenty years at Moody's with assignments in New York, London, Tokyo, and Hong Kong. After retiring in 2004, in 2015 Young completed an MA in the History of Warfare at King's College London, returning to King's College to complete a PhD in History in 2020.

Young has written 22 books and monographs on aviation and military history, including *Aerial Nationalism: A History of Aviation in Thailand* published by the Smithsonian Institution and *The Tenth Air Force in World War II: Strategy, Command, and Operations 1942-1945* published by Schiffer Publishing. He is the co-author with Louis Eltscher of *Curtiss-Wright: Greatness and Decline*. He lives with his wife in Seattle, Washington.

www.ingramcontent.com/pod-product-compliance
Lightning Source LLC
LaVergne TN
LVHW071342210926
848486LV00003B/208